S0-EEK-578

11-30-76

SEMICONDUCTOR ELECTRONICS DESIGN

FRED K. MANASSE

Professor of Electrical Engineering
University of New Hampshire

PRENTICE-HALL, INC., *Englewood Cliffs, New Jersey 07632*

Library of Congress Cataloging in Publication Data

Manasse, Fred K
Semiconductor electronics design.

Includes bibliographical references and indexes.
1. Semiconductors. 2. Electronic circuit design. I. Title.
TK7871.85.M33 621.3815'3'042 76-13638
ISBN 0-13-806273-0

Englewood Cliffs, New Jersey 07632

This text was previously published in 1967, titled *Modern Transistor Electronics Analysis and Design* by Fred K. Manasse, John A. Ekiss, and Charles R. Gray.

10 9 8 7 6 5 4 3 2 1

Printed in the United States of America

Prentice-Hall International, Inc., *London*
Prentice-Hall of Australia Pty. Limited, *Sydney*
Prentice-Hall of Canada, Ltd., *Toronto*
Prentice-Hall of India Private Limited, *New Delhi*
Prentice-Hall of Japan, Inc., *Tokyo*
Prentice-Hall of Southeast Asia Pte. Ltd., *Singapore*
Whitehall Books, Limited, *Wellington, New Zealand*

Contents

Preface

This is more than just a revision of a somewhat out-of-date text. In the eight years since its publication, the solid state electronics revolution *has* arrived. The integrated circuit, in its MSI and LSI forms, has made memory, logic, even complete calculators and computer microprocessors for military and consumer electronics available on a single chip of silicon. The original book, *Modern Transistor Electronics Analysis and Design*, which had called these circuits SIC's (Silicon Integrated Circuit) was then very up-to-date and even today still contains completely valid material. The processing procedure is essentially unchanged but has become much more technologically advanced. However, we have gone so far from there in reducing costs, increasing yields, shrinking size, reducing power consumption, and increasing complexity of functions and component densities that extensive new material and several chapters had to be added to the text. We do this at peril, however, since the acceleration of progress in the field is so rapid that even what we write today may be superceded by the time you read this book and thus require another extensive revision.

However, it should be made clear that the fundamental premises have not changed so that what has occurred to permit cheap and reliable electronic watches, calculators, radios, color televisions, automobile ignitions as well as automatic braking and anti-skidding computer systems, etc. can still be understood on a component function basis. That is to say, it is just as important today to know how a simple bipolar transistor amplifier operates, is biased, and has temperature and frequency limitations as it was ten years ago; this is so since each IC chip merely reflects the interconnection and assembly of large numbers of these discrete components in a more cost effective and space saving arrangement.

Changes have therefore been made in each chapter to reflect what has occurred in the last decade. New devices, such as the MOSFET, LED, and CCD are described as well as new logic arrangements such as T^2L and I^2L. The integrated operational amplifier or op-amp has such wide applicability that an entirely new chapter has been added on its operation and use in circuits. Extensive new material has been included on complete circuits and systems. An example which illustrates this involves complete modulator and demodulator functions on a chip, which now make modems and other communication circuits small and cheap. Along with coders and decoders for BCD and ASCII, these all permit, with microprocessor and memory chips, a complete 4 or 8 Kbit minicomputer to fit on one small PC card. Semiconductor memories such as RAM, ROM, PROM, etc., made on a single chip form MOS or bipolar transistors, are also described, since most new electronic systems now incroporate some memory in them.

The recent advances in device physics are introducing solid state electronics to new areas. No longer is the designer confronted only with p-n junction diodes and pnp or npn transistors (along with passive elements) for solving electronic circuit design problems. He must now be capable of designing a complete circuit on a monolithic substrate the size of the head of a pin. Thus, circuit design in many areas, especially that of digital circuits for use in computers, now entails putting together many such microcircuits into a subsystem and many of these subsystems into a complete system. The circuit designer must therefore become systems-oriented, and the scope of his task is correspondingly widened.

Since, in digital circuits, the microcircuit rather than the device is the fundamental building block, one might feel that the ability of a design engineer to develop the circuit might less important than the ability to put the circuits together into a system. Considerations such as tolerances with respect to temperature, bias, humidity, vibration, etc., are even more important in the proper operating characteristics of the device and circuit than in conventional design since mass production and linking of these subassemblies into a working system are essential. Thus, an elementary understanding of the limitations of the basic amplifying device and of the fundamental circuit building blocks is not only essential but crucial to the entire field. Ground plane problems, shielding to reduce interaction between circuits, power-supply filtering to reduce unwanted mixing of signals, tolerances on components, etc., are just a few of the complicating problems which the circuit system designer must now consider. Each of the problems that are encountered in designing a simple circuit such as a multivibrator, an inverter, or a logical gate, must now be multiplied a hundredfold. Unless the designer has the ability to master the design of simple and basic circuit structures with lumped elements, he cannot hope to ever design a microcircuit or to put them together into a subsystem. We see, therefore, that even in this age of miniaturization and integrated electronics, knowledge of the design of simple lumped circuits is essential.

The microelectronics industry is today primarily involved in developing digital types of circuits for use in computers, military electronics, and space vehicles. The major area of consumer use of electronics—such as communications, television, radio, high fidelity equipment, heating, lighting and cooling controls, etc.—where reusability or repairability is essential and size and weight are not at a premium is still relatively untouched by microelectronics. However, this area is one of great and promising activity. In consumer applications, large power-handling capability, ease of maintenance, basic components designed to avoid obsolescence, and most importantly, cost, are factors which will insure that lumped element design will continue to be important to the electronics industry. Thus, microcircuit techniques and lumped element circuitry will probably be combined in all these areas for a long time. Again we see the need for engineers who can design circuits with lumped elements.

Another major reason for the need of such a text as this concerns engineering education. The student must be able to understand simple amplifying circuits utilizing a single transistor before he can appreciate the design complexities of a multistage amplifying circuit using five or six transistors contrived in a single chip of semiconducting material. He must appreciate the difficulties inherent in bias stabilization for the single transistor stage before he can comprehend the intricacies of stabilizing an entire microcircuit where the individual stages can interact and where the rise in temperature of one circuit indirectly affects all the others. He must clearly understand the workings of a single multivibrator before he can successfully design a miniature integrated-circuit shift register.

The study of electronic design incorporates not only circuit and network theory and electronic devices, but also ties together all of electrical engineering. Study in any part of this field requires an understanding of circuitry, whether in analog or digital computers, control systems, power generation and transmission, communications, instrumentation, or in all the rest of the vast domain of the electrical engineering technology. This unification is fundamental and must be made an important part of an electrical engineering education.

So much for the whys of the text. As to the hows, we have attempted to include under one cover a sufficient scope of topics, but with enough detail, to satisfy both the undergraduate and the graduate student, as well as the practicing engineer. We have assumed that the reader is familiar with the basic theory of operation of semiconducting devices, and therefore have included only one chapter on the theory. This first chapter is not rigorous, and serves mainly to refresh the memory of the reader, or to introduce simply the physical ideas which are important to an understanding of the operation of junction diodes and transistors. While not greatly detailed, this chapter serves to unify the topics covered in the bulk of the text by clearly laying the groundwork for the understanding of concepts such as temperature stability, frequency dependence, etc., which are important considerations in proper design.

Chapter 2 considers equivalent circuits for the transistor with emphasis on both the equivalent-circuit approach—the conventional matrix parameters such as h, y, or z parameters—as well as the physical parameters derived from the fundamental equations, such as the equation of continuity. Chapter 3 introduces the methods of biasing transistors, primarily to compensate for temperature variation in the operating parameters. However, the techniques of feedback, both shunt and series, as well as compound, are described and illustrated. Analytic techniques for evaluating and designing low-frequency untuned amplifiers are developed in Chapter 4.

Tuned amplifiers, which are still important in many consumer electronics are taken up in Chapter 5. The design of this type of circuit is crucial to the understanding of communications, tuned oscillators, detectors, etc., which are discussed in later chapters. The automatic gain control (AGC) feedback circuit so essential in many different types of equipment is discussed and analyzed.

Video or wide-band amplifiers, useful in television and FM receivers, are discussed in Chapter 6. These design techniques are exactly those necessary in the development of pulse amplifiers since their Fourier spectrum includes components with very large frequencies. Thus, this chapter serves to bridge the gap in the design of analog and digital circuitry. Chapter 7 discusses choppers and is a valuable addition to the text, as it clearly illustrates the advantages of semiconductor devices over mechanical devices (relays). The conversion of dc to ac necessary to enable the efficient amplification, modification, transmission, etc., of ac is readily accomplished by these circuits.

This leads directly to operational amplifiers discussed in Chapter 8. This is an all new chapter and is very design oriented. Again, this material is not generally available in electronics texts at this level, but is essential in the modern approach to design.

The operation and design of switching circuits are taken up in Chapter 9. These circuits include multivibrators, logical circuits (useful in digital computer development), and differing circuit schemes for connecting fundamental building blocks, such as binary counters (modified binary multivibrator) and gates, into ring counters, adders, and clock systems. The emphasis here is on the use of the bistable nature of the transistor in circuitry. However, since several simple circuits are put together to form a small system, this is the beginning of a look into the motivation for, and design, of integrated circuits.

In Chapter 10, we return to analog circuitry and discuss some aspects of the nonlinear behavior of the transistors. Here, the nonlinearity is utilized to effect control of a desirable oscillation, to mix two signals together, or to recover an information-carrying signal from an amplitude- or frequency-modulated carrier. This is still one of the most active areas for engineering designers and will continue to require engineers skilled in circuit analysis.

Discussion of the transistor operating in its nonlinear range is continued in Chapter 11 as the question of high-power amplifiers and frequency converters is taken up. Using the earlier discussion of tuned amplifiers as a background, we now go into the details of obtaining large power output in a narrow frequency range suitable for transmitters, etc. Noise behavior of transistors and their circuits is the subject of Chapter 12. The concept of noise figure and its measurement, fundamental in communications, is discussed and illustrated. In Chapter 13 we discuss two-terminal devices, similar to transistors, which can be used in certain applications. These devices can exhibit gain and are useful in both linear and nonlinear operation. They can be used as oscillators, amplifiers, switches, etc., and under certain circumstances are much to be preferred over their three-terminal counterparts. The use of the tunnel diode, backward diode, Gunn "diode," and avalanching pn junction diode are discussed. Because of their nature, these devices can be used at much higher frequencies than even the most modern transistors, and thereby extend the range of frequencies over which the circuit designer can function.

Integrated circuits are discussed in the two concluding chapters. This important topic is of an extremely specialized nature. No two circuits are identical, and since all devices within the subsystem or microcircuit are formed on the same chip, many new problems in design arise. These complications only add to the normal ones inherent in lumped element design, and require a volume of their own to be fully expounded. However, an attempt is made to introduce the circuit designer to the new demands and capabilities of the monolithic circuitry. By this means, we hope to convince him that certain considerations in lumped circuit design can be applied and that others must be modified in order to design in this new domain.

A few exercises have been included at the end of each chapter to stimulate the interest of the reader and to provide some real design problems which can test his understanding of the material. No simple "plug-in" exercises have been used; the exercises are designed primarily to *amplify* the text material and to suggest real industrial designs.

The references at the end of each chapter are not meant to be exhaustive, but merely indicate other readily available text and journal material dealing with the subject. Since much of the subject matter in many of the individual chapters deals with material previously developed by other authors whose works contain adquate bibliographies, it was felt that no extensive referencing was necessary.

In summary then, considerable amounts of material have been added on modern transistor electronic devices and systems. However, we have maintained the direct approach to design, and continue to illustrate the use of these circuits with practical examples taken from widely used industrial applications. We have also eliminated some material which we felt was redundant, or no longer of wide interest, as well as reducing several of the mathematically tedious developments in the interest of improving readability.

We believe this is now a better and more modern book and we encourage your critical comments to improve it still further. Since my two co-authors for the earlier edition are now heavily committed to managerial functions in geographically widely separated regions of our nation, it has not been expedient to continue the collaboration for this major revision. However, the base of this text still amply represents the fruits of that earlier unity, and I am greatly indebted to both Charles Gray and John Ekiss for their willingness to allow me to continue to use much of their creative effort intact here. The original preface listed the many other contributors to that work and my continued indebtedness to these individuals is hereby acknowledged. It may also be of interest to the reader to note that my own career has changed direction somewhat, being less involved with solid state research and more with higher education and its management. I trust this does not detract from the text but perhaps even enhances its teachability.

Finally, I would like to thank all my students, colleagues, and former associates at Princeton University, Dartmouth College, and Drexel University who helped to guide my efforts and who provided useful critical reviews of the material by teaching or learning from it. My secretary of almost 2 years, Miss Connie Wallgren who typed and corrected the MS is to be especially commended for her untiring fortitude. I would especially like to thank Mr. James Chege of Prentice-Hall, Inc., for his invaluable assistance throughout the production of the book. To my wife who has put up with me for more than 20 years, I can only say gratefully, thank you.

FRED K. MANASSE

Philadelphia, Pa.

1

Overview of Appropriate Semiconductor Diode and Transistor Theory

INTRODUCTION

Since the invention of the transistor at the Bell Telephone Laboratories in 1948, it has found its way into varied applications in the commercial, industrial, and military fields. It is an active electrical device and because of input-output decoupling it can exhibit gain. Like the vacuum tube, the transistor finds its largest application as a control element, where a small input signal is used to control a large output signal as well as being useful as an on-off switch. Actually, in some applications, the gain is less important than the capacity for remote control of a given signal. In these so-called switching applications, the transistor has almost completely replaced the vacuum tube, gas tube, and relay because of its almost ideal characteristics.

This chapter does not present a rigorous treatment of semiconductor physics, because this is the subject of many books. Rather, we have attempted to present the information in a manner useful to circuit designers. Those interested in a more detailed discussion of semiconductor physics should consult the list of references.

PROPERTIES OF SEMICONDUCTOR MATERIALS

A *transistor* can loosely be said to be composed of two semiconductor diodes on a common piece of semiconductor material. These diodes are in very close proximity to each other, which affords a mutual coupling between the diodes. This is loosely referred to as *transistor action*. With this in mind, it seems only natural that the properties of semiconductor diodes should be the first subject of interest in studying transistors.

By way of review, a semiconductor is material that is neither a good conductor nor a good insulator, hence, the name *semiconductor*. An ideal diode is an electrically unilateral device; that is, current flows through it in only one direction. A practical diode approaches this ideal and under many applications can be said to approximate it quite closely. Why, one may ask, does the diode conduct in only one direction? In order to answer this question, we must first determine why any material does, or does not, conduct current.

BAND THEORY

In order for conduction to take place there must be, within a material, free electrical charges or carriers. A *free charge* is one that is not confined to any particular atom within the material. A material is classified as a conductor, semiconductor, or insulator depending on the number and mobility of these free carriers it contains per unit volume. The *mobility* of the carrier is, as its name implies, related to how mobile a carrier is in moving from one point to another.

In materials which have a regular, repetitive, atomic arrangement which we denote by the term *crystalline*, the outermost electrons from each atom find themselves bound to their nucleus more weakly than the inner electrons. Because of the Pauli exclusion principle, which can loosely be taken to imply that no two electrons can ever be in the same energy and momentum state, not all electrons can be at the lowest energy state possible. For a single atom, and for 0°K, energy levels can be schematized as shown in Fig. 1-1(a).* If we bring a second identical atom near the first we will now have an energy level scheme as shown in Fig. 1-1(b). The energy difference between pairs of levels is much larger than the energy difference within the pair itself. If we put many more identical atoms together, the energy structure becomes as shown in Fig. 1-1(c) where, because of the large numbers of discrete levels in each band, we have almost a continuous energy distribution within the band. Depending on the interatomic spacing of the crystal lattice structure, we will then have three rather distinct possibilities, as shown in Fig. 1-2.† These are shown in more detail in Fig. 1-3(a,b,c). Figure 1-3(a) shows a schematic band diagram of a metal where the interatomic spacing is large and the two outermost bands overlap so that electrons can freely go from one band to the other if we give them some thermal energy by raising the temperature. Because of this we will have many "free" electrons which can be moved in an external field thereby leading to a large current. Since the generalized Ohm's law between the cur-

*Here we plot energy as ordinate and momentum as abscissa (proportional to the inverse of interatomic spacing).

†Note that we now plot energy as a function of spacing and not momentum!

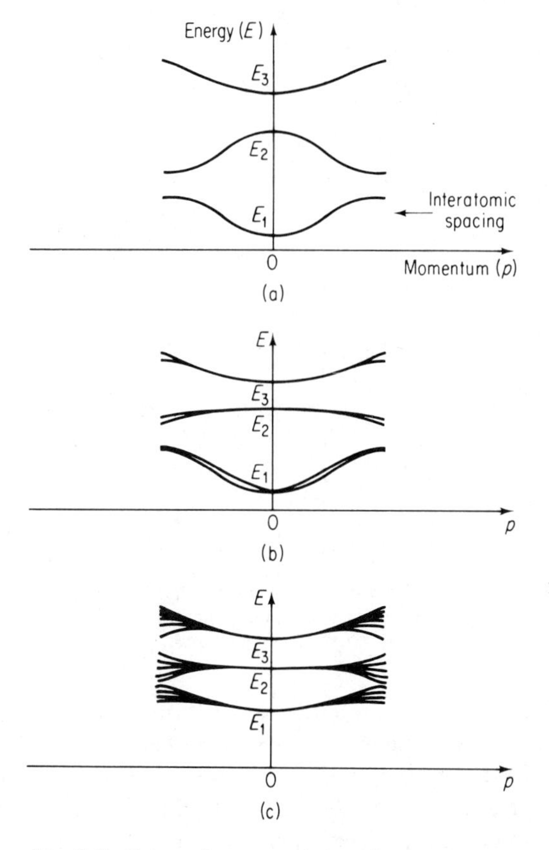

Fig. 1-1. Schematic representation of energy-momentum levels in solids. (a) energy levels for one atom; (b) energy levels for two atoms in proximity; (c) energy levels (bands) for many atoms in proximity.

rent density J and the electric field $\mathcal{E}$ is given by

$$J = \sigma \mathcal{E} \tag{1-1}$$

where σ is the conductivity, and is a constant for uniform isotropic material, we see that a metal has a large conductivity and is therefore a good conductor. If we change the temperature, we shall not change matters very much since even at room temperature we have a rather large number of free carriers (essentially all the so-called valence electrons of each atom are free in a metal at room temperature). If the interatomic spacing is increased we shall have a situation as shown in Fig. 1-3(b) where the bands are separated by a large energy of the order of 3 ev or more (an ev or electron volt, is 1.6×10^{-19} joule or the energy an electron acquires on falling through a potential difference of 1 volt). Thus at 0°K we have the case where the lower band (or

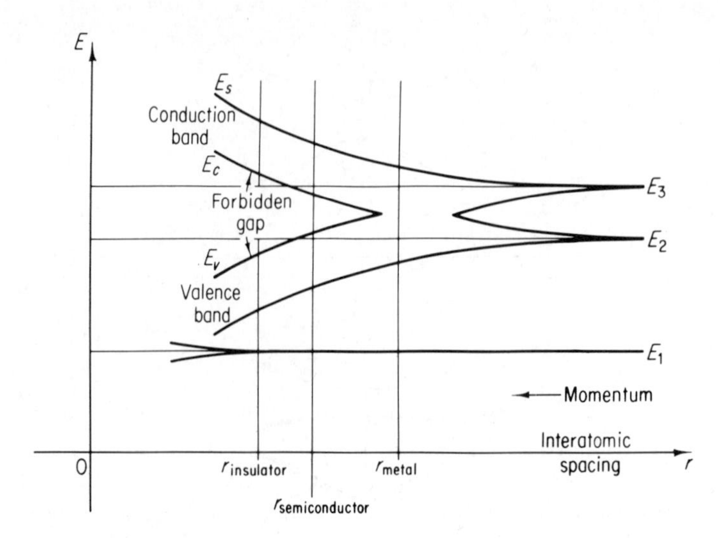

Fig. 1-2. Schematic representation of energy-interatomic spacing bands in solids, where

E_s = energy level of top of conduction band (vacuum level)
E_c = energy level of bottom of conduction band
E_v = energy level of top of valence band.

valence band) is filled, say, and the upper band (or conduction band) is empty. If we raise the temperature to room temperature (297°K) we have an average energy of the order of 0.025 ev per electron, which is not sufficient to raise very many electrons from the valence to the conduction band. Thus an applied field would not cause very much current and the conductivity would be low. We call this material an *insulator* for that reason. If we position our interatomic spacing somewhere between the two extremes we get the band picture of Fig. 1-3(c) where we still have an energy gap between the two bands but this gap is of the order of 1 ev or less. At room temperature we will not have as many carriers as in a metal but considerably more than in an insulator, although the conductivity of this material is still rather low. We call this a *semiconductor*.*

*We thus see that, because of the nature of the crystalline state of matter, different substances form materials of varying conductivity, with the interatomic spacing the controlling factor. Thus since, for example, the temperature causes this spacing to vary (pressure will also do similar things), we can modify the properties of a material in this manner. We see that an insulator can be made semiconducting by raising the temperature and thereby reducing the interatomic spacing (for example, carbon in its various states of diamond, graphite, etc.), and metallic by going still further in temperature. We also see that if we increase the temperature sufficiently we can distort the crystal structure and eventually break it down into the liquid or gas phase where all pure materials are basically insulators again.

If we raise the temperature somewhat, we increase the energy of the electrons in the crystal and increase the number of electrons which go from valence to conduction bands, thereby increasing the conductivity. In a metal, the phenomenon of resistance owing to the collisions the free electrons make with the lattice atoms and with each other is such as to increase with temperature. That is, the number of free electrons is large and more or less independent of temperature (at moderate temperatures), but as their energy goes up (so does the energy of the fixed nuclei) they collide more often and therefore the conductivity goes down. In a semiconductor, on the other hand, because the number of free carriers increases rapidly with temperature, and since this is the controlling factor, the conductivity then increases with temperature. This fact is employed in some devices, such as thermistors, for temperature regulation of circuits.

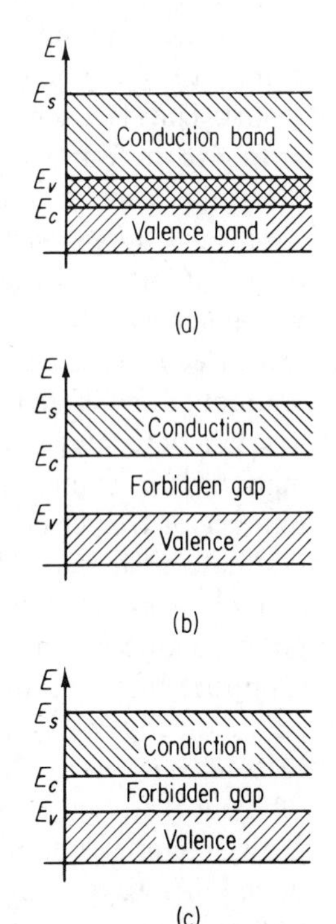

Fig. 1-3. Band structure for materials. (a) Metal with overlapping bands; (b) insulator with wide gap; (c) semiconductor (intrinsic).

In taking an electron from the valence to the conduction band, a gap in the electronic structure between lattice atoms is left in the valence band. This gap can be filled by another electron from a neighboring atom in the lattice and the gap can move within the valence band. Thus, under the action of an external field, if an electron goes from the left to the right to fill a gap, the gap has in a sense moved from right to left. We call this gap a *hole* and it can also contribute to the conductivity. It behaves somewhat like a positively charged electron (since it moves oppositely in an electric field) although it is somewhat more "massive" than an electron because of differences in the statistical properties of bound electrons moving into a gap as opposed to free electrons moving among themselves and a lattice.

IMPURITY SEMICONDUCTORS

If we had only intrinsic semiconductors (so called because they are pure materials) and for every electron created we also created a hole (since the material has all its electrons bound in the lattice at 0°K), the use of semiconductors in circuitry would indeed be limited, since at room temperature these materials behave almost like insulators and conduct only nanoamperes (10^{-9} amp) of current. We can, however, modify their properties by adding controlled amounts of impurities to the pure material. If we have materials such as germanium or silicon, which have four electrons in their outermost bands, a crystalline solid is formed by sharing electrons between neighboring atoms. That is, each atom would like either to have four additional electrons or to give up four electrons to form a completely filled band as in an insulator. Since this can be done by sharing electrons among four neighbors, called *covalent bonding*, the materials can form a crystal and do so in a so-called diamond lattice. By adding a small amount of material, such as phosphorus which has five electrons in its outermost band, a bond can be formed with an extra electron for each bond formed available for conduction. So long as the amount added is small, the crystal lattice is not disturbed very much and even a small percentage (one atom of phosphorus for each 10^7 atoms of germanium, for example, raises the conductivity by a factor of 10^3) will raise the number of free electrons by a very large factor. Actually we can *dope* (contaminate with a controlled amount of selected impurity) an intrinsic semiconductor so heavily that we can actually cause it to become a reasonably good conductor and still maintain its temperature variation characteristic (opposite that of a normal conductor). The energy band structure is shown schematically in Fig. 1-4(a). An extra "level" (really a very narrow band) is located very near the bottom of the conduction band (typically 0.01 ev difference) so that a very small thermal energy is required to ionize the impurity completely and make its electron available for conduction. A material doped this way is called *n-type* to indicate that primarily an electron conduction mechanism is responsible for current flow.

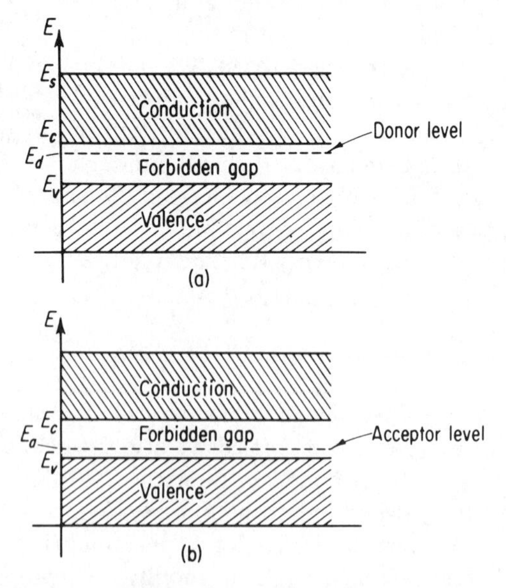

Fig. 1-4. Extrinsic or doped semiconductors. (a) n-type semiconductor; (b) p-type semiconductor.

Correspondingly, a material such as aluminum, which has but three electrons in its outer band, can be used to dope germanium, thereby leaving a hole. The energy band structure is as shown in Fig. 1-4(b) where the dopant is called an *acceptor* since it accepts electrons or creates a hole and the material is called *p-type* to indicate that hole conduction in the valence band is the main reason for current flow. Both of these doped semiconducting materials are called *extrinsic semiconductors*. By suitable combination of these types of dopants, many different characteristics can be developed for semiconductors for different applications.

In order to obtain conduction, the charges must flow from one point to another, but if the carrier has many collisions along the way, its average velocity will be decreased. At each collision, the carrier will be deflected in a different direction so that its path will be greatly lengthened. A more formal definition of *mobility* follows: The mobility of carriers within a material is defined to be the average velocity which the carriers assume per unit applied electric field.

$$\mu = \frac{v_c}{\mathcal{E}_a} \tag{1-2}$$

where μ = mobility
v_c = average velocity of the carriers in cm/sec
$\mathcal{E}_a$ = applied electric field in volts/cm.

A simple expression for the conductivity of a material is shown in Eq. 1-3. Notice that the conductivity is dependent upon both the mobility and the number of free carriers as stated earlier.

$$\sigma = q\mu_n n + q\mu_p p \tag{1-3}$$

where σ = conductivity in mho/cm
μ_n = mobility of negatively charged carriers in cm²/v-sec
μ_p = mobility of positively charged carriers in cm²/v-sec
q = electronic charge of the carrier in coulombs
n = number of negatively charged carriers/cm³
p = number of positively charged carriers/cm³.

In a semiconductor, we have seen that the negatively charged current carriers are electrons and those positively charged are holes. The free carriers within a material are thermally generated; thus, the conductance of the material, which is dependent upon the number of carriers, is a function of temperature. So, as indicated earlier, some materials which are insulators at room temperature become conductors at high temperatures. Likewise, materials which are conductors at room temperatures may become insulators at very low temperatures. In fact, another way of specifying whether a material is a conductor, semiconductor, or insulator is by defining the amount of thermal energy it takes to generate a free carrier. For example, in intrinsic silicon it takes 1.1 ev of energy to create a carrier pair (electron and hole) whereas in germanium it takes only 0.7 ev to generate a pair thermally.

Some Effects of Temperature in Semiconductors

Since both the mobility and the number of free carriers are a function of temperature, it is interesting to show how the conductivity of a semiconductor material varies with temperature—Fig. 1-5(a). Notice the several phases which the conductivity goes through as the temperature is increased. Let us examine this curve in detail and explain how the variation of conductivity in a semiconductor crystal differs from that of a metal—shown in Fig. 1-5(b). In most of our discussions extrinsic semiconductors are used so that some impurities are present within the semiconductor material—Fig. 1-5(a). These impurities may be residual, or they may have been added intentionally to alter the material properties in a prescribed way. In the case of a metal, the number of free carriers is large and essentially independent of temperature. As previously mentioned, however, the conductivity of a material is dependent not only upon the number of free carriers but also upon the carrier mobility.

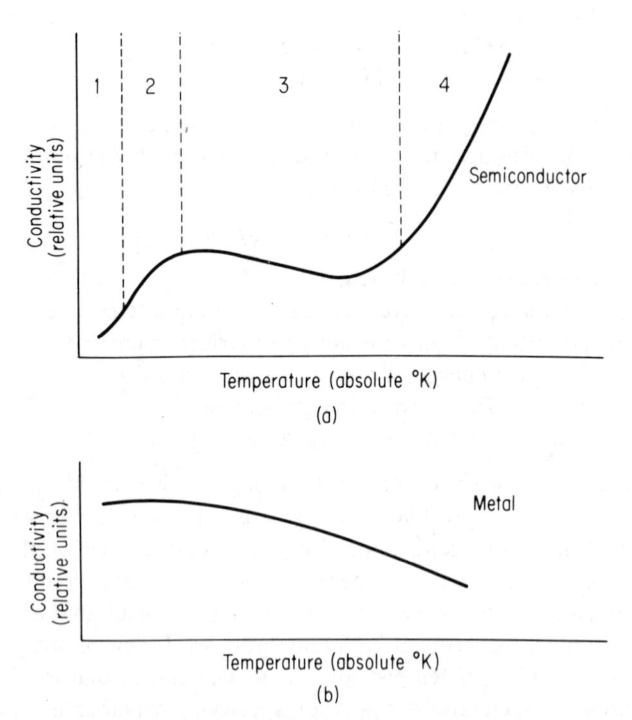

Fig. 1-5. Temperature dependence of conductivity in semiconductors and metals. (a) Conductivity of semiconductor (extrinsic); (b) conductivity of metal.

In a metal it is the mobility which falls with rising temperature. Thus metals exhibit a continuous fall of conductivity over a wide range of temperatures. This is shown in Fig. 1-5(b). In a semiconductor crystal, however, the number of free carriers as well as the mobility varies with temperature. Let us examine the various regions of Fig. 1-5(a).

Region 1: At very low temperatures, the number of free carriers is very, very small so that no matter what the mobility, the conductivity is very low.

Region 2: As the temperature is increased, carriers from any impurities which may be present will be freed within the crystal, thus increasing the number of free carriers. As the number of free carriers in increased, the conductivity increases.

Region 3: As the temperature is increased further, the increase in the total number of carriers levels off. At the same time, the increasing temperature of the material causes the crystal lattice vibrations to increase, thus increasing the probability of a collision between the free carrier and fixed atoms of the crystal. This, of course, lowers the mobility of the free carrier. Thus, for some temperature range, the number of free carriers is increasing slightly while the mobility is decreasing slightly so that the conductivity may increase or decrease, depending on which of these effects dominates. It is interesting to note that Region 3 occurs at approximately room temperature.

Region 4: At the high-temperature end of Region 3, essentially all the impurity carriers are free. If the temperature is raised still further, the valence bonds of the crystal break. When this occurs, the number of free carriers increases rapidly. In this region, the semiconductor starts behaving like an ordinary conductor.

Although this explanation may seem superfluous, note that almost all transistor parameters depend upon the conductivity of the semiconductor materials which are used to make the transistor. From the foregoing discussion, we may observe a most important fact concerning semiconductors—their characteristic properties are temperature sensitive. This is one of the big disadvantages of the transistor and other semiconductor devices.

Current Flow in Semiconductors

In contrast to metals, a semiconductor has *two* important current flow mechanisms: drift and diffusion. The first of these, the drift current, is the familiar form of current flow in which charge carriers move through a material owing to the force created by an applied electric field. This is the type of current flow which follows Ohm's law. It is referred to as *drift* because the applied electric field forces the mobile carriers to drift in a certain direction.

There is, however, another form of current flow which is not usually studied because of its limited transmission distance. This second form is called *current flow by diffusion* and is analogous to the diffusion of gas molecules. Consider what happens if one places an open perfume bottle in the corner of a closed room. After some time the perfume diffuses to other parts of the room. The concentration of perfume is dependent upon the distance one stands from the open bottle. A similar situation exists if one injects a large number of electrical charges (carriers) into a piece of solid material, such as a semiconductor crystal. Consider what happens if a very large concentration of electrons is instantaneously injected into a solid crystal at point A in Fig. 1-6. For simplicity, we shall study a two-dimensional model; however, the effect is three-dimensional.

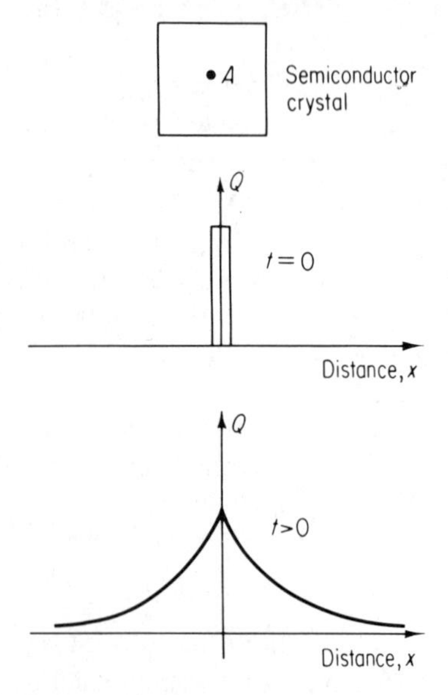

Fig. 1-6. Diffusion of electrons injected into a semiconductor crystal.

Figure 1-6 shows a plot of the electron concentration versus distance from point A, both initially and after a short period of time. Note that the curve is an exponential (as it is for the diffusion of gases) and is symmetrical in both directions. Since it is obvious that the electrons have moved through the material and thus traveled some distance, we may talk about a current flow. This charge flow will always flow from a region of high carrier concentration to a region of low carrier concentration. The amount of current which flows is given by

$$I_D = -DA\frac{dQ}{dx} \tag{1-4}$$

where A = cross-sectional area
D = diffusion constant (a constant of the material)
dQ/dx = charge gradient.

As one might expect, the steeper the charge gradient, the higher the current flow. The minus sign takes care of the current direction since electrons move opposite to the conventional current direction. After a long time, the electron concentration, due to the diffusion current flow, will be equal throughout the crystal, and the current will cease because there will be no charge gradient.

Carrier Lifetime and Diffusion Length

Before leaving the subject of current flow by diffusion, we shall define two other terms of interest. Any semiconductor material has a certain number of imperfections (for example, foreign material and broken covalent bonds) which will, if given the opportunity, capture free electrons or holes. Thus, if we plot the concentration of injected electrons versus time we will get a curve similar to Fig. 1-7. This curve is again exponential. The time required for 63 per cent $(1 - 1/e)$ of the carriers to recombine with crystal imperfections is called the *lifetime*, τ, of the material. The lifetime is a measure of the purity of the crystal. For some purposes (high transistor current gain) a large value of τ is desirable, for other purposes (reducing certain switching times in pulse circuits) a low value of τ is desirable. Typical values of τ range from 10 nanosec to 10 μsec.

Along with τ, a similar quantity may be defined, the diffusion length, L. The *diffusion length* is defined as the average net displacement of a carrier through the crystal before being trapped by a crystal imperfection. Thus, the average carrier will travel one diffusion length, L, in τ sec. Typical values of L range from 10^{-4} cm to 10^{-1} cm. The illustration, Fig. 1-8, shows the importance of the diffusion length. At distances from the point $x = 0$ greater than the diffusion length, the carrier concentration has decreased to a value too low to be useful. For all practical purposes, the crystal material, a distance of $3L$ from the point of injection, does not even know that the high concentration exists at the injection point.

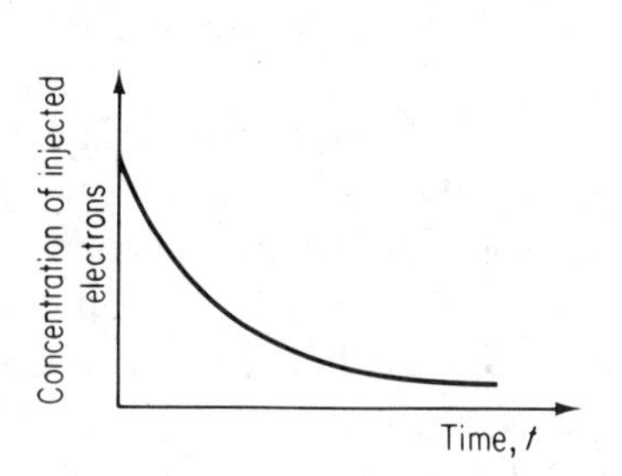

Fig. 1-7. Exponential decay of injected electrons.

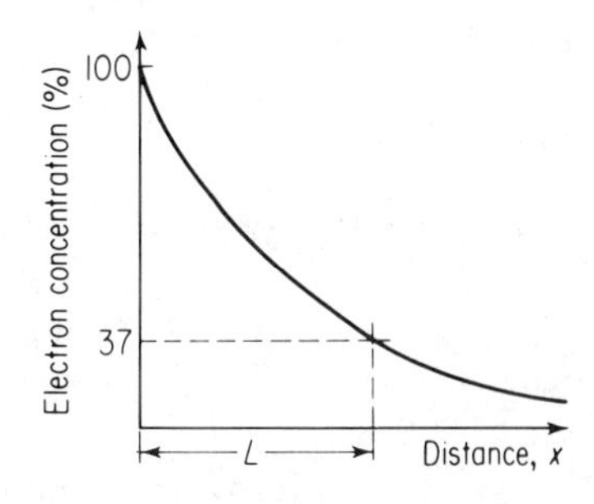

Fig. 1-8. Definition of the diffusion length L.

The diffusion constant, lifetime, and diffusion length are simply related by the expression

$$L = \sqrt{D\tau} \tag{1-5}$$

In a typical sample of germanium at room temperature, with a lifetime of 10 μsec, the diffusion length for holes would be about 2×10^{-2} cm.

THE PN JUNCTION

Junction Formation

A semiconductor diode may be formed, in theory, by fusing together two pieces of semiconductor material of opposite conductivity type. A p-type semiconductor material is formed by doping (adding to) the semiconductor crystal an elemental material which will contribute a positive charge carrier. Examples of materials which are p-type dopants are boron, indium, gallium, and aluminum. An n-type semiconductor material is similarly formed by doping with materials such as phosphorus, antimony, or arsenic, which contribute negative charge carriers.

In the most general sense a junction may be said to exist wherever there is a variation in the charge density throughout a material. Thus for every variation or discontinuity, whether it be an imperfection in the crystal lattice of the semiconductor or a variation of a chemical impurity (dopant), there will exist an electric field. This is true because of Poisson's law which in a one-dimensional form is:

$$\frac{d\mathcal{E}}{dx} = -\frac{\rho}{\epsilon} \tag{1-6}$$

where ρ is the charge density and ϵ is the dielectric constant of the material.

The PN junction is a special type of junction in which the charge variation is attributed to two regions of opposite conductivity type in a host crystal.

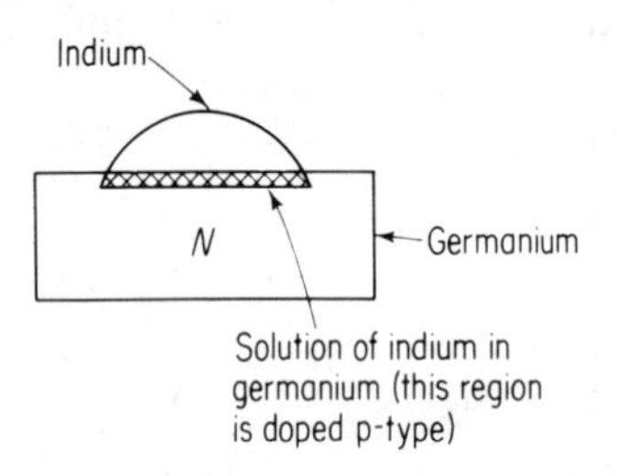

Fig. 1-9. Schematic representation of an alloy junction diode.

We may symbolically fabricate a PN junction by joining the faces of two semiconductor crystals of opposite conductivity type. This cannot, however, be done mechanically. One practical way of forming the junction is to alloy the materials of opposite conductivity type together. Such a PN junction diode is shown in Fig. 1-9.

The junction is formed by heating the indium and germanium structure until the two materials alloy together. The host germanium crystal is doped n-type. Where the indium alloys into the germanium, the semiconductor is doped p-type. This p-type region is shown as the crosshatched area in Fig. 1-9 and represents a solid solution of indium in germanium.

The PN Junction under Equilibrium Conditions

A PN junction is shown schematically in Fig. 1-10(a). In the vicinity of the junction, holes from the P material have diffused across the junction to the N material and electrons from the N material have diffused across the junction to the P material. Hence, on either side of the junction there exists a charge layer which is depleted of mobile charge carriers. The result is a space charge layer or depletion layer on either side of the junction—Fig. 1-10(b) and (c).

Associated with this charge density plot is a resultant electric field in the depletion layer, and hence, a potential difference across the junction. This

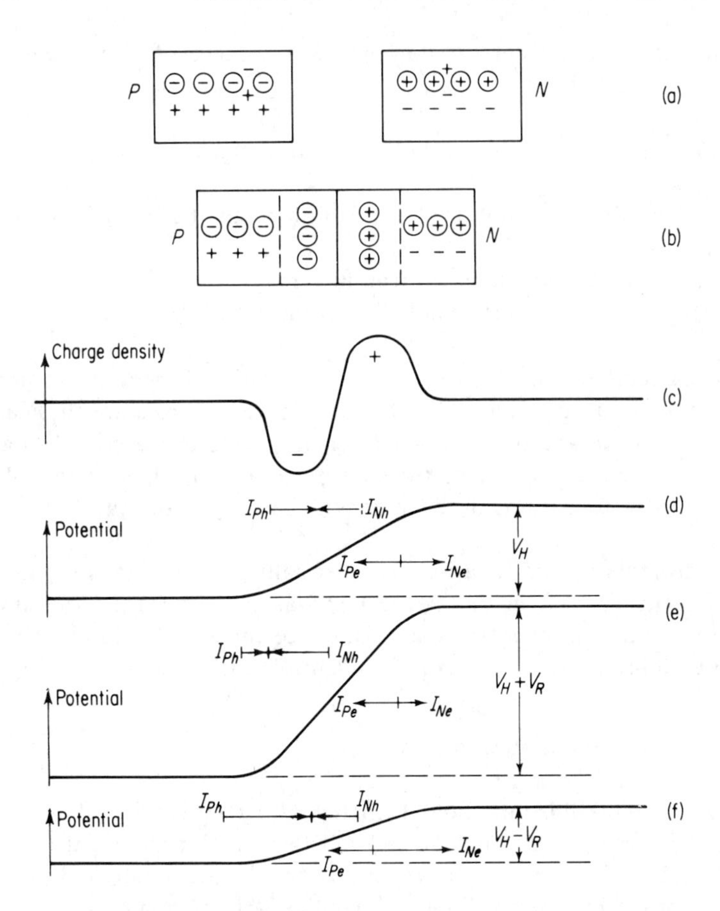

Fig. 1-10. The PN junction. (a) Two crystals of opposite conductivity; (b) formation of depletion layer; (c) a plot of the net charge density through the crystal showing the charge dipole at the junction; (d) with no external applied voltage there is a potential barrier at the junction and the net current flow is zero; (e) with a reverse voltage applied, the potential barrier is increased in height and current flow is reduced; (f) with a forward voltage applied, the potential barrier is reduced and a large current flows.

potential distribution across the junction is plotted in Fig. 1-10(d). The potential difference across the junction serves as a barrier to further hole migration from p-type to n-type. At the same time, the relatively few holes in the n-type material can be easily swept across the junction by the electric field. Similar comments can be made for the effect of the barrier on electron flow across the junction. We see that the initial migration of carriers across the junction sets up a barrier to further migration. Since the requirements for equilibrium in the system are that no net steady-state current flows across the junction, the potential barrier height adjusts to fulfill this condition.

When considering current flow across the junction, four components must be considered:

1. I_{Ph}, current flow due to hole flow from the P to the N side of the junction
2. I_{Ne}, current flow due to electron flow from the N to P side of the junction
3. I_{Pe}, current flow due to electron flow from P to N
4. I_{Nh}, current flow due to hole flow from N to P

The currents I_{Ph} and I_{Ne} arise from the majority charge carriers in the P and N regions, respectively. In the p-type material, holes are the majority carriers. Electrons are the majority carriers in the n-type material. Not all the majority carriers contribute to these currents, as only those with sufficient kinetic energy to surmount the potential barrier at the junction can be involved.

The currents I_{Pe} and I_{Nh} arise from thermally generated minority carriers. The minority carriers in the p-type material are electrons. Holes are the minority carriers in the n-type material. These currents due to the thermally generated minority carriers are independent of barrier height.

Reverse Conduction or Reverse Bias

Figure 1-10(e) illustrates how the potential barrier is raised if an external potential is applied in a direction to aid the junction potential. With the barrier higher, the current components I_{Ph} and I_{Ne} are greatly reduced since they are primarily due to diffusion and the carriers cannot overcome the increased barrier height without added energy.

For moderate values of applied voltage they are insignificant compared to I_{Nh} and I_{Pe}. The latter components, I_{Nh} and I_{Pe}, are, however, independent of the barrier height since they do not fall under the influence of the barrier until they diffuse to the edge of the depletion layer. When they reach the edge of the potential barrier, they are swept down since they are attracted by the potential. Consequently, reverse conduction is characterized by a saturation current I_{Nh} and I_{Pe} which is in proportion to the minority carrier density—since both I_{Nh} and I_{Pe} are due to minority carriers. At the same time one must remember that the number of available minority carriers is quite small so that the current which flows in reverse conduction is quite small (10^{-11} to 10^{-3} amp, depending on material, junction area, and temperature). Because these minority carriers are thermally generated, the saturation current would be expected to rise with increased temperature.

Forward Conduction or Forward Bias

Figure 1-10(f) illustrates how the potential barrier is affected if an external potential is applied with a polarity to oppose the junction potential. It is easier for positive carriers in the p-type material to be injected into the n-type material. Thus, the flow of majority carriers, components I_{Ph} and I_{Ne}, is greatly increased while the saturation component of the current, due to the minority carriers, remains constant. Notice that it is now easy for a large number of majority carriers to flow, resulting in a relatively large current. It is interesting to note that the majority carriers are not all thermally generated. In fact, most of them were added to the semiconductor on purpose, in the form of an impurity dopant; hence the current which flows in forward conduction is not nearly so temperature sensitive as the reverse conduction current.

PN Junction Current-Voltage Characteristics

The relationship between current and voltage in an ideal PN junction diode has been shown to be exponential in character (5).* Below the diode breakdown voltage and ignoring diode series resistance

$$I = I_S(e^{qV/kT} - 1) \tag{1-7}$$

in which I_S = saturation current
V = applied junction voltage
q = charge of an electron
k = Boltzmann's constant
T = temperature in degrees Kelvin.

The value of q/kT is about 40 at room temperature (25°C). Note that the equation predicts the slope of the V-I characteristics for both forward- and reverse-bias conditions. For the reverse-bias condition, V negative, the exponential term is negligible and the current becomes $-I_S$ (the minus sign signifies the reverse direction). For forward-bias voltages, V positive, the equation predicts an exponential relationship between the applied voltage and the current through the diode.

The nonideal diode may be statically characterized by the ideal diode with the addition of resistances. These resistances are termed the *spreading* resistance (R_S) and the *leakage* resistance (R_L). As shown in Fig. 1-11, the spreading resistance is primarily the resistance of the bulk material between the terminal and the actual junction and is dependent on the length and

*Numbers in parentheses refer to list at the end of the chapter.

cross-sectional area of the bulk material. Contact resistance between the external leads and the bulk material must also be included, if appreciable. The leakage resistance shown in Fig. 1-11 constitutes a current path around the diode. This current is a bilateral current and is primarily a surface phenomena. Since the leakage current is dependent on the junction surface condition, one would expect its variation to be erratic. This is usually the case. If the leakage component can be represented by a resistor as shown in Fig. 1-11, the current should be, and is, voltage sensitive. This suggests that the leakage and saturation currents may be separated in a diode by measuring at different voltages. Manufacturers sometimes show these data with a low voltage (1–2 volts) leakage current measurement (saturation component) and a higher voltage measurement for the leakage component. Because the leakage component is primarily due to surface states, it is not usually temperature dependent. Thus, the method just described also separates the component which is temperature sensitive. A practical diode may be represented as shown in Fig. 1-12. The effect of these two resistances on the diode characteristic curves is shown in Fig. 1-13.

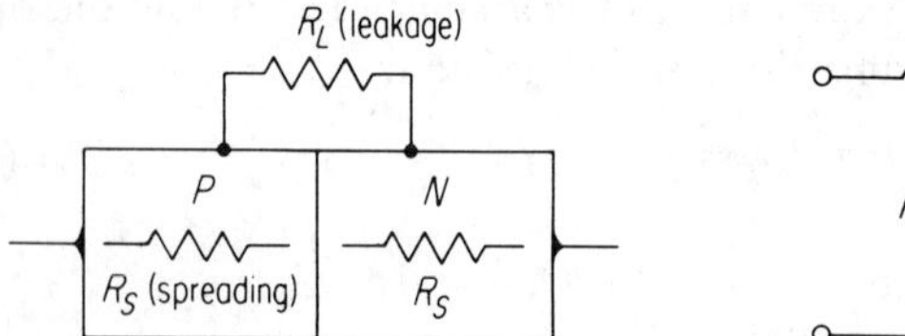

Fig. 1-11. Parasitic resistances in a PN junction diode.

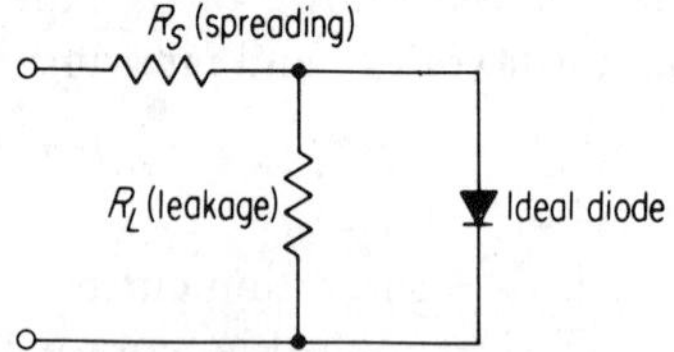

Fig. 1-12. Equivalent circuit of a practical diode.

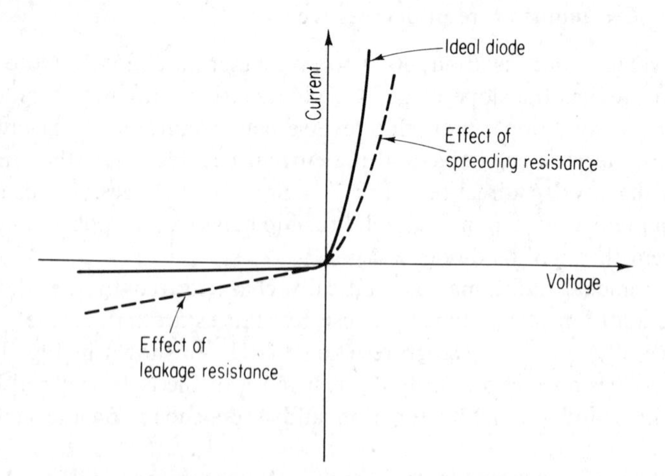

Fig. 1-13. The effect of parasitic resistances on the V-I characteristics of the ideal diode.

Another especially interesting property of the PN junction is the phenomenon of voltage breakdown. Two accepted mechanisms can cause voltage breakdown: (1) the Zener effect, (2) the avalanche effect. The Zener effect occurs when the electric field within a material becomes very large. When the electric field across the junction becomes large enough to overcome the binding force of a covalent bond, the bond will be broken, thus generating free carriers (a hole-electron pair). When this occurs, the current through the depletion layer increases rapidly. Zener breakdown is characterized by a very sharp break in the *V-I* characteristic as shown in Fig. 1-14. It is relatively free from noise and is very stable with temperature.

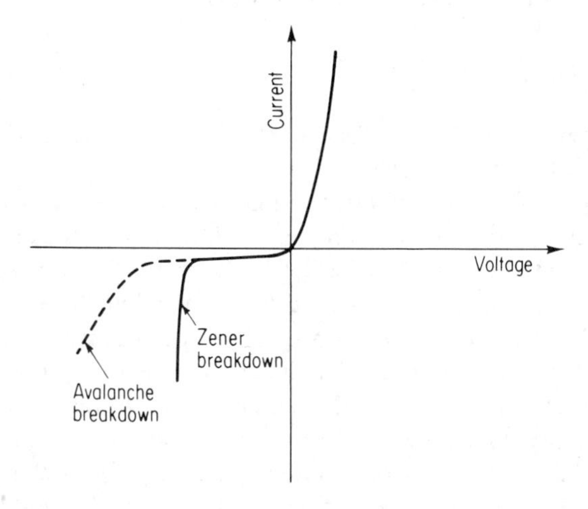

Fig. 1-14. The V-I characteristics of avalanche and zener breakdown.

The value of the electric field necessary to cause this in germanium is given by

$$\mathcal{E} = 2 \times 10^4 \sqrt{\sigma V} \tag{1-8}$$

where V is the applied voltage and σ the conductivity. As a typical example for $V = 50$ volts, $\sigma = 0.5$ mho/cm, we get $\mathcal{E} = 10^5$ volts/cm.

Avalanche breakdown, as its name implies, is similar to the avalanche of a snow-covered mountain. Consider a carrier at the edge of the depletion layer. Owing to the electric field in the depletion layer, the carrier is accelerated through the depletion layer. Now, if the electric field is high enough, the carrier will be accelerated and obtain a large amount of kinetic energy. If this high-energy carrier collides with a stable atom, it may free other carriers, thus generating hole-electron pairs. These newly generated carriers will themselves be accelerated, causing further collisions and the generation of more hole-electron pairs. Since the distance between collisions (mean free path) is small, the electric field must be large enough to impart the proper amount of kinetic energy to the carrier within a distance equal to one mean free path. Avalanche breakdown is characterized by a high noise content (because of the many collisions) and the relatively soft knee of the *V-I* characteristic. This is exactly the same mechanism as occurs in the glow discharge of a gas diode.

Since the avalanche process is dependent upon the probability of making collisions, it is apparent that one way to vary the breakdown voltage of the diode is to change the amount of dopant added to the pure semiconductor material. The more heavily doped the crystal, the lower the breakdown voltage. This is, of course, also true of transistors.

Typical AC Properties of the Diode

Up to this point, we have characterized the dc properties of the diode by a *V-I* curve in both the forward and reverse regions. Such a curve, although quite necessary, is far from sufficient to tell us all we need to know about the diode as a circuit element. For example, the *V-I* curve tells us nothing about the high-frequency behavior of the diode. It would be extremely useful to obtain a lumped parameter equivalent circuit of the diode which characterizes its ac performance as the circuit of Fig. 1-12 characterizes the dc performance.

Figure 1-15(a) shows a normal PN junction with connecting leads. Figure 1-15(b) shows the same diode with the various parasitic resistances of the diode shown approximately where they appear physically in a practical diode. Starting from the left we encounter the resistance of the connecting leads (R_{wp}) which is almost always negligible. Next, there is the contact resistance of the solder. This resistance is not always negligible because the solder (R_{cp}) is between a conductor and a semiconductor. Next, there is the series resistance of the bulk semiconductor material (R_{sp}), in this case the p-type

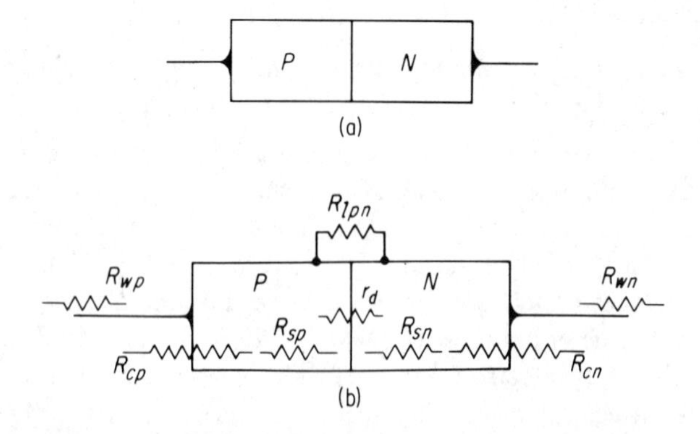

Fig. 1-15. Location of resistances in a PN junction diode.

material. At the actual junction there is associated with the diode a small-signal resistance (r_d) which may be obtained from the slope of the V-I curve in either the forward or reverse directions. This resistance is referred to simply as the *internal diode resistance* and is dependent upon the amount of current flow through the diode. For example, $r_d \approx 26\ \Omega$ for 1 ma current flow.

Electrically in parallel with the internal junction is a leakage resistance, R_{lpn}. This leakage resistance is composed of two components representing the saturation current and surface leakage current as described earlier. The leakage resistance primarily affects operation in the reverse-biased region. Of course, there are also spreading and contact resistance in the n-type material. Figure 1-16 shows the lumped parameter circuit representation of the various resistances of the diode.

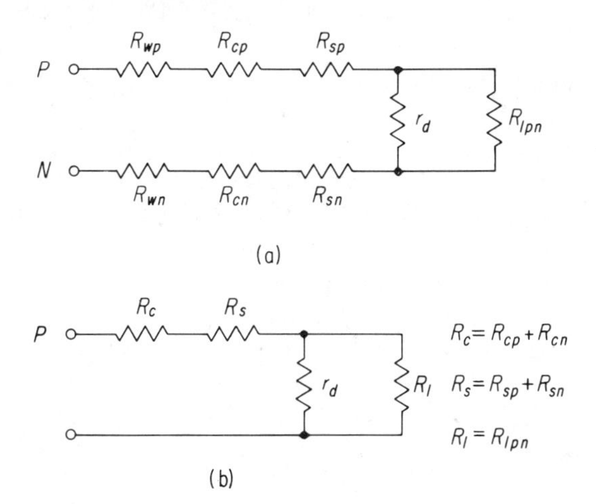

Fig. 1-16. (a) Lumped parameter equivalent circuit; (b) simplified equivalent circuit.

Besides these resistances two capacitances are associated with the diode. The first is the junction capacitance which is a result of the charge stored on both sides of the depletion layer. In fact, it represents the charge which must be displaced in order that an applied voltage will change the width of the depletion layer. As such, the junction capacitance is quite similar to an ordinary parallel plate capacitor in which the P and N materials act as the plates and the depletion layer as the dielectric. In fact, the value of the junction capacitance is given by the familiar equation

$$C_j = \frac{\epsilon A}{W} \tag{1-9}$$

where ϵ = dielectric constant
A = junction cross-sectional area
W = width of the depletion layer.

Typical values of C_j run from less than 1 pf for high-frequency diodes to 500 pf for power diodes. Since the value of the junction capacity depends on the width of the depletion layer, it follows that its value is dependent upon the applied voltage. A curve of C_j versus voltage is shown in Fig. 1-17 for both forward- and reverse-bias conditions.

The second capacitance associated with the PN junction is termed the *diffusion capacitance*. From Eq. 1-4 we see that the current is dependent upon the slope of the charge versus distance curve. Figure 1-18 shows the distribution of holes which are injected into the N region from the P region. The charge density at the junction is a function of the applied voltage and is quite large, depending on how much current is flowing through the diode. A large distance from the junction, the charge concentration must be small because of the exponential nature of the charge versus distance curve. It is important to note that, in order to maintain a certain current flow, a charge distribution must be set up and maintained such that the slope of the charge versus distance curve has the required value. Integrating I_d over the distance x will provide the charge, Q, which must be stored in the bulk material of the diode in order to maintain a specified current flow. This total charge is represented by the cross-hatched area of Fig. 1-18. This charge is related to a lumped capacitance known as the *diffusion capacitance*. Let us consider what happens when the diode is conducting a certain amount of current and a small signal is applied to the diode in such a direction as to increase the current through the diode. In order to change the diffusion current, we must change the term dQ/dx. In fact, to increase the current, the slope of the charge distribution curve must also increase. This is accomplished by increasing the voltage across the junction. This case is shown in Fig. 1-19. The area between the

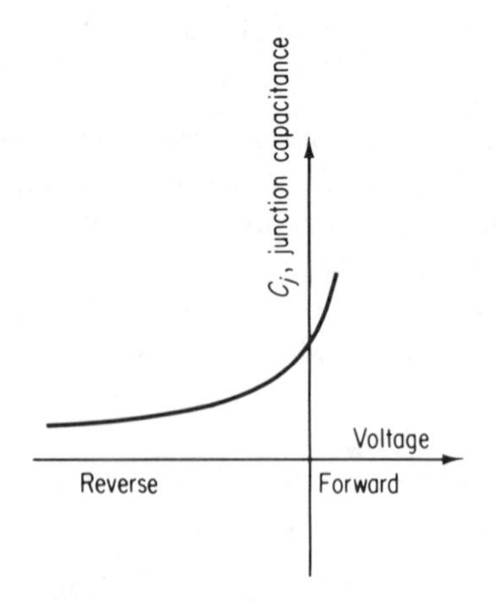

Fig. 1-17. Voltage dependence of junction capacitance.

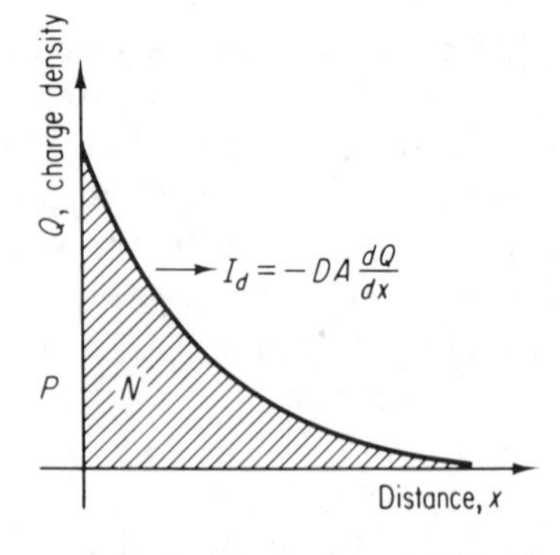

Fig. 1-18. Distribution of injected holes which leads to a diffusion current.

original curve and the curve for increased current represents the additional amount of charge which must be stored within the bulk material of the diode in order that the current be increased. Again, this increased amount of stored charge may be represented by a small signal capacitance which we term the *diffusion capacitance*. Another way of saying the same thing is that an increase in stored charge was realized owing to an increase in applied voltage. Thus, since

$$C = \frac{dQ}{dV} \tag{1-10}$$

we may represent this action by a capacitance. It is apparent that the value of the diffusion capacitance is dependent upon the dc current through the diode.

With the addition of the junction and diffusion capacitances, a more complete equivalent circuit for the junction may be drawn, as shown in Fig. 1-20. This circuit provides the designer with a method of predicting the high-frequency performance of the diode.

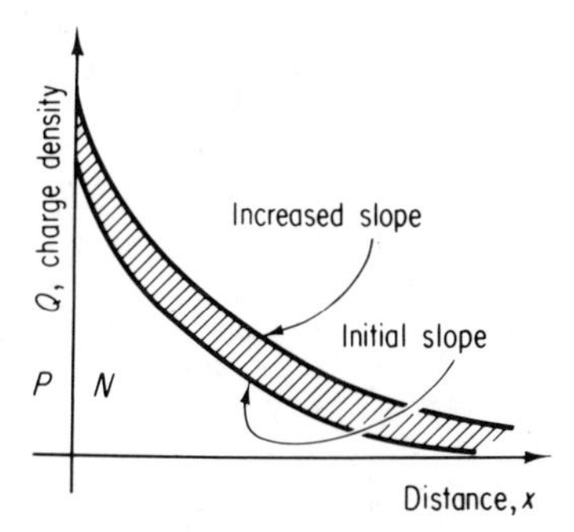

Fig. 1-19. The origin of the diffusion capacitance.

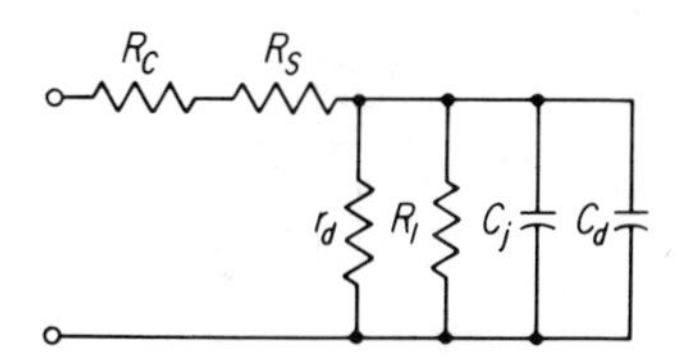

Fig. 1-20. A complete ac equivalent circuit for the PN junction diode.

In addition to the elements shown in Fig. 1-20, various parasitic elements, such as series lead inductance and stray capacitance, will also be found in any practical diode. The elements of this circuit when added to the inherent characteristics of an ideal diode provide a good approximation of the performance of a practical diode.

THE TRANSISTOR (BIPOLAR)

If the operation of the PN junction is thoroughly understood, the operation of a bipolar transistor may be comprehended by the simple superposition of the characteristics of two PN junctions.* Earlier in this chapter, it was stated

*Later in this chapter we will discuss unipolar devices generally called FET's and will reserve the term transistor for bipolar devices in the balance of the text.

that the transistor was composed of two PN junctions on a common piece of semiconductor material in very close proximity to each other. Electrically connecting two diodes with wires will not work, nor will placing two diodes in direct mechanical contact. The two PN junctions must be formed very close to each other and on a common piece of semiconductor material. Specifically, they must be within one diffusion length (L) of each other so that the carriers which flow through one diode may reach the second diode before they recombine and disappear.

When conventionally biased, the transistor has one diode forward-biased (the emitter-base) and one diode reverse-biased (collector-base). Consider the single forward-biased PN junction of Fig. 1-21(a) which shows the physical structure of an alloyed PN junction. Figure 1-21(b) shows the positive charge distribution from the emitter out into the n-type region which, for simplicity, is extended indefinitely. The diffusion length is denoted by L. Figure 1-21(c) illustrates the diagram of the PN junction potential and the fact that positive carriers (holes) are being injected into the n-type material.

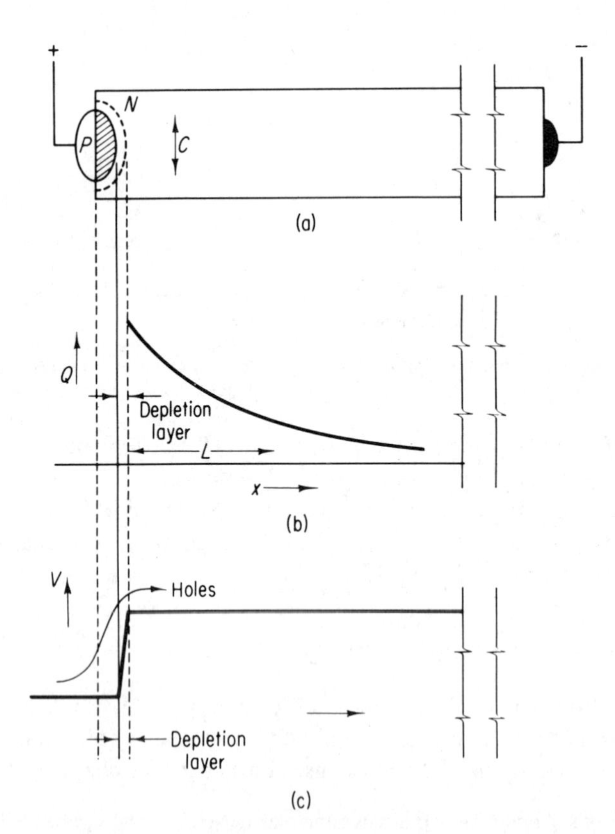

Fig. 1-21. Injection of holes into the base of a PNP transistor.

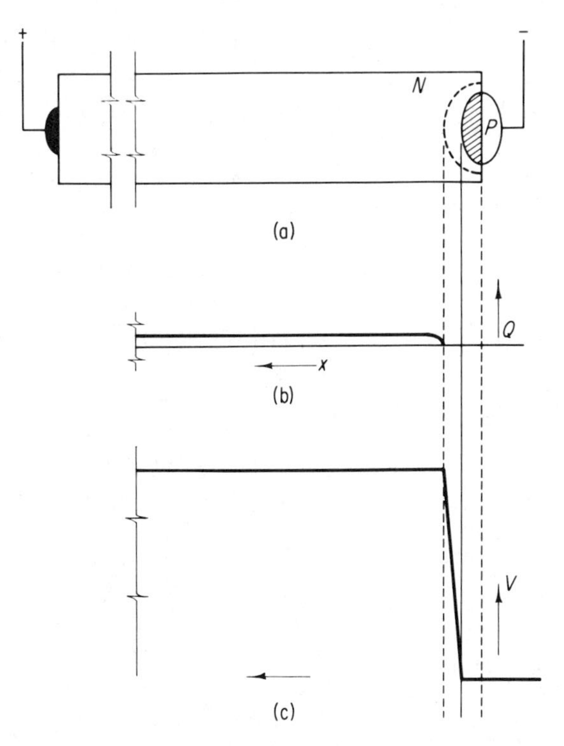

Fig. 1-22. Collection of holes at the collector in a PNP transistor.

Now consider the reverse-biased NP junction at the right in Fig. 1-22(a). This shows the physical structure of an alloyed NP junction. Figure 1-22(b) shows the positive hole charge distribution from the collector into the bulk material of the N region. The positive charge density is zero right at the edge of the depletion layer (electrical junction) since any positive carrier which reaches the edge of the junction is immediately swept down the potential hill and into the p-type material. This hole current is, of course, nothing more than the small number of holes which exist in the n-type material. If we suppose that this small number of holes in the n-type material were suddenly increased, for example, by having a large number of holes injected into the n-type region, then the reverse current through the diode would increase and would be a function of the number of holes injected into the n-type material. Referring back to Fig. 1-21, if we formed our NP junction at line *C* which is within a diffusion length of the forward-biased PN junction, the forward-biased diode would then act as a source of holes for the reverse-biased NP junction. In other words, holes which were injected (or emitted) by the PN junction would be randomly diffused through the n-type material and collected by the NP junction. Since the n-type material is common to both diodes, we have obtained a transistor from two diodes connected through a common region. These are the basic concepts explaining transistor action. Carriers are emitted by a forward-biased junction, in an amount controlled by the applied forward bias voltage, and collected by a second junction which is placed within a diffusion length (L) of the emitting junction. Almost all the emitted carriers will be collected if the collector junction is placed very close ($L/100$) to the emitter diode (recombination will remove only a few of the charges). The number of collected carriers will be nearly independent of the collector voltage since all that is required is that it be reverse-biased.

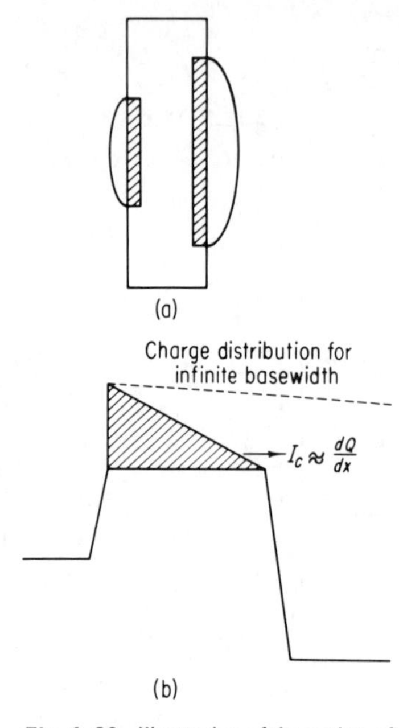

Fig. 1-23. Illustration of the action of a simple transistor.

Figure 1-23 illustrates the action of a simple transistor. Figure 1-23(a) shows the physical PNP junctions. Figure 1-23(b) shows the potential barrier diagram with the addition of the charge distribution in the base region. The charge distribution curve is the initial portion of an exponential and thus is almost linear. Note that the flow of current carriers through the n-type material (base) is entirely by diffusion. There is no voltage across the base region in the lateral direction.

With this analysis of transistor action we can look into the question of its operation. A positive carrier at the emitter must overcome a potential of approximately 0.5 volt in Ge* before it can be injected into the base. Once there it diffuses to the collector edge of the base with no change in potential. When the charge falls over the potential hill into the collector, a potential of many volts (typically of the order of 10 volts), it does useful work. Since approximately the same current is collected, the power gain is approximately equal to the ratio of the voltages and this is of the order 20. Thus, we can get significant gain and have constructed an extremely useful device.

BIPOLAR TRANSISTOR CHARACTERISTICS

One way of characterizing a device is by a set of input-output curves. Let us consider the transistor connected in the common-base configuration which means that the input and output are connected to the base as a common lead.

*About 0.7 v in Si.

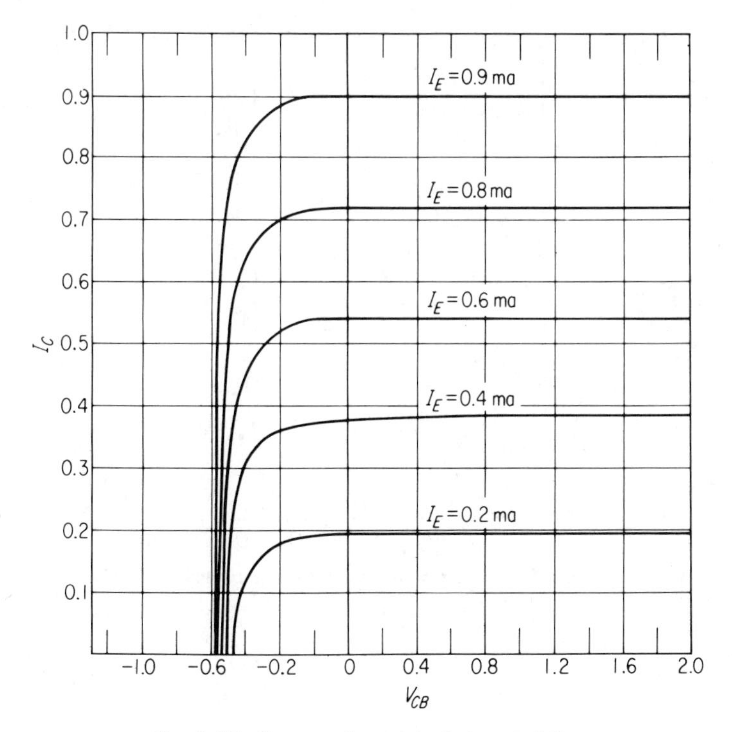

Fig. 1-24. Common-base output characteristics.

If we plot a *V-I* characteristic of the collector with the emitter current set at various constant values we will obtain Fig. 1-24. Notice that the curves are extremely flat, indicating that the collector is approximately a current source. Notice also that the collector must be slightly forward-biased in order to reduce the current to zero. Now let us see whether we can explain this curve from what we know of the transistor. Perhaps the simplest explanation can be obtained from the potential hill diagram of Fig. 1-23(b). Since the diffusion current is proportional to the slope of the charge gradient in the base, and since the charge density at the collector must be zero, there is no reason for the slope of the charge distribution to change significantly as the collector voltage is increased from zero. Instead the collector current is relatively independent of voltage. In order for the collector current to become zero, the collector must inject as much current into the base as it receives from the base. As should be expected, the collector current is slightly smaller than the emitter current by an amount equal to the base current; that is:

$$I_E = I_C + I_B \tag{1-11}$$

The current gain (actually a loss) is the ratio of the collector current to the

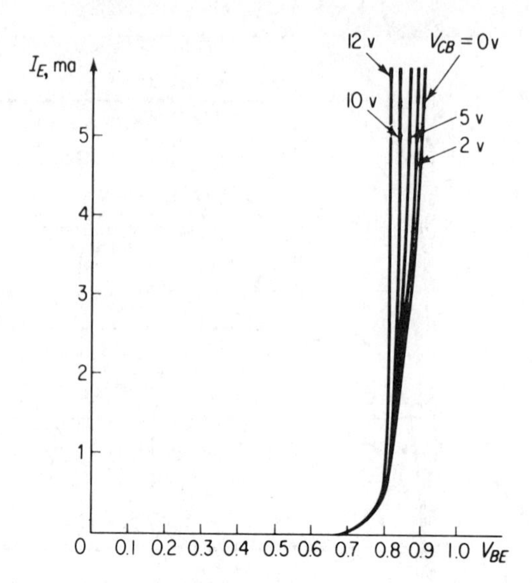

Fig. 1-25. Common-base input characteristics.

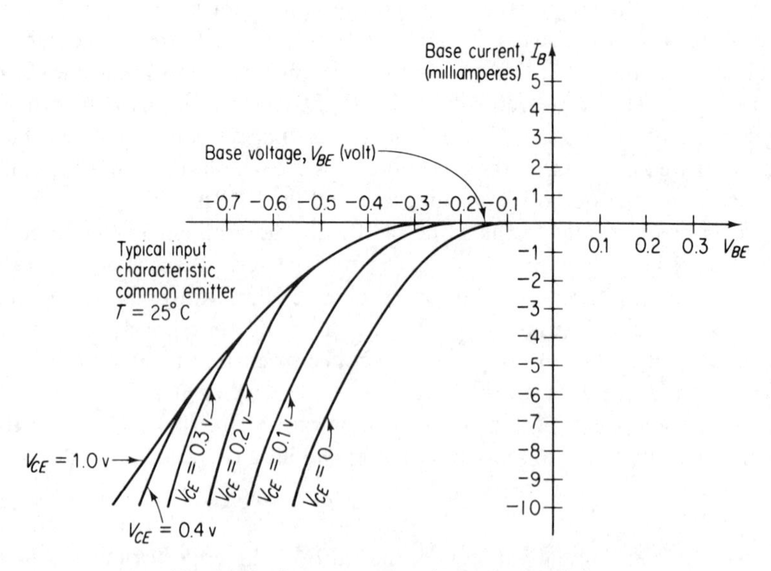

Fig. 1-26. Common-emitter input characteristic.

emitter current and is denoted by α where

$$\alpha = \frac{I_C}{I_E} \tag{1-12}$$

A plot of the input characteristic for the common-base connection is shown in Fig. 1-25. As expected, we see the exponential curve of a forward-biased diode. Again, the *V-I* characteristic is almost independent of variations in collector voltage. The slight dependence on collector voltage occurs because the collector voltage sets the width of the collector depletion layer and thus varies the base width. The change in base width changes the slope of the charge distribution and thus the diffusion current. Similar curves may be obtained for the common-emitter and common-collector connection. Since most applications of the transistor are in the common-emitter connection, a discussion of these curves is indicated.

The common-emitter characteristic is quite similar to that for the common base, and again takes on the form of an exponential curve (Fig. 1-26). The base-emitter voltage required to obtain a certain base current is quite dependent upon the collector-emitter voltage for small values of V_{CE} but nearly independent of V_{CE} for large values of this voltage because the movement of the collector depletion layer varies approximately as the square root of the applied voltage. Again, the voltage required to produce a certain current is dependent upon the type of material used and averages 0.3 volt for germanium and 0.8 volt for silicon. The output characteristic (Fig. 1-27) shows a plot of collector current versus collector-to-emitter voltage for various base

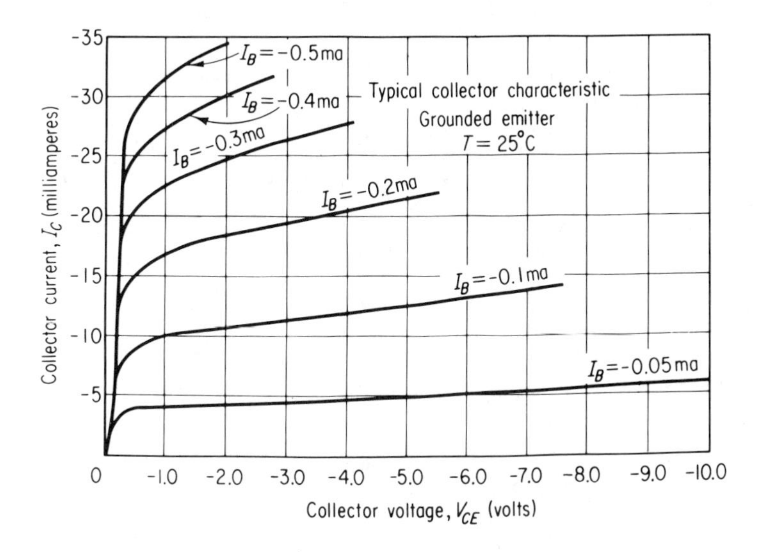

Fig. 1-27. Common-emitter output characteristic.

currents. Unlike the common-base output characteristic, all the curves lie to the right of the axis. The left-hand region of the curve is called the *saturation region* and represents a point where an increase in base current yields virtually no increase in collector current. In this region both emitter and collector are forward biased. The output curves are less flat than the common-base curves because the output impedance is many times less than in the common base connection. Notice also that the base current is very small compared to the collector current. The current gain in the common-emitter connection, β, is defined as the ratio of the collector current to the base current and is given as

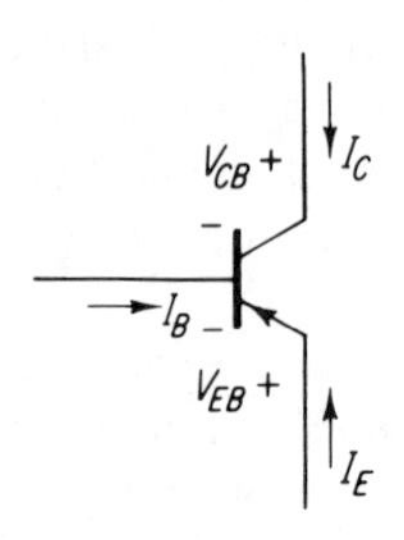

Fig. 1-28. Definition of voltage and current polarities.

$$\beta = \frac{I_C}{I_B} \tag{1-13}$$

Common-emitter and common-base current gains can be readily related. With the assumed current polarities of Fig. 1-28

$$\beta = \frac{I_C}{I_B} = \frac{\alpha I_E}{(1 - \alpha)I_E} = \frac{\alpha}{1 - \alpha} \tag{1-14}$$

Note that as α approaches 1, β approaches ∞. Typical values of β range from 20 to 300. This high value of current gain is the most desirable characteristic of the common-emitter connection. Whereas the common-base connection yields only voltage gain, the common-emitter connection yields both current and voltage gain.

Another method of characterizing the transistor is a set of equations derived by Ebers and Moll. Sometimes called the *symmetrical equations*, they relate the forward and reverse current gains, voltages, and currents of the transistor. The conventions of Fig. 1-28 must be used in conjunction with these equations. Ebers and Moll have shown that (4)

$$I_E = \frac{-I_{EBO}}{1 - \alpha_N \alpha_I}(e^{qV_{EB}/kT} - 1) + \frac{\alpha_I I_{CBO}}{1 - \alpha_N \alpha_I}(e^{qV_{CB}/kT} - 1) \tag{1-15}$$

$$I_C = \frac{\alpha_N I_{EBO}}{1 - \alpha_N \alpha_I}(e^{qV_{EB}/kT} - 1) - \frac{I_{CBO}}{1 - \alpha_N \alpha_I}(e^{qV_{CB}/kT} - 1) \tag{1-16}$$

where I_{CBO} = collector-base saturation current
I_{EBO} = emitter-base saturation current
α_N = normal common-base current gain
α_I = inverted common-base current gain (collector and emitter inverted).

These equations are very useful, for much can be learned from them. They predict to a reasonable degree of accuracy the dc characteristic curves shown in Fig. 1-24–1-27. The application of these equations will be discussed in Chap. 7.

TRANSISTOR FABRICATION TECHNIQUES

It is generally not necessary for the circuit designer to know the type of construction process used for a given transistor or integrated circuit (IC) if the specifications of the device are known. It is sometimes helpful, however, to know what properties of the device are optimized by a particular fabrication technique. Before IC's, greater justification for considering production methods was the increased understanding gained by the designer of the physical mechanisms involved. The following section then can still be useful but has now lost much of its importance due to the greater use of IC chips and functional blocks in design. The processing of IC's will be discussed fully in a later chapter but does use many of the individual steps outlined here. In general, six basic techniques exist for forming junction devices:

1. Point contact
2. Alloyed
3. Grown junction
4. Diffused
5. Electrochemical
6. Epitaxial

Although other fabrication methods exist, these represent the most important techniques at the present time.

Point-Contact Junctions

The earliest technique, which has been known since the days of the crystal radio, consists of forming a rectifying junction by pressing together two materials, one of which has a small area at the point of contact. This is usually accomplished by bringing a small whisker of one material into contact with a wafer of the other material. Hence the name *point-contact*. The exact theory of point-contact diodes has been debated for many years.* It is known, however, that when a small metallic whisker is pressed against the surface of an n-type wafer of semiconductor material, a rectifying junction may be formed.

Some type of spring tension is used to hold the whisker against the wafer. Often this is accomplished by placing a kink in the whisker wire. A cross section of a point-contact diode is shown in Fig. 1-29. Point-contact transistors are now obsolete, although

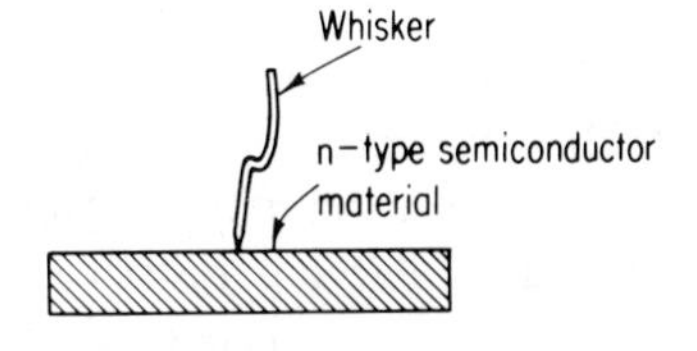

Fig. 1-29. Cross-sectional sketch of a point-contact diode.

*The most current theory which is generally accepted is that a metal/semiconductor junction or Schottky barrier is formed in the process.

this technique is still used in many very high-frequency diodes because it is possible to realize small junction areas and, therefore, low capacitance.

Alloy Junctions

The formation of PN junctions by alloying techniques produces the majority of all low- and medium-frequency transistors. This technique consists of taking a blank of one type of semiconductor material, which ultimately forms the base, and then alloying two pellets into opposite sides of the blank. [See Fig. 1-30a.] The pellet material must be capable of doping the semiconductor material. The two pellets are held in place in a fixture (usually

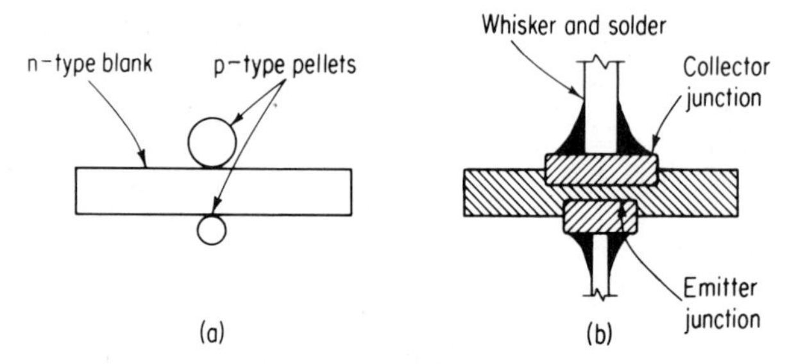

Fig. 1-30. Sketch of alloy junction transistor. (a) Before alloying, and (b) after attaching whiskers. Sketch is idealized in that junctions are not flat but rounded in an actual transistor.

a carbon boat); then the whole assembly is passed through a furnace. Under the heat and pressure applied by the fixture, the pellets soften and alloy into the n-type blank. When cooled, the material recrystallizes and forms PN junctions as shown in Fig. 1-30(b). Contacts are made to the three elements to form the PNP transistor as shown in Fig. 1-30(b). A similar process is used for an NPN device except that n-type pellets are alloyed into a p-type blank.

Diffused Junctions

Diffused regions are used in almost all modern high-frequency transistors. Diffusion processes are used in device fabrication for two reasons: first, as a means of forming a PN junction and second, as a technique for forming a graded or varying resistivity region within the device. In either case the diffusion process is essentially the same.

A source of the diffusant atoms is introduced at the surface of the semiconductor material which is heated to a relatively high temperature. For germanium, diffusion is carried out at material temperatures of 600–800°C; temperatures in the range of 100–1300°C are used for diffusion in silicon. The depth of penetration of the diffusant (dopant) depends upon material temperature, diffusion time, and the concentration of the diffusant atoms at the semiconductor surface.

A typical diffusion process is carried out in a furnace in which the semiconductor material is heated in one temperature zone and a source of the dopant is heated in another temperature zone (Fig. 1-31). A diffusion system such as this is often used, for example, in diffusion of phosphorus into silicon. In this case the dopant source is phosphorus pentoxide (P_2O_5). The source and material temperatures, T_1 and T_2, are typically 230°C and 1000°C, respectively.

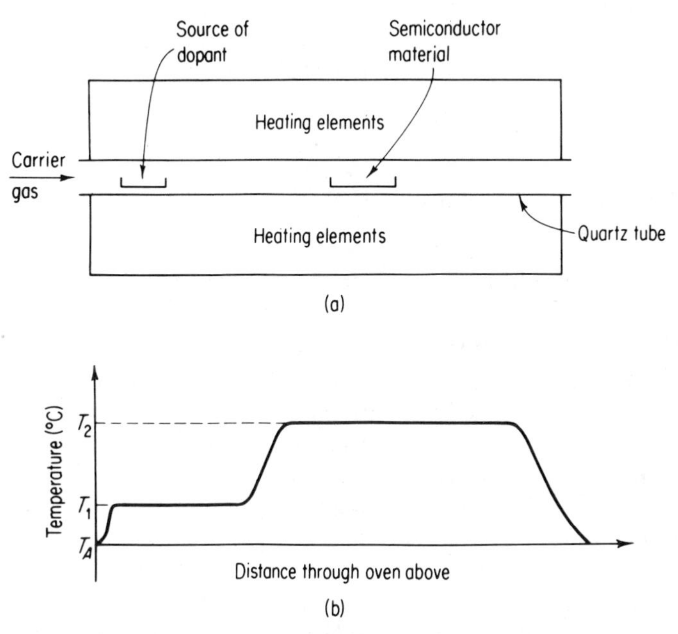

Fig. 1-31. Sketch of typical diffusion process. (a) Relative location of dopant and semiconductor material; (b) oven temperature profile.

If diffusion is being used to form a junction, then the diffusant is such that it will dope the material to the opposite conductivity type. For instance, n-type semiconductor material is diffused with a p-type dopant. Figure 1-32(a) shows a cross section of an n-type blank which has been diffused with a p dopant. Note that the dopant diffuses into all sides of the blank. Where the concentration of n- and p-type dopants is equal, a PN junction is formed as shown by the inner rectangle.

Figure 1-32(b) shows how the dopant concentration varies across the blank. Note that the resistivity is low at the surface owing to the high impurity concentration and decreases through the blank until the bulk resistivity of the

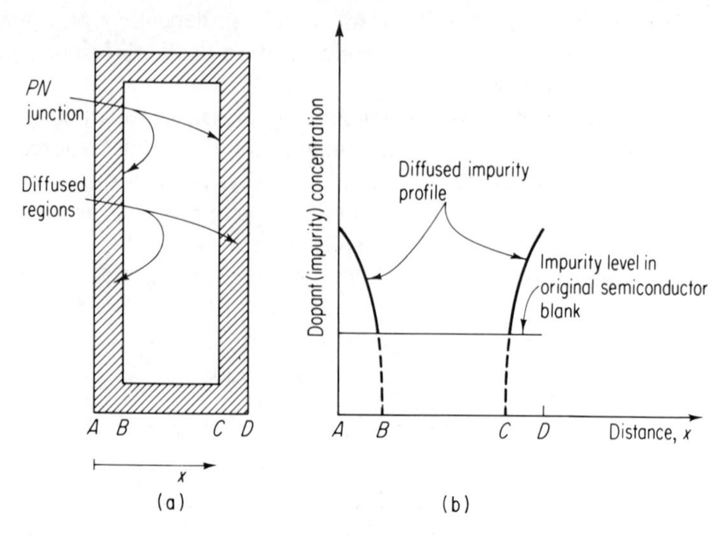

Fig. 1-32. (a) Sketch of the cross section of a diffused blank; (b) a typical impurity concentration diagram for a diffused blank.

original blank is reached. The material resistivity is inversely proportional to the dopant concentration.

If diffusion is used to produce only a graded region then the same type of dopant is used for the diffusant. That is, an n-type dopant is diffused into n-type material or p-type dopant into p-type material. The diffusant causes a higher impurity concentration at the surface than in the original blank. Thus, the dopant concentration as a function of the distance through the blank would be similar to Fig. 1-32(b) for any diffused material. Note, however, that no junction would be formed in this case.

Common diffusants for germanium transistors are antimony and arsenic as n-type dopants and gallium and indium as p-type dopants. For silicon, phosphorus is often used as an n-type dopant and boron and aluminum are often used as p-type diffusants.

Electrochemical Junctions

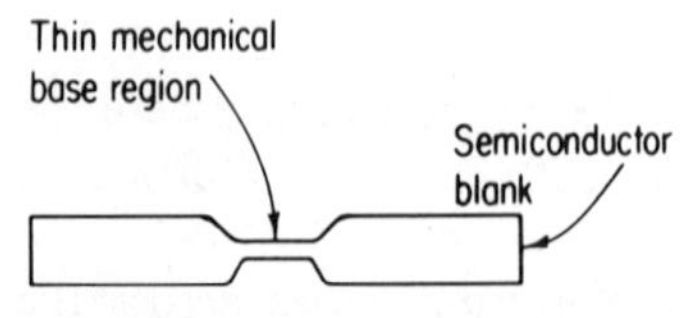

Fig. 1-33. Cross section of a semiconductor blank which has been etched to produce a thin mechanical base width.

Electrochemical techniques may be combined with any of the other techniques to form a transistor. Basically, electrochemical techniques consist of forming the active transistor region electrochemically by a combination of etching and plating. In the simplest case, shown in Fig. 1-33, the basic blank is etched to produce a narrow mechanical base region. The pits, which are circular, are formed by directing a jet of etching solution at the surface of the blank. Etching time is controlled to produce a very thin web (about 0.2 to 0.5 mils) between the two pits. The etching process is then essentially reversed and contacts are electrochemically plated onto the bottoms of the pits. This is rarely used today for forming junctions but still may be important for device formation.

Expitaxial Junctions

Epitaxial layer growth is a relatively new technique for PN junction formation. With this technique, a semiconductor film or layer is grown or deposited on a host substrate of the same material (for example, silicon deposited on silicon). The deposited material assumes the same crystal lattice orientation as the host substrate, a condition required to produce high-quality junctions.

In a typical process the semiconductor material is deposited via a chemical reaction which "frees" the semiconductor from a host compound. The process is carried out at a high temperature (typically 1100–1200°C for silicon).

As an example, the epitaxial growth of silicon may be accomplished by the chemical reduction of silicon tetrachloride ($SiCl_4$) with hydrogen, yielding elemental silicon and hydrochloric acid (HCl). The deposited silicon may be doped n-type or p-type by adding a suitable dopant compound to the silicon tetrachloride. By growth, for example, of an n-type layer on a p-type substrate, a PN junction may be formed.

In the manufacture of transistors several of these techniques may be combined. For instance, the PADT® (Post Alloy Diffused Transistor) combines diffusion and alloying techniques. The MADT® (Micro Alloy Diffused Transistor) combines diffusion, alloying, and electrochemical techniques.

FIELD EFFECT TRANSISTORS— UNIPOLAR TRANSISTORS

In recent years the search for semiconductor devices with higher input impedance, and hence lower current drain and power consumption, for use in large electronic systems such as computers has led to the development of integrated circuits. These use not only bipolar transistors and devices but also in the last few years field effect devices such as the Junction Field Effect Transistor (JFET) and the Insulated Gate Field Effect Transistor (IGFET) have been developed. We discuss their use more fully in a later chapter on IC's but discuss here briefly their physical basis as discrete devices since they are often used by the designer almost interchangeably with their bipolar cousins. Since they involve only one type of carrier, they are often called unipolar transistors.

Junction FET

Consider the structure of Fig. 1-34 which is an idealized version of the JFET originally conceived by Shockley (12). An n-type channel is between two end electrodes which are called source and drain. On either side of this channel step p-n junctions are fabricated and are operated with reverse bias so that the depletion region formed is wide. The actual bias across the junction is a function of position in the channel since the drain-source voltage is chosen to be such as to draw electrons from the source through the channel to the drain contact. By controlling the reverse voltage on the p-n junction, the gate-source voltage, we can narrow or widen the channel to carrier flow. We thus have a mechanism for modulating the drain current by varying the gate voltage. An ideal and almost typical output characteristic is shown in Fig. 1-35. We see that for a given V_{GS}, initially I_D increases linearly with V_{DS} but soon saturates and reaches a constant value known as the *saturation* or *pinchoff current*. This occurs when the channel has been narrowed as far as it can be or pinched off. The actual situation is such that a finite slope exists

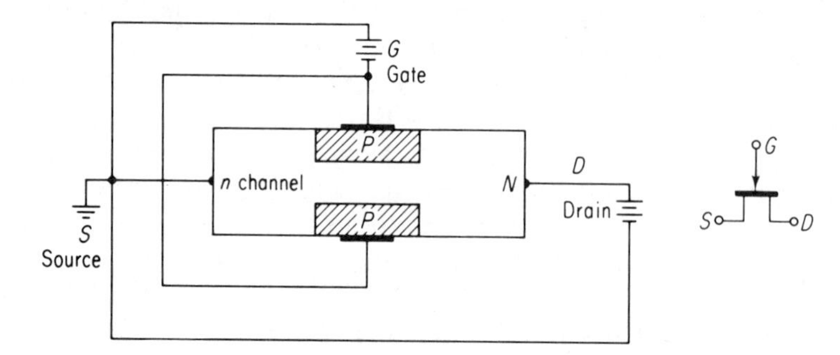

Fig. 1-34. The junction FET: (a) Schematic diagram showing bias; (b) circuit symbol n channel.

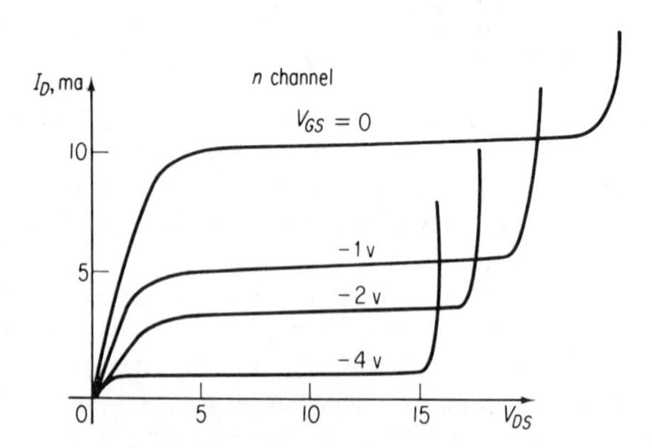

Fig. 1-35. JFET output characteristics.

beyond the initial pinchoff point before breakdown, almost exactly analogous to that for a reverse biased p-n diode. The larger the reverse bias on the gate, the lower the saturation or pinchoff current, which is consistent with our understanding of the reverse bias action on a depletion layer width. Because the device is normally operated with the source as common to the input and output circuits, with the gate as input and drain as output terminals, and since the gate-source junction is reverse biased, very small currents flow and the input impedance of the device is very high. This is analogous to the reverse biased p-n junction discussed earlier.

Insulated Gate FET

A second and far more generally useful device, since its integrated circuit form is easily fabricated and large arrays are convenient to make, is the Metal-Insulator-Semiconductor version of the Insulated Gate FET. The most well known of these is when the insulator layer is made of Silicon Dioxide or the MOSFET (13). A simplified schematic of such a device is shown in Fig. 1-36. Here the intrinsic or p-silicon substrate has photolithographically been prepared with heavily doped n regions, (n+) which constitute the source and

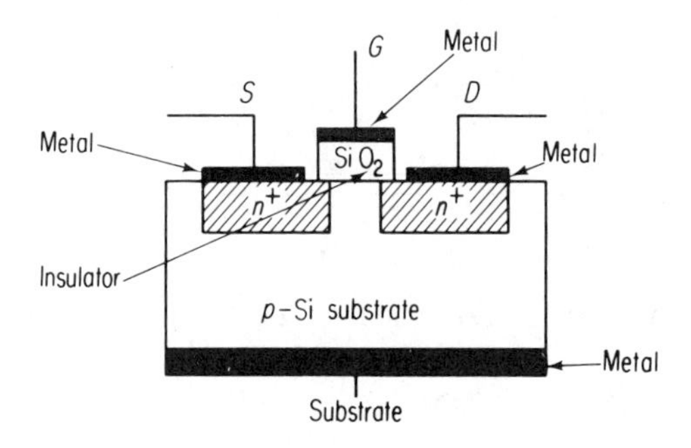

Fig. 1-36. MOSFET.

drain. These effectively form ohmic contacts with the metal connections and higher resistivity channel so that the flow of current between them is controlled by the width of the channel. By growing a thin insulating layer (~1000A°) above the p-substrate which separates the metal gate electrode from the silicon, a capacitive structure is developed. When a voltage is applied between gate and source, charges are realigned within the insulator and cause surface charges to occur on the top of the channel. In the enhancement type of device which is shown in the figure no channel exists unless a large positive voltage is applied to the gate. This causes a movement of holes away from the channel and thereby makes essentially an n-channel device by the creation of an inversion layer. The width of the channel is controlled by the voltage and hence the conductivity of the channel can be varied. If too small a positive gate bias is applied or we have a negative gate voltage, holes are attracted to the surface, and a reverse biased p-n$^+$ junction between source and drain always exists regardless of drain voltage thereby preventing current flow. Because of the insulator, no conduction current can ever flow in the gate unless the breakdown field is exceeded. A p-channel device can similarly be made on an n-substrate with opposite polarities of gate and drain voltages. The output characteristic is shown in Fig. 1-37. Note that no current flows until the gate voltage exceeds a threshold value. A second operating mode can also be achieved by diffusing* in a thin n-channel between source and drain. This device is called a *depletion-type MOSFET* since it will operate even with zero gate bias. A positive gate will cause increased conduction just

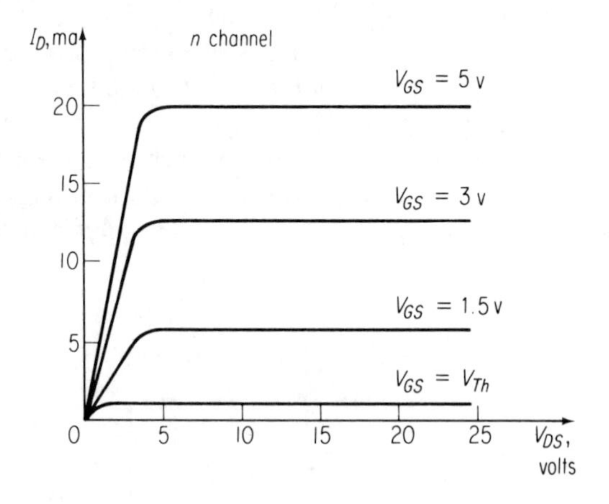

Fig. 1-37. Enhancement MOSFET output characteristic.

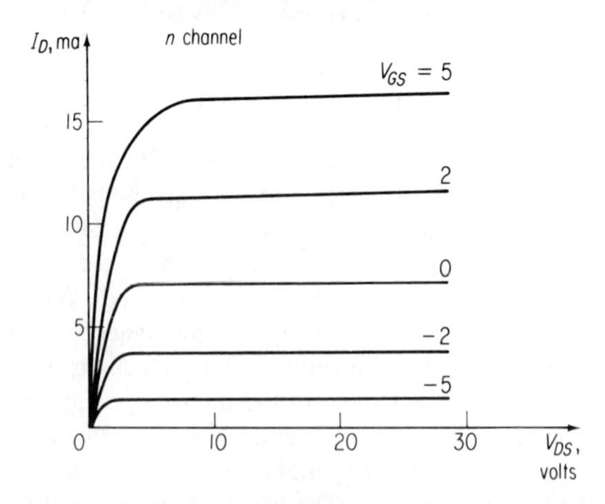

Fig. 1-38. Depletion mode MOSFET output characteristic.

*Currently done more often by ion implantation and used as depletion loads.

as in an enhancement device, while a negative voltage on the gate will reduce the current as shown in Fig. 1-38. Thus we can get a controlled output current for both positive or negative gate inputs in this device, without inversion, and with extremely high input impedance. This gives us an excellent device for large-scale integration and is currently heavily in use for computer and other systems (16). A later chapter will deal with these applications in greater detail.

EXERCISES

1. Assume that the excess electrons due to doping 100 grams of germanium (Ge) with 3.0 μg of pentavalent antimony (Sb) atoms are fully excited at room temperature. If the mobility of the electrons is $\mu_n = 3600$ cm²/volt-sec, calculate:
(a) the conductivity of the n-type semiconductor;
(b) the resistance of a block of Ge 1 cm long and 1 mm square in cross section.

2. Suppose 0.90 μg of trivalent gallium (Ga) is added to the material in exercise 1. If the hole mobility is $\mu_p = 1700$ cm²/volt-sec, calculate:
(a) the conductivity;
(b) the type of semiconductor (n- or p-type).

3. If a crystal were subjected to large pressure, what would you expect to happen to material it was made of assuming it was:
(a) an insulator;
(b) a metal
(c) a semiconductor?
Discuss your answer mainly from the point of view of conductivity changes.

4. (a) Calculate the current flowing through an ideal diode at room temperature (25°C) when $I_s = 1$ μa and a forward bias of 0.15 v is applied.
(b) If the temperature is lowered to such a value that this current is to double what will its value now be?
(c) What changes in forward bias will restore this current to its original value at the new temperature?

5. (a) For a Silicon diode doped with $N_d = 10^{16}$ cm^{-3}, Na $= 10^{15}$ cm^{-3}, and area of 0.015 cm² calculate the current carried by electrons and holes for a forward bias of 0.15 v.
(b) What is the total current?

REFERENCES

1. W. SHOCKLEY, *Electrons and Holes in Semiconductors.* Princeton, N.J.: D. Van Nostrand Company, Inc., 1950.

2. E. SPENKE, *Electronic Semiconductors.* New York: McGraw-Hill Book Company, 1958.

3. C. KITTEL, *Introduction to Solid State Physics*. New York: John Wiley & Sons, Inc., 1953.

4. J. J. EBERS and J. L. MOLL, "Large Signal Behavior of Junction Transistors," *Proc. IRE*, **42**, December, 1954.

5. D. LECROISSETTE, *Transistors*. Englewood Cliffs, N.J.: Prentice-Hall, Inc., 1963.

6. J. LINDMAYER and C. Y. WRIGLEY, *Fundamentals of Semiconductor Devices*. Princeton, N.J.: D. Van Nostrand Company, Inc., 1965.

7. A. NUSSBAUM, *Semiconductor Device Physics*. Englewood Cliffs, N.J.: Prentice-Hall, Inc., 1962.

8. L. P. HUNTER, *Introduction to Semiconductor Phenomena and Devices*. Reading, Mass.: Addison-Wesley Publishing Co., Inc., 1966.

9. J. L. MOLL, *Physics of Semiconductors*. New York: McGraw-Hill Book Company, 1964.

10. R. B. ADLER *et al.*, *Introduction to Semiconductor Physics*, SEEC, Vol. 1. New York: John Wiley & Sons, Inc., 1964.

11. M. J. MORANT, *Introduction to Semiconductor Devices*. Reading, Mass.: Addison-Wesley Publishing Co., Inc., 1964.

12. W. SHOCKLEY, "A Unipolar Field Effect Transistor," *Proc. IRE*, **40** Nov. 1952.

13. S. R. HOFSTEIN and F. P. HEIMAN, "The Silicon Insulated-Gate Field Effect Transistor," *Proc. IEEE*, September, 1963.

14. D. H. NAVON, *Electronic Materials and Devices*. Boston, Mass.: Houghton Mifflin Company, 1975.

15. P. D. ANKRUM, *Semiconductor Electronics*. Englewood Cliffs, N.J.: Prentice-Hall, Inc., 1971.

16. M. F. UMAN, *Introduction to the Physics of Electronics*. Englewood Cliffs, N.J.: Prentice-Hall, Inc., 1974.

17. T. L. MARTIN, JR. and W. F. LEONARD, *Electrons and Crystals*. New York: Brooks/Cole Publishing Company, 1970.

2

Transistor Equivalent Circuits

INTRODUCTION

A transistor equivalent circuit is a model, derived empirically and/or from basic physical principles. Because the operation of the transistor is so complex, most equivalent circuits are necessarily approximations to the true behavior of the device. Even though the models are approximate, equivalent circuits are useful analytical tools. In the design of any electronic equipment it is desirable to have models of the components which make up the individual circuits being designed. Use of models for the various active and passive components allows prediction of performance and optimization of that performance with respect to the many variables normally present. Hence two different categories of equivalent circuits are considered in this chapter.

The first form of equivalent circuit considered is that of the transistor as a "black box," the electrical behavior of which can be characterized by knowing the relationship between the terminal currents and voltages. We shall find that the relationships between these terminal currents and voltages are specified in terms of sets of matrix parameters. Hence these matrix parameters form one broad category of equivalent circuit. This approach is only useful at the particular frequency where the black box parameters are measured.

Another approach is based on a detailed analysis of the flow of minority and majority carriers within the emitter, base, and collector of the transistor. Equivalent circuits in this category consist of collections of passive elemnts and active elements (current generators and/or voltage generators) which, in a sense, have a one-to-one correspondence with the physical operation of the transistor. This form of equivalent circuit has the advantage of being useful over a wide frequency range.

Equivalent Circuits Using Matrix Parameters

We now consider equivalent circuits based on four-terminal, linear network theory. It is assumed that the transistor has been biased to a dc operating point and that we wish an equivalent circuit for the transistor for small ac signals. The "small" signal condition is imposed in order that the transistor may be considered a linear element. Figure 2-1 defines the assumed polarities of the terminal currents and voltages for the transistor considered as a black box. The parameters p_1, p_2, p_3, and p_4 signify that, under the assumption of linearity, the terminal currents and voltages may be related by four parameters. These parameters are, in general, complex quantities of the form $p_r + jp_i$; that is, they consist of a real part, p_r and an imaginary part, p_i.

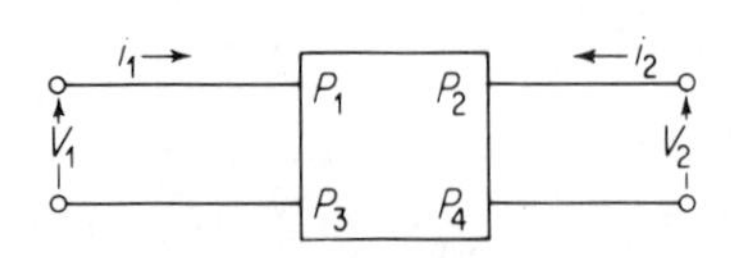

Fig. 2-1. "Black box" representation of the transistor.

Of the four terminal voltages and currents, two may be chosen as independent and two as dependent variables, thus yielding six pairs of linear equations. Three sets of equations used in analysis of transistor circuits are:
(Impedance equations)

$$V_1 = z_{11}i_1 + z_{12}i_2 \tag{2-1}$$

$$V_2 = z_{21}i_1 + z_{22}i_2 \tag{2-2}$$

(Hybrid equations)

$$V_1 = h_{11}i_1 + h_{12}V_2 \tag{2-3}$$

$$i_2 = h_{21}i_1 + h_{22}V_2 \tag{2-4}$$

(Admittance equations)

$$i_1 = y_{11}V_1 + y_{12}V_2 \tag{2-5}$$

$$i_2 = y_{21}V_1 + y_{22}V_2 \tag{2-6}$$

The subscripts of the coefficients refer merely to the position of that coefficient in the equations. The h's, y's, and z's are called *matrix parameters* since they lend themselves conveniently to matrix algebra. Any four h, y, or z parameters will describe the black box. In order for these equations to be valid, the dc operating point must be fixed and the transistor operating as a linear element for small variations in signal (small-signal operation). *Remember that matrix parameters are valid only at the specified operating point and frequency.*

Consider the hybrid equations. If the output is ac short-circuited, Eqs. 2-3 and 2-4 become

$$h_{11} = \left.\frac{V_1}{i_1}\right|_{V_2=0} = \text{imput impedance} \tag{2-7}$$

$$h_{21} = \left.\frac{i_2}{i_1}\right|_{V_2=0} = \text{forward current gain} \tag{2-8}$$

If the input is open-circuited, then Eqs. 2-3 and 2-4 become

$$h_{12} = \left.\frac{V_1}{V_2}\right|_{i_1=0} = \text{reverse voltage gain} \tag{2-9}$$

$$h_{21} = \left.\frac{i_2}{V_2}\right|_{i_1=0} = \text{output admittance} \tag{2-10}$$

If the input and output terminals of the black box are open- or short-circuited in the proper combinations, then similar expressions are obtained for the y and z parameters.

An example of the calculation of the z parameters of a simple passive network will be illustrated. Consider the parallel passive network shown in Fig. 2-2.

If i_1 and i_2 are separately made zero, the z parameters become:

$$\begin{aligned} z_{11} &= \left.\frac{V_2}{i_1}\right|_{i_2=0} = R_1 + R_3 \qquad & z_{12} &= \left.\frac{V_1}{i_2}\right|_{i_1=0} = R_3 \\ z_{21} &= \left.\frac{V_2}{i_1}\right|_{i_2=0} = R_3 \qquad & z_{22} &= \left.\frac{V_2}{i_2}\right|_{i_1=0} = R_2 + R_3 \end{aligned} \tag{2-11}$$

In a passive network it is always true that the forward and reverse matrix parameters are equal. This is a consequence of the principle of reciprocity which is discussed in many texts on networks (7, 8).* This technique of using the simplifying and reduced equations that result from a simple choice of the independent parameters is extremely useful and is often employed.

In order to standardize notation, the small-signal parameters have been given letter subscripts rather than numerical subscripts. The letter subscripts completely identify the parameters and are

$h_{11} = h_i$ (i represents input)
$h_{12} = h_r$ (r represents reverse)
$h_{21} = h_f$ (f represents forward)
$h_{22} = h_o$ (o represents output)

Since there are three possible transistor† connections, common-base, common-emitter, and common-collector (where by *common* we imply that input and output connections have a common terminal) a second letter subscript, b, e, or c, is needed to identify the way in which the transistor is connected. Hence, h_{fe} is the forward, short-circuit, current gain of a transistor in the common-emitter connection. One complete set of h, y, or z parameters will completely describe the transistor for all possible connections.

Assume the h parameters for the common-base connection are given and that we desire to find the h parameters for the common-emitter connection. Figure 2-3 shows an equivalent circuit for the grounded-base connection and

*Numbers in parentheses refer to list at end of chapter.

†Whenever the term *transistor* is used alone it shall mean bipolar transistor.

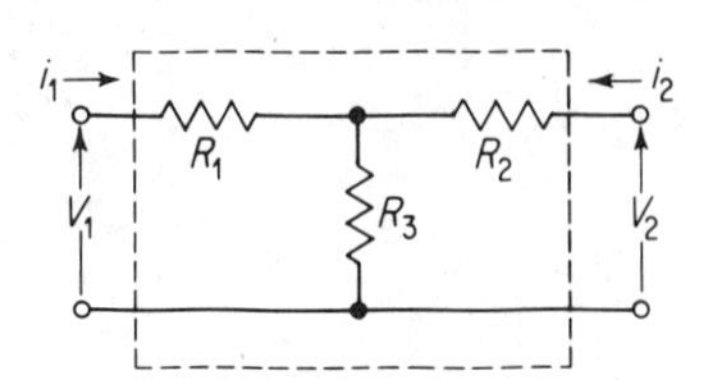

Fig. 2-2. A passive network whose terminal parameters may be described by the z parameters.

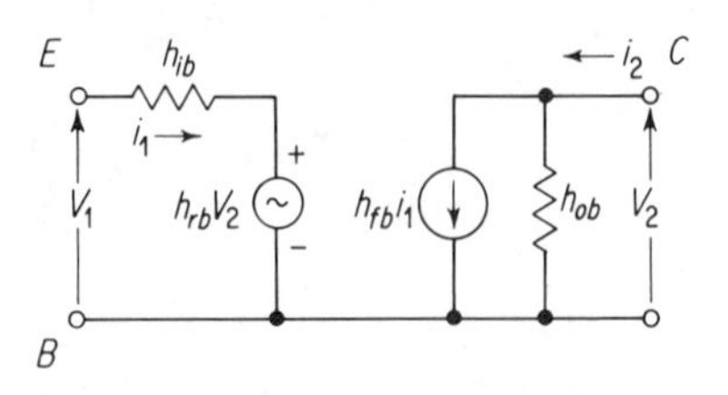

Fig. 2-3. A common-base equivalent circuit using h parameters.

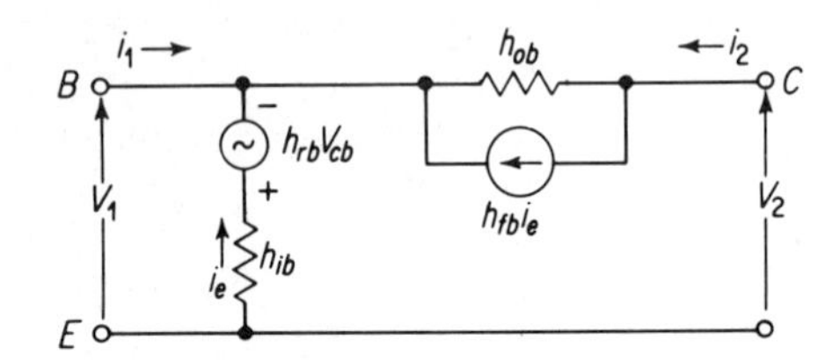

Fig. 2-4. The common-base equivalent circuit of Fig. 2-3 rearranged for analysis of the common-emitter connection.

Fig. 2-4 is a rearrangement of the common-base equivalent circuit which is suitable for the grounded-emitter connection. Note that we have merely redrawn the circuit without changing any connections. Since the input impedance and forward current gain of a transistor in the common-emitter connection are

$$h_{ie} = \left.\frac{V_1}{i_1}\right|_{V_2=0} \tag{2-12}$$

$$h_{fe} = \left.\frac{i_2}{i_1}\right|_{V_2=0} \tag{2-13}$$

we need to find V_1 in terms of i_1, and i_2 in terms of i_1 (for $V_2 = 0$), in order to determine the equivalent values for these h parameters when the transistor is connected in the common-emitter connection. Figure 2-5 shows the circuit of Fig. 2-4 with the output short-circuited ($V_2 = 0$). The loop equations are

$$V_1 = -h_{rb}V_{cb} - i_e h_{ib} \tag{2-14}$$

$$V_1 = \frac{i_3}{h_{ob}} \tag{2-15}$$

and

$$i_3 = i_e h_{fb} - i_2 \tag{2-16}$$

$$i_e = -(i_1 + i_2) \tag{2-17}$$

and where we have obviously

$$V_{cb} \equiv -V_1 \tag{2-18}$$

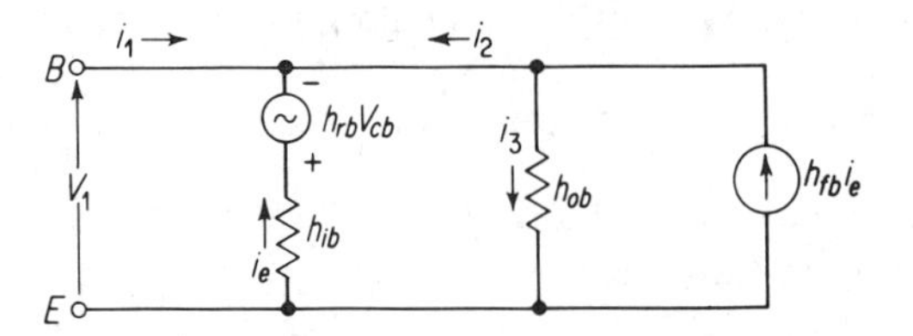

Fig. 2-5. The common-emitter connection with the output short-circuited.

Making these substitutions into Eqs. 2-14 and 2-15, the simultaneous equations become

$$V_1 = \frac{h_{ib}}{1 - h_{rb}}(i_2 + i_1) \tag{2-19}$$

and

$$V_1 = -\left(\frac{h_{fb}}{h_{ob}} + \frac{1}{h_{ob}}\right)i_2 - \frac{h_{fb}}{h_{ob}}i_1 \tag{2-20}$$

Solving these equations simultaneously, the resulting expression for the input current is

$$i_1 = V_1\frac{[1 + h_{fb} - h_{rb}h_{fb} + h_{ib}h_{ob} - h_{rb}]}{h_{ib}} \tag{2-21}$$

Often, the symbol Δ_{hb} is used to indicate the determinant of the h parameters for the common-base configuration:

$$\Delta_{hb} = \begin{vmatrix} h_{ib} & h_{rb} \\ h_{fb} & h_{ob} \end{vmatrix} = h_{ib}h_{ob} - h_{fb}h_{rb} \tag{2-22}$$

This quantity, as well as h_{rb} the reverse voltage "feedback," is generally small, so that quite useful and accurate simplifying approximations can be made.

In such practical applications then, it may be assumed that the value of $\Delta_{hb} - h_{rb} \equiv h_{ib}h_{ob} - h_{rb}h_{fb} - h_{rb} \ll 1 + h_{fb}$ so that Eq. 2-21 becomes:

$$i_1 \approx V_1\frac{1 + h_{fb}}{h_{ib}} \tag{2-23}$$

Therefore, the input impedance in the common-emitter connection in terms of the common-base h parameters is

$$h_{ie} = \left.\frac{V_1}{i_1}\right|_{V_2=0} = \frac{h_{ib}}{1 + h_{fb}} \tag{2-24}$$

To find h_{fe}, solve Eqs. 2-19 and 2-20 for i_2. Hence

$$i_2 = V_1\frac{h_{ib}h_{ob} + h_{fb}(1 - h_{rb})}{h_{ib}h_{fb} - h_{ib}(1 + h_{fb})} \tag{2-25}$$

This can be simplified by using the foregoing assumption to

$$i_2 = -V_1 \frac{h_{fb}}{h_{ib}} \tag{2-26}$$

Substituting Eqs. 2-23 and 2-26 into Eq. 2-13 by using the preceding simplifying assumption, we have, finally,

$$h_{fe} \approx \frac{-h_{fb}}{1 + h_{fb}} = \frac{\alpha}{1 - \alpha} \tag{2-27}$$

Referring back to Fig. 2-4, which is the equivalent circuit for the common-emitter connection with an open-circuit input, the output voltage is

$$V_2 = \frac{i_3}{h_{ob}} - V_{cb}h_{rb} - h_{ib}i_e \tag{2-28}$$

where

$$i_3 = i_2 - h_{ib}i_e \tag{2-29}$$

$$i_e = -i_2 \tag{2-30}$$

and

$$V_{cb} = \frac{i_2 - i_e h_{fb}}{h_{ob}} \tag{2-31}$$

Using these substitutions, Eq. 2-28 yields

$$V_2 = \left[\left(\frac{1 + h_{fb}}{h_{ob}}\right) - \left(\frac{1 + h_{fb}}{h_{ob}}\right)h_{rb} + h_{ib}\right]i_2 \tag{2-32}$$

Taking the ratio of i_2 to V_2, with $i_1 = 0$, we get

$$h_{oe} = \left.\frac{i_2}{V_2}\right|_{i_1=0} = \frac{h_{ob}}{\Delta_{hb} - h_{rb} + h_{fb} + 1} \tag{2-33}$$

As before, assume that

$$\Delta_{hb} - h_{rb} \equiv (h_{ib}h_{ob} - h_{rb}h_{fb} - h_{rb}) \ll (1 + h_{fb}) \tag{2-34}$$

so that

$$h_{oe} \approx \frac{h_{ob}}{1 + h_{fb}} \tag{2-35}$$

Referring to Fig. 2-4, V_1 is

$$V_1 = \left[h_{ib} - \frac{h_{rb}(1 + h_{fb})}{h_{ob}}\right]i_2 \tag{2-36}$$

The ratio of V_1 to V_2, with $i_1 = 0$, is

$$h_{re} = \left.\frac{V_1}{V_2}\right|_{i_1=0} = \frac{h_{ib}h_{ob} - h_{rb}h_{fb} - h_{rb}}{1 + h_{fb} - h_{rb} + h_{ib}h_{ob} - h_{rb}h_{fb}} \tag{2-37}$$

Using the same assumptions as before, Eq. 2-37 becomes

$$h_{re} \approx \frac{\Delta_{hb} - h_{rb}}{1 + h_{fb}} \tag{2-38}$$

where $\Delta_{hb} = h_{ib}h_{ob} - h_{rb}h_{fb}$ as before.

By similar methods any of the 12 h, y, and z parameters may be found by knowing only four parameters of a particular connection. This can be a tedious procedure, and in many cases an odious one, so that an empirical or tabular data approach is thus a very useful tool in design. Table 2-1 gives a summary of these interrelations for the h parameters.

TABLE 2-1 Interrelations between h Parameters using approximation that $\Delta_{hb} - h_{rb} \ll 1 + h_{fb}$

CB Parameter		*CE* Parameter	*CC* Parameter
Numbered	Lettered		
h_{11}	h_{ib}	$h_{ie} = \dfrac{h_{ib}}{1 + h_{fb}}$	$h_{ic} = \dfrac{h_{ib}}{1 + h_{fb}}$
h_{12}	h_{rb}	$h_{re} = \dfrac{\Delta_{hb} - h_{rb}}{1 + h_{fb}}$	$h_{rc} = 1$
h_{21}	h_{fb}	$h_{fe} = \dfrac{-h_{fb}}{1 + h_{fb}}$	$h_{fc} = \dfrac{-1}{1 + h_{fb}}$
h_{22}	h_{ob}	$h_{oe} = \dfrac{h_{ob}}{1 + h_{fb}}$	$h_{oc} = \dfrac{h_{ob}}{1 + h_{fb}}$

Interrelations between h, y, and z Parameters

We have shown that it is possible to find all the 12 small-signal parameters in a group if a set of four parameters is known. We now consider the case where it is desired to convert to a group of matrix parameters from another set of matrix parameters.

Considering Eqs. 2-1 and 2-2, and 2-5 and 2-6, it is possible to convert from y parameters to z parameters. Defining the y parameters as

$$y_{11} = \frac{i_1}{V_1}\bigg|_{V_2=0} \qquad y_{12} = \frac{i_1}{V_2}\bigg|_{V_1=0} \tag{2-39}$$

$$y_{21} = \frac{i_2}{V_1}\bigg|_{V_2=0} \qquad y_{22} = \frac{i_2}{V_2}\bigg|_{V_1=0} \tag{2-40}$$

From Eq. 2-2, we obtain

$$i_2 = \frac{V_2 - z_{21}i_1}{z_{22}} \tag{2-41}$$

Using the substitution, Eq. 2-1 becomes

$$V_1 = z_{11}i_1 + z_{12}\left[\frac{V_2 - z_{21}i_1}{z_{22}}\right] \tag{2-42}$$

Let $V_2 = 0$ in Eq. 2-42. Then we get

$$V_1 = \left[z_{11} - \frac{z_{12}z_{21}}{z_{22}}\right]i_1 \tag{2-43}$$

Taking the ratio of i_1 to V_1, y_{11} is

$$y_{11} = \frac{i_1}{V_1}\bigg|_{V_2=0} = \frac{z_{22}}{\Delta_z} \tag{2-44}$$

where $\Delta_z = z_{11}z_{22} - z_{12}z_{21}$ is now the determinant of the z parameters.

From Eq. 2-2,

$$i_1 = \frac{V_2 - z_{22}i_2}{z_{21}} \tag{2-45}$$

Substituting Eq. 2-45 into Eq. 2-1, we have

$$V_1 = z_{11}\left[\frac{V_2 - z_{22}i_2}{z_{21}}\right] + z_{12}i_2 \tag{2-46}$$

When $V_2 = 0$, V_1 becomes

$$V_1 = \left[\frac{z_{12}z_{21} - z_{11}z_{22}}{z_{21}}\right]i_2 \tag{2-47}$$

Taking the ratio of i_2 to V_1, y_{21} is

$$y_{21} = \frac{i_2}{V_1}\bigg|_{V_2=0} = \frac{-z_{21}}{\Delta_z} \tag{2-48}$$

From Eq. 2-1

$$i_1 = \frac{V_1 - z_{12}i_2}{z_{11}} \tag{2-49}$$

Using this substitution in Eq. 2-2, we have

$$V_2 = z_{21}\left[\frac{V_1 - z_{12}i_2}{z_{11}}\right] + z_{22}i_2 \tag{2-50}$$

When $V_1 = 0$, Eq. 2-50 becomes

$$V_2 = \left[\frac{z_{22}z_{11} - z_{12}z_{21}}{z_{11}}\right]i_2 \tag{2-51}$$

Taking the ratio of i_2 to V_2, y_{22} becomes

$$y_{22} = \frac{i_2}{V_2}\bigg|_{V_1=0} = \frac{z_{11}}{\Delta_z} \tag{2-52}$$

From Eq. 2-1

$$i_2 = \frac{V_1 - z_{11}i_1}{z_{12}} \tag{2-53}$$

Using this substitution, Eq. 2-1 becomes

$$V_2 = z_{21}i_1 + z_{22}\left[\frac{V_1 - z_{11}i_1}{z_{12}}\right] \tag{2-54}$$

When $V_1 = 0$,

$$V_2 = \left[\frac{z_{12}z_{21} - z_{11}z_{22}}{z_{12}}\right]i_1 = \frac{-\Delta_z}{z_{12}}i_1 \tag{2-55}$$

Taking the ratio of i_1 to V_2, y_{12} is

$$y_{12} = \left.\frac{i_1}{V_2}\right|_{V_1=0} = -\frac{z_{12}}{\Delta_z} \tag{2-56}$$

The preceding derivation was carried out for y-to-z conversion. Similar derivations are available for h-to-z conversions, z-to-y conversions, etc. It has been shown that, given any four parameters of a set, the remaining parameters of that set as well as the other related parameters may be easily found. Table 2-2 summarizes the important relations.

TABLE 2-2 Conversions among *z*, *h*, and *y* Parameters

Matrix	y		z		h	
y	y_{11}	y_{12}	$\frac{z_{22}}{\Delta_z}$	$\frac{-z_{12}}{\Delta_z}$	$\frac{1}{h_{11}}$	$\frac{-h_{12}}{h_{11}}$
	y_{21}	y_{22}	$\frac{-z_{21}}{\Delta_z}$	$\frac{z_{11}}{\Delta_z}$	$\frac{h_{21}}{h_{11}}$	$\frac{\Delta_h}{h_{11}}$
z	$\frac{y_{22}}{\Delta_y}$	$\frac{-y_{12}}{\Delta_y}$	z_{11}	z_{12}	$\frac{\Delta_h}{h_{22}}$	$\frac{h_{12}}{h_{22}}$
	$\frac{-y_{21}}{\Delta_y}$	$\frac{y_{11}}{\Delta_y}$	z_{21}	z_{22}	$\frac{-h_{21}}{h_{22}}$	$\frac{1}{h_{22}}$
h	$\frac{1}{y_{11}}$	$\frac{-y_{12}}{y_{11}}$	$\frac{\Delta_z}{z_{22}}$	$\frac{z_{12}}{z_{22}}$	h_{11}	h_{12}
	$\frac{y_{21}}{y_{11}}$	$\frac{\Delta_y}{y_{11}}$	$\frac{-z_{21}}{z_{22}}$	$\frac{1}{z_{22}}$	h_{21}	h_{22}

where $\Delta_y = y_{11}y_{22} - y_{12}y_{21}$; $\Delta_z = z_{11}z_{22} - z_{12}z_{21}$
$\Delta_h = h_{11}h_{22} - h_{12}h_{21}$

Measurement of Matrix Parameters

If a set of parameters is to be useful, they must be measurable under conditions that allow application of dc operating conditions. Remember that the small-signal parameters are obtained with the quiescent conditions fixed and the ac components of the input and output signals open- or short-circuited.

By definition, the z parameters are most directly determined by a series of open-circuit measurements, y parameters are determined by a series of short-circuit measurements, and h parameters are determined by a combination of open- and short-circuit measurements.

As discussed in Chap. 1, an important feature of the transistor is its high output impedance and low input impedance. Generally speaking, it may be said that at low frequencies it is relatively difficult to obtain an ac short circuit with respect to the input impedance of the transistor, and relatively difficult to obtain an ac open circuit with respect to the output impedance of a transistor. This leads to the conclusion that h parameters are more widely used since this measurement is dependent upon short-circuit output and open-circuit input measurement conditions which are more readily obtained. Such is indeed the case in practical commercial devices.

The measurement of h_{ie} and h_{fe} is performed with the output short-circuited. This is easily accomplished by applying the dc bias supply to the output with the output bypassed to present a small impedance at the ac signal frequency compared to the transistor output impedance as shown in Fig. 2-6. The constant dc input current may be supplied by a current source consisting of a battery in series with a resistance which is very large compared to the input impedance of the transistor. The constant ac input current can be supplied by a signal generator in series with the same resistor. By establishing the dc operating conditions and the ac input current, h_{ie} and h_{fe} are found by measuring V_1 and i_2.

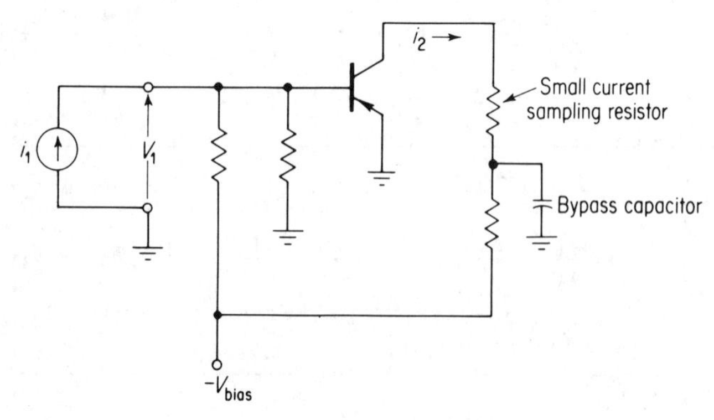

Fig. 2-6. This circuit may be used for measurement of h_{ie} and h_{fe}.

In comparison to the method of measuring $h_{ie}(h_{11})$ and $h_{fe}(h_{21})$, the method of measuring z_{11} and z_{21} will be considered. This measurement calls for an open-circuit output. To accomplish this a high-impedance network can be connected in series with a battery in order to establish a dc operating condition and an ac open circuit. For this impedance to be large in comparison to the transistor output impedance, it must be in the order of many megohms. This is not desirable, since using a large resistance will dissipate a large amount of power. If a tuned circuit is used, it must be carefully tuned to the signal frequency.

The measurement of h_{oe} and h_{re} requires an open-circuit input and a source of constant-output sine wave voltage. The quiescent input conditions are readily obtained in the same manner as prescribed in the measurement of h_{ie} and h_{fe}. With the application of a constant-output voltage, $h_{oe}(h_{22})$ and $h_{re}(h_{12})$ may be obtained by measuring V_1 and i_2. This method may be compared to the method of measuring y_{22} and y_{12} which calls for a shorted input circuit. In order to satisfy this shorted condition and supply a dc current to the input it is necessary to bypass the input circuit with a capacitor which has a small impedance at the signal frequency compared to the input impedance of the transistor. In order for this to be true, a large value of capacitance must be used.

It can be seen from this discussion that the h parameters have become the most popular of the small-signal, low-frequency parameters because of the relative ease of measurement. Since we can always relate one set to another, ease of measurement is an important consideration. A listing of typical values of the h parameter is given in Table 2-3.

TABLE 2-3 Typical h Parameters for Transistors at 1 KHz

Parameter	*CB* Values	*CE* Values	*CC* Values
$h_{11} = h_i$	$h_{ib} = 40\Omega$	$h_{ie} = 2000\Omega$	$h_{ic} = 2000\Omega$
$h_{12} = h_r$	$h_{rb} = 4 \times 10^{-4}$	$h_{re} = 16 \times 10^{-4}$	$h_{rc} \approx 1$
$h_{21} = h_f$	$h_{fb} = -0.98$	$h_{fe} = 49$	$h_{fc} = -50$
$h_{22} = h_o$	$h_{ob} = 10^{-6}℧$	$h_{oe} = 5 \times 10^{-5}℧$	$h_{oc} = 5 \times 10^{-5}℧$
Δ_h	$\Delta_{hb} = 0.432 \times 10^{-3}$	$\Delta_{he} = 0.0218$	$\Delta_{hc} = 50.1$

Application of Matrix Parameters

The z and y parameters, although not as popular as the h parameters, do have suitable applications from the standpoint of formal circuit analysis. They are often used, particularly for high-frequency analysis. The matrix parameters take on great significance when they are applied to complex multistage and multiloop networks. By use of matrix methods, a network consisting of active and/or passive elements whose individual matrix parameters are known may be analyzed and produces a result in the form of a composite matrix network. For example, a multistage amplifier circuit may be analyzed by using nodal or mesh equations, whose result is in the form of complex simultaneous equations. The network may also be analyzed by treating each stage separately, knowing only the matrix parameters of the circuit, and combining these parameters by the use of matrix algebra, the result being in the form of a single network whose parameters are a composite of the known individual parameters.

Before one may attempt to analyze a complex multistage or multiloop network by matrix methods, the analysis of a single stage must be complete.

If the important quantities, such as power gain, input and output resistances, and current and voltage gain, can be derived for a single network in terms of small-signal parameters, then these techniques may be extended to the complete network. Small-signal parameters may be used, whatever the connection (common-emitter, base, or collector), in deriving the desired quantity. The derived expression is then applicable to any configuration if the proper parameters are inserted into the expression. As an example, the voltage gain of a grounded-base amplifier will be derived.

The *voltage gain* of an amplifier is defined as the ratio of the output voltage to the input voltage. Figure 2-7 shows an ac equivalent circuit for a grounded-base amplifier, in terms of its h parameters.

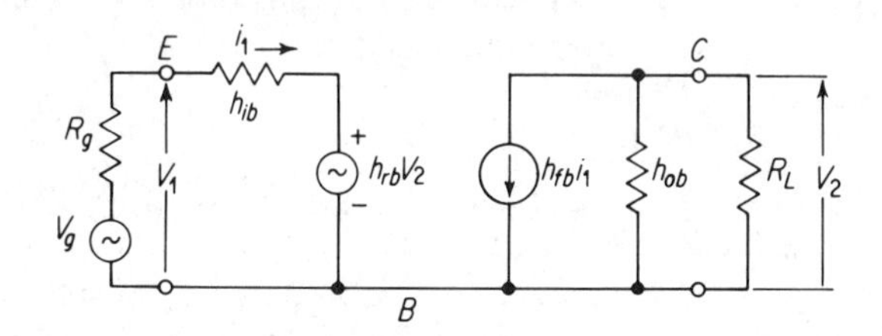

Fig. 2-7. Common-base equivalent circuit used for calculation of voltage gain.

For this calculation, we need the ratio of V_2 to V_1; the simultaneous equations are

$$V_1 = h_{ib}i_1 + h_{rb}V_2 \tag{2-57}$$

$$i_1 h_{fb} = \left(\frac{1}{R_L} + h_{ob}\right)V_2 \tag{2-58}$$

Solving for V_2, we get

$$V_2 = \frac{h_{fb}R_L V_1}{h_{ib}(1 + h_{ob}R_L) - h_{fb}h_{rb}R_L} \tag{2-59}$$

Taking the ratio of V_2 to V_1, the voltage gain is

$$A_{vb} = \frac{h_{fb}R_L}{R_L\Delta_{hb} + h_{ib}} \tag{2-60}$$

where $\Delta_{hb} = h_{ib}h_{ob} - h_{rb}h_{fb}$.

Equation 2-60 is the expression for the voltage gain of a grounded-base amplifier, in terms of grounded-base h parameters. If it is desirable to obtain the voltage gain of a grounded-emitter amplifier, then all that is necessary is to substitute the corresponding common-emitter h parameters into Eq. 2-60, since the form of the circuit is identical. Thus, we obtain

$$A_{ve} = \frac{h_{fe}R_L}{R_L\Delta_{he} + h_{ie}} \tag{2-61}$$

If only the common-base h parameters were available, we would have to convert using the appropriate factors in Table 2-1. A tabular representation of results for an amplifier is given in Table 2-4.

TABLE 2-4 External Characterization of an h Parameter Equivalent Circuit Representation for a Transistor

Characteristic	Symbol	Value
Input resistance	R_i	$\dfrac{\Delta_h R_L + h_i}{1 + h_o R_L}$
Output resistance	R_o	$\dfrac{h_i + R_g}{\Delta_h + h_o R_g}$
Current gain	A_i	$\dfrac{h_f}{1 + h_o R_L}$
Voltage gain	A_v	$\dfrac{h_f R_L}{\Delta_h R_L + h_i}$
Power gain	$A_p = A_v A_i$	$\dfrac{h_f^2 R_L}{(\Delta_h R_L + h_i)(1 + h_o R_L)}$
where $\Delta_h = h_i h_o - h_f h_r$		

Complex Multistage Networks

Transistor stages may be coupled to succeeding stages by five different methods: series-coupled, parallel-coupled, series-parallel coupled, parallel-series coupled, and cascade-coupled. Figure 2-8 is a diagram of two four-terminal networks coupled in parallel. It is desired to find the parameters of the composite network (indicated by the dashed lines). The choice of the small-signal parameters used in an analysis is arbitrary and somewhat dependent upon personal preference. Of the possible interstage couplings mentioned, certain small-signal parameters lend themselves more easily to a

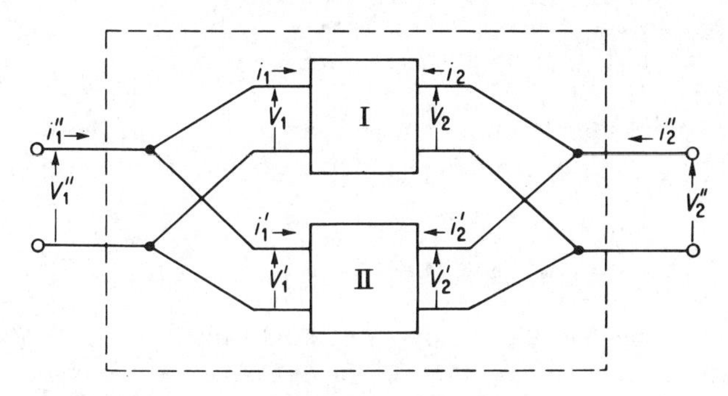

Fig. 2-8. Two four-terminal networks connected in parallel.

network solution for a particular coupling. In addition to h, y, and z parameters already discussed, three other matrix parameters are compatible with the aforementioned parameters. These small-signal parameters are a, b, and g parameters and have useful application for certain interstage couplings but are not generally useful in transistor circuit design.

To analyze the circuit of Fig. 2-8 we will have

$$\begin{aligned} i''_1 &= i_1 + i'_1 \\ i''_2 &= i_2 + i'_2 \\ V''_1 &= V_1 = V'_1 \\ V''_2 &= V_2 = V'_2 \end{aligned} \tag{2-62}$$

Using the y parameters to analyze the network (these are the most convenient when common voltages at input and output exist) we can write the matrix equations for each network:

$$\begin{aligned} i_1 &= y_{11}V_1 + y_{12}V_2 \\ i_2 &= y_{21}V_1 + y_{22}V_2 \end{aligned} \tag{2-63}$$

$$\begin{aligned} i'_1 &= y'_{11}V'_1 + y'_{12}V'_2 \\ i'_2 &= y'_{21}V'_1 + y'_{22}V'_2 \end{aligned} \tag{2-64}$$

Combining these equations, we have

$$\begin{aligned} i''_1 &= (y_{11} + y'_{11})V''_1 + (y_{12} + y'_{12})V''_2 \\ i''_2 &= (y_{21} + y'_{21})V''_1 + (y_{22} + y'_{22})V''_2 \end{aligned} \tag{2-65}$$

The matrix equation of the composite network is obviously:

$$\begin{aligned} i''_1 &= y''_{11}V''_1 + y''_{12}V''_2 \\ i''_2 &= y''_{21}V''_1 + y''_{22}V''_2 \end{aligned} \tag{2-66}$$

Comparing these equations, it is obvious that each of the y parameters of the composite network is the sum of the appropriate y parameters of the individual networks. In matrix notation, they are

$$\begin{pmatrix} y''_{11} & y''_{12} \\ y''_{21} & y''_{22} \end{pmatrix} = \begin{pmatrix} y_{11} + y'_{11} & y_{12} + y'_{12} \\ y_{21} + y'_{21} & y_{22} + y'_{22} \end{pmatrix} \tag{2-67}$$

Similar composite matrix parameters can be found for the other types of coupling by applying the same theory used for finding the preceding matrix.

SIMPLIFIED EQUIVALENT CIRCUITS

In the case of the most common transistor connection used, that of the common-emitter, the values of two of the h parameters in comparison to the rest of the circuit values suggest some obvious simplifications. These are mainly useful in order to obtain order of magnitude values for external characteristics, such as gain, input impedance, etc., of multistage networks. If more accurate results are desired, the entire equivalent circuit must be used, but the solution can be guided and done more efficiently if some idea of the expected answer is known.

The output admittance h_{oe} is generally quite small in comparison with the load admittance $(1/R_L)$ and can in first approximation be neglected.

Similarly, owing to the smallness of h_{re} and the resulting small "feedback" voltage into the input circuit, this is often neglected. The equivalent circuit is now merely that illustrated in Fig. 2-9 where the input impedance is simply given by the value of h_{ie} and the current gain is h_{fe}. An example of the usefulness of this simplified circuit in computation is given in a later chapter, in the discussion of the two-stage amplifier.

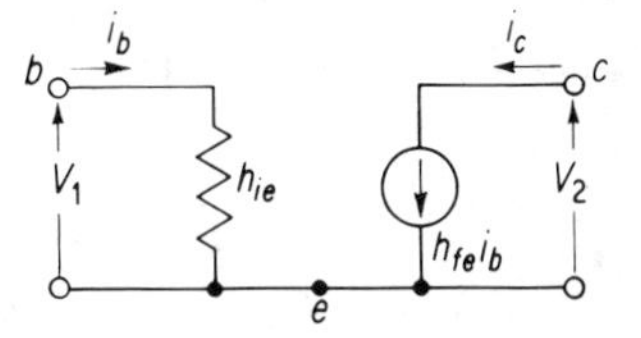

Fig. 2-9. Simplified equivalent circuit for common-emitter transistor useful for multistage calculations.

PHYSICAL EQUIVALENT CIRCUITS

The Physical Basis for Equivalent Circuits

Transistor equivalent circuits may be derived from an analysis of the flow of minority and majority carriers within the emitter, base, and collector regions of the transistor. The analysis is usually carried out by considering the intrinsic transistor and its current-voltage terminal characteristics and adding to this intrinsic model, extrinsic elements which are not truly a result of "transistor action." Figure 2-10 illustrates a complete transistor model made up of the intrinsic transistor, described by y'_{ij}, and the extrinsic parameters, consisting of spreading resistances r'_b, r'_e, and r'_c, transistor barrier capacitances C_{TE} and C_{TC}, plus stray capacitances and inductance.

In discussing equivalent circuits in this chapter, we shall ignore all extrinsic elements except r'_b, C_{TE}, and C_{TC}. In general, this will be a valid assumption, except when considering transistor behavior at very high frequencies.

To obtain an equivalent circuit we must determine the y'_{ij} of the physical parameters of the device. These may be found by analyzing the differential equations which describe the flow of current carriers within the transistor. Here one must solve the equation of continuity with appropriate boundary conditions for the base region and interface region. This is a tedious procedure which is found in many standard works. Since the equations that describe the physical operation of the transistor are differential, as opposed to algebraic, the equivalent circuits are distributed in nature, in much the same way as the equivalent circuit of a transmission line is distributed. The common-base y' parameters, obtained from the aforementioned differential equations, are

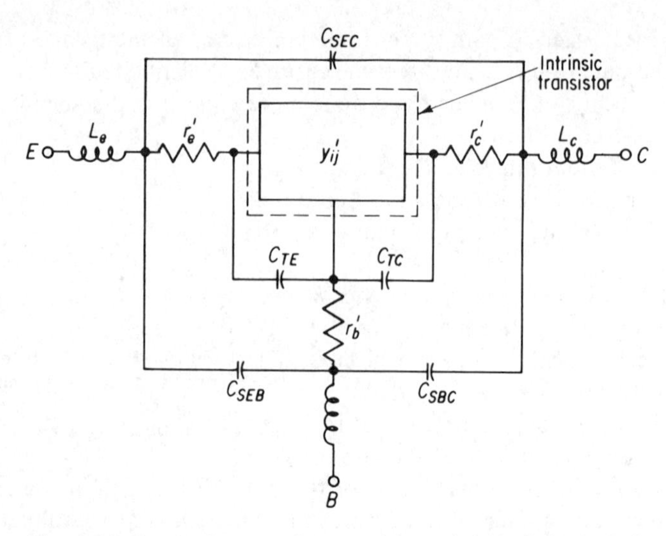

Fig. 2-10. A complete transistor model consisting of the intrinsic transistor described by parameters y'_{ij} and external parasitic elements.

shown in order to present some idea of the distributed nature of the transistor (1)

$$y'_{11} = \frac{qI_E}{kT}\frac{W}{L_p}\sqrt{1 + j\omega\tau_p}\coth\frac{W}{L_p}\sqrt{1 + j\omega\tau_p} \tag{2-68}$$

$$y'_{12} = \frac{1}{K}\frac{qI_E}{kT}\frac{W}{L_p}\sqrt{1 + j\omega\tau_p}\,\text{cosech}\,\frac{W}{L_p}\sqrt{1 + j\omega\tau_p} \tag{2-69}$$

$$y'_{21} = \frac{-qI_E}{kT}\frac{W}{L_p}\sqrt{1 + j\omega\tau_p}\,\text{cosech}\,\frac{W}{L_p}\sqrt{1 + j\omega\tau_p} \tag{2-70}$$

$$y'_{22} = \frac{1}{K}\frac{qI_E}{kT}\frac{W}{L_p}\sqrt{1 + j\omega\tau_p}\coth\frac{W}{L_p}\sqrt{1 + j\omega\tau_p} \tag{2-71}$$

where

$$\frac{1}{K} = \frac{kT}{q}\frac{1}{W}\frac{\partial W}{\partial V_{C'B'}} \tag{2-72}$$

The various quantities used in these equations are defined as follows:

W = electrical base width, the distance between adjacent edges of the emitter and collector depletion layers.
L_p = diffusion length for minority carriers in the base.
τ_p = lifetime for minority carriers in the base.
ω = angular frequency.
q = electronic charge.
k = Boltzmann's constant.
T = temperature, °K.
I_E = dc emitter current.
$V_{C'B'}$ = dc collector-base voltage.

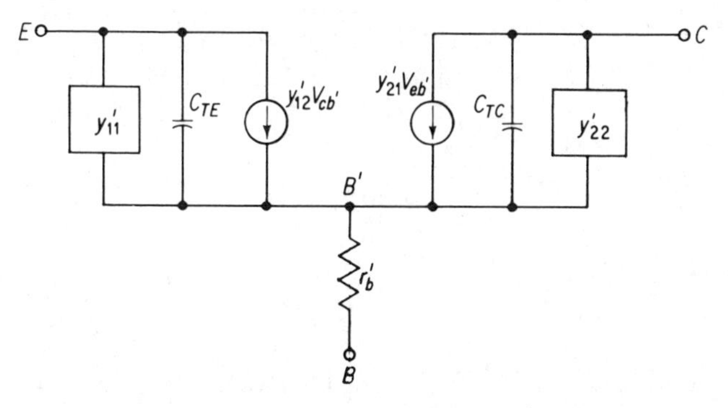

Fig. 2-11. An equivalent circuit based on the y' parameters.

With these y' parameters calculated in terms of the physical parameters of the transistor, we may combine these intrinsic parameters with the extrinsic parameters to formulate a complete equivalent circuit as shown in Fig. 2-11.

If the y'_{ij} of Eqs. (2-68)–(2-71) are used in this equivalent circuit, it becomes very difficult to work with the resulting model. Hence, the trigonometric functions and square root functions are expanded, assuming that $W/L_p \ll 1$ (always met in practice). Consequently, over some finite frequency range, each of the y'_{ij} may be represented by an equation of the form,

$$y'_{ij} = g'_{ij} + j\omega C'_{ij} \tag{2-73}$$

where g'_{ij} and C'_{ij} are in general dependent on frequency. With these equations put in this form, the equivalent circuit of Fig. 2-11 may be altered to the form shown in Fig. 2-12.

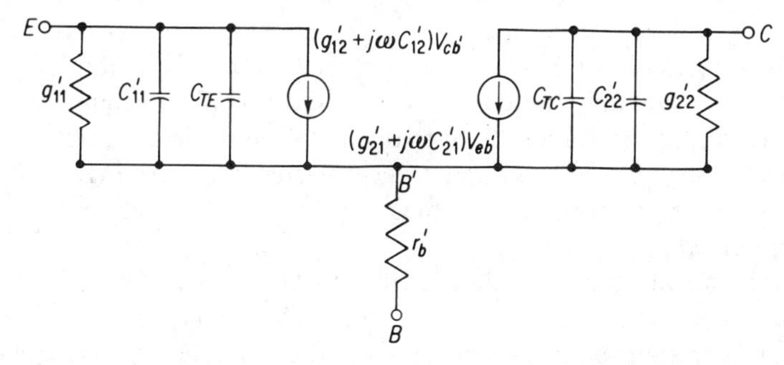

Fig. 2-12. An alternative form of the equivalent circuit shown in Fig. 2-11 in which the y' parameters are shown as parallel g'-C' parameters.

The g'_{ij} and C'_{ij} may be obtained from Eqs. (2-68)–(2-71) by expansion. Expressions for g'_{ij} and C'_{ij} valid in the range $0 \leq \omega < \omega_\alpha$ (ω_α is the frequency at which the common-base current gain is 0.707 of its low-frequency value) follow:

$$g'_{11} = \frac{qI_E}{kT} \tag{2-74}$$

$$C'_{11} = 0.81\frac{qI_E}{kT}\frac{1}{\omega_\alpha} \tag{2-75}$$

$$g'_{22} = \frac{1}{K}\frac{qI_E}{kT} \tag{2-76}$$

$$C'_{22} = 0.81\frac{1}{K}\frac{qI_E}{kT}\frac{1}{\omega_\alpha} \tag{2-77}$$

$$g'_{12} = -\frac{1}{K}\frac{qI_E}{kT}\cdot\left[1 + 0.17\left(\frac{\omega}{\omega_\alpha}\right)^2\right]^{-1/2} \tag{2-78}$$

$$C'_{12} = \frac{0.41}{K}\frac{qI_E}{kT}\frac{1}{\omega_\alpha}\cdot\left[1 + 0.17\left(\frac{\omega}{\omega_\alpha}\right)^2\right]^{-1/2} \tag{2-79}$$

$$g'_{21} = \frac{qI_E}{kT}\cdot\left[1 + 0.17\left(\frac{\omega}{\omega_\alpha}\right)^2\right]^{-1/2} \tag{2-80}$$

$$C'_{21} = 0.41\frac{qI_E}{kT}\frac{1}{\omega_\alpha}\left[1 + 0.17\left(\frac{\omega}{\omega_\alpha}\right)^2\right]^{-1/2} \tag{2-81}$$

These are given in order to demonstrate the frequency dependence and order of magnitude of the conductances and capacitances which become quite important in developing equivalent circuits at higher frequencies. Because of their complexity, however, simpler equivalent circuits which approximate the true situation are usually employed.

Practical Equivalent Circuits

Based on the foregoing analysis, a variety of equivalent circuits may be derived. By suitable transformation the T-equivalent circuit of Fig. 2-12 containing two active generators may be simplified to a single-generator T-equivalent circuit (2). This is shown in Fig. 2-13. The form of this equivalent circuit is considerably different from that normally encountered in the literature under the name "high-frequency T-equivalent circuit." At high frequencies, the equivalent circuit reduces to that shown in Fig. 2-14. This is the form normally seen, except that the "collector resistance" 0.3 Kr_e is usually termed r_c.

In this equivalent circuit, the activity* of the transistor is represented by a current generator αi_1, in shunt with the collector junction. The α is the common-base short-circuit current gain of the intrinsic transistor. This may be calculated from a knowledge of the intrinsic y'_{ij} and may be accurately represented by (3)

$$\alpha = \alpha_0 \operatorname{sech}\sqrt{j\frac{\omega}{0.82\omega_\alpha}} \tag{2-82}$$

*By *activity* we mean the action of the base-emitter and base-collector junctions as well as of the base region in transferring carriers from the emitter to the collector. That is, the mechanism of action and carrier flow in a transistor.

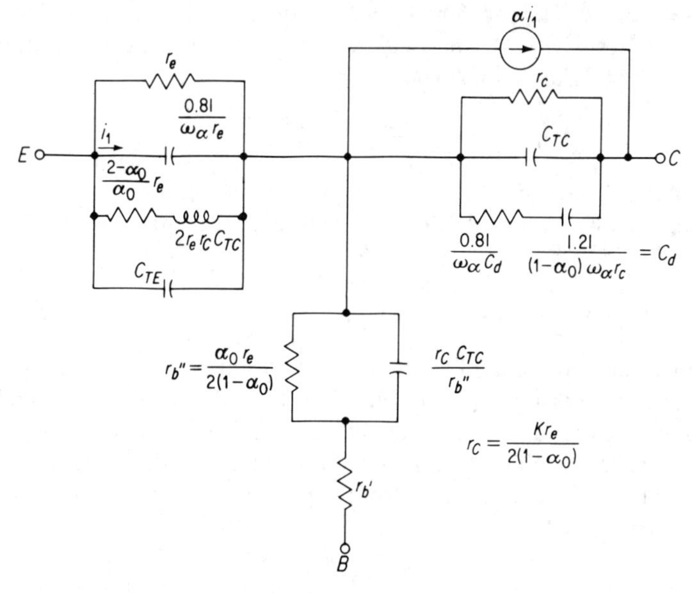

Fig. 2-13. A single-generator *T*-equivalent circuit.

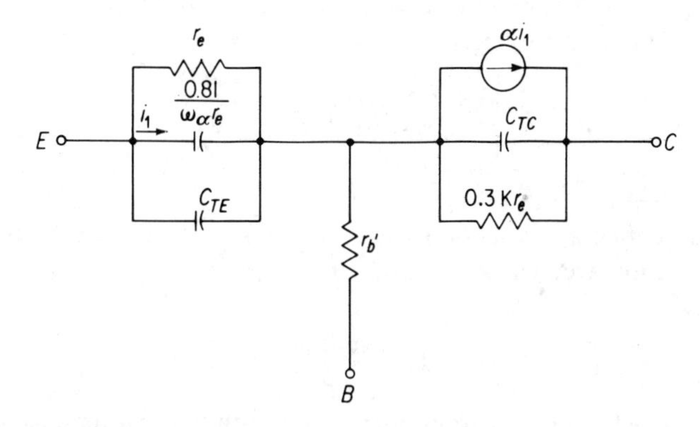

Fig. 2-14. The *T*-equivalent circuit applicable at high frequencies.

Here α_0 is the low-frequency value of α. This, however, is valid only for nondiffused base transistors (such as alloy junction types), since the y' parameters from which it was derived were obtained from an analysis of the homogeneous-base transistor. For diffused-base transistors the expression for α is much more complex. Both may be approximated very accurately in amplitude and phase by the expression (3)

$$\alpha = \frac{\alpha_0 \exp\left[j\left(\frac{K_0 - 1}{K_0}\right)\frac{f}{f_\alpha}\right]}{1 + j\frac{f}{f_\alpha}}, \qquad (f < f_\alpha) \tag{2-83}$$

where $f_\alpha = \omega_\alpha/2\pi$. The constant K_0 is dependent upon the degree of "grading" of the base resistivity. In particular, for a homogeneous-base transistor $K_0 = 0.82$. The factor in the exponential,

$$\left(\frac{K_0 - 1}{K_0}\right)\frac{f}{f_\alpha} \tag{2-84}$$

is called *excess phase shift* because it gives the phase shift in α over and above that associated with a simple *RC*-type roll-off of the form:

$$\alpha = \frac{\alpha_0}{1 + j\dfrac{f}{f_\alpha}} \tag{2-85}$$

This expression is often used as an approximation to α. For homogeneous-base transistors, it is a good approximation; for diffused-base transistors, it is rather poor because of the effect of the built-in field in the base.

An interesting result may be obtained if we calculate the intrinsic common-emitter short-circuit current gain from Eq. 2-83. If extrinsic parameters are negligibly small,

$$\beta = \frac{\alpha}{1 - \alpha} \tag{2-86}$$

where β is the common-emitter current gain. Substituting, we obtain, after a good deal of manipulation:

$$\beta \approx \frac{\beta_0 \exp j\left(\dfrac{K_0 - 1}{\sqrt{K_0}}\right)\dfrac{f}{f_\alpha}}{1 + j\dfrac{\beta_0 f}{\alpha_0 K_0 f_\alpha}}, \qquad (f < f_\alpha) \tag{2-87}$$

We may define a *beta cut-off frequency* (f_β) as the frequency at which β is 0.707 of its low-frequency value, β_0. This is easily obtained from Eq. 2-87 as

$$f_\beta = \frac{\alpha_0 K_0 f_\alpha}{\beta_0} \tag{2-88}$$

A more widely used "figure of merit" for common-emitter current gain is the frequency at which $|\beta|$ is unity. This is commonly called the *common-emitter current gain-bandwidth poduct*, f_T, or *gain-bandwidth product* for short. The variation of β with frequency is sketched in Fig. 2-15. Note that the current gain decreases at the rate of 6 db/octave increase in frequency. From this graph we see that:

$$f_T = \beta_0 f_\beta \tag{2-89}$$

and thus

$$f_T = \alpha_0 K_0 f_\alpha \tag{2-90}$$

This result shows that the excess phase shift (as measured by K_0) acts to decrease the frequency response of the common-emitter connection. Equation

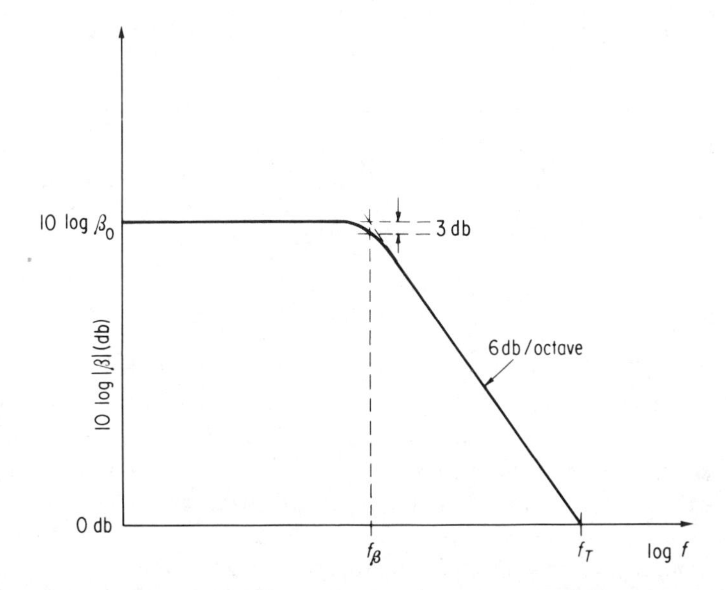

Fig. 2-15. Common-emitter current gain vs frequency defining the gain-bandwidth product f_T.

2-90 shows the origin of the "rule of thumb" often used, that in graded-base transistors f_T is "about" one-half the alpha cut-off frequency. A more correct statement would be, f_T is "about" one-half the intrinsic alpha cut-off frequency. Because of the effects of the extrinsic elements r_b' and C_{TC}, some transistors do not have a measurable extrinsic alpha cut-off frequency.

The magnitude of the common-emitter current gain decreases at the rate of 6 db/octave; this provides a relatively easy way to measure f_T. For frequency in excess of the beta cut-off frequency f_β,

$$f|\beta| = f_T \tag{2-91}$$

A measurement of the magnitude of the common-emitter current gain at a known frequency thus yields f_T directly. Measuring the alpha cut-off frequency is much more difficult because it must be done at a higher frequency and because a variable-frequency, constant-amplitude frequency generator is required. Ease of measurement is one reason that f_T is preferred over f_α as a figure of merit for frequency response.

Common-Emitter Equivalent Circuits

By using the approach outlined previously it is possible to derive a common-emitter equivalent circuit in much the same way that the common-base T-equivalent circuit was done. This procedure was followed by L. J. Giacoletto (5) in deriving a single-generator, pi-type equivalent circuit

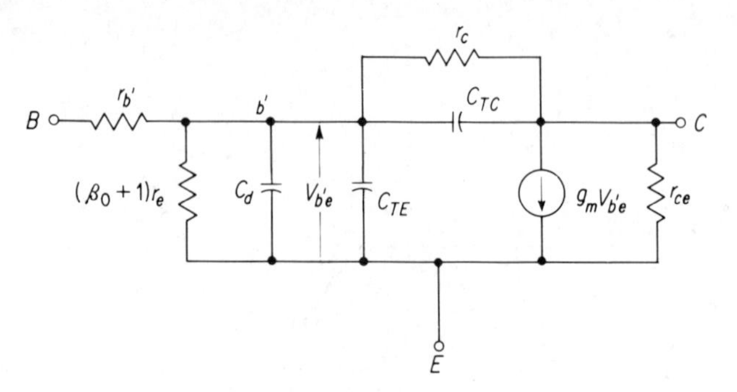

Fig. 2-16. Single-generator pi-type equivalent circuit (hybrid-pi).

(shown in Fig. 2-16). This equivalent circuit is an accurate model for many types of transistors, including alloy junction and electrochemical types. It is not as accurate a representation for certain mesa and planar transistors because of the distributed nature of the base resistance in these devices. For most applications, however, it is a reasonable approximation. The form of the circuit shown in Fig. 2-16 is somewhat in error because of the neglect of the excess phase shift factor that appears in Eq. 2-87. The short-circuit current gain may be calculated (neglecting extrinsic elements) as

$$\beta = \frac{g_m\beta_0 r_e}{1 + j\dfrac{\omega}{\omega_T}} \tag{2-92}$$

where $1/\omega_T = r_{b'e}C_{b'e}$. In order that Eqs. 2-87 and 2-92 agree we must replace g_m by an effective transconductance, g'_m given by

$$g'_m = g_m \exp j\left(\frac{K_0 - 1}{\sqrt{K_0}}\right)\frac{f}{f_\alpha} \tag{2-93}$$

with $g_m = 1/r_e$. With these substitutions, the correct expression for current gain is then obtained.

Two very useful simplifications may be made to the hybrid-pi circuit of Fig. 2-16. Before introducing these, however, note that this circuit may be simplified somewhat, provided that high-frequency operation is considered ($f > 10$ MHz) and relatively low load impedances are to be used ($Z_L <$ 20 KΩ). The simplified hybrid-pi circuit is shown in Fig. 2-17.

In this equivalent circuit, $r_{b'}$ is the base-spreading resistance. This represents the actual ohmic resistance from the external base contact to a point under the emitter junction (the internal base contact).

The resistance, $r_{b'e}$ between the internal base contact and the emitter represents the dynamic resistance of the base-emitter junction as seen from the base. The diode dynamic resistance $r_e = kT/qI_e$ and $r_{b'e}$ are simply

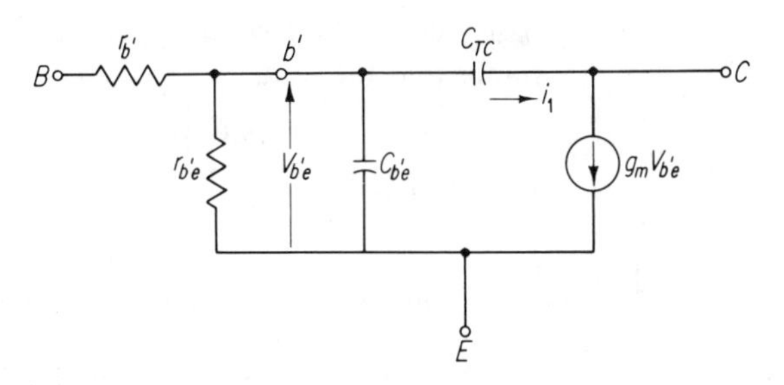

Fig. 2-17. Simplified version of the hybrid-pi circuit valid for high frequency and relatively low load impedances.

related by

$$r_{b'e} = (\beta_0 + 1)r_e \tag{2-94}$$

The capacitance shunted across $r_{b'e}$ consists of two components: a diffusion capacitance, C_d, and the emitter-base junction barrier (or transition) capacitance, C_{TE}.

The capacitance, C_{TC}, between the collector contact and the internal base contact is the collector-base junction barrier capacitance.

The activity of the transistor is represented by the current generator $g_mV_{b'e}$ which here represents the current multiplication or transfer conductivity of the device.

The presence of the collector capacitance between internal base and collector constitutes a feedback path between output and input. In other words, the transistor is not unilateral. In many practical situations, the transistor is operated with such a small load resistance that the effect of the feedback via the collector capacitance, C_{TC}, is negligible except to reflect a capacitance from the output to the input. We wish to replace the circuit of Fig. 2-17 by a unilateral one which has approximately the same terminal characteristics. Figure 2-18 shows the form of the circuit we desire. The desired circuit may be obtained by use of the Miller effect transformation. For the circuit of Fig. 2-19 we must calculate the impedance at the internal

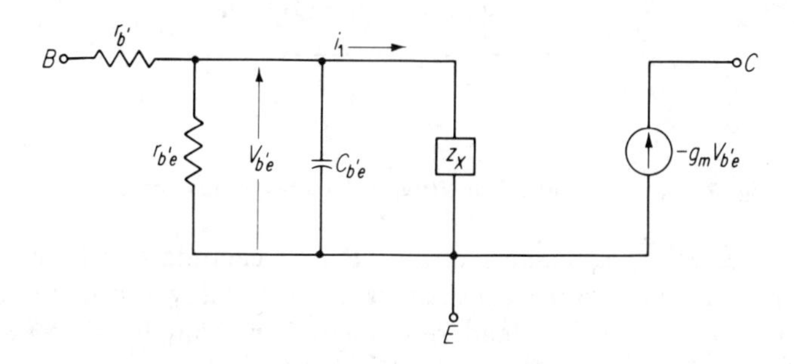

Fig. 2-18. A unilateral equivalent circuit.

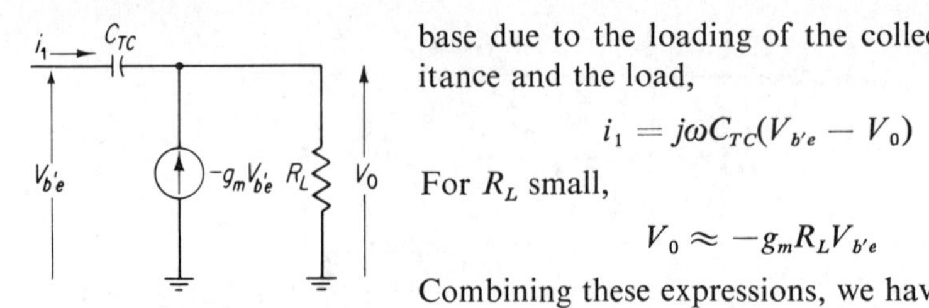

Fig. 2-19. Calculation of the Miller effect capacitance.

base due to the loading of the collector capacitance and the load,

$$i_1 = j\omega C_{TC}(V_{b'e} - V_0) \tag{2-95}$$

For R_L small,

$$V_0 \approx -g_m R_L V_{b'e} \tag{2-96}$$

Combining these expressions, we have

$$Z_X = \frac{V_{b'e}}{i_1} = j\omega(1 + g_m R_L)C_{TC} \tag{2-97}$$

The desired unilateral equivalent circuit is shown in Fig. 2-20. The effect of a complex load (R, L, C) may also be analyzed in a similar way. For the case of a parallel RC load (6), the unilateral equivalent circuit obtained after applying the Miller effect transformation is shown in Fig. 2-21. Use of these unilateral equivalent circuits will often simplify analysis greatly.

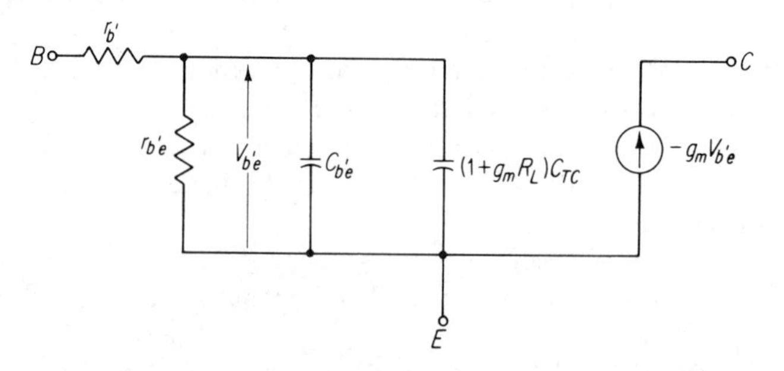

Fig. 2-20. Unilateral equivalent circuit showing the effect of the Miller capacitance.

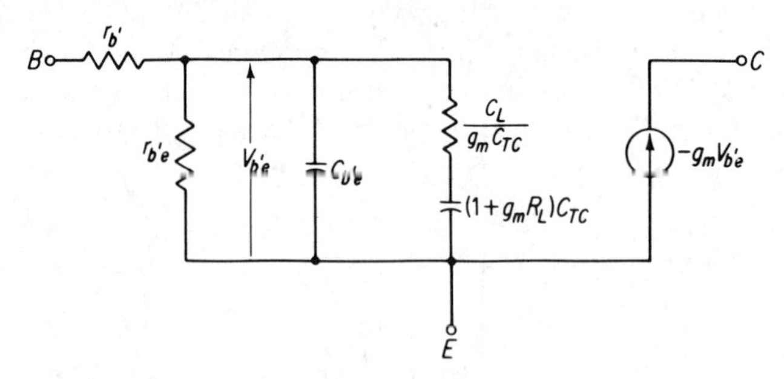

Fig. 2-21. Equivalent circuit showing the effect of load capacitance.

The Miller effect capacitance $(1 + g_m R_L)C_{TC}$ can add significantly to the total input capacitance of the transistor. This is especially true at low collector current and high values of load resistance. For example, a typical high-frequency transistor may have $C_{b'e} = 10$ pf at $I_E = 1.0$ ma and $C_{TC} = .5$–1 pf. In an amplifier with $R_L = 1000\ \Omega$, the Miller capacitance will be (1 +

$1000/27)^2$ pf $\approx$ 80 pf. In this case, the Miller capacitance almost completely controls the frequency response of the transistor.

A Common-Collector Equivalent Circuit

For completeness, we shall present a high-frequency equivalent circuit for the common-collector connection. The equivalent circuit (2) shown in Fig. 2-22 may be derived in the same way as the common-base T-equivalent circuit. Other common-collector equivalent circuits may be obtained by using the hybrid-pi or T-equivalent circuit and interchanging the terminal connections.

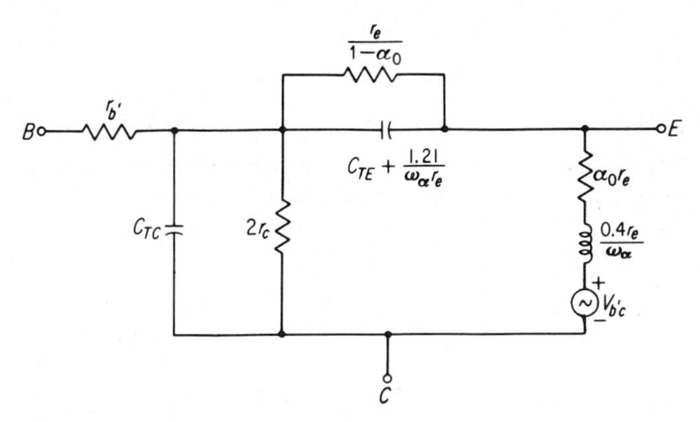

Fig. 2-22. A common-collector equivalent circuit.

The usefulness of these equivalent circuits will be demonstrated in succeeding chapters. Having discussed these so-called physical models of the transistor, we can thus obtain a complete characterization of the device from a purely phenomenological or electrical circuit point of view as well as from a physical approach. One can thus easily make the interrelation of the parameters involving the physical capacitances due, for example, to the diffusion of charge, and the resistances due to carrier flow in real materials, with the externally measurable quantities describing the terminal voltage-current characteristic of the device.

EXERCISES

1. In the accompanying circuit, evaluate the z parameters.

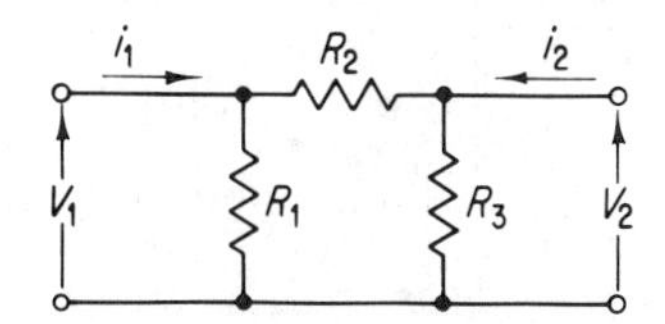

2. For the same circuit evaluate the y parameters, then, using the table, convert to the z parameters. (This illustrates the ease of evaluating one set of parameters from another and indicates that some particular set is often easier to evaluate from the circuit.)

3. Using the simplified equivalent circuit of a transistor, evaluate the current gain, voltage gain, and output resistance of the equivalent circuit shown below:

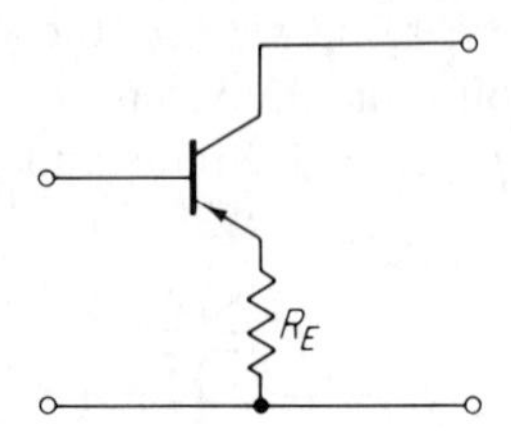

These should be evaluated with respect to the 50Ω-load resistance terminals.

4. (a) What parameters for the networks shown would be best to use in setting up an equivalent circuit at the input of the multistage network shown below? Why?
   (b) At the output?

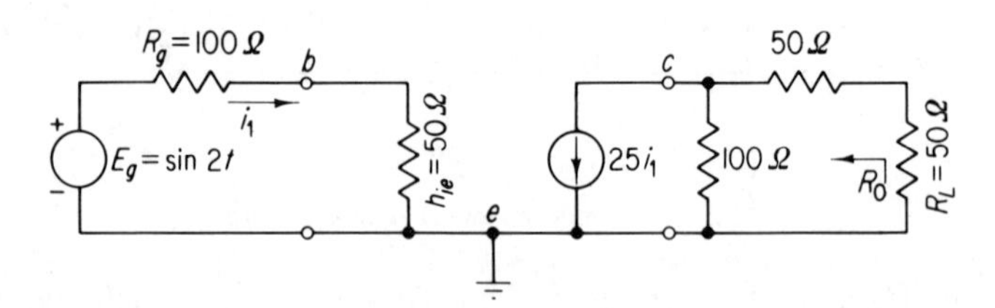

5. What is the voltage gain of an amplifier with parameters given in Table 2-3 driving a 10 KΩ amplifier. The transistor is connected
   (a) common emitter
   (b) common base
   (c) common collector

6. Evaluate the h parameters for the physical circuit of the transistor given in Fig. 2-17 at frequencies such that all capacitances are negligible.

7. Derive expressions for the equivalent h parameters of a common-emitter amplifier with emitter resistance R_E as shown.

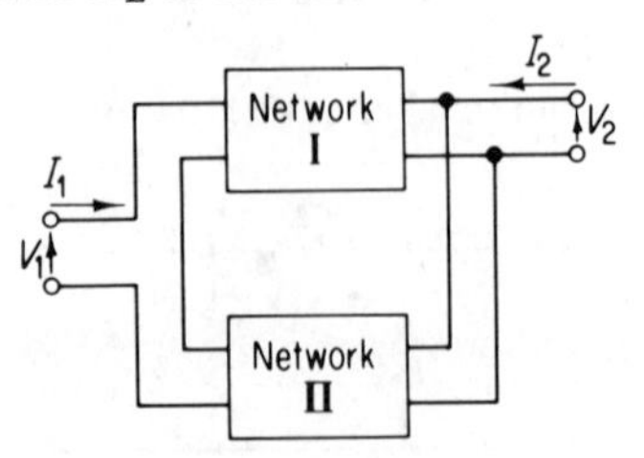

Show that

$$h_i = \frac{h_{ie} - R_E(1 + \Delta_{he} + h_{fe}) - h_{re}}{1 + h_{fe}R_E}$$

$$h_r = \frac{h_{re} + R_E h_{oe}}{1 + R_E h_{oe}}$$

$$h_f = \frac{h_{fe} - R_E h_{oe}}{1 + R_E h_{oe}}$$

$$h_o = \frac{h_{oe}}{1 + R_E h_{oe}}$$

8. Derive the equivalent circuit shown in Fig. 2-21. Use the Miller effect transformation as outlined in Eqs. (2-95)–(2-97).

REFERENCES

1. J. ZAWELS, "Physical Theory of New Circuit Representation for Junction Transistors," *J. Appl. Phys*, **25** (August, 1954), 976–81.
2. R. M. SCARLETT, "Some New High-Frequency Equivalent Circuits for Junction Transistors," Technical Report No. 103, Stanford University, Stanford Electronics Laboratory, Stanford, Calif., 1955.
3. C. A. LEE, "A High-Frequency Diffused Base Germanium Transistor," *Bell Syst. Tech. J.*, **35** (January, 1956), 23–24.
4. D. E. THOMAS, and J. L. MOLL, "Junction Transistor Short-Circuit Current Gain and Phase Determination," *Proc. IRE*, June 1958, pp. 1177–84.
5. L. J. GIACOLETTO, "Study of PNP Alloy Junction Transistors from D-C through Medium Frequencies," *RCA Rev.* (December, 1954), 506–62.
6. J. A. EKISS, "Calculation of the Small Signal Transient Response of a Grounded Emitter Amplifier Having a Parallel *RC* Load Using the Giacoletto Hybrid-Pi Equivalent Circuit," Application Lab Report 549. Philco Corp., September, 1958.
7. E. BRENNER, and M. JAVID, *Analysis of Electric Circuits*. New York: McGraw-Hill Book Company, 1959.
8. F. F. KUO, *Network Analysis and Synthesis*. New York: John Wiley & Sons, Inc., 1962.
9. R. D. MIDDLEBROOK, *Introduction to Junction Transistor Theory*. New York: John Wiley & Sons, Inc., 1957.
10. P. E. GRAY, *et al.*, *Physical Electronics and Circuit Models of Transistors*, SEEC, Vol. II. New York: John Wiley & Sons, Inc., 1964.
11. J. L. MOLL, *Physics of Semiconductors*. New York: McGraw-Hill Book Company, 1964.

12. P. E. GRAY and C. L. SEARLE, *Electronic Principles—Physics, Models and Circuits*. New York: John Wiley & Sons, Inc., 1969.

13. E. R. JONES, *Solid State Electronics*. New York: Intext Educational Publishers, 1971.

14. F. C. FITCHEN, *Electronic Integrated Circuits and Systems*. New York: D. Van Nostrand Reinhold Company, 1970.

3

Techniques for Biasing Transistors

INTRODUCTION

This chapter discusses techniques for establishing the dc operating point (biasing) for transistors in a variety of circuit applications. A dc operating point is determined, for example, when dc collector-emitter voltage and dc emitter current are specified. Selection of an operating point is based on particular performance criteria that the circuit must meet.

Factors Influencing Selection of the Operating Point

The following factors are among those which influence selection of the operating point:

1. The ac and dc parameters of the transistor vary with operating point.
2. The fundamental performance characteristics, such as power gain, input impedance, voltage gain, etc., are functions of the operating point.
3. The signal level to be handled must be considered, particularly in power stages.
4. The power dissipated within the device is a function of the operating point and must be kept within allowable limits.
5. The ambient temperature must be considered for two reasons: first, most of the parameters just mentioned also vary with junction temperature; second, the operating point must be stabilized by careful external circuit design.

Let us now discuss each of these considerations in detail that we may better understand the importance of selecting the correct operating point.

Variations of AC and DC Parameters

Figures 3-1, 3-2, and 3-3 illustrate typical variations of both dc and small-signal h parameters with both current and voltage. Note that variations of several hundred per cent are not uncommon. Although the effect of vari-

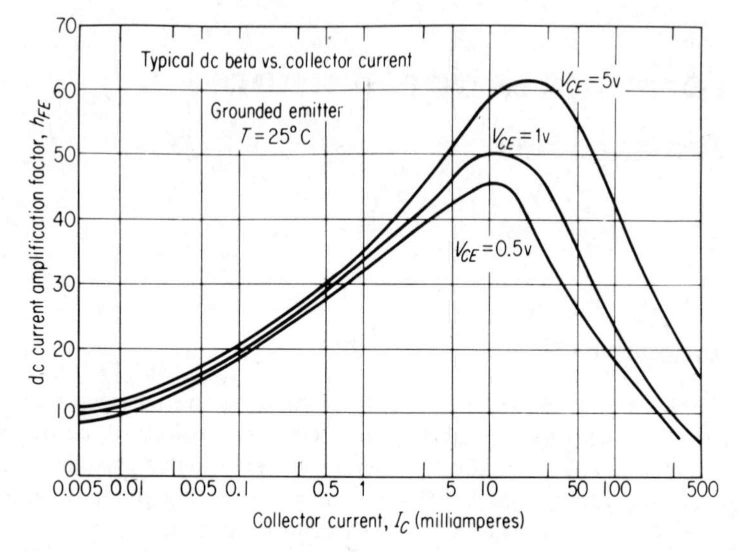

Fig. 3-1. Typical variation of dc beta (h_{FE}) with current and voltage for the 2N708.

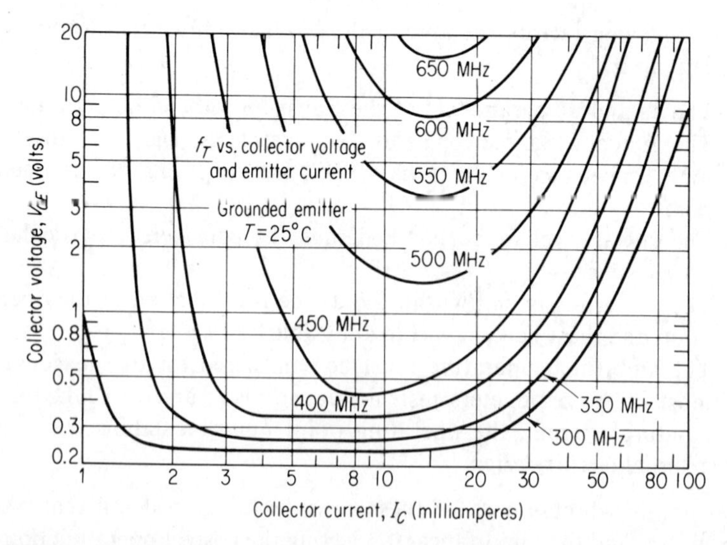

Fig. 3-2. Typical variation of high-frequency h_{FE} (f_t) with current and voltage for the 2N708.

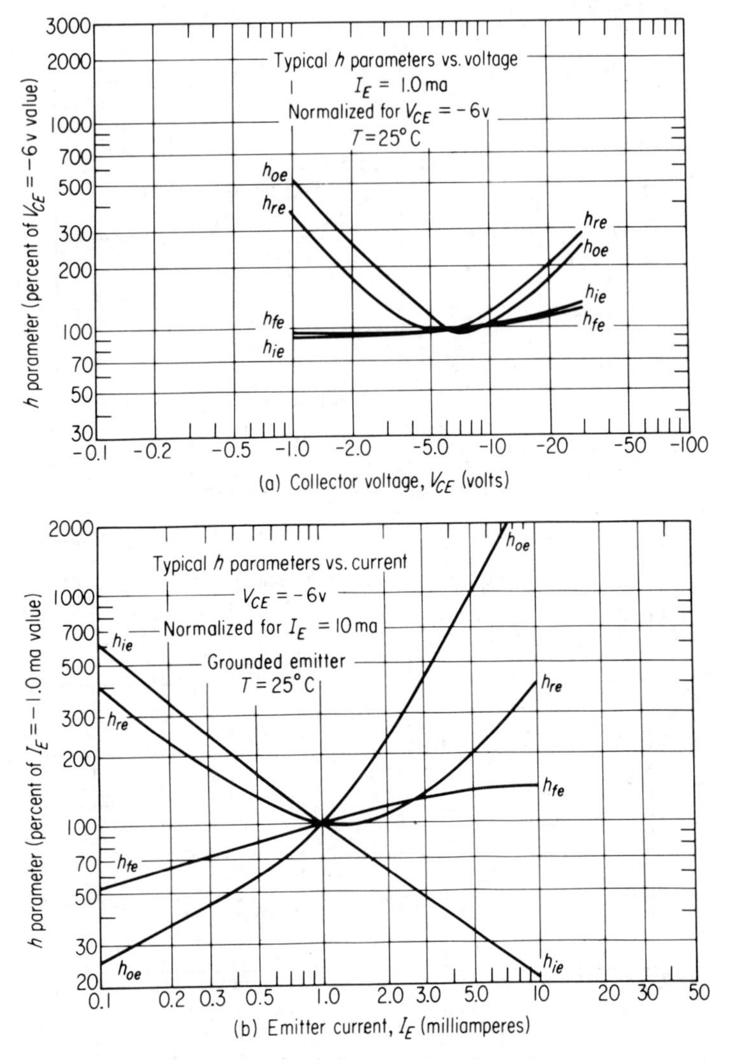

Fig. 3-3. Typical variation of 2N1747 small-signal h parameters with (a) voltage and (b) current.

ations in all the h parameters must be considered in designing the circuit, the dc beta (h_{FE}) is probably the most critical. As will be shown later, not only does h_{FE} vary with the applied bias, but also the collector current in an actual circuit varies from unit to unit as the h_{FE} changes.

Variations in Performance Characteristics

Figures 3-4 and 3-5 illustrate how gain, noise figure (an indication of purity of output frequency response or of the spurious noise contribution of

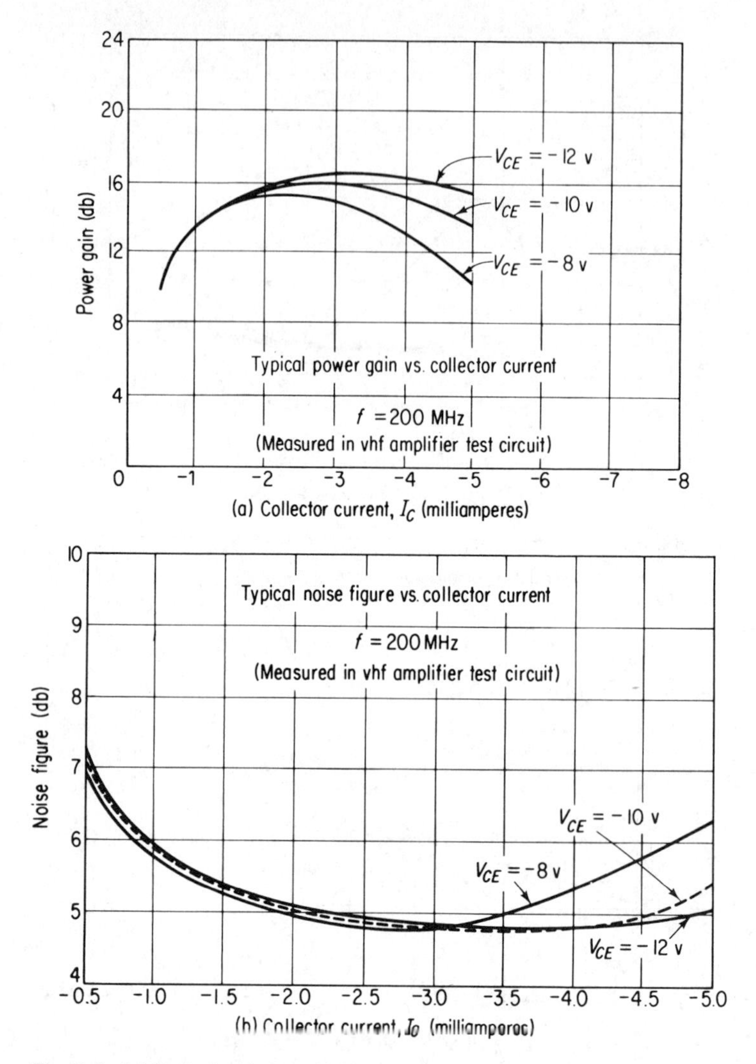

Fig. 3-4. (a) Typical variation of 200 MHz power gain and (b) noise figure with operating point for the 2N1742.

the device as discussed in a later chapter), and switching times (time to turn on and turn off a transistor acting as a switch) vary as a function of operating point. Note that sometimes one must compromise the performance of one parameter for another. For instance, in Fig. 3-4 it is not possible to obtain highest gain and lowest noise at the same operating point. Thus, the designer must choose what is most important for his particular application.

The Effect of Signal Level on Operating Point

In most low-level stages where the signal level at the collector is a few hundred millivolts, the operating point is chosen not only to obtain sufficient

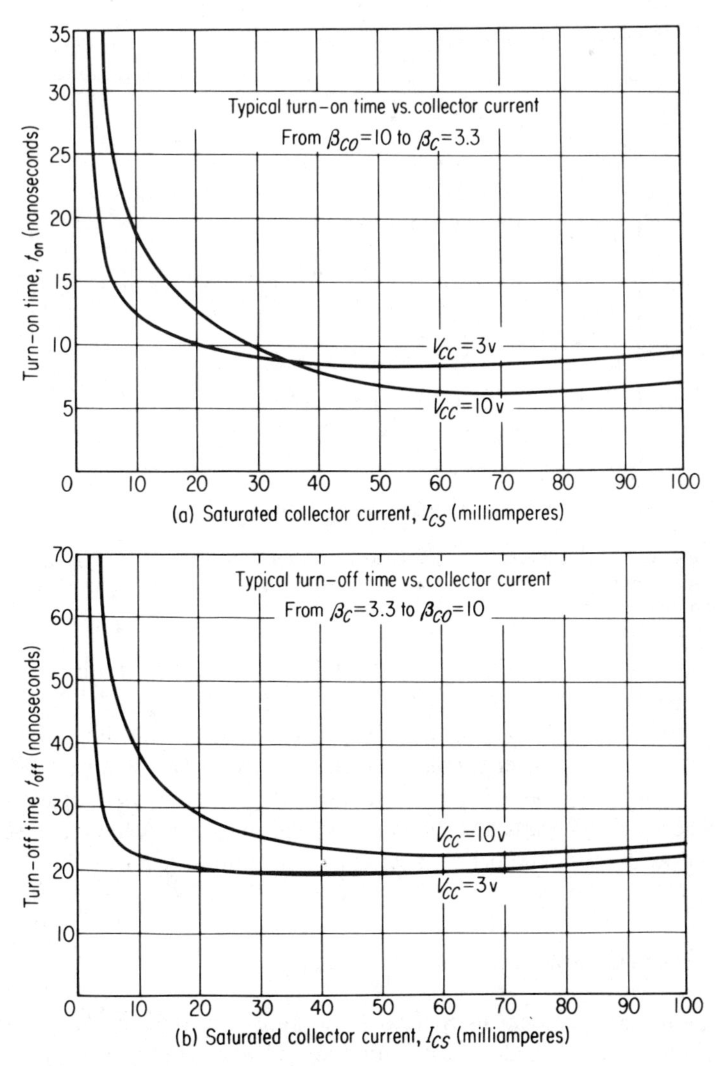

Fig. 3-5. Typical turn-on and turn-off times as a function of operating point for the 2N708.

output but also for other reasons. As the signal level increases, however, the proper output can be obtained only if two factors are considered. First, the maximum output voltage or current swing without distortion, and second, the output power, are important and essentially govern the placement of the operating dc quiescent point. Figure 3-6 illustrates a set of collector characteristics with a load line superimposed. If the quiescent operating point is at Q_1 (about -2 volt and 4 ma), note that the linear swing along the load line is limited to approximately ± 2 volts. Any further increase in signal will cause clipping (cutting off the top of the waveform and consequent distortion) during the low-voltage, high-current part of the waveform. If the operating

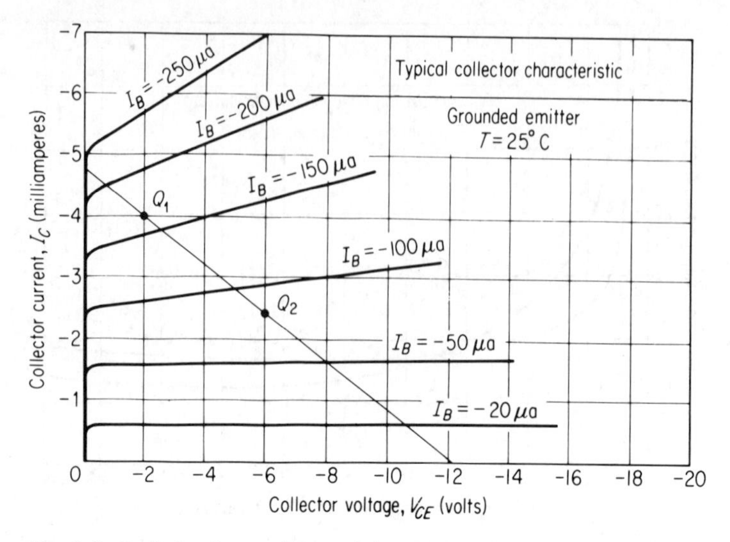

Fig. 3-6. Typical collector characteristic with load line showing quiescent point for maximum output (Q_2) and Point Q_1 which will result in less linear output.

point is moved to Q_2 (about -6 volt, 2.5 ma) it will be possible to obtain nearly a ± 6-volt difference or swing in the output.

A similar analysis holds for power stages, and as will be shown in a later chapter, maximum power available from a Class A stage (one where the output faithfully reproduces the input waveform) occurs when biased at Q_2. Thus, where large signals are encountered, the output power or undistorted signal output may be the prime factor in determining the particular point of operation.

The Effect of Operating Point on Power Dissipation

The allowable power dissipation of a transistor is a maximum rating which must not be exceeded. It is a function of many things, including junction area, material, internal construction, and the package in which the transistor is mounted. When a transistor is biased at an appropriate operating point, the dc voltages and currents set up dissipate power within the transistor. Care should be taken to assure that this dissipation is within rated limits at all times. The dissipation in a transistor may be easily calculated from either Fig. 3-7(a) or Fig. 3-7(b).

In Fig. 3-7(a),

$$P_D = I_B V_{BE} + I_C V_{CE}, \tag{3-1}$$

where P_D means power dissipated. In Fig. 3-7(b),

$$P_D = I_E V_{BE} + I_C V_{CB} \tag{3-2}$$

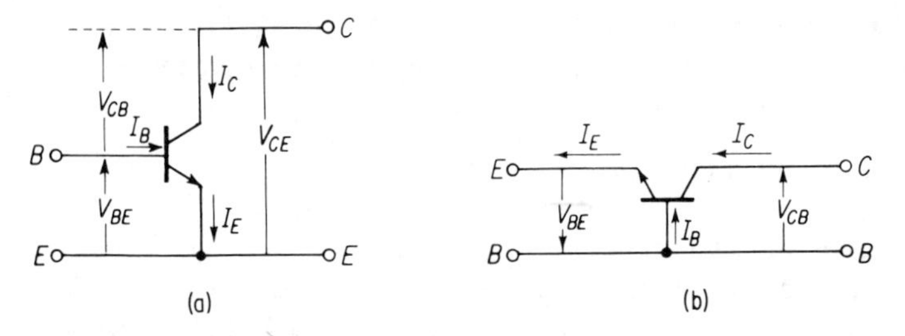

Fig. 3-7. (a) Common-emitter and (b) common-base dc voltage and current conventions for calculating power dissipation.

However,

$$V_{CE} = V_{CB} + V_{BE} \tag{3-3}$$

and

$$I_E = I_B + I_C \tag{3-4}$$

Thus, Eq. 3-2 becomes

$$P_D = (I_B + I_C)V_{BE} + I_C(V_{CE} - V_{BE}) \tag{3-5}$$

$$= V_{BE}I_B + I_C V_{CE} \tag{3-6}$$

which is the same as Eq. 3-1.

In many applications the dissipation in the base-emitter diode ($V_{BE}I_B$) is small and may be neglected. Also, when V_{CB} and V_{CE} are large, their difference is small and the power dissipation is often calculated as the product of I_C times the voltage across the transistor (V_{CB} or V_{CE}). In those cases where ac power is being either extracted or dissipated within the device, the actual dissipation must be adjusted accordingly.

As noted earlier, the power dissipation of a transistor is dependent on many factors. Most transistors, however, are rated under the assumption that junction temperature is the limiting factor. The junction is heated from two sources: first, the dissipation of power within the device causes heating; second, there is a contribution due to the ambient temperature in which the transistor must operate. To a first approximation it may be assumed that the junction cannot distinguish between heating caused by ambient temperature rise over its rated value and heating from dissipation. Thus, storing a transistor at 100°C should have the same effects as operating for the same period of time at a dissipation which creates 100°C junction temperature. Although there is considerable evidence to show this is not exactly correct, most manufacturers design on this basic principle and run life tests to establish the validity of the ratings.

In any event, most transistors are rated such that the maximum allowable dissipation decreases with ambient temperature. Thus the bias point must be such that the dissipation rating is not exceeded over the operating temperature range. This reduction in allowable dissipation with temperature is often expressed in terms of thermal resistance (Θ), in °C/watt. It also relates the junction temperature to the ambient temperature by the expression:

$$T_J = T_A + \Theta P_D \tag{3-7}$$

where T_J = junction temperature of the device in °C
T_A = ambient temperature in °C
Θ = thermal resistance in °C/watt
P_D = power dissipated in the device in watts.

In general, most transistor data sheets either show Θ, or give data from which it may be calculated, namely, the maximum junction temperature, rated dissipation, and the temperature at which the rated dissipation applies. It has become common practice to use also the reciprocal of the thermal resistance (watts/°C). This is often referred to as *derating factor*. Since the two terms are sometimes used interchangeably, one must be careful to observe the units before calculating the derating. (An example of a typical derating curve showing allowable dissipation as a function of temperature is shown in Fig. 3-8.)

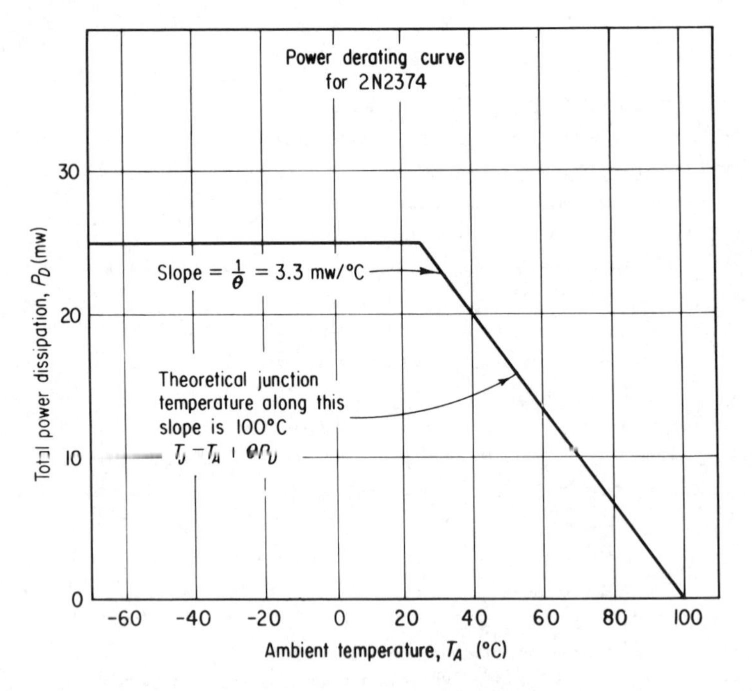

Fig. 3-8. Power derating curve for 2N2374.

The Effect of Temperature on Operating Point

The preceding discussion has indicated the need to consider a number of factors in selecting the operating point. We have also shown that the operating point affects the junction temperature. To complicate matters further, almost all transistor parameters are temperature dependent. Thus, we must consider not only the variations of parameters with operating point, but also the manner in which these same critical parameters vary with junction temperature over the expected ambient temperature range. Figures 3-9 and 3-10 show typical variations of transistor parameters with temperature.

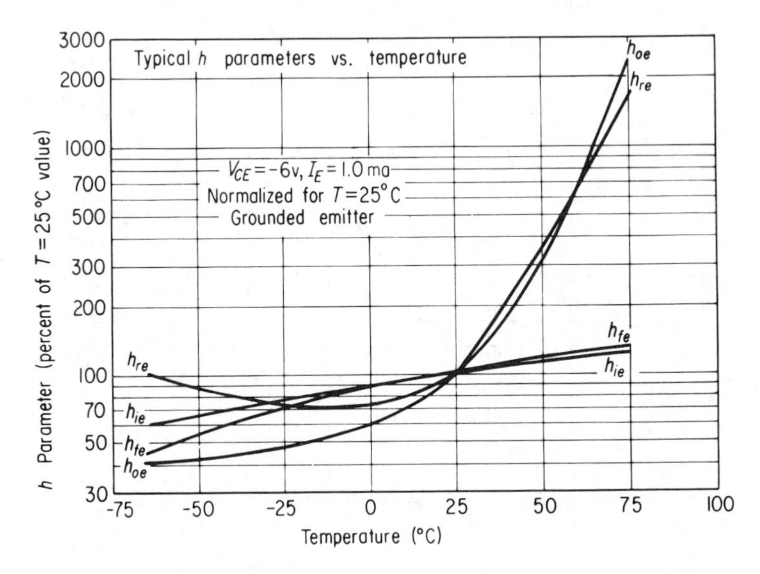

Fig. 3-9. Typical variation of h parameters with temperature for the 2N1747.

The change in leakage currents with temperature probably creates more problems for the circuit designer than any other parameter. The leakage currents across the back-biased collector base (I_{CBO}) or emitter base (I_{EBO}) diodes can be considered as consisting of two components: the diode saturation current (I_S) and the surface leakage current (I_{SL}). This latter component is highly variable, unpredictable, and relatively independent of temperature. It is usually neglected when calculating the temperature variation of leakage current. In order to separate the two components, the leakage currents are often specified at two voltages on manufacturers' data sheets. A low-voltage specification (usually 1 volt or less) measures I_S, the temperature-dependent portion. A higher voltage specification (usually about 50–75 percent of the diode breakdown rating) measures the total leakage $I_S + I_{SL}$. Sometimes they are specified at a third point (at high voltage) to define the diode breakdown. Be careful not to confuse this with the leakage specification ($I_S + I_{SL}$).

The diode saturation current varies exponentially with temperature as follows:

$$I_S = I_{SO}e^{K_1(T-T_0)} \tag{3-8}$$

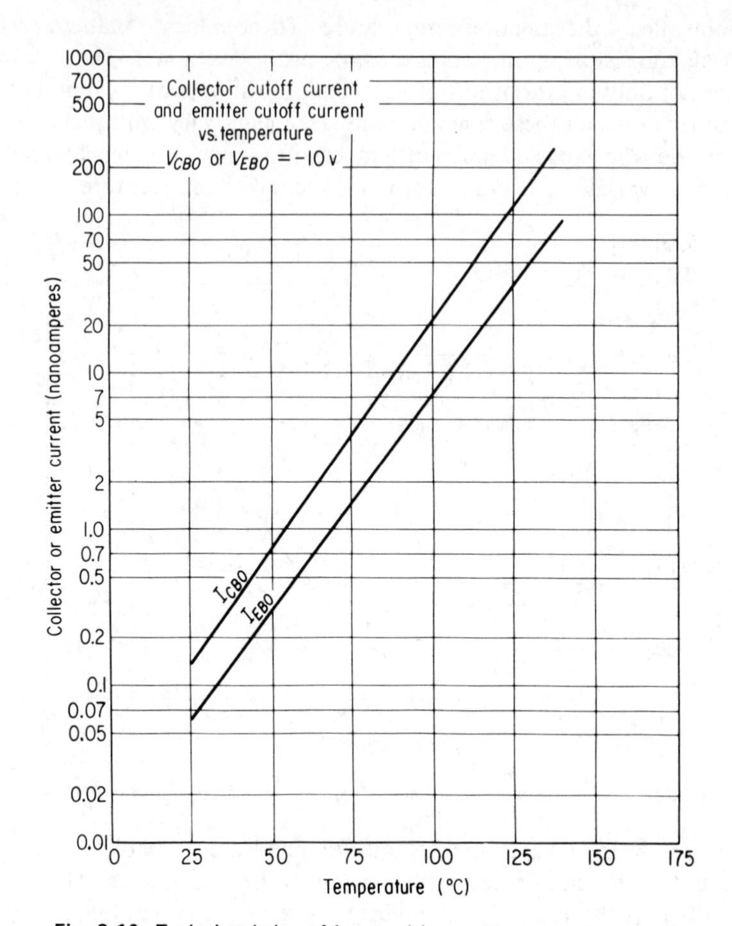

Fig. 3-10. Typical variation of I_{CBO} and I_{EBO} with temperatures for the 2N2278.

where I_{SO} = diode saturation current at T_0
K_1 = constant, ≈ 0.07 for germanium and ≈ 0.09 for silicon
T = temperature of operation, °C
T_0 = room temperature, usually considered to be 25°C.

If K_1 is 0.07, as it is in germanium, then I_S will approximately double for each 10°C increase in temperature, and if K_1 is 0.09, as it is in silicon, then I_S will approximately double for every 8°C. Figure 3-11 shows how I_S changes with temperature for germanium and silicon (they are normalized for better visualization). Note that although the change in I_S with temperature is less for germanium transistors, silicon transistors will, in general, have a lower total leakage current over the temperature range.

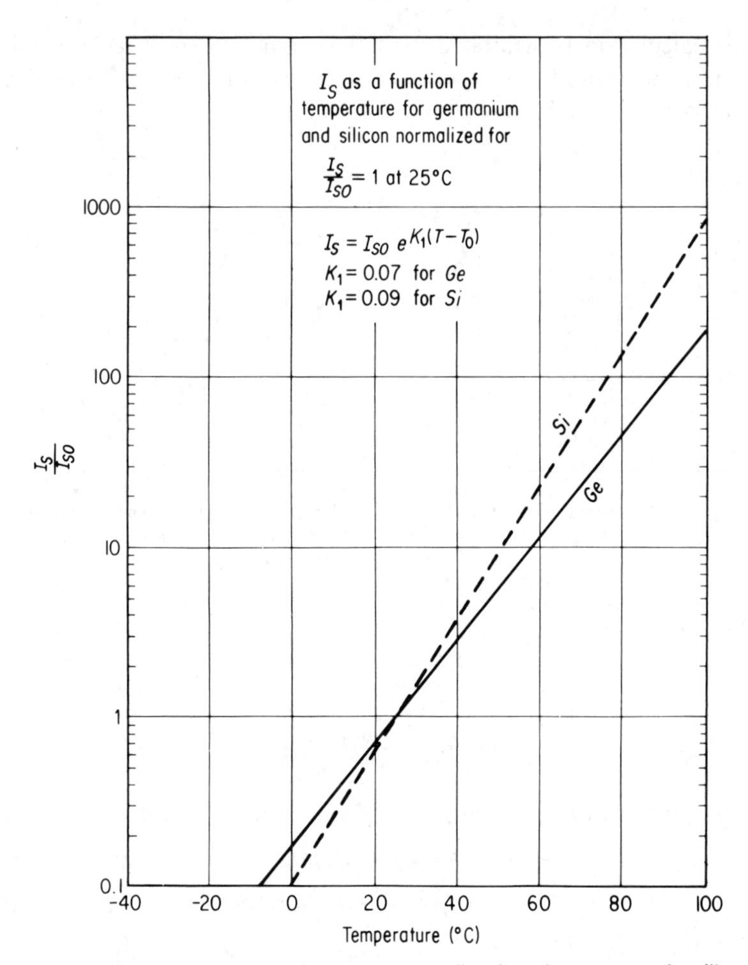

Fig. 3-11. Normalized saturation current as a function of temperature for silicon and germanium.

Let us now look at the tools available to the circuit designer to compensate for some of the variation with operating point and temperature.

GENERAL BIAS CIRCUIT

A generalized bias circuit has been derived which incorporates most of the circuits in present use. This circuit is shown in Fig. 3-12.

Given V_1, V_2, R_1, R_2, R_L, R_E, h_{FE}, and I_{CBO} it is possible to calculate the operating point as specified by the collector current, I_C, and the collector-emitter voltage, V_{CE}.

This calculation is facilitated if the base bias circuit is replaced by the Thevenin equivalent of V_1, R_1, and R_2 as shown in Fig. 3-13. The equivalent base bias voltage and base resistance are V_3 and R_B,

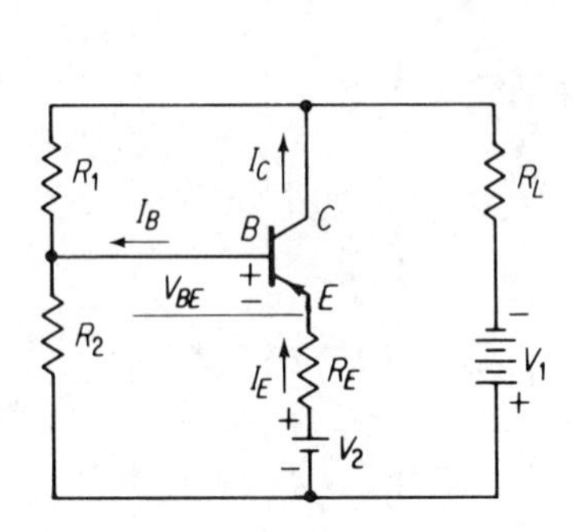

Fig. 3-12. General bias circuit.

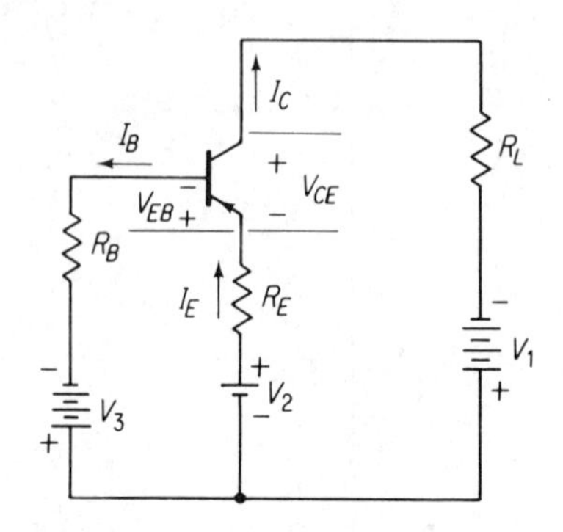

Fig. 3-13. General bias circuit with base bias circuit replaced with the Thevenin equivalent.

where

$$V_3 = \frac{V_1 R_1}{R_1 + R_2} \tag{3-9}$$

and

$$R_B = \frac{R_1 R_2}{R_1 + R_2} \tag{3-10}$$

Now, by writing two loop equations, we have (assuming $V_2 + V_3 \gg V_{EB}$)

$$V_2 + V_3 \approx (R_E + R_B)I_B + R_E I_C \tag{3-11}$$

$$V_1 + V_2 + V_{CE} = R_E I_E + R_L I_C \tag{3-12}$$

where we have used the relation,

$$I_E = I_B + I_C \tag{3-13}$$

These are the two dc loop equations of the basic grounded emitter circuit.

The collector leakage current does not appear in any of the preceding equations, but since the relationship between I_C and I_{CBO} is desired for reasons we shall see later, an equation containing I_{CBO} is necessary. This equation is

$$I_{CBO} + h_{FB} I_E = I_C \tag{3-14}$$

where I_{CBO} = grounded-base cutoff current
h_{FB} = grounded-base short-circuit dc current gain.

This equation can be obtained by putting into the Ebers-Moll equations, expressions 1-15 and 1-16, the actual dc operating conditions for the transistor, where V_{CB} is a large negative bias, etc. The resulting equation can be simplified to Eq. 3-14. The output equation then is

$$I_{CBO} = I_C - h_{FB} I_E \tag{3-15}$$

which is in terms of I_C and I_E, and the input equation 3-11 becomes

$$V_2 + V_3 = -I_C R_B + I_E(R_E + R_B) \tag{3-16}$$

Equations 3-15 and 3-16 are simultaneous equations from which I_C can be determined as

$$I_C = \frac{(I_{CBO})(R_E + R_B) + h_{FB}(V_2 + V_3)}{R_E + R_B - h_{FB}(R_B)} \tag{3-17}$$

or

$$I_C = \frac{(R_E + R_B)(I_{CBO}) + h_{FB}(V_2 + V_3)}{R_E + R_B(1 - h_{FB})} \tag{3-18}$$

However,

$$h_{FE} = \frac{h_{FB}}{1 - h_{FB}} \tag{3-19}$$

and Eq. 3-18 becomes

$$I_C = \frac{(R_E + R_B)(I_{CBO})}{R_E + R_B/(h_{FE} + 1)} + \frac{h_{FE}(V_3 + V_2)}{(h_{FE} + 1)R_E + R_B} \tag{3-20}$$

This specifies the dc collector current as a function of the bias network parameters and two transistor parameters h_{FE} and I_{CBO}. The collector operating voltage is $V_1 - I_C R_L$.

It might be well to point out here that the general bias circuit can be applied just as well in a transformer-coupled amplifier as shown in Fig. 3-14. The big advantage of using this type of bias network is that R_B is limited only by the maximum allowable current drain. The bias network has no effect on the ac signal if the base biasing resistance R_B is bypassed for ac signals by a large capacitor.

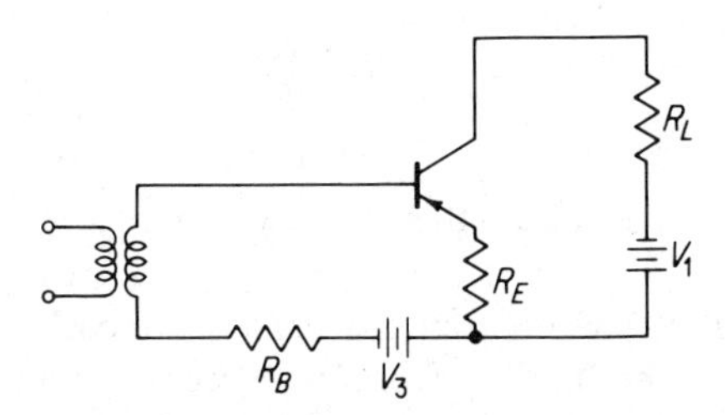

Fig. 3-14. Application of general bias circuit to transformer-coupled amplifier.

TEMPERATURE DEPENDENCE OF THE BASE-EMITTER VOLTAGE

Where the base is biased from a low impedance source, the base-emitter voltage and its temperature dependence cannot be ignored. The relationship between I_E and V_{BE} for $|V_{CB}| \gg kT/q$ is (from Chap. 1, Eqs. 1-15, 1-16)

$$I_E = \frac{I_{EBO}}{1 - \alpha_N \alpha_I}[e^{qV_{EB}/kT} - 1] - \frac{\alpha_I I_{CBO}}{1 - \alpha_N \alpha_I} \tag{3-21}$$

If V_{EB} is positive, as it will be in a forward-biased diode, the first term in Eq. 3-21 is much larger than the second term, and V_{EB} can be given approximately as

$$V_{EB} \approx \frac{kT}{q}\left[\ln\left(\frac{I_E(1-\alpha_N\alpha_I)}{I_{EBO}}\right) - K_1(T - T_0)\right] \tag{3-22}$$

where I_{EBO} is measured at temperature T_0, and K_1 is a constant, usually considered to be 0.07 for germanium and 0.09 for silicon.

The change in V_{EB} with temperature, then, is approximately

$$\frac{\partial V_{EB}}{\partial T} = -\frac{2kTK_1}{q} + \frac{kT_0K_1}{q} \tag{3-23}$$

which is approximately -2 mv/°C at 25°C ($T = T_0$).

THERMAL STABILITY

Thermal stability is a measure of the stability of the dc bias point with variations in temperature. For a properly designed bias network the change in collector current with temperature is due to the increase of collector diode leakage current, I_{CBO}, with temperature.

A convenient measure of the thermal stability of a particular bias network is the thermal stability factor, S, defined by

$$S \equiv \left.\frac{\partial I_C}{\partial I_{CBO}}\right|_{\text{constant voltage}} \tag{3-24}$$

From Eq. 3-20,

$$S = \frac{R_E + R_B}{R_E + \dfrac{R_B}{h_{FE} + 1}} \tag{3-25}$$

Note that R_L does not affect the stability, since I_C is almost independent of V_{CE}.

From Eq. 3-24 it is apparent that if the change in collector current is equal to the change in I_{CBO}, the stability factor will become equal to 1, which is the best stability factor possible. Equation 3-25 can be made to approach 1 if R_E is increased and R_B is decreased. On the other hand, if the change in collector current is equal to the change in I_{CEO} where

$$I_{CEO} = (h_{FE} + 1)I_{CBO} \tag{3-26}$$

then

$$S = h_{FE} + 1 \tag{3-27}$$

Equation 3-25 can be made to equal $(h_{FE} + 1)$ if R_E is made equal to zero. This, then, is the worst stability attainable.

Normally, however, we have a stability factor somewhere between 1 and $(h_{FE} + 1)$. Since R_E dissipates power and since the current flowing through it causes a voltage drop, it cannot be increased without limitations. The base resistance, R_B, cannot be decreased to zero since it may shunt the signal and increase the battery drain. In a transformer-coupled amplifier, the battery drain is the only consideration in the selection of R_B. In a common-base circuit, when V_2 is not zero, R_B may be made zero.

DC BETA STABILITY

Since a particular type of transistor may have a wide beta spread, it is easy to see that the operating point might vary from transistor to transistor. It is not practical to use a different bias circuit for every transistor. Therefore, a bias circuit should be designed to constrain the collector current within reasonable limits for the expected beta range. At low temperatures, I_{CBO} is generally negligible and I_C from Eq. 3-20 is

$$I_C \approx \frac{h_{FE}(V_2 + V_3)}{(h_{FE} + 1)R_E + R_B} \tag{3-28}$$

With a minimum beta transistor, then, at low temperatures:

$$I_C = I_{C(\text{min})} = \frac{h_{FE(\text{min})}(V_2 + V_3)}{(h_{FE(\text{min})} + 1)R_E + R_B} \tag{3-29}$$

and with a maximum beta transistor at high temperatures, from Eq. 3-20 we obtain, where we now must include I_{CBO},

$$I_C = I_{C(\text{max})} = \frac{h_{FE(\text{max})}(V_2 + V_3)}{(h_{FE(\text{max})} + 1)R_E + R_B} + \frac{(R_E + R_B)I_{CBO(\text{max})}}{R_E + [R_B/(h_{FE(\text{max})} + 1)]} \tag{3-30}$$

But from Eq. 3-25

$$S = \frac{R_E + R_B}{R_E + R_B/(h_{FE} + 1)} \tag{3-25}$$

and thus

$$I_{C(\text{max})} = \frac{h_{FE(\text{max})}(V_2 + V_3)}{R_E(h_{FE(\text{max})} + 1) + R_B} + SI_{CBO(\text{max})} \tag{3-31}$$

Now the change in the collector current, ΔI_C, with beta and temperature is

$$\Delta I_C = I_{C(\text{max})} - I_{C(\text{min})} \tag{3-32}$$

This may be evaluated for the two extreme cases, $S = 1$ and $S = h_{FE} + 1$. For $R_E = 0$, $S = h_{FE} + 1$, and

$$\Delta I_C = \frac{(h_{FE(\text{max})} - h_{FE(\text{min})})(V_2 + V_3)}{R_B} + (h_{FE} + 1)I_{CBO(\text{max})} \tag{3-33}$$

This may be rewritten as

$$\Delta I_C = I_{C(\text{max})}\left(1 - \frac{h_{FE(\text{min})}}{h_{FE(\text{max})}}\right) + (h_{FE} + 1)I_{CBO(\text{max})} \tag{3-34}$$

where

$$I_{C(\max)} = \frac{h_{FE(\max)}(V_2 + V_3)}{R_B} \tag{3-35}$$

On the other hand, if R_E is made large, $I_{C(\min)}$ from Eq. 3-29 approaches $(V_3 + V_2)/R_E$ and $I_{C(\max)}$ from Eq. 3-31 approaches $[(V_3 + V_2)/R_E] + SI_{CBO(\max)}$, from which I_C approaches $SI_{CBO(\max)}$. Since $S \longrightarrow 1$ in a case where R_E is large

$$\Delta I_C \simeq I_{CBO(\max)} \tag{3-36}$$

It can be seen from the foregoing equations for ΔI_C that beta variation from unit to unit in the same circuit produces the same general effect as temperature variation. Therefore, stabilizing for temperature variations also stabilizes for unit to unit variations in h_{FE}.

A PRACTICAL BIAS DESIGN

The general bias circuit of Fig 3-12 is an example of dc current feedback and is redrawn as a frequently used circuit in Fig. 3-15.

For many designs a good rule of thumb is that the ratio of R_B/R_E be approximately 3 or less. Once the parameters desired to optimize the design are known, the following steps should be taken to design the bias circuit.

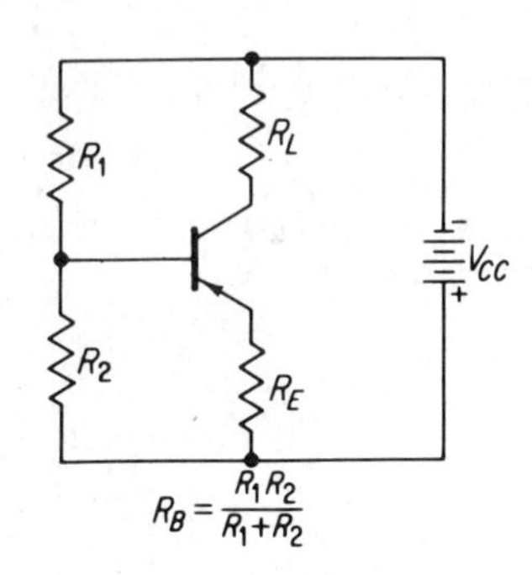

Fig. 3-15. Typical common-emitter bias circuit.

1. Determine I_E and V_{CE} to produce the desired operating characteristics.
2. Determine R_L from the ac power requirements. Maximum power transfer occurs when the load impedance (R_L) is equal to the transistor output impedance, $1/h_{oe}$. This is not always practical with a resistive load, but it may be with a transformer load (in which case $R_L = 0$ from a dc standpoint).
3. Select an emitter resistance (R_E), keeping in mind that you would like it to be $\frac{1}{3}$ the value of R_B and that R_B may shunt the input signal. Also R_E must be small enough so that the dc drop across it does not lower V_{CE} below the desired value, consistent with supply voltage limitations.
4. Determine V_{CC}.
5. Select R_2. It should be at least 10 times larger than the input resistance of the transistor so as not appreciably to shunt the input signal (necessary except when connected in transformer-coupled stages).
6. Calculate the voltage drop across R_E: ($V_E = I_E R_E$).

7. Add approximately 0.3 volt, which is approximately V_{BE} in most germanium transistors. Add about 0.7 volt for silicon transistors. The sum is the approximate voltage drop that is desired across R_2 (V_{R_2}).
8. Determine R_1 from the voltage-divider relationship

$$\frac{V_{CC}}{R_1 + R_2} R_2 = V_{R_2}.$$

The bias network is now designed and ready to operate at the desired point with reasonable stability. If it is desirable to calculate the actual stability and current changes, then a more exact design can be made from Eqs. 3-20 and 3-34. For most cases where R_E is reasonably large, however, the preceding approximation is sufficiently accurate. This technique is extremely useful in designing initial circuits and also in estimating the current present in a designed circuit which shows the values of the components but not the operating point.

EXAMPLE 1

Suppose we desire to operate a transistor amplifier to drive a load with an impedance of 100 Ω with a circuit similar to Fig. 3-15. Assume a transistor whose value of $h_{FE} \approx 200$ and which must operate over an output voltage range of ± 1 volt.

Since we need to drive ± 1 volt into 100 Ω, we require at least a current change of ± 10 ma from the quiescent operating point. We bias the transistor, therefore, so that approximately 10 ma flow in the collector with at least 2 volts across the device for linearity. Thus our operating point is approximately $V_{CE} = -2$ volts, $I_E = 10$ ma. Since R_E must be such that it is approximately $\frac{1}{3}R_B$ and R_B must be large compared to the source impedance and input impedance of the transistor which is of the order of 300 Ω, choose $R_E = 100$ Ω. This will give, with a quiescent current of 10 ma, a drop of 2 volts across both the transistor and the dc resistance of $R_L + R_E = 100\ \Omega + 100\ \Omega$. Thus a supply voltage of 4 volts is required. Also this gives $R_B \approx 300$ Ω. If we assume a silicon transistor with $V_{BE} \approx 0.7$ volt, we obtain $V_{R_2} \approx 1.7$ volts, and since R_2 must be about 10 times the input impedance, we choose $R_2 = 600$ Ω. Using voltage division, we obtain for R_1:

$$\frac{1.7}{4} = \frac{0.6\text{K}}{R_1 + 0.6\text{K}} \longrightarrow R_1 \approx 0.8\text{K}$$

Thus we have $R_1 = 800\ \Omega$, $R_2 = 600\ \Omega$ which gives an R_B of $(800 \times 600)/1400 \approx 340\ \Omega$. To check this result, we compute the base current and compare it with $I_C/\beta = 10\ \text{ma}/200 = 50\mu\text{a}$. Since this current will be the difference current between that in R_1 and that in R_2, we have

$$\frac{2.3}{800} - \frac{1.7}{600} = (2875 - 2830) \times 10^{-6} \approx 45\mu\text{a}$$

This agrees quite closely with the value of 50μa required. Since we must take resistors in standard values anyhow, the operating point will probably change sufficiently so that our computation will be adequate. We would now check the stability, etc., to see whether our design was reasonable.

SOME LINEAR BIAS CIRCUITS

The general bias circuit shown in Fig. 3-12 can now be used to show some of the more frequently used bias circuits. In order to determine the quiescent point on a curve of the output characteristics, it is necessary to determine the base current, I_B. Equation 3-11 may be used to find I_B, and since

$$I_C \approx h_{FE} I_B$$

$$I_B = \frac{V_2 + V_3}{R_E(h_{FE} + 1) + R_B} \tag{3-37}$$

For the circuits which follow, equations will be given for finding the stability factor, S, and base current, I_B.

Figure 3-15 shows essentially the same circuit as the general bias circuit, except that $V_2 = 0$. Therefore, the same equations are used for the stability factor and base current as for the general bias circuit with $V_2 = 0$.

Figure 3-16 is another conventional bias circuit in which R_2 is infinite.

The stability factor equation for this circuit is the same as for the general bias circuit, except that

$$R_B = R_1 \quad \text{and} \quad V_3 = V_{CC} \tag{3-38}$$

The equations for the stability factor and base current are

$$S = \frac{R_E + R_1}{R_E + R_1/(_{FE} + 1)} \tag{3-39}$$

and

$$I_B = \frac{V_{CC}}{R_E(h_{FE} + 1) + R_1} \tag{3-40}$$

In this circuit, however, R_1 is generally very much greater than R_B would be in the general bias network; hence a higher stability factor is obtained.

In the grounded-base circuit shown in Fig. 3-17, $R_B = 0$, $V_3 = 0$, so that

$$S = \frac{R_E}{R_E} = 1 \tag{3-41}$$

and

$$I_B = \frac{V_2}{R_E(h_{FE} + 1)} \tag{3-42}$$

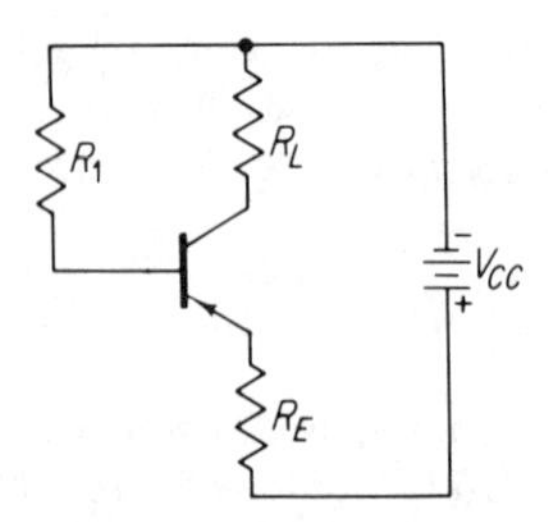

Fig. 3-16. Simple bias circuit often used where economy is important.

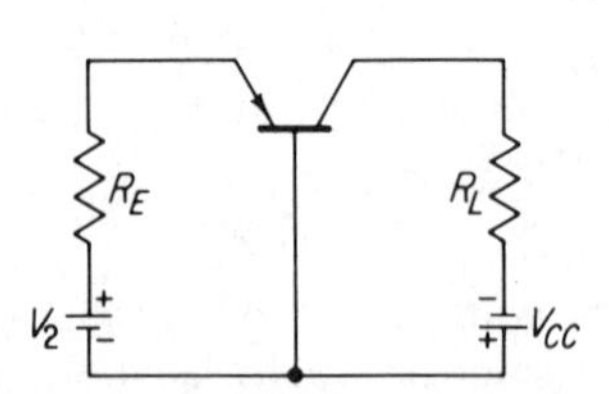

Fig. 3-17. Grounded-base bias circuit having excellent thermal stability.

In a grounded collector circuit, the emitter resistor becomes the load. The stability factor and base current for this circuit would be the same as those found for the general bias circuit since R_L does not influence the stability factor.

All the foregoing circuits are examples of linear bias circuits which utilize dc current feedback. Another method of bias stabilization is the use of dc voltage feedback. This means that part of the output voltage is fed back to the input.

The simplest dc voltage feedback circuit is shown in Fig. 3-18. When collector current flows through R_L, part of the voltage developed across R_L is fed back to the base via R_F. This circuit is similar to that of Fig. 3-16, where R_F and R_L act as R_1. This circuit also produces ac feedback.

The stability for this circuit can be found in the same manner as in the general bias circuit when V_{BE} is neglected.

$$I_C = I_{CBO} + \frac{(h_{FE})I_E}{h_{FE} + 1} \tag{3-43}$$

and

$$V_{CC} = -R_F I_C + (R_F + R_L)I_E \tag{3-44}$$

from which

$$I_C = \frac{I_{CBO}(R_F + R_L)}{R_L + R_F/(h_{FE} + 1)} + \frac{h_{FE}V_{CC}}{R_F + R_L(h_{FE} + 1)} \tag{3-45}$$

and

$$I_B = \frac{V_{CC}}{R_F + R_L(h_{FE} + 1)} \tag{3-46}$$

The stability factor is

$$S = \frac{R_F + R_L}{R_L + R_F/(h_{FE} + 1)} \tag{3-47}$$

The preceding development was a simple example of dc voltage feedback. A more useful example is shown in Fig. 3-19, where R_F is broken into two parts and one part is bypassed so there is no ac feedback. The stability

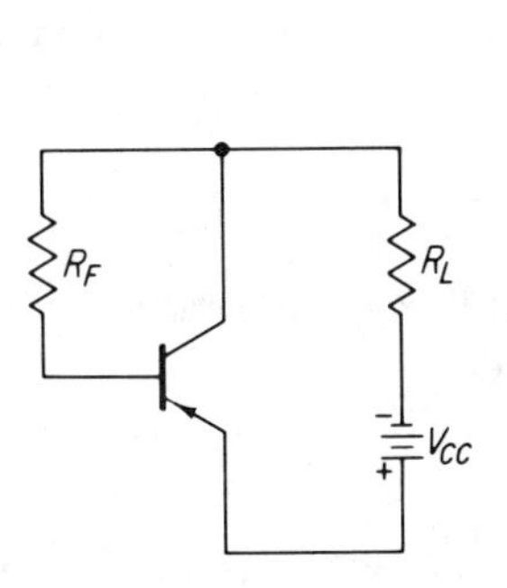

Fig. 3-18. Bias circuit with ac and dc voltage feedback.

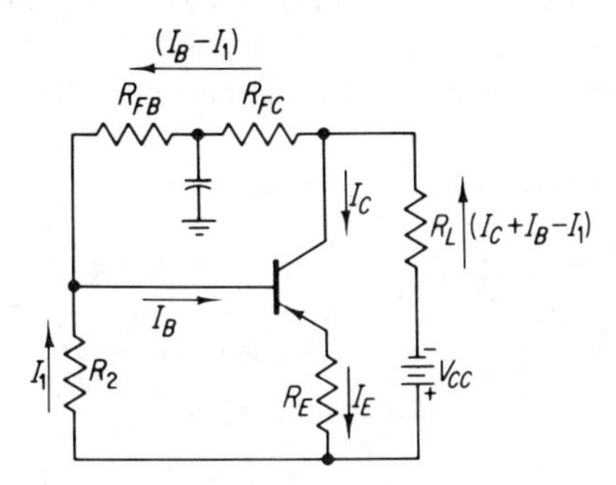

Fig. 3-19. Bias circuit with dc voltage feedback. R_2 added for better stability.

and base current equations remain the same, however, except that wherever R_F appears it should be replaced by $R_{FB} + R_{FC}$ (assume R_2 is infinite). For the general case where R_2 has a finite value, S may be found by using the following equations:
(Assume $V_{BE} = 0$.)

$$I_C = I_{CBO} + \frac{h_{FE}I_E}{h_{FE} + 1} \tag{3-48}$$

$$V_{CC} = -R_F I_C + I_E(R_F + R_i) - I_1(R_2 + R_F + R_i) \tag{3-49}$$

$$R_E I_E + R_2 I_1 = 0 \tag{3-50}$$

Solving these, we have

$$I_C = \frac{R_E(R_2 + R_F + R_L)(I_{CBO}) + R_2(R_F + R_L)(I_{CBO})}{R_E(R_2 + R_F + R_L) + R_2R_L + R_2R_F/(h_{FE} + 1)} + \frac{h_{FE}R_2V_{CC}}{(h_{FE} + 1)[R_E(R_2 + R_F + R_L) + R_2R_L] + R_2R_F} \tag{3-51}$$

from which

$$S = \frac{R_E(R_2 + R_F + R_L) + R_2(R_F + R_L)}{R_E(R_2 + R_F + R_L) + R_2R_L + R_2R_F/(h_{FE} + 1)} \tag{3-52}$$

and

$$I_B = \frac{V_{CC}R_2/(h_{FE} + 1) - I_{CBO}(R_ER_2 + R_ER_F + R_ER_L + R_2R_L)}{R_2R_F/(h_{FE} + 1) + R_E(R_2 + R_F + R_L) + R_2R_L} \tag{3-53}$$

Equation 3-53 is a general equation which can be altered when some of the resistances go to zero or infinity to describe the stability factor for any voltage feedback condition.

NONLINEAR BIASING TECHNIQUES

In most cases, adequate biasing of circuits can be achieved by linear techniques discussed in the preceding section. Some special applications may arise, however, where very accurate control of the operating point is required over wide temperature variations or power supply fluctuations. To achieve close control of the transistor operating point in these cases, nonlinear temperature compensating elements may be used.

Temperature-Sensitive Resistances

Thermal stability can be achieved by utilizing temperature-sensitive elements, such as thermistors and junction diodes.

Thermistors have a resistance which changes with temperature according to the relationship

$$R = R_0 \exp B\left(\frac{1}{T} - \frac{1}{T_0}\right) \tag{3-54}$$

where R_0 = resistance at T_0

B = temperature constant on the order of $\pm 10^3$, depending on whether a negative or positive coefficient is used. As the temperature increases, the resistance changes in a manner shown in Fig. 3-20.

Figure 3-21 is a circuit which uses a thermistor (R) in its bias network to retain I_E constant as temperature increases. At room temperature R has a value of R_0 from which the operating point can be determined.

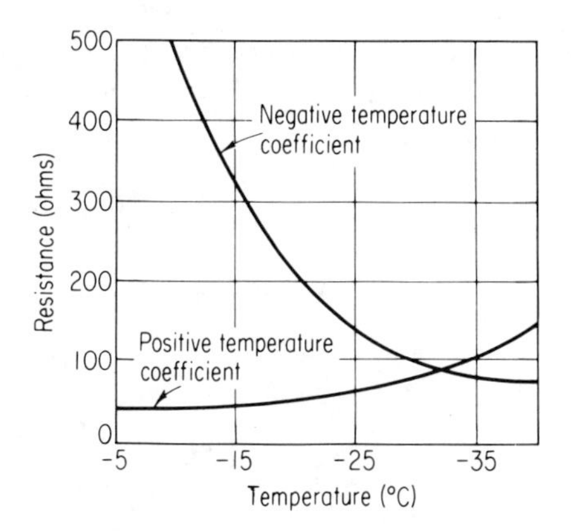

Fig. 3-20. Typical variation of resistance with temperature for both positive and negative coefficient thermistors.

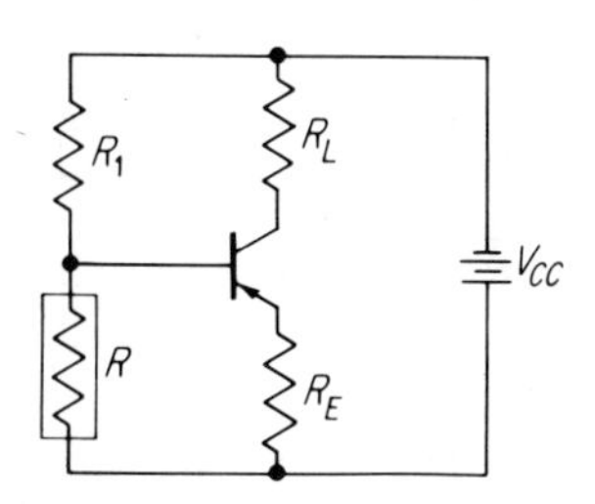

Fig. 3-21. Typical bias circuit using thermistor for stabilization.

As the temperature increases, R gets smaller, tending to decrease the bias which has risen because of the increased I_{CBO} with increased temperature. It can be seen from Eq. 3-25 that as the temperature increases, the stability increases since R_B decreases. If the temperature characteristics of the thermistor are properly chosen, the emitter current can be made to remain nearly constant over a wide temperature range.

It can be seen in Fig. 3-21 that, as R decreases, more and more of the ac signal is shunted, making the circuit useful only for large values of R at the increased temperature. A better circuit might be the circuit shown in Fig. 3-22. In this circuit R is at ac ground potential, and since its resistance decreases with increasing temperature the increase in I_C is absorbed by R if it is chosen properly.

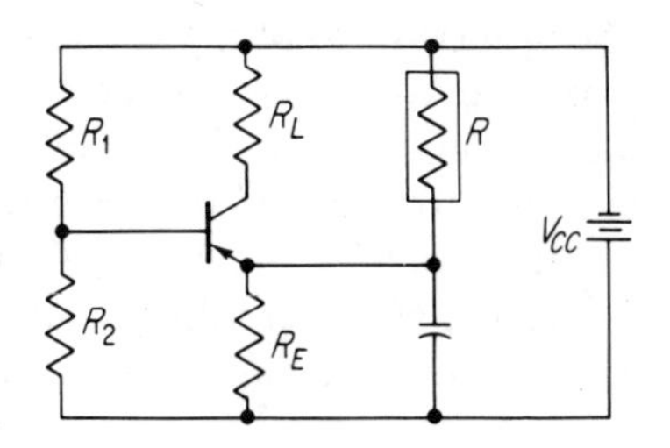

Fig. 3-22. Bias circuit using thermistor which does not shunt the ac signal.

Junction diodes are extremely useful as temperature-sensitive elements in transistor circuits since the transistor is itself two PN junctions and has similar temperature-dependent properties. As the temperature is increased, the forward resistance of the diode decreases, reducing R_B in Eq. 3-25 and improving the stability.

In order to maintain a constant collector voltage, a Zener diode may be used. A Zener diode is biased into the breakdown region so that it has a very low dynamic resistance. In this region, the voltage drop across the diode is essentially constant for large variations in diode current. Figure 3-23 shows the current-voltage characteristics of such a diode. Zener diodes, however, have a positive temperature coefficient, and in order to compensate for this, it may be necessary to operate the diode in series with a negative temperature coefficient element, such as a forward-biased diode.

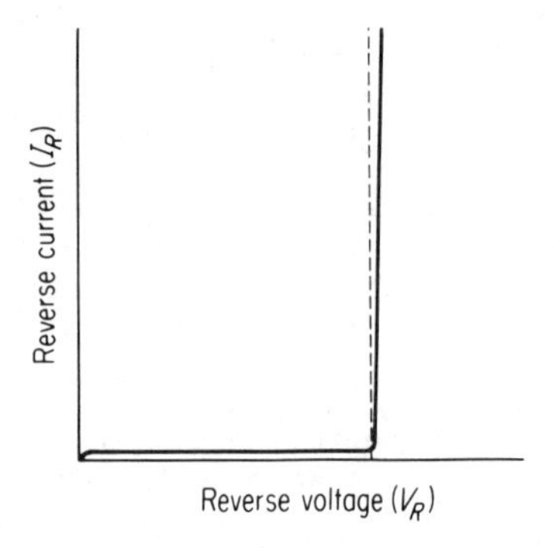

Fig. 3-23. Typical *V-I* characteristic of Zener diode.

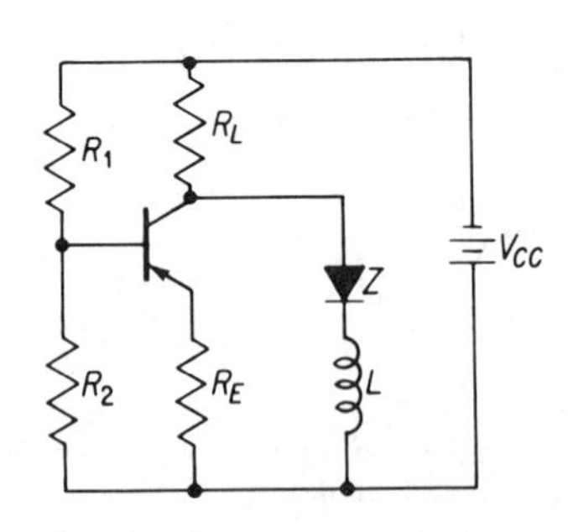

Fig. 3-24. Temperature compensation using a Zener diode.

Figure 3-24 shows a temperature-compensated stage using a Zener diode. The stabilizing effect is to maintain the collector voltage constant. R_L in this circuit serves two purposes: as a regulator and as an ac load. The inductance L is used to prevent the ac signal from being shunted through the Zener diode. This circuit will retain the collector voltage constant with changes in supply voltage or changes in temperature.

Zener diodes are also useful for deriving more than one supply voltage from a major power supply of a higher voltage. For example, a 12-volt supply can be maintained in an aircraft whose primary supply is 28 volts.

An Example of a Transistor as a Bias Stabilization Element

Several methods of transistor compensation can be used for bias stabilization. One of these methods is to utilize a PNP transistor and an NPN transistor and derive the bias voltage for the PNP transistor amplifier from the base-emitter voltage of the NPN transistor as shown in Fig. 3-25(a).

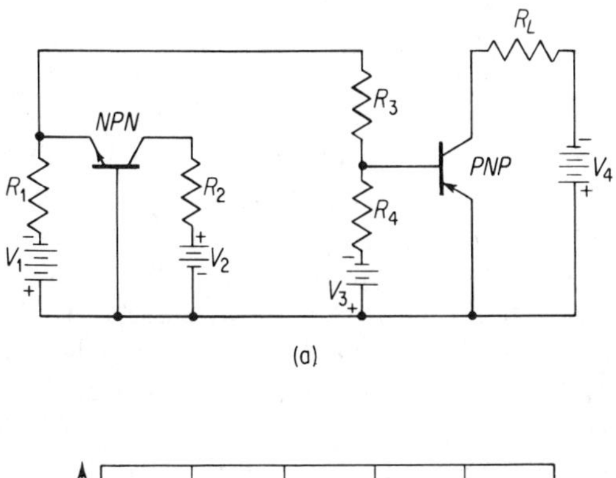

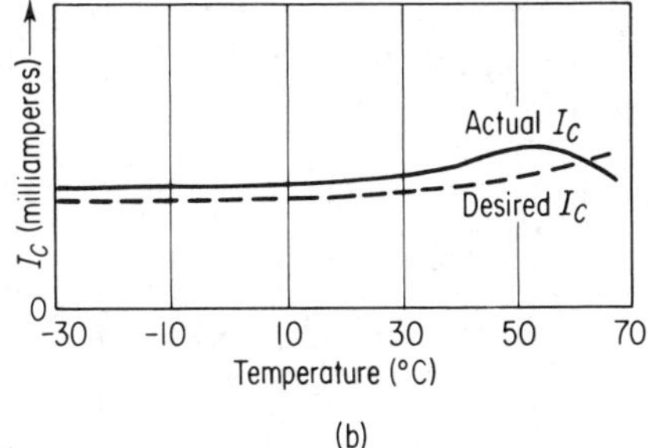

Fig. 3-25. Example of (a) transistor compensation and (b) resulting collector current versus temperature characteristic.

In this circuit, the NPN transistor is biased for constant-emitter current by means of a voltage source and a high emitter resistance R_1. The PNP is biased by the base-emitter voltage of the NPN, the resistors R_3 and R_4 and the voltage V_3. When the temperature is increased, the current in the NPN transistor increases, causing the base voltage of the PNP to decrease, thereby decreasing its emitter current. Figure 3-25(b) shows how the collector current is maintained over a wide temperature range.

We can develop other types of bias stabilization circuitry which have particular usefulness depending on the application. The combination of simple shunt compensation and series compensation, for example, results in a compound structure which has better properties than either one alone. In most instances, however, the design equations become quite involved, and practical analytical designs in lieu of breadboarding techniques are tedious and often unproductive. This is often true of active network synthesis, and unfortunately, especially true in devices, such as transistors, where not only temperature, but voltage and other environmental changes can be the controlling factors in operational characteristics. Many other texts and articles indicate the proper type of circuit to use in a given application and the reader may turn to these works given in the list of references.

INTEGRATED CIRCUIT BIASING

The popular emitter-bias scheme for bipolar transistors and the FET self-bias circuit both generally use a large bypass capacitor to achieve high ac gain. The unavailability of large capacitors in monolithic circuits and the lack of integrated inductors require that special techniques be used to establish bias currents for integrated amplifiers. Differential stages and complex feedback circuits are often used to obtain the correct bias in the integrated circuit amplifier. Since differential and feedback amplifiers are discussed in later chapters, only the simplest integrated circuit biasing schemes will be considered at this point.

The circuit of Fig. 3.26 shows one method of achieving proper bias on the transistor T_2. The transistor T_1 is used strictly for bias current generation and does not take part in signal amplification.

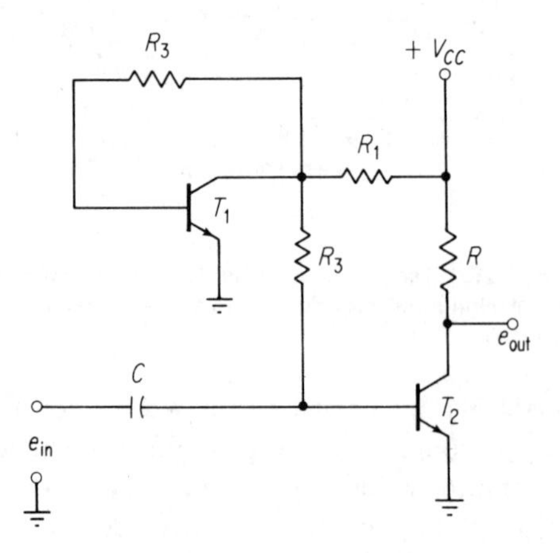

Fig. 3-26. Diode-biased stage.

While a coupling capacitor C is required for this biasing scheme, the value of this capacitor will be two orders of magnitude smaller than the bypass capacitor used in the emitter-bias circuit. The diode-biased current sink of Fig. 3.27 is closely related to the network of Fig. 3.26 and can be used to explain this type of bias technique.

We note that this configuration has the base terminals in parallel and the emitters both return to ground; thus $V_{BE1} = V_{BE2}$. Assuming that the transistors are integrated with the same geometry and are perfectly matched, the base currents, the emitter currents, and the collector currents of both transistors will be equal. The collector current of T_2 can be expressed as

$$I_2 = I_1 - 2I_B \tag{3-55}$$

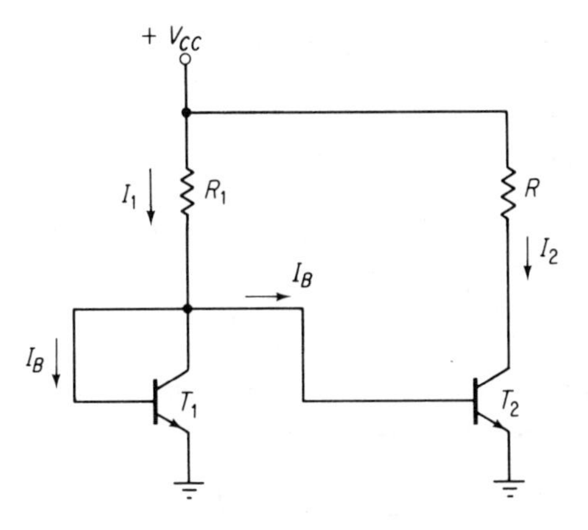

Fig. 3-27. Diode-biased current sink.

The current I_1 is given by

$$I_1 = \frac{V_{CC} - V_{BE}}{R_1} \approx \frac{V_{CC}}{R_1} \tag{3-56}$$

allowing us to write

$$I_2 = \frac{V_{CC}}{R_1} - 2I_B \approx \frac{V_{CC}}{R_1} \tag{3-57}$$

Control of collector bias current for T_2 is then accomplished by controlling the ratio of V_{CC} to R_1.

Although bias current for T_2 can easily be controlled with this configuration, it is difficult to apply the signal to the base of T_2 because this is a low impedance point. The transistor T_1 is near saturation and the low impedance path from collector to emitter of this transistor appears from the base of T_2 to ground. Insertion of the resistors R_3 as shown in Fig. 3.26 overcomes the impedance problem while maintaining a balanced dc circuit. Once the desired collector current of T_2 is selected, the values of R_1 and R_3 are found from Eq. (3.55) which is still valid and

$$I_1 = \frac{V_{CC} - V_{C1}}{R_1} \tag{3-58}$$

The collector voltage of T_1 is

$$V_{C1} \approx \frac{V_{CC}R_3 + \beta_0 V_{BE} R_1}{\beta_0 R_1 + R_3} \tag{3-59}$$

If $\beta_0 R_1$ is much larger than R_3, then the bias current will be relatively independent of the value of β_0 and the collector current of T_2 will again be given by Eq. (3.57). The input impedance from the base of T_2 to ground will now consist of the resistance R_3 in parallel with the input impedance to the transistor.

EXERCISES

1. The common-emitter class-A amplifier shown is being designed as a low-level audio amplifier. It is desired that the voltage gain variations with temperature and from unit to unit be held to a minimum. For such an amplifier the voltage gain is approximated by

$$G_V = \frac{R_L I_E}{kT/q}$$

Given that $V_{CC} = 10$ volts and the required nominal $G_V = 50$, find R_1, R_2, and R_E such that the stability factor $S \approx 5$. Assume V_{EB} of the transistor is zero.

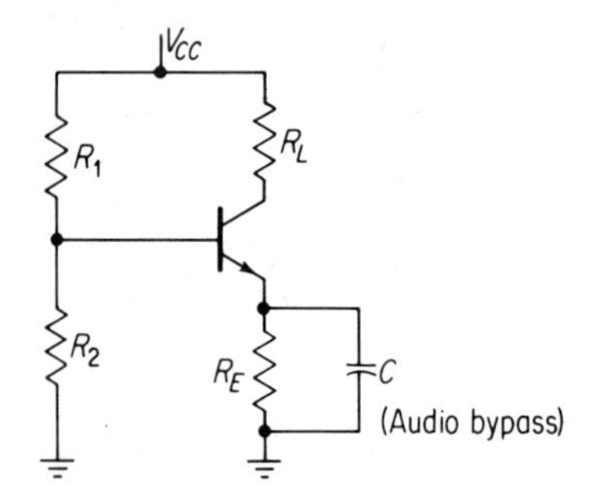

Calculate the variation in gain with temperature and from unit to unit assuming $h_{FE(\text{max})} = 300$, $h_{FE(\text{min})} = 100$, and h_{FE} increases with temperature at the rate of 1 percent/°C.

2. The accompanying biasing scheme is sometimes used in semiconductor integrated circuits.

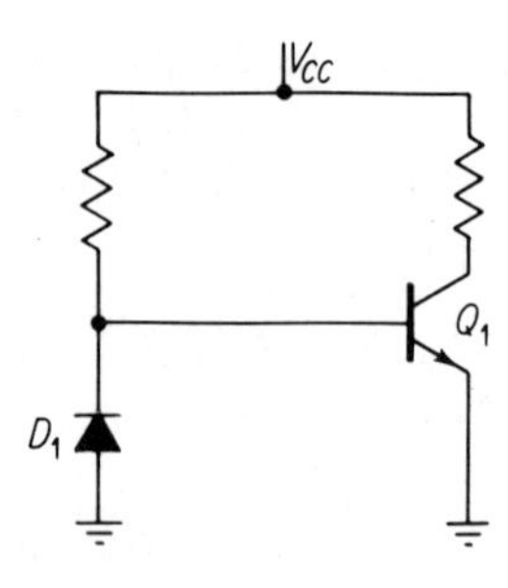

Discuss the pros and cons of such a biasing scheme for use with discrete components. Under what conditions is transistor emitter current independent of temperature? What is the effect on the emitter current of a temperature difference between D_1 and Q_1?

3. (a) For the circuit shown, derive an expression for I_c in terms of V_{CC}, R_B, R_C, V_{BE}, and β.

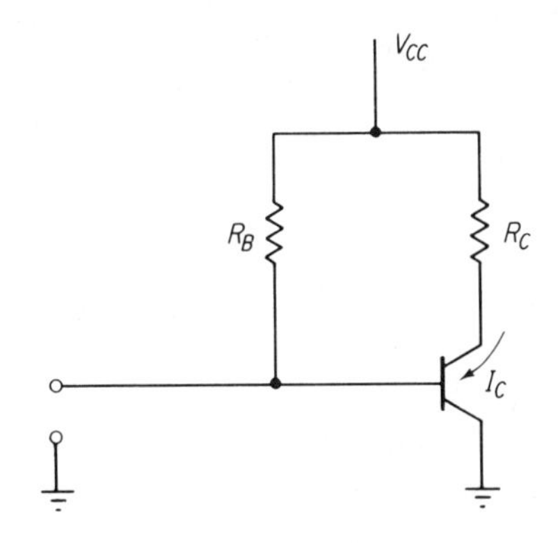

(b) Find S.

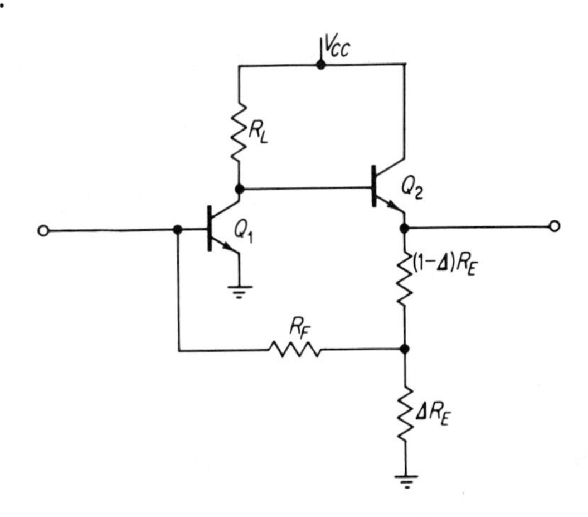

4. A bias circuit useful in single-ended dc amplifiers is shown. Calculate a general expression for the collector current of Q_1. Show quantitatively the variation in this current with temperature.

REFERENCES

1. R. F. SHEA, *Transistor Circuit Engineering*. New York: John Wiley & Sons, Inc., 1957.

2. D. LECROISSETTE, *Transistors*. Englewood Cliffs, N.J.: Prentice-Hall, Inc., 1963.

3. J. F. PIERCE, *Transistor Circuit Theory and Design*. New York: Charles E. Merrill Books, Inc., 1963.

4. A. W. LO, *et al.*, *Transistor Electronics*. Englewood Cliffs, N.J.: Prentice-Hall, Inc., 1961.

5. M. V. JOYCE and K. K. CLARKE, *Transistor Circuit Analysis.* Reading, Mass.: Addison-Wesley Publishing Company, Inc., 1961.

6. E. J. ANGELO, JR., *Electronic Circuits.* New York: McGraw-Hill Book Company, 1964.

7. R. R. WRIGHT and H. R. SKUTT, *Electronic Circuits and Devices.* New York: The Ronald Press Company, 1965.

8. F. C. FITCHEN, *Electronic Integrated Circuits and Systems.* D. Van Nostrand Reinhold Company, 1970.

9. M. P. RISTENBATT, *Semiconductor Circuits, Linear and Digital.* Englewood Cliffs, N.J.: Prentice-Hall, Inc., 1975.

10. C. BELOVE, H. SCHACHTER and D. L. SCHILLING, *Digital and Analog Systems, Circuits and Devices: An Introduction.* New York: McGraw-Hill Book Company, 1973.

11. H. E. STEWART, *Engineering Electronics.* Boston, Mass.: Allyn & Bacon, Inc., 1969.

4

Low-Frequency Untuned Amplifiers

INTRODUCTION

In studying the practical use of transistors in amplifiers, it is convenient to divide the scope of application into several operating ranges. In this chapter, we consider both small- and large-signal, low-frequency untuned amplifiers. Small-signal implies that the signal swing from the normal operating point is small enough so that the parameters of the transistor may be considered essentially constant over the signal swing. Conversely, large signal swings exceed the range over which the parameters are constant. The term *low-frequency* covers that range of operations over which the transistor parameters are relatively independent of frequency. Note that the upper limit of this range cannot be specified as a single frequency since the frequency-independent range depends to a great extent on the particular transistor type employed, ranging from perhaps 10 KHz for audio transistors to several megahertz for high-frequency units.

Amplifiers of the type under consideration may be used singly or in cascaded stages. In either case it is convenient to consider a single stage as consisting of a source, an amplifier, which can be considered as a black box for the present, and a load. In the case of a cascaded amplifier, the load actually represents the input circuit of the following stage, and the source can be the output of a previous amplifier stage.

The amplifier may have to meet many requirements. The primary one is to provide the specified level of signal at the load when the input is at a certain level. Other considerations are distortion, gain and temperature stability, input and output impedances, and cost.

GAIN DEFINITIONS AND SPECIFICATIONS (1)*

One of the primary considerations in an amplifier is the required gain between input and output. This gain can be expressed in several different ways, each of which is useful under different conditions. Consider first Fig. 4-1 that shows a source, a load, and a matching factor. The matching factor (*MF*) does not represent a matching network. It is the ratio between the power actually delivered to the load, and the maximum power available from the source.

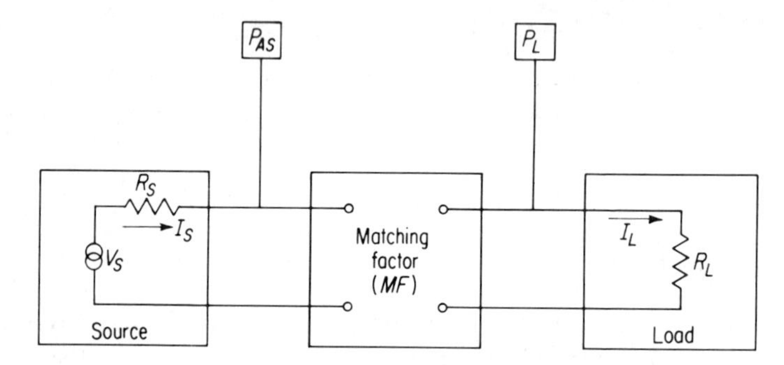

Fig. 4-1. Source and load with matching factor.

The power delivered to the load is simply $I_L^2 R_L$, and is designated as P_L. The maximum power available from the source into a real resistive load is obtained under matched conditions; that is, when the resistance of the load is equal to the resistance of the source. Then P_{AS}, the power available from the source, is equal to $V_S^2/4R_S$ or $I_S^2 R_S/4$.

When R_L is equal to R_S, then P_L is equal to P_{AS}, and the *MF* (matching factor) is P_L/P_{AS} or 1 for the matched condition. Figure 4-2 shows the power

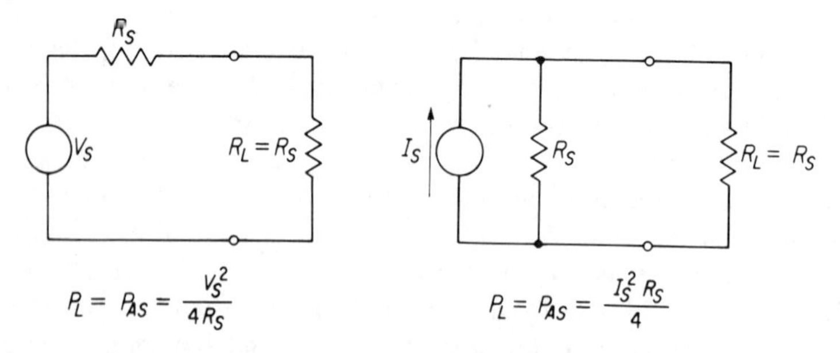

Fig. 4-2. Load matched to voltage and current sources.

*Numbers in parentheses refer to material listed at the end of the chapter.

into the load under matched conditions. When R_L is not equal to R_S, the MF is easily shown to be

$$MF = \frac{P_L}{P_{AS}} = \frac{4R_S R_L}{(R_S + R_L)^2} \tag{4-1}$$

The matching factor varies between 0 and 1 as the degree of matching is changed.

Consider now the practical case shown in Fig. 4-3 where the source and load are connected by a device X. This device X can be either an amplifier or merely a matching network between the source and the load. Here, P_{AS} and P_L have the same connotation previously discussed, P_{in} is the actual power delivered to the device and P_{AX} is the power available from the device under matched conditions. MF_{in} is the ratio of P_{in}/P_{AS}, and MF_{out} is P_L/P_{AX}.

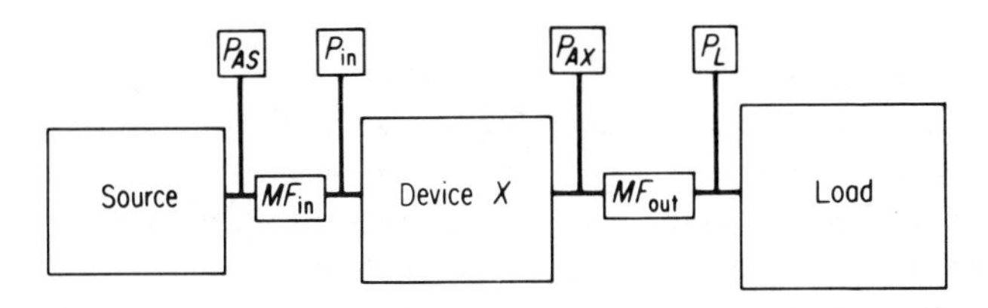

Fig. 4-3. Power levels between source and load.

Several different expressions can be written for the gain in the block diagram of Fig. 4-3, depending at which point the input and output levels are measured. The four most important ratios are as follows:

Transducer gain

$$A_T = \frac{P_L}{P_{AS}} \tag{4-2}$$

Available gain

$$A_A = \frac{P_{AX}}{P_{AS}} \tag{4-3}$$

Power gain

$$A_P = \frac{P_L}{P_{\text{in}}} \tag{4-4}$$

Maximum available gain

$$A_{MA} = \frac{P_{AX}}{P_{\text{in}}} \tag{4-5}$$

The transducer gain, A_T, is a ratio that compares the performance of device X in transferring to the load the power which is available from the source, to the performance of a perfect transformer which is used to match R_L to R_S. It is useful as a figure of merit for comparing various coupling devices. Transducer gain, A_T, includes the effect of the mismatching and coupling network losses which exist at the input and output of device X.

Available gain, A_A, previously defined as P_{AX}/P_{AS} includes the effect of the mismatch at the input to the device, but not the effect of the output mismatch.

The power gain, A_P, which is P_L/P_{in}, takes into consideration the output mismatch, but ignores the effect of any mismatch at the input. Notice that if the degree of input matching is varied so as to affect P_{in}, then P_L and P_{in} change in the same proportions, and their ratio A_P remains constant.

These three gain definitions are interrelated by the matching factors MF_{in} and MF_{out}, as follows:

$$A_T = A_P \cdot MF_{\text{in}} = A_A \cdot MF_{\text{out}} \tag{4-6}$$

$$A_A = \frac{A_T}{MF_{\text{out}}} = A_P\left(\frac{MF_{\text{in}}}{MF_{\text{out}}}\right) \tag{4-7}$$

$$A_P = \frac{A_T}{MF_{\text{in}}} = A_A\left(\frac{MF_{\text{out}}}{MF_{\text{in}}}\right) \tag{4-8}$$

The foregoing equations illustrate an important point, namely, as the matching factors at the input and output approach 1, the three expressions approach the same value. This value is sometimes called the *maximum available gain*, A_{MA}, although it could just as readily and correctly have been named *maximum transducer gain*, or *maximum power gain.*

The maximum available gain is independent of the input and output mismatch, and so characterizes the device itself. As such, there are applications in which it is a useful measure of the performance of the device. In practical circuits, however, it is usually not realizable, particularly at low frequencies, and one should not use A_{MA} as the sole criterion in predicting useful circuit gain. Since A_{MA} represents the case of matched input and output, it is sometimes referred to as *matched power gain.* This matched power gain is frequently specified as a parameter of high-frequency transistors.

Yet another gain parameter is frequently encountered. This is the *insertion gain*, A_N, that is defined as the ratio of the load powers with the device in and out of the circuit. That is,

$$A_N = \frac{P_L}{P_{LO}} = \frac{\text{load power with device in use}}{\text{load power with device out of circuit}} \tag{4-9}$$

Figure 4-4 shows the conditions existing for the measurement of insertion gain.

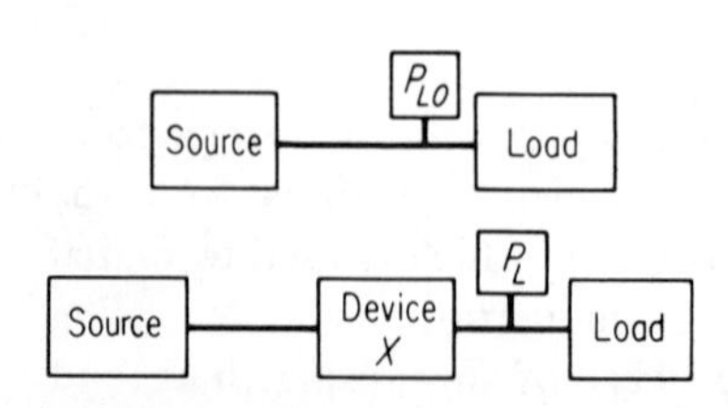

Fig. 4-4. Measurement of insertion gain.

Insertion gain, A_N, is a function of both the input and output device mismatches, and of the mismatch that exists when the load is directly connected to the source. It is related to the transducer gain, A_T, by

$$A_N = \frac{A_T}{MF} \tag{4-10}$$

where, again, MF represents the matching between the load and the source.

A_T compares the actual load power to that which would be delivered to the load from the source through an ideal matching transformer; A_N compares the actual load power to the power delivered from source to load in a direct connection.

The idea of insertion gain is frequently applied to the evaluation of an interstage transformer. Here, the efficiency of the transformer as a coupling device is compared to that of a perfect transformer which might be used to connect the source to the load. In this application, it is more convenient to think of the term *insertion loss*, since, regardless of the design, a transformer will certainly not yield power gain. Indeed, nothing has been said that prevents any of the aforementioned gain parameters from being used to measure the performance of a stage having less than unity gain.

Now that we have several ways of expressing the amplifier gain and have considered the requirements on individual stages for maximum over-all performance, we shall look at the optimum design method for each stage.

SMALL-SIGNAL AMPLIFIERS

Characteristics of Single-Stage Amplifiers

Three possible ac connections of the transistor will give gain. These are illustrated in Fig. 4-5, together with their elementary biasing circuits.

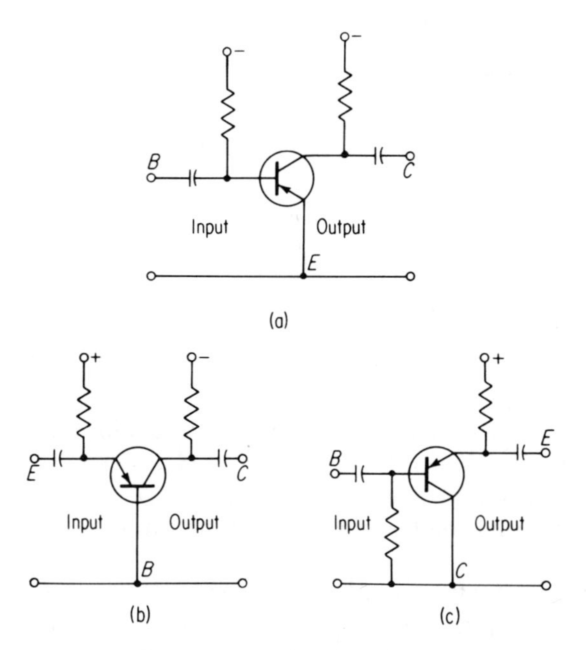

Fig. 4-5. Three basic transistor connections. (a) Common-emitter; (b) common-base; (c) common-collector.

The term *common* in Fig. 4-5 refers to the terminal which is common to the input and output signals. In general, the common-emitter stage provides the greatest power gain, and is best adapted to direct-coupled stages. The common-base stage provides a large impedance stepup, and the common-collector (or emitter-follower) provides an impedance stepdown.

The input and output impedances of a transistor were shown earlier and are given by the expressions

$$R_{\text{in}} = h_i - \frac{h_f h_r}{h_o + \dfrac{1}{R_L}} \tag{4-11}$$

and

$$\frac{1}{R_{\text{out}}} = h_o - \frac{h_f h_r}{h_i + R_S} \tag{4-12}$$

These two equations can be used to calculate the input impedance and output impedance for any of the three transistor connections, if we know the *h* parameters for that connection. It is only necessary to know the *h* parameters for one connection since each set can be derived from any other, as was shown in Chap. 2.

The values of R_{in} and R_{out} can also be calculated by using the low-frequency equivalent circuit shown in Fig. 4-6. This circuit is useful in the low-frequency region which we are considering.

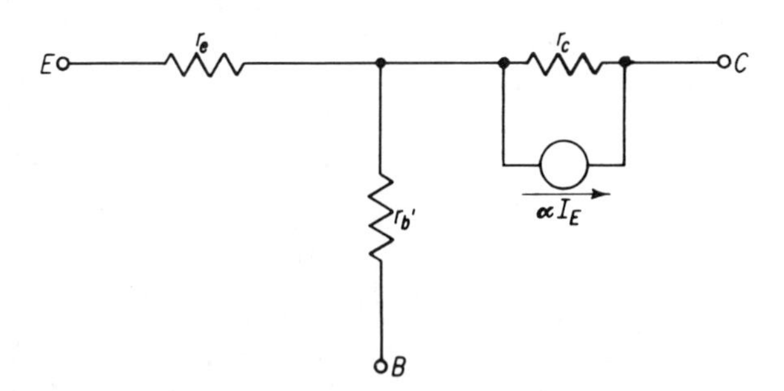

Fig. 4-6. A low-frequency equivalent circuit.

The expressions for input and output resistances for the three transistor connections are shown in Table 4-1. Two simplifying assumptions have been made in these equations, namely $r_b \ll r_c$, and $r_e \ll r_c(1 - \alpha)$. These assumptions are quite valid for the bipolar transistors usually employed in low-frequency amplifiers.

From the equations in Table 4-1, it is seen that R_{in} in all cases is a function of the transistor parameters and the load resistance R_L. On the other hand, R_{out} is a function of the transistor parameters and the source resistance R_S.

TABLE 4-1 Input and Output Resistances for Three Connections

	Input Resistance (R_{in})	Output Resistance (R_{out})
Common base	$R_{\text{in}(b)} = r_e + r_{b'}\dfrac{r_c(1-\alpha)+R_L}{r_c+R_L}$	$R_{\text{out}(b)} = r_c\dfrac{r_e+r_{b'}(1-\alpha)+R_S}{r_e+r_{b'}+R_S}$
Common emitter	$R_{\text{in}(e)} = r_{b'} + r_e\dfrac{r_c+R_L}{r_c(1-\alpha)+R_L}$	$R_{\text{out}(e)} = r_c(1-\alpha) + r_e\dfrac{\alpha r_c+R_S}{r_e+r_{b'}+R_S}$
Common collector	$R_{\text{in}(c)} = r_{b'} + r_c\dfrac{r_e+R_L}{r_c(1-\alpha)+R_L}$	$R_{\text{out}(c)} = r_e + r_c(1-\alpha)\dfrac{R_S+r_{b'}}{R_S+r_c}$

R_L and R_S are load and source resistance, respectively.

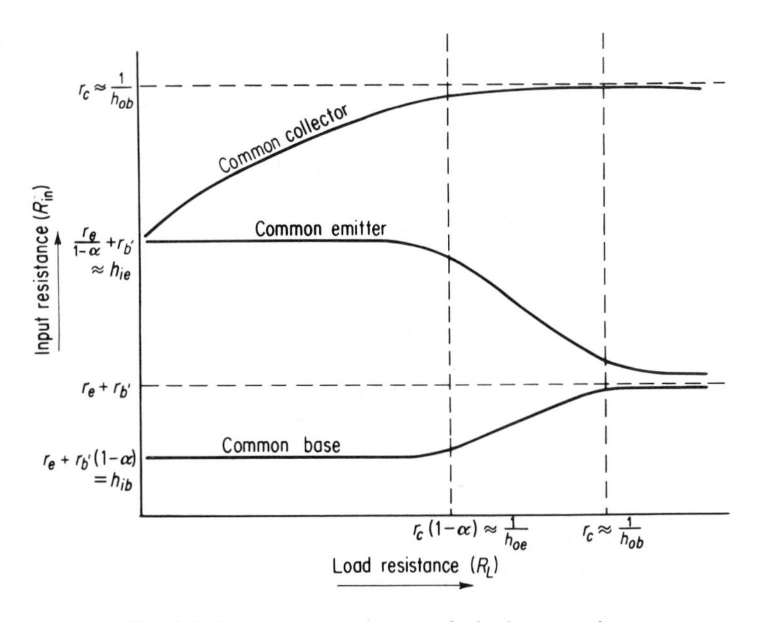

Fig. 4-7. Input vs output resistance for basic connections.

Figure 4-7 shows how R_{in} varies with R_L for the three connections. The curves are meant to show only the trend of the input resistance, since the vertical dimension is not to scale. Figure 4-8 shows the same type of curve for the output resistance as a function of R_S.

Figures 4-7 and 4-8 illustrate that only certain resistances are attainable and that each connection has its own possible values which do not duplicate those of any other connections.

The equations for input and output resistance shown in Table 4-1 can be further simplified by imposing some additional conditions. First, let the load be such that

$$R_L \ll r_c(1-\alpha) \tag{4-13}$$

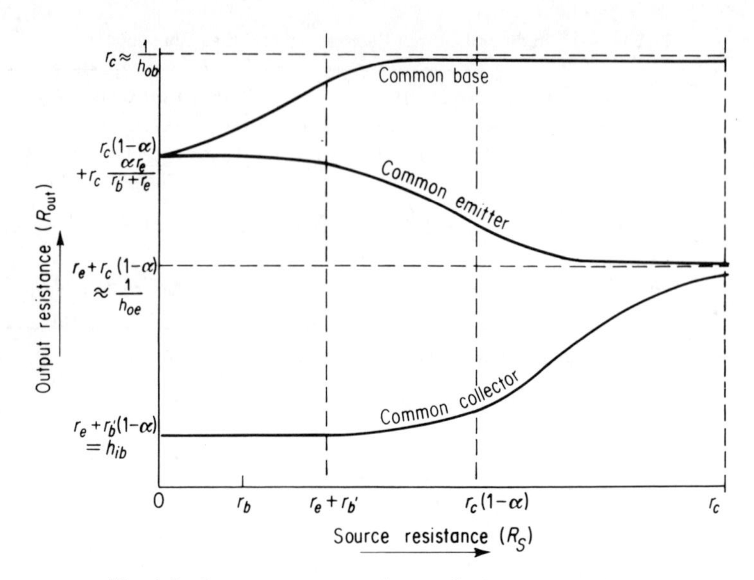

Fig. 4-8. Output vs source resistance for basic connections.

Then, the common-base input resistance becomes

$$R_{\text{in}(b)} = r_e + r_{b'}(1 - \alpha) = h_{ib} \tag{4-14}$$

This is the common-base input resistance with the output short-circuited. The common-emitter input resistance becomes

$$R_{\text{in}(e)} = r_{b'} + \frac{r_e}{1 - \alpha} = h_{ie} \tag{4-15}$$

which is the common-emitter input resistance with the output short-circuited. For the common-collector connection, we have

$$R_{\text{in}(c)} = r_{b'} + \frac{r_e}{1 - \alpha} + \frac{R_L}{1 - \alpha} \tag{4-16}$$

Making the additional assumption that

$$R_L \gg r_{b'}(1 - \alpha) + r_e = h_{ib} \tag{4-17}$$

then, $R_{\text{in}(c)}$ reduces to the following:

$$R_{\text{in}(c)} \approx \frac{R_L}{1 - \alpha} \approx \frac{\alpha R_L}{1 - \alpha} = \beta R_L \tag{4-18}$$

which is a very useful approximation, since the imposed conditions are frequently met in practice.

The expressions for output resistance can be similarly simplified. If

$$R_S \ll r_c \tag{4-19}$$

then the common-collector output impedance is

$$R_{out(c)} = r_e + (1 - \alpha)(R_S + r_{b'}) \tag{4-20}$$

Further, if

$$R_S \gg r_{b'} + \frac{r_e}{1 - \alpha} = h_{ie} \tag{4-21}$$

then this may be further simplified to

$$R_{out(c)} \approx (1 - \alpha)R_S \approx \frac{R_S}{\beta} \tag{4-22}$$

which again is the condition usually met in practice.

Looking now at the current gain for each of the three connections, we can use the equivalent circuit of Fig. 4-6 to obtain

Common base

$$A_{ib} = \frac{\alpha}{1 + (R_L/r_c)} \tag{4-23}$$

Common emitter

$$A_{ie} = \frac{\alpha}{(1 - \alpha) + \dfrac{R_L}{r_c}} = \frac{\beta}{1 + \dfrac{R_L}{r_c(1 - \alpha)}} = \frac{\beta}{1 + R_L h_{oe}} \tag{4-24}$$

Common collector

$$A_{ic} = \frac{1}{(1 - \alpha) + \dfrac{R_L}{r_c}} \tag{4-25}$$

Here we are still using the assumption made for Table 4-1. The current gain expressions are plotted versus R_L in Fig. 4-9. We see that the current gain of

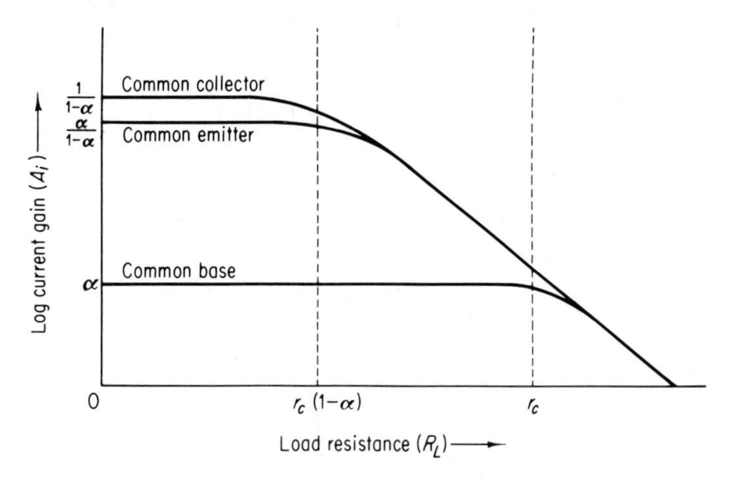

Fig. 4-9. Current gain vs load resistance.

the common-emitter and common-collector stages are almost identical, since in fact, $A_{ie} = \alpha A_{ic}$. When R_L is larger than r_c, the three converge.

COUPLED AMPLIFIERS (MULTISTAGE)

An analysis of the previous results show the common-emitter connection to be generally optimal for all but the last stage of a cascaded amplifier. This results from the fact that power gain and impedance matchups for this connection give the best intermediate stage results for maximum gain. The first stage is occasionally chosen to be other than the common-emitter when special impedance considerations warrant it. The common-collector connection is almost always the last stage since it provides a low impedance power amplifier for driving a load.

RC-Coupled Amplifiers

One common method of interstage coupling between low frequency amplifiers is the RC network shown schematically in Fig. 4-10. Here, R_1 represents the effective load resistance on the left of coupling capacitor C_1, and R_2 represents the effective resistance on the right side of C_1, thereby shunting the input of T_2.

The ac equivalent circuit for the *RC*-coupled stage is given in Fig. 4-11. The signal current to be amplified by T_2 is i_2, which is that portion of i_1 that is

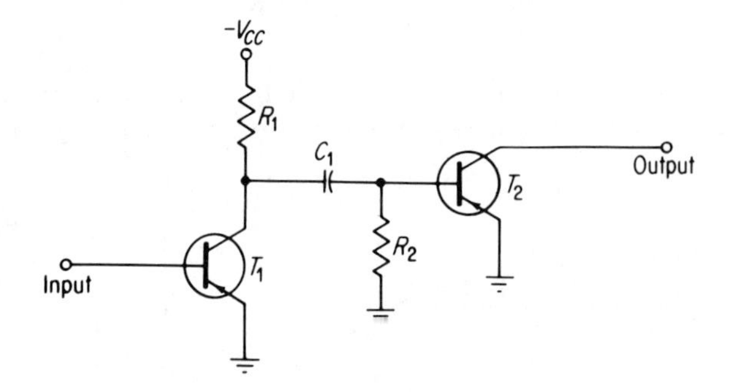

Fig. 4-10. Circuit of *RC*-coupled amplifier.

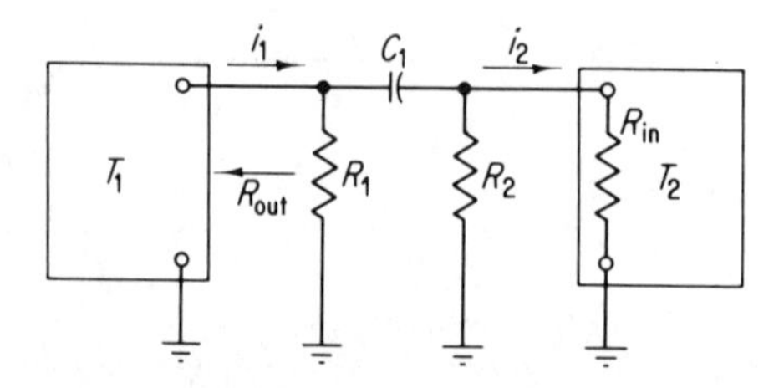

Fig. 4-11. Low frequency ac circuit of *RC*-coupled amplifier.

not shunted to ground by R_1 and R_2. The current gain of the coupling networ (actually a loss) is given by

$$A_i = \frac{i_2}{i_1} \tag{4-26}$$

Over the frequency range where the reactance of C_1 is very small, the current gain is

$$A_i = \frac{\dfrac{R_1 R_2}{R_1 + R_2}}{\dfrac{R_1 R_2}{R_1 + R_2} + R_{in}} \tag{4-27}$$

Using R_p to represent the parallel combination of R_1 and R_2, then Eq. 4-27 becomes

$$A_i = \frac{R_p}{R_p + R_{in}} \tag{4-28}$$

For $A_i = 1$, R_p should be large relative to R_{in}. The value of R_p is dictated by temperature stability considerations, the available supply voltage, and the desired quiescent collector current.

The R_{in} of the common-emitter stage is much smaller than that for the common-collector stage, another point in favor of using common-emitter amplifiers. Figure 4-12 shows the variation in A_i with the ratio R_p/R_{in}. From this curve, the loss for any R_p/R_{in} ratio can be found.

The value of C_1 can be selected from the desired low-frequency response of the amplifier. If R_1 is small compared to the output impedance of T_1, and if

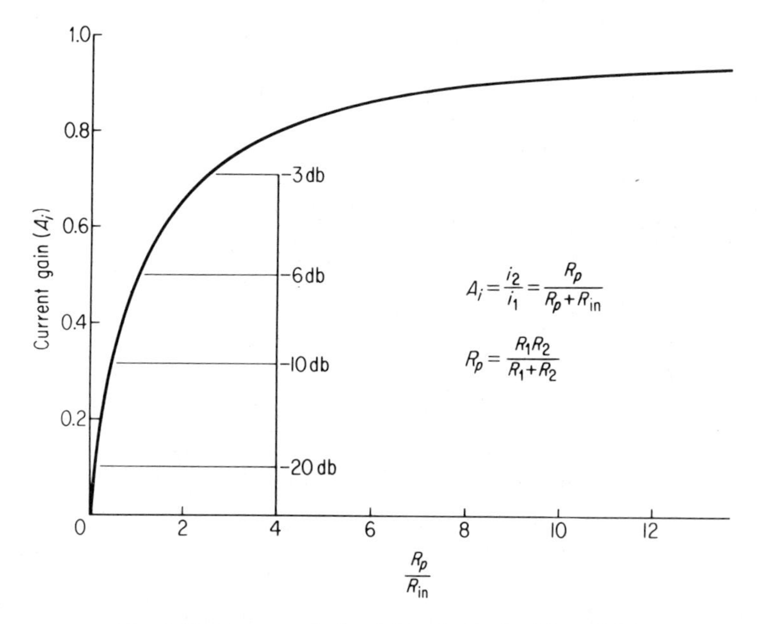

Fig. 4-12. Current gain (loss) A_i vs R_p/R_{in} for RC amplifier.

R_2 is large compared to R_1, the low-frequency cut-off will be

$$f_{co} = \frac{1}{2\pi R_1 C_1} \tag{4-29}$$

If R_1 is not small compared to T_1's output impedance, but R_2 is large compared to R_{in}, the cut-off frequency becomes

$$f_{co} = \frac{1}{2\pi C_1 \dfrac{R_1 R_{out}}{R_1 + R_{out}}} \tag{4-30}$$

The power gain per stage of a cascade of identical common-emitter RC-coupled amplifiers is a complex function of load and biasing resistances. However, this gain is always less than h_{fe}^2.

Transformer-Coupled Amplifiers

From the concept of maximum available gain, A_{MA}, we know that the gain of a transistor is maximum when the source is matched to the input resistance of the transistor, and when the load is matched to the output resistance.

Transformers can be used as coupling devices between transistor stages to match the resistances and thus yield maximum gain, as shown in Fig. 4-13.

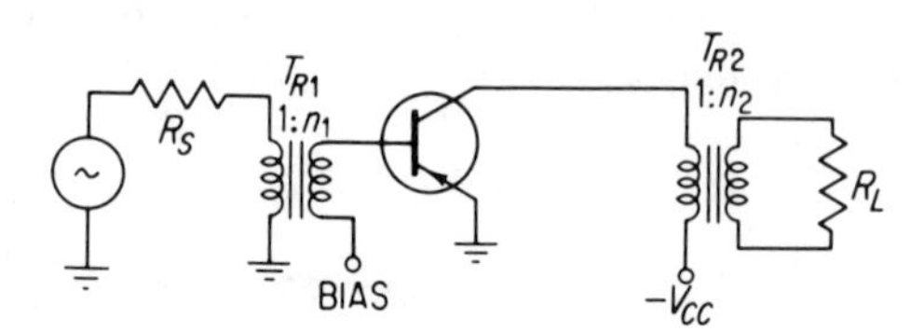

Fig. 4-13. Transformer-coupled amplifier.

The expressions for matched input and output resistances have been derived in the literature, and are listed (7) herewith:

$$R_{S(\text{matched})} = h_i\sqrt{1 - \frac{h_f h_r}{h_i h_o}} \tag{4-31}$$

$$R_{L(\text{matched})} = \frac{1}{h_o\sqrt{1 - \dfrac{h_f h_r}{h_i h_o}}} \tag{4-32}$$

To satisfy the matched conditions, the transformer turns-ratio requirements, n_1 and n_2, are such that $n_1^2 R_S = R_{in}$, and $(1/n_2)^2 R_L = R_{out}$. This yields

$$n_1 = \sqrt{\frac{h_i}{R_S}\sqrt{1 - \frac{h_f h_r}{h_i h_o}}} \tag{4-33}$$

and

$$n_2 = \sqrt{R_L h_o \sqrt{1 - \frac{h_f h_r}{h_i h_o}}} \tag{4-34}$$

as the design equations for the transformers involved. The expression for matched power gain (A_{MA}) of the stage is

$$A_{MA} = \frac{h_f^2}{h_i h_o\left(1 + \sqrt{1 - \dfrac{h_f h_r}{h_i h_o}}\right)} \tag{4-35}$$

PRACTICAL DESIGN OF A TRANSFORMER-COUPLED AMPLIFIER

To illustrate the use of the equations that have been derived, consider the following problem: It is desired to build a transformer-coupled amplifier to provide the maximum gain from a 600 Ω source to a 600 Ω load. The available input power is 0.5×10^{-8} watt. The 2N223 transistor is selected and its parameters are given in Table 4-2. The basic circuit is shown in Fig. 4-14. As a starting operating point, we select 4.5 volts and 2 ma.

TABLE 4-2 Typical h Parameters for 2N223

Parameter	Common Base	Common Emitter	Units
h_i	15	1500	ohms
h_r	2.5×10^{-4}	12.5×10^{-4}	
h_f	-0.99	110	
h_o	1×10^{-6}	1×10^{-4}	mhos
Δ_h	2.68×10^{-4}	125×10^{-4}	

$V_{CE} = -4.5\ V$, $I_C = -2$ ma.

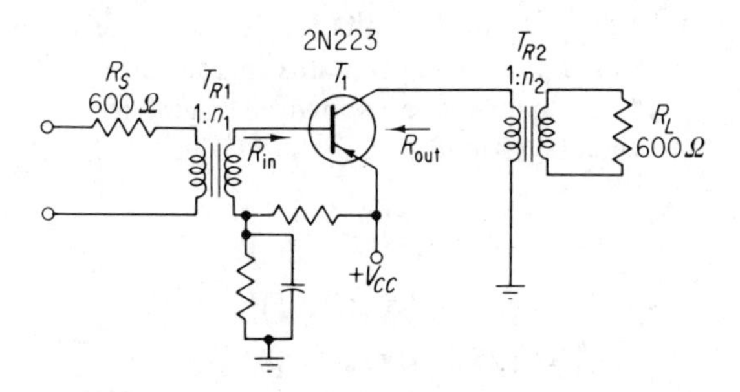

Fig. 4-14. Design example of transformer matching.

For a single stage, $A_P = i_2^2R_L/i_1^2R_{in}$. This expression has a maximum value when R_L is equal to $R_{out\,(matched)}$ of expression 4-32. Using the parameters shown, we obtain $R_{L(matched)} = 35{,}400\ \Omega$. R_{in} can now be computed either by using the expression for $R_{in\,(matched)}$ from Eq. 4-31, or from the general expression for R_{in} given by Eq. 4-11. Using Eq. 4-11 we obtain

$$R_{in} = h_i - \frac{h_f h_r}{h_o + \dfrac{1}{R_L}}$$

$$= 1500 - \frac{110 \times 12.5 \times 10^{-4}}{10^{-4} + \dfrac{1}{35.4 \times 10^3}}$$

$$= 420\ \Omega$$

Since we want matched conditions for maximum gain, we set the secondary resistance of the input transformer T_{R1} equal to R_{in} or 420 Ω. From this we can determine the turns ratio of T_{R1} as

$$n_1 = \sqrt{\frac{R_{in}}{R_S}} = \sqrt{\frac{420}{600}} = 0.84$$

The turns ratio of the output transformer T_{R2} is determined similarly, as

$$\frac{1}{n_2} = \sqrt{\frac{R_{out}}{R_L}} = \sqrt{\frac{35{,}400}{600}} = 7.7$$

The voltage drop across the dc resistance of the primary winding of T_{R2} must be subtracted from the supply voltage to get the actual no-signal collector voltage. If we assume this resistance to be 1000 Ω (a reasonable value for the small transformers generally employed), then the supply voltage should be 4.5 volts + 2 ma × 1000 Ω, or 6.5 volts.

The input signal current, assuming a lossless transformer, is given by

$$i_{in} = \sqrt{\frac{P_{in}}{R_{in}}} = \sqrt{\frac{0.5 \times 10^{-8}\ \text{watt}}{420\ \Omega}} \quad \text{or} \quad 3.5 \times 10^{-6}\ \text{amp}$$

The current gain as given by expression 4-24 is 24.5, so that ac collector current is $24.5 \times 3.5 \times 10^{-6}$, or 85 μa, rms. The peak collector voltage is $35{,}400 \times 85 \times 10^{-6} \times \sqrt{2}$, or 4.2 volts. Since the selected operating point was 4.5 volts, the peak signal swing remains in the linear operating region. The operating point and load line are shown in Fig. 4-15.

The power into the 600 Ω load is

$$P_L = \left(\frac{1}{n_2} \times I_c\right)^2 \times R_L$$

$$= (7.7 \times 85 \times 10^{-6})^2 \times 600$$

$$= 425 \times 10^{-9} \times 600$$

$$= 255 \times 10^{-6}\ \text{watt}$$

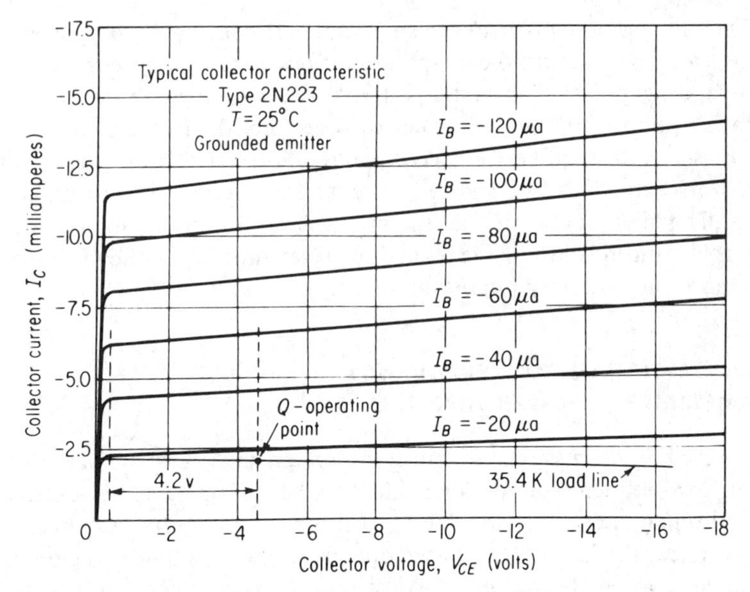

Fig. 4-15. Collector characteristics showing operating point and load line.

Then, the power gain is

$$A_P = \frac{P_{\text{out}}}{P_{\text{in}}} = \frac{255 \times 10^{-6}}{0.5 \times 10^{-8}} = 51{,}000 \quad \text{or} \quad 47.1 \text{ db}$$

From this power gain must be subtracted any loss in the coupling transformers which, as mentioned previously, will be small for the type of transformers normally employed in this application.

Direct Coupling

A third coupling technique is sometimes employed with transistors. This is *direct coupling*, where the collector of one stage is tied directly to the base of the following stage, as shown in Fig. 4-16.

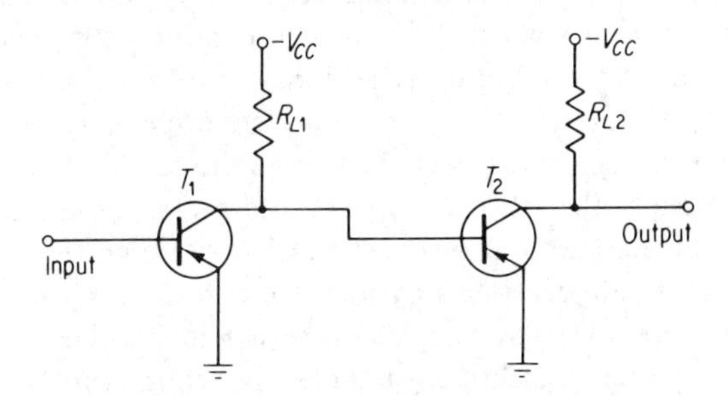

Fig. 4-16. Two-stage direct-coupled amplifier.

The obvious advantage in such a circuit is the saving in components that results from eliminating the coupling network. There is a design difficulty: the operating point of each transistor in the amplifier depends upon the operating point of the preceding stage. Therefore, the temperature stabilization of such an amplifier is exceedingly complex. For this reason, silicon transistors with their low leakage currents are often used in direct-coupled stages. The main application as an amplifier is in subminiature applications, such as hearing aids, where the reduction in size and weight due to elimination of components is a very important virtue.

LARGE-SIGNAL, LOW-FREQUENCY AMPLIFIERS (POWER AMPLIFIERS)

The preceding discussion presented the design and performance of small-signal, low-frequency amplifiers. Small-signal parameters, however, apply over a limited operating range of a transistor. When several amplifier stages are cascaded, the latter stages often operate in their nonlinear region. When this occurs, as in the output stage of an audio amplifier, the small-signal parameters of the transistor are no longer valid and the designer must resort to other forms of analysis. In most low-frequency amplifier systems, the last stage is usually the only stage operated in its nonlinear region. For this reason, one usually thinks of the large-signal, low-frequency amplifier as being the power output stage.

Design Criteria for Low-Frequency Power Amplifiers

In the amplifier stages preceding the output stage, power gain is one of the most important performance characteristics desired. In the output stage, however, maximum power output is usually of prime importance. Power amplifiers are classified according to their operating point as Class A, B, C, or AB. In Class A operation, the zero signal operating point is chosen so that the transistor will conduct over the entire range of input signal swing. In Class B operation, the transistor conducts for one-half of the input cycle. Class C operation occurs when the transistor conducts for less than one-half of the input cycle. Class AB allows the transistor to conduct more than one-half of the input cycle, yet not for the complete cycle. Choice of the class of operation will be influenced by such factors as the required power output, dc power requirement (efficiency), distortion, and power gain.

Any of the three configurations may be used in a high-level amplifier. The common-emitter configuration provides the greatest power gain and is generally used. The linear transfer characteristic exhibited by the common-base stage is, however, attractive for those applications requiring minimum distortion. Power gain of the common-collector stage is somewhat lower than for the common-emitter stage, except under extremely low load resistances. Also, in the common-collector stage, the load resistance acts as the emitter stabilization resistance.

Since power output is one of the important performance characteristics of a high-level amplifier stage, the excursions of the collector voltage and current may approach the absolute maximum ratings. Careful consideration must be given to these limits as well as the power dissipation capabilities of the transistor. Figure 4-17 shows a typical collector output characteristic for a transistor. Region A is the only region in which the transistor can be safely operated as an amplifier. If the bias excursions cause the operating point of the transistor to swing into Region B, the maximum collector-to-emitter voltage rating of the transistor will be exceeded. If the excursion of the output signal extends into Region D, the maximum collector current of the transistor is exceeded, and if the operating point is chosen such that the signal swing extends into Region C, then the maximum power dissipation of the transistor has been exceeded. All these parameters are very important in high-level stages, and the load line must be constructed such that it does not enter into either Region B, C, or D.

It can be seen in Fig. 4-17 that an infinite number of load lines may be drawn within the limits of load line L_1 and load line L_2. The values for these two load lines were obtained by constructing lines tangent to the power

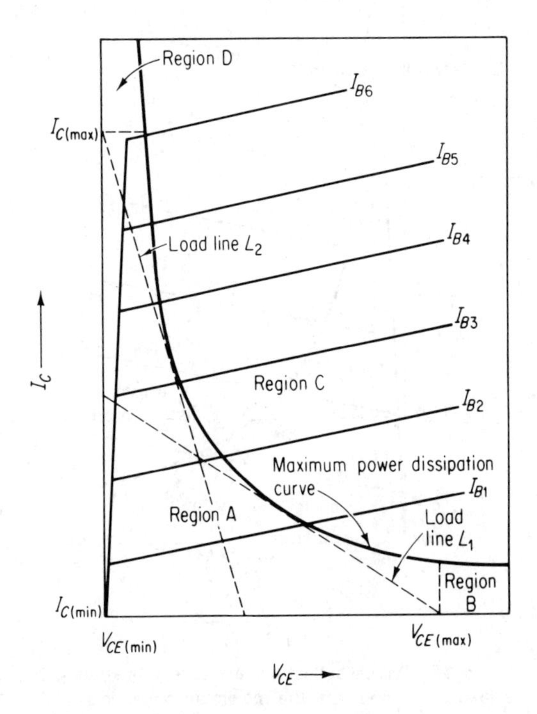

Fig. 4-17. Typical collector characteristic showing limit load lines for power amplifiers.

dissipation curve from the maximum allowable voltage and current ratings, respectively. Any value of load resistance between these two curves will be capable of delivering approximately the same power output. The resistance represented by load line L_1 is the maximum value, and usually more nearly matches the output resistance of the transistor; hence, load resistance values close to the maximum permissible value are normally used. Excursions of the operating point are also confined to the more linear regions of the transfer characteristic curve. Additionally, the losses due to operation in the saturation region of the transistor are lower when the larger load is used.

ANALYSIS AND DESIGN OF CLASS A POWER AMPLIFIERS

Design Equations

Figure 4-18 again shows the typical collector output characteristics of a transistor. Any load line constructed tangent to the maximum dissipation curve will be bisected by this curve. That is, if the quiescent operating point

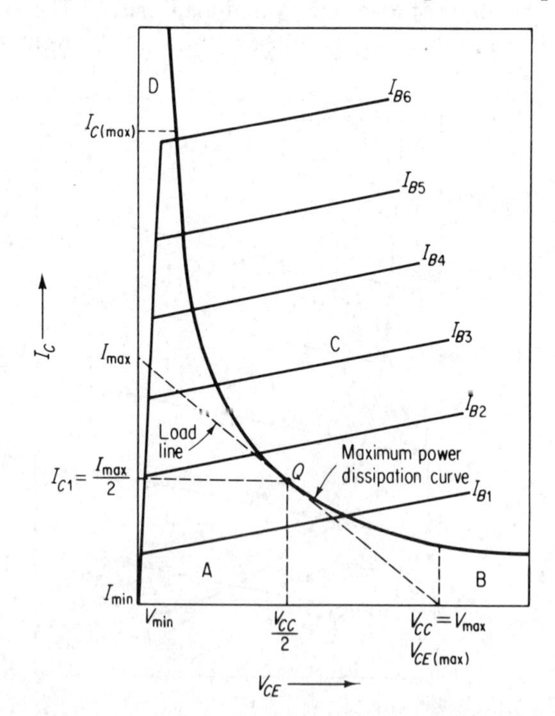

Fig. 4-18. Typical collector characteristic illustrating that a load line tangent to the maximum power dissipation curve is bisected at point of tangency (Q).

for the transistor is chosen to be at this intersection, equal voltage and current swings will be possible when the input signal drives the transistor into cutoff ($I_C = 0$ and $V_{CE} = V_{CC}$), and into saturation ($I_C = I_{\max}$ and $V_{CE} = 0$). These definitions of *cutoff* and *saturation*, which will be used throughout this chapter, neglect collector leakage current, I_{CBO}, and the saturation voltage $V_{CE(\text{sat})}$.

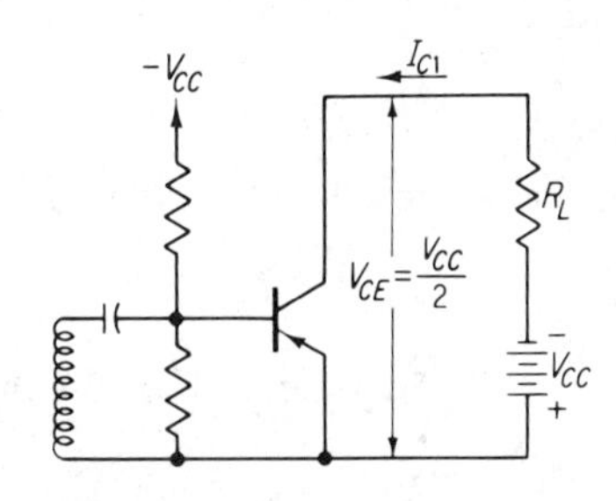

Fig. 4-19. Typical Class A power amplifier with resistive load.

To prove that the load line tangent to the maximum dissipation curve is bisected, we note that the maximum dissipation for a Class A stage, shown in Fig. 4-19, occurs when $V_{CE} = V_{CC}/2$ and $I_C = V_{CC}/2R_L$. Then the equation for maximum dissipation is

$$P_{D(\max)} = V_{CE}I_{C1} = \frac{V_{CC}^2}{4R_L} \tag{4-36}$$

The load current is

$$I_C = -\frac{V_{CE}}{R_L} + \frac{V_{CC}}{R_L} \tag{4-37}$$

These equations are solved simultaneously for V_{CE} to give

$$V_{CE} = -\frac{V_{CC}}{2} \tag{4-38}$$

Class A Amplifier with Resistive Load

Figure 4-18 showed a load line tangent to the maximum power dissipation curve and passing through $V_{CE(\max)}$. This load line was the maximum allowable value of load resistance to give maximum power output. The basic circuit for a Class A amplifier was shown in Fig. 4-19 with a resistive load determined by the load line. This value of load resistance dictates that the supply voltage must be equal to the maximum collector-emitter voltage rating of the transistor. This choice is not always practical, but it gives the maximum power gain, since it more nearly matches the output impedance of the transistor.

The transistor is biased at the operating point Q, which is at a collector-emitter voltage of $V_{CC}/2$, and a collector current I_{C1}. The collector current excursion will be along the load line and will vary from some maximum value, $I_{\max}$, to a minimum value, $I_{\min}$. Likewise, the collector voltage excursion will be from some value, $V_{\max}$ to $V_{\min}$ (see Fig. 4-18).

Therefore, for a sinusoidal time variation of the collector current

$$I_{\text{rms}} = \frac{I_{\max} - I_{\min}}{2\sqrt{2}} \tag{4-39}$$

and

$$V_{\text{rms}} = \frac{V_{\max} - V_{\min}}{2\sqrt{2}} \tag{4-40}$$

The ac power output is then

$$P_{\text{ac}} = V_{\text{rms}} I_{\text{rms}} = \frac{(V_{\max} - V_{\min})(I_{\max} - I_{\min})}{8} \tag{4-41}$$

The dc power will be the product of the supply voltage and current, or

$$P_{\text{dc}} = V_{CC} I_{C1} \tag{4-42}$$

The efficiency of the collector circuit may be defined as the ratio of the ac power delivered to the load to the dc power obtained from the collector power supply or:

$$\eta_C = \text{collector efficiency} = \frac{P_{\text{ac}}}{P_{\text{dc}}} 100\% \tag{4-43}$$

and, therefore,

$$\eta_C = \frac{(V_{\max} - V_{\min})(I_{\max} - I_{\min})}{8 V_{CC} I_{C1}} 100\% \tag{4-44}$$

Maximum Class A collector efficiency occurs at maximum power output. This occurs when the excursion of the operating point is a maximum, or from Fig. 4-18,

$$(V_{\max} - V_{\min}) = V_{CC} \tag{4-45}$$

and

$$(I_{\max} - I_{\min}) = 2 I_{C1} \tag{4-46}$$

The maximum collector efficiency of a Class A amplifier with a resistive load is, then,

$$\text{maximum collector efficiency} = \eta_{C(\max)} = \frac{(V_{CC})(2I_{C1})}{8 V_{CC} I_{C1}} 100\% = 25\% \tag{4-47}$$

In addition to the collector efficiency and power output, the power dissipated in the transistor is also a design consideration, since the maximum allowable transistor dissipation should not be exceeded. The power dissipated in the transistor is given by

$$P_D = P_{\text{dc}} - P_{\text{ac}} - I_{C1}^2 R_L \tag{4-48}$$

where $I_{C1}^2 R_L$ is the dc power dissipated in the load resistor. Maximum power dissipation occurs when the power output is zero, or

$$P_{D(\max)} = P_{\text{dc}} - I_{C1}^2 R_L = V_{CC} I_{C1} - I_{C1}^2 R_L \tag{4-49}$$

The ratio of Eq. 4-49 to Eq. 4-41 for maximum power output is, then

$$\frac{P_{D(\max)}}{P_{\text{ac}(\max)}} = \frac{(V_{CC} I_{C1} - I_{C1}^2 R_L)}{\dfrac{2 V_{CC} I_{C1}}{8}} = \frac{4\left(V_{CC} I_{C1} - \dfrac{V_{CC} I_{C1}}{2}\right)}{V_{CC} I_{C1}} = 2 \tag{4-50}$$

Thus, the transistor in a Class A power amplifier with a resistive load must be capable of dissipating at least twice the desired ac power output, a distinct disadvantage.

Class A Transformer-Coupled Amplifier

Consider now the transformer-coupled Class A amplifier. The basic circuit is shown in Fig. 4-20 and the load line and operating point are shown on the collector characteristics of Fig. 4-21. Considering the dc resistance of the output transformer to be negligible, the total supply voltage will appear across the transistor. Therefore, the maximum supply voltage must not exceed one-half the collector voltage rating of the transistor. The maximum excursion of the operating point is now

$$V_{\max} - V_{\min} = 2V_{CC} \tag{4-51}$$

and

$$I_{\max} - I_{\min} = 2I_{C1} \tag{4-52}$$

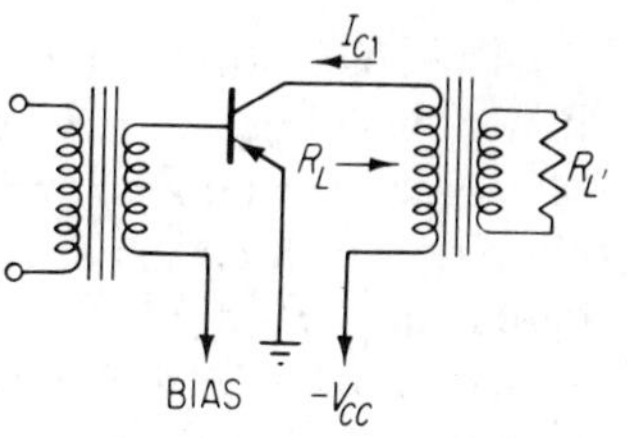

Fig. 4-20. Basic transformer-coupled Class A power amplifier.

The maximum collector efficiency for this Class A circuit is

$$\eta_{C(\max)} = \frac{2V_{CC}(2I_{C1})100\%}{8V_{CC}I_{C1}} = \frac{100\%}{2} = 50\% \tag{4-53}$$

as compared to 25 per cent for the resistive load. However,

$$\frac{P_{D(\max)}}{P_{ac(\max)}} = \frac{(V_{CC}I_{C1})}{\frac{(2V_{CC})(2I_{C1})}{8}} = 2 \tag{4-54}$$

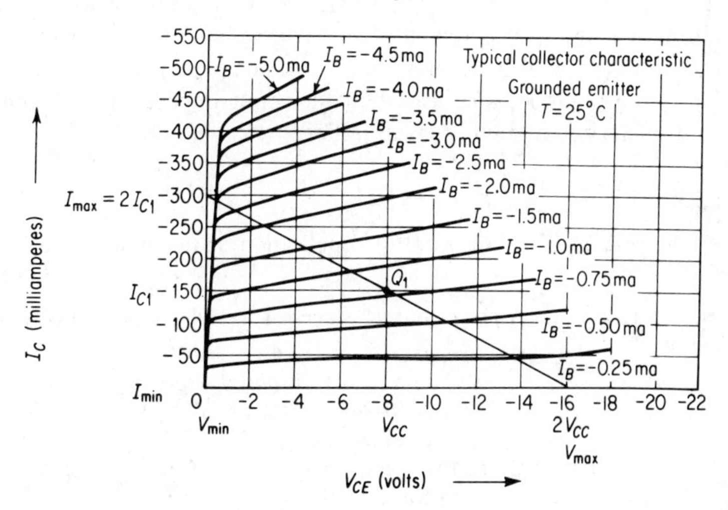

Fig. 4-21. Collector characteristic and load line for typical transformer-coupled Class A amplifier. Notice $V_{\max} = 2V_{CC}$ and $I_{\max} = 2I_{C1}$.

which is the same as for the resistive load ($I_{C1}^2 R_L = 0$). Even though a transformer-coupled amplifier is more efficient, the high dissipation remains a problem. In fact, the requirement that the transistor must be capable of dissipating twice the maximum available output power is the major disadvantage with Class A circuits.

Distortion in Class A Amplifiers

The application of large signals to transistor amplifiers necessarily results in nonlinear operation. Distortion in Class A amplifiers is caused by a number of factors, including nonlinear current gain and input resistance, and the clipping which occurs at high signal levels as the transistor is driven toward cutoff and/or saturation. If clipping is avoided, distortion may usually be limited in Class A amplifiers to less than 5 per cent.

Current gain in a common-base amplifier configuration is more linear than that of the common-emitter connection. In fact, the current gain linearity of the common-base connection is so good that distortion due to this cause may be considered negligible. Because of this excellent current gain linearity, distortion in the common-base connection is minimized if driven from a high-impedance source. This, of course, results in a decrease in gain, obviating a compromise between gain and distortion.

In the common-emitter amplifier a compromise must be made in the selection of source impedance for minimizing distortion. Distortion caused by the relatively nonlinear current gain characteristic is minimized by driving the amplifier from a low-impedance source. Distortion resulting from the nonlinear input impedance characteristics is minimized by a high-source impedance. A compromise for the source impedance must be made that will best minimize the distortion and result in adequate gain. In practice, we find that a source impedance less than the input impedance will minimize distortion of the common-emitter amplifier.

DESIGN EXAMPLE

The following example illustrates the application of the design equations just derived:

Problem: Design a Class A power output stage for an auto radio to the following specifications:

Audio power output = 4 watts
Supply voltage = 12 volts
Speaker resistance = 3.2 Ω
Driver transistor output resistance = 5000 Ω
Transformer efficiencies = 80%
Operating temperature = 45°C

Solution: A transistor must first be selected. Since the speaker resistance is only 3.2 Ω, a transformer will have to be used, which means the maximum collector voltage rating of the transistor must be at least 2 × 12 or 24 volts. From expression 4-54, the power dissipation must be twice the power output delivered to the output transformer primary, or 2 × 4/0.80 = 10 watts. The 2N176 meets these two requirements and has a frequency response suitable for the application.

Plotting a 10-watt dissipation curve on the collector characteristic curve for the 2N176 as shown in Fig. 4-22(a), a load line is now constructed tangent to the dissipation curve at the 12-volt intercept. Notice that this line crosses the voltage axis at 24 volts and the current axis at 1.6 amp.

The slope of the load line is 24/1.6 = 15 Ω, the value of the load resistance required to be reflected back to the transistor. The turns ratio of the output transformer may now be calculated to be, therefore

$$\left(\frac{n_1}{n_2}\right)^2 = \frac{15}{3.2}$$

or

$$\frac{n_1}{n_2} = 2.15$$

The operating point is taken from the curve to be 12 volts and 0.8 amp. A suitable bias network, as discussed in an earlier chapter, would now be designed.

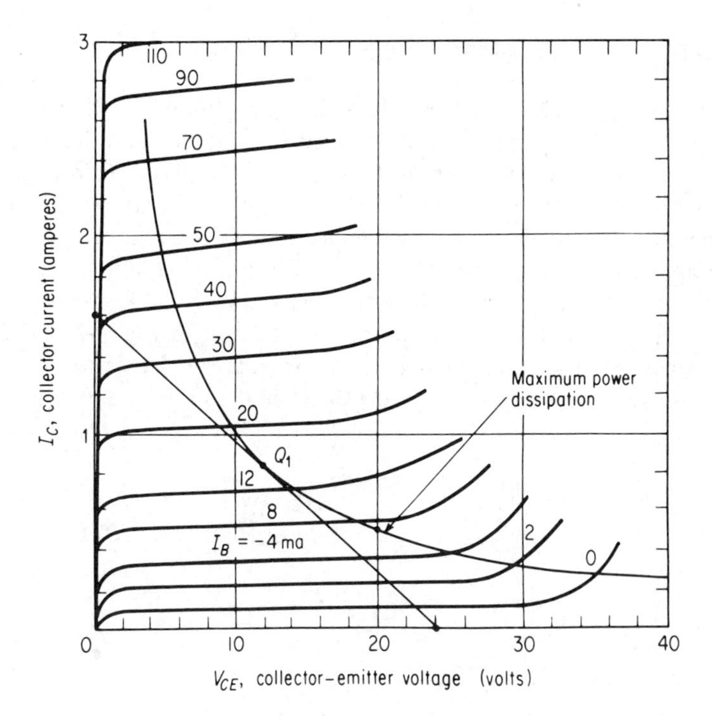

Fig. 4-22a. Typical 2N176 collector characteristic showing 15 ohm load line tangent to maximum power dissipation curve at V_{CE} = 12 volts.

Typical common-emitter input resistance is given on the data sheet as 15 Ω. A source impedance of 10 Ω has been found experimentally to give reasonable linearity. Thus, the turns ratio of the input transformer is

$$\left(\frac{n_1}{n_2}\right)^2 = \frac{5000}{10} \quad \text{or} \quad \frac{n_1}{n_2} = 22.3$$

The designed amplifier schematic (neglecting bias circuitry) is shown in Fig. 4-22(b).

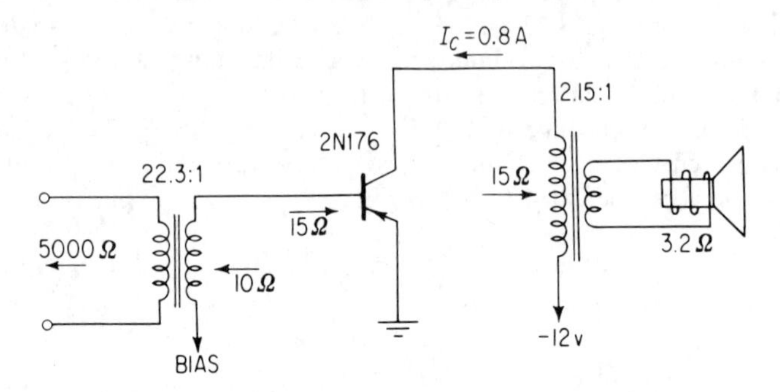

Fig. 4-22b. Power amplifier to meet design requirements of Class A design example.

CLASS B POWER AMPLIFIERS

A transistor operated in Class B is biased to the point of cutoff ($I_C = 0$) with zero input signal. Consequently, it conducts for only one-half the input cycle. A single-transistor Class B amplifier would not be practical in an untuned amplifier because of the extreme distortion introduced by the one-half cycle of nonoperation. A second transistor operating Class B can, however, be added to the circuit, to conduct when the first transistor is not conducting in order to eliminate this distortion. Such operation is generally called *push-pull operation*.

The basic Class B push-pull amplifier is shown in Fig. 4-23. The input transformer secondary is center-tapped to provide the phase reversal between the transistors necessary for conduction over the entire input signal cycle.

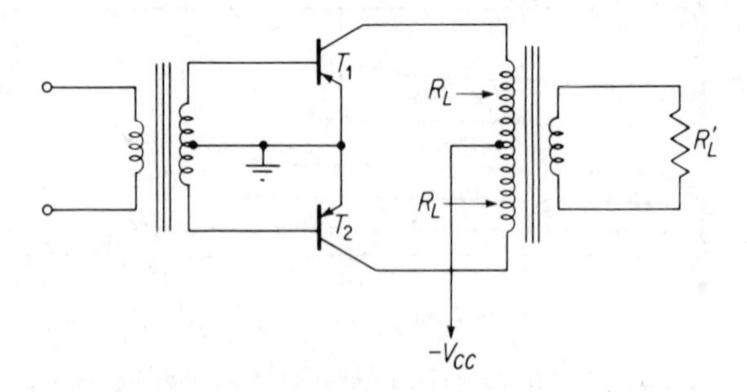

Fig. 4-23. Basic Class B push-pull amplifier.

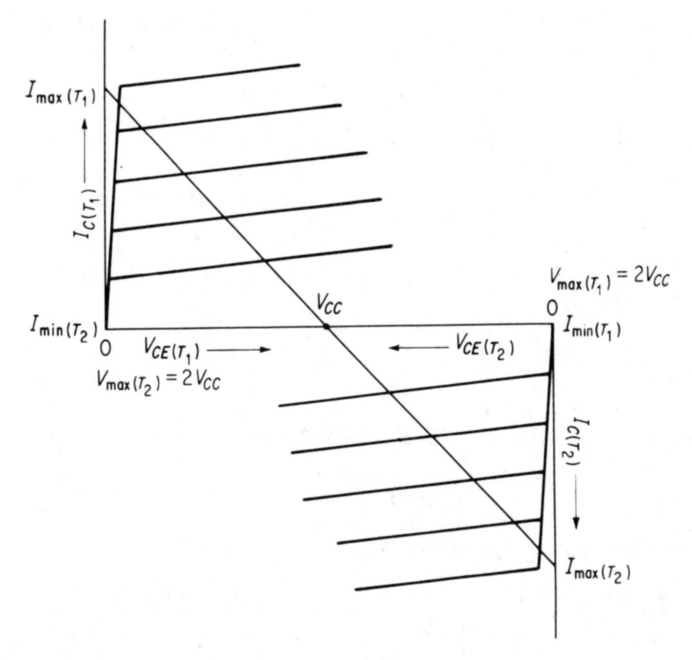

Fig. 4-24. Typical collector characteristics with one inverted and offset so they coincide at V_{CC} to illustrate push-pull operation.

The collector characteristics shown in Fig. 4-24 are arranged so as to show the excursion of the operating point over an entire input cycle. As is the case for a transformer-coupled Class A amplifier, the maximum supply voltage that may be used is one-half the maximum collector-emitter voltage rating of the transistor. Since only one transistor is in operation at one time, equations can be derived on the basis of single-transistor operation. The collector current for a sine wave input signal will be half cycles of a sine wave as shown in Fig. 4-25. The dc value of this current is

$$I_{dc} = \frac{I_{max}}{\pi} \tag{4-55}$$

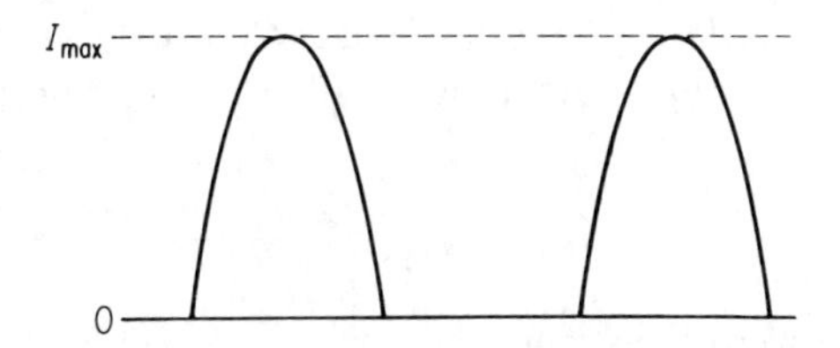

Fig. 4-25. Collector-current wave form for one transistor during push-pull operation with sine wave input.

The total dc power required will be twice that required by one transistor, or

$$P_{dc} = 2\left(V_{CC}\frac{I_{max}}{\pi}\right) \tag{4-56}$$

Combining this equation with Eq. 4-41 gives an expression for the collector efficiency

$$\eta_C = \frac{(V_{max} - V_{min})(I_{max} - I_{min})\pi 100\%}{16(V_{CC}I_{max})} \tag{4-57}$$

The excursion of the operating point for maximum power output is

$$V_{max} - V_{min} = 2V_{CC} \tag{4-58}$$

$$I_{max} - I_{min} = 2I_{max} = \frac{2V_{CC}}{R_L} \tag{4-59}$$

The maximum collector efficiency for a Class B push-pull power amplifier is

$$\eta_{C(max)} = \frac{(2V_{CC})(2I_{max})\pi 100\%}{16V_{CC}I_{max}} = \frac{\pi \times 100\%}{4} = 78\% \tag{4-60}$$

The maximum power output is

$$P_{ac(max)} = \frac{(V_{max} - V_{min})(I_{max} - I_{min})}{8} = \frac{(2V_{CC})\left(2\frac{V_{CC}}{R_L}\right)}{8} = \frac{V_{CC}^2}{2R_L} \tag{4-61}$$

This equation states that the optimum value of load resistance for a given supply voltage and power output is given by

$$R_L = \frac{V_{CC}^2}{2P_{ac}} \tag{4-62}$$

This value of resistance is the load presented to one transistor. The total primary impedance of the output transformer would, therefore, be four times this amount.

Since the dc resistance of the output transformer is assumed to be zero, the power dissipation for both transistors is $P_D = P_{dc} - P_{ac}$, or

$$P_D = \frac{2V_{CC}I_{max}}{\pi} - \frac{V_{CC}^2}{2R_L} = \frac{2V_{CC}I_{max}}{\pi} - \frac{I_{max}^2 R_L}{2} \tag{4-63}$$

The load resistance is restricted by the power dissipation for a given V_{CC} and the dissipation varies with input signal level. Maximum power dissipation may be found by taking the derivative of Eq. 4-63 and is found to be a maximum when

$$I_{max(signal)} = \frac{2}{\pi}\frac{V_{CC}}{R_L} \tag{4-64}$$

The maximum power dissipation, then, is

$$P_{D(max)} = \frac{2V_{CC}^2}{\pi^2 R_L} \tag{4-65}$$

This occurs when the power output is

$$P_{ac} = \frac{I^2_{max(signal)} R_L}{(\sqrt{2})^2} \tag{4-66}$$

where the $(\sqrt{2})^2$ is inserted to change the peak value to rms. The ratio of the ac power output at maximum dissipation to the maximum output power is found by combining Eqs. 4-61, 4-64, and 4-66 and is

$$\frac{P_{ac} \times 100\%}{P_{ac(max)}} = \left(\frac{2}{\pi}\right)^2 \times 100\% = 40.6\% \tag{4-67}$$

Thus, the maximum power dissipation in a Class B push-pull amplifier occurs when the ac output power is 40.6 per cent of the maximum obtainable.

In designing any amplifier for a given power output, one must consider the maximum power dissipation. From Eqs. 4-61 and 4-65,

$$\frac{P_{ac(max)}}{P_{D(max)}} = \frac{V^2_{CC}/2R_L}{2V^2_{CC}/\pi^2 R_L} = \frac{\pi^2}{4} = 2.47 \tag{4-68}$$

In other words, a Class B push-pull amplifier (assuming the ideal circuit, $\eta_C = 78$ percent) may be designed for a power output of 2.47 times the combined rated power dissipation of both the transistors, or approximately five times the rating of one of the units.

From Eq. 4-65, the maximum power dissipation per transistor is

$$P_{D(max-one\ transistor)} = \frac{V^2_{CC}}{\pi^2 R_L} \tag{4-69}$$

or the maximum output power in a Class B amplifier is obtained for a given transistor power dissipation rating and a given voltage supply when

$$R_L = \frac{V^2_{CC}}{\pi^2 P_{DM}} \tag{4-70}$$

where P_{DM} is the maximum power dissipation rating of one transistor.

Distortion

The transistor characteristics that contribute to the distortion in a Class A amplifier also contribute to distortion in Class B amplifiers. Nonlinearity of current gain and input resistance and, at high signal levels, clipping due to saturation and/or cutoff of the transistor, occur in Class B as well as Class A amplifiers. Hence, the same driving impedance considerations apply. That is, the common-emitter amplifier is usually driven from a low-impedance source, and the common-base amplifier from a high-impedance source. Class B push-pull operation has the additional advantage that even-order harmonics cancel. If the current transfer characteristics of the two transistors used in a push-pull amplifier differ considerably, distortion may also result. Transistors for push-pull amplifiers are, therefore, usually matched by the manufacturers for current gain (h_{fe}).

In a Class B push-pull amplifier, the output current is supplied first by one transistor and then the other. In the region where the output current is switched from one transistor to the other, considerable distortion may occur owing to the nonlinearity in the collector transfer characteristic near the cut-off region of the transistor. Figure 4-26(a) is a plot of the input voltage versus the output current. It can readily be seen that considerable nonlinearity exists near the origin. If a sine wave signal were applied to a push-pull Class B amplifier from a voltage source, the output current wave form would be highly distorted.

If the transistors of a Class B push-pull amplifier were slightly forward-biased to some point, V_{BE1}, as shown in Fig. 4-26(b), the linearity of the transfer characteristic would be greatly improved. This type of operation would no longer be Class B but Class AB. This is the method normally used to reduce cross-over distortion. A small amount of collector current will flow when no signal is present in the amplifier, thus reducing the collector efficiency of the amplifier. Figure 4-26(b) illustrates how a value of bias voltage may be arrived at for a minimum cross-over distortion and maximum collector efficiency. The higher the value of V_{BE1}, the lower the collector efficiency will be. It can be seen, however, that once a certain value of V_{BE1} is reached, very little improvement in cross-over distortion will result. The curve as shown in Fig. 4-26 will be different for each transistor type. For most

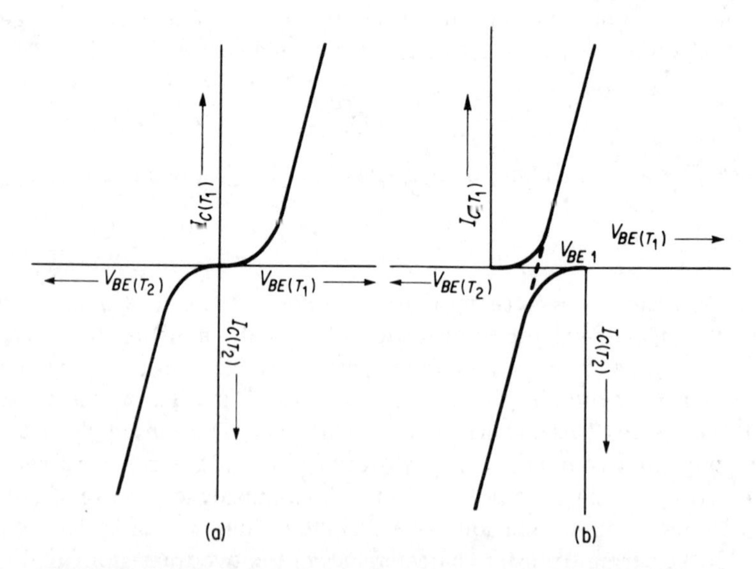

Fig. 4-26. Transfer characteristics of typical push-pull transistor amplifier. (a) Without forward bias; (b) a method of determining proper forward bias.

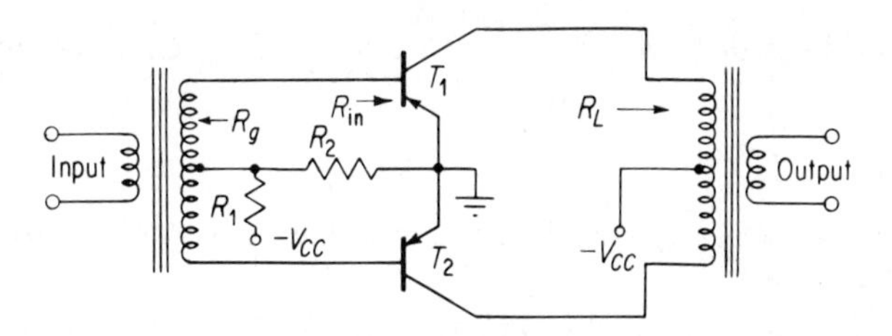

Fig. 4-27. Push-pull amplifier showing bias network to reduce cross-over distortion. Note bias network should not be bypassed.

germanium transistors, however, the amount of bias will be about 0.1 to 0.2 volt, and for silicon devices about 0.6 to 1.0 volt. Figure 4-27 shows how bias may be applied to a Class AB amplifier to improve cross-over distortion. The parallel combination of R_1 and R_2 should be as low as the power supply will permit. These resistors must *not be bypassed with a capacitor*, for it would charge through the base-emitter junction of the transistor during the conduction portion of the signal cycle and then discharge through the bias resistors during nonconduction of the transistors. The discharge current would reverse-bias the transistors, driving the amplifier into Class C operation. Since the use of bypass capacitors is not possible, some power gain in the amplifier must be sacrificed. Equation 4-71 shows the power gain that will be lost with the addition of the bias resistors:

$$P_{G(\text{loss})\text{db}} = 10 \log \left(\frac{R_{\text{in}} + R_B + R_g}{R_{\text{in}} + R_g}\right)^2 \tag{4-71}$$

where R_{in} = input resistance of the transistor
R_B = parallel combination of R_1 and R_2
R_g = secondary impedance of the input transformer.

Nonlinearity near the zero operating point of the input current versus output current characteristic is not as great as for the input voltage versus output current characteristic. Figure 4-28 shows a plot of input current versus output current for the common-emitter connection. Note here that the distortion is not as great at the zero point, but linearity at the higher currents is not as good. The distortion caused by the nonlinearity near the cut-off point of the transistor is called *cross-over distortion.* Cross-over distortion can be improved somewhat by driving the push-pull amplifier from a current source.

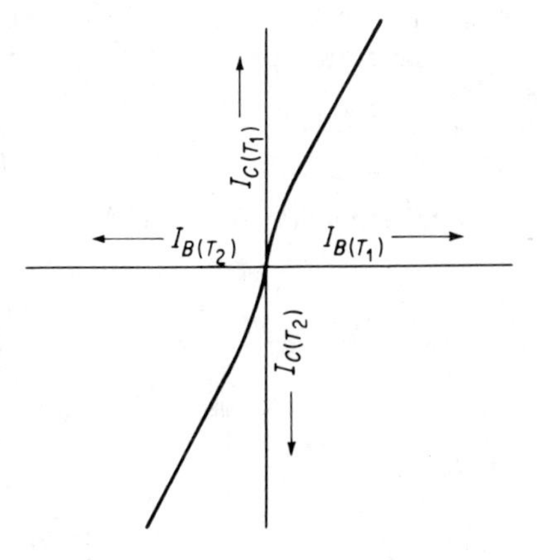

Fig. 4-28. Typical current transfer characteristic of push-pull amplifier.

Higher-level distortion will, however, be worse in a common-emitter amplifier.

DESIGN EXAMPLE

Design a Class B push-pull amplifier to meet the same conditions as those given for the Class A stage.

Solution: Again, the power that must be delivered to the output transformer primary is 4/0.8 = 5 watts. The maximum load resistance then is given by Eq. 4-62 as

$$R_{L(\max)} = \frac{V_{CC}^2}{2P_{ac}} = \frac{(12)^2}{2 \times 5} = 14.4\ \Omega$$

This maximum value will be used more nearly to match the transistor output resistance, and therefore,

$$I_{\max} = \frac{V_{CC}}{R_L} = \frac{12}{14.4} = 0.833 \text{ amp}$$

From Eq. 4-68, maximum power dissipation is 1/2.47 times the power output, or 5/2.47 = 2 watts, or, each transistor must be capable of dissipating 1 watt. Again, the 2N176 easily meets the voltage, current, and power dissipation requirements.

Notice that the power dissipation requirement of each transistor is only one-tenth of that required by the Class A amplifier.

The calculated value of R_L is the load resistance across one-half the primary; hence, the total primary resistance must be 4 × 14.4 = 57.6 Ω. The turns ratio of the output transformer is then

$$\left(\frac{n_1}{n_2}\right)^2 = \frac{57.6}{3.2}$$

or

$$\frac{n_1}{n_2} = 4.2$$

with the primary center-tapped.

The source resistance should be low compared to the input resistance of the transistor (15 Ω typical), and again 10 Ω is chosen. The total secondary resistance of the input transformer is then 4 × 10 = 40 Ω. The turns ratio of this transformer is

$$\left(\frac{n_1}{n_2}\right)^2 = \frac{5000}{40} \quad \text{or} \quad \frac{n_1}{n_2} = 11.1$$

with the secondary center-tapped. The circuit is shown in Fig. 4-29.

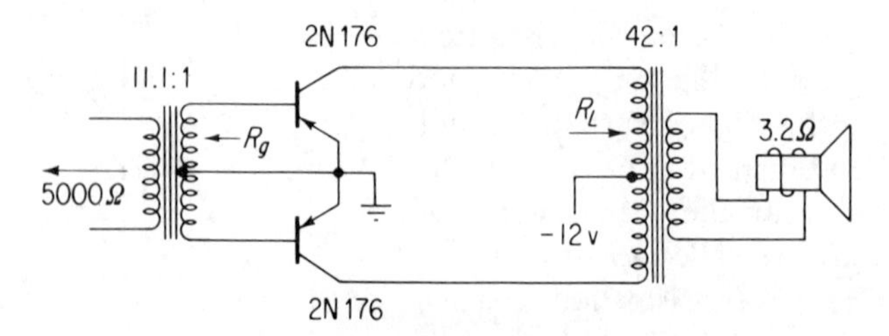

Fig. 4-29. Circuit for Class B amplifier of design example.

Frequency Response

The frequency response of an audio amplifier, whether Class A or Class B, is determined by the transformers, the capacitors, and the transistors. The low-frequency response is determined by the primary inductance of the transformers and also the value of the bypass capacitors, if used. The high-frequency response is determined by the leakage reactance and winding capacitance of the transformer, and the frequency response of the transistor.

OTHER CLASS B CIRCUITS

Phase reversal of the Class B amplifier in the preceding discussion was accomplished by the use of a center-tapped input transformer. This is one of the more common means of obtaining the phase reversal. Figure 4-30 shows an *RC*-coupled amplifier providing phase reversal to a push-pull amplifier.

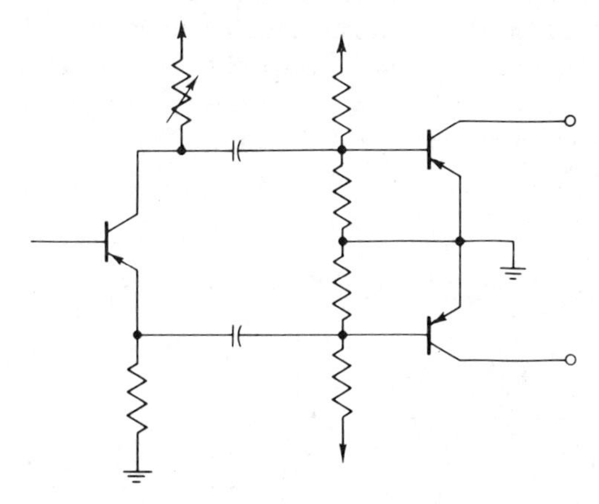

Fig. 4-30. Phase splitting transistor amplifier to provide phase reversal for a push-pull output stage.

The signal for one side of the amplifier is derived from the collector side of the driver transistor, and the signal for the other side of the push-pull circuit is derived from the emitter circuit of the driver stage. Owing to the phase reversal of a common-emitter amplifier, the signal taken from the collector circuit of the driver stage is 180° out of phase with the signal taken from the emitter circuit. Some means is usually provided in the driver stage to balance the amplitude of the two output signals to prevent an unbalanced condition in the push-pull amplifier. Figure 4-30 shows the collector resistor of the driver stage variable for this purpose.

If a PNP and an NPN transistor having similar transfer characteristics are connected in a Class B push-pull amplifier, automatic phase reversal is accomplished. This circuit connection is often referred to as a *complementary*

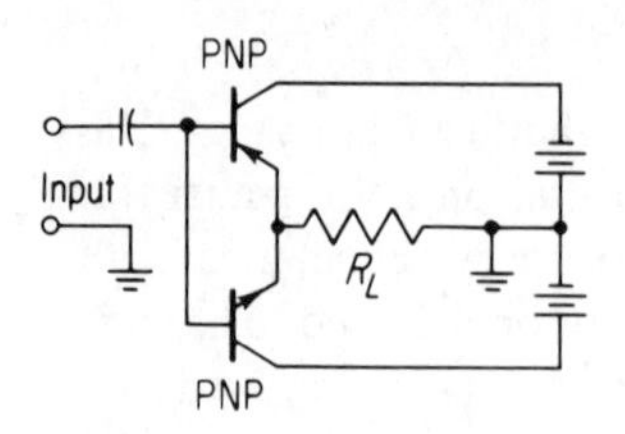

Fig. 4-31. Typical complementary symmetry push-pull amplifier with resistive load.

symmetry amplifier. Figure 4-31 is a circuit of a push-pull amplifier using a PNP and an NPN transistor. In the Class B condition, the PNP transistor will conduct for one-half of the input cycle and the NPN transistor will be cutoff. During the other half of the input cycle the PNP transistor will be cutoff and the NPN transistor will conduct.

Since the output transformer in any power amplifier contributes to the cost, weight, and volume of the amplifier, an attempt is sometimes made to produce a speaker of a proper load impedance for the amplifier. In a Class A amplifier, dc flowing in the collector circuit must also pass through the voice coil of the speaker. This causes the voice coil to be offset from its normal operating position and distortion may result. In a Class B amplifier, a center-tapped speaker is sometimes coupled directly to the amplifier. A speaker of this type results in poor electrical-to-acoustical efficiency, because only one-half of the winding of the voice coil is in use for each half cycle. The circuit for such an amplifier stage is shown in Fig. 4-32.

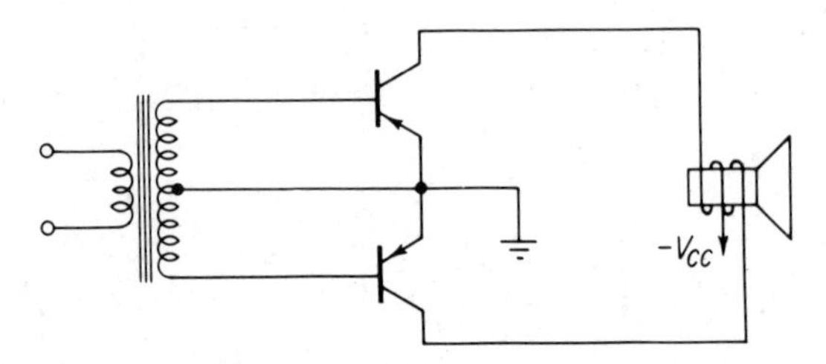

Fig. 4-32. Typical push-pull amplifier with voice coil of speaker used as the output transformer.

If the speaker of a push-pull amplifier were arranged in a bridge circuit, as shown in Fig. 4-33(a), no current would flow through the speaker when the bridge was in balance. If the bridge were to become unbalanced with the application of an input signal, then only signal current would flow through the speaker. Figure 4-33(b) is a circuit of a bridge-type push-pull amplifier with a speaker connected in the same manner as shown in Fig. 4-33(a). Under no-signal conditions, the potential at Point A is the same as at Point B and no current flows through the speaker. Upon the application of an input signal, one of the transistors is driven toward saturation and the other transistor toward cutoff. The potential then at Point A shifts either negative or positive

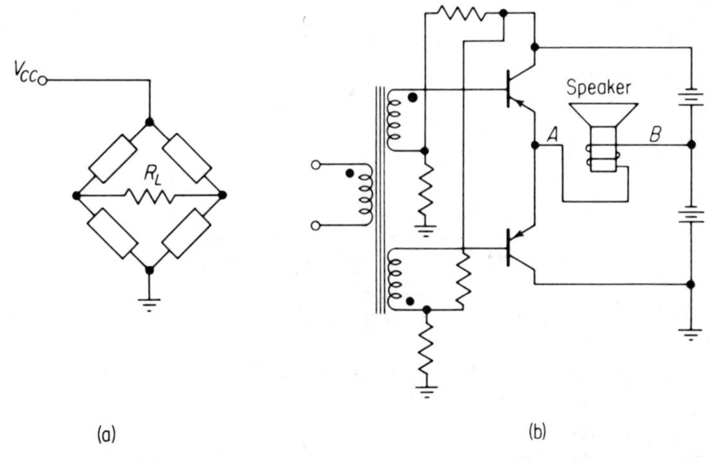

Fig. 4-33. Bridge-type push-pull amplifier showing (a) bridge and (b) complete circuit.

with respect to Point *B*, depending on which transistor saturates, and current then flows through the speaker. In the other half of the cycle, the condition is reversed and the opposite transistor is driven toward saturation and the current reverses in the speaker.

Notice that a center-tapped input transformer cannot be used in a circuit of this type because of bias conditions. A circuit of this type is very practical if a center-tapped voltage supply and a speaker of the proper impedance are available. The design procedure used in circuits of this type is the same as that used for the basic Class B circuit provided that the proper values are given to the load impedance and the supply voltage. Figure 4-34 is a comparison of a circuit of this type and the conventional Class B circuit.

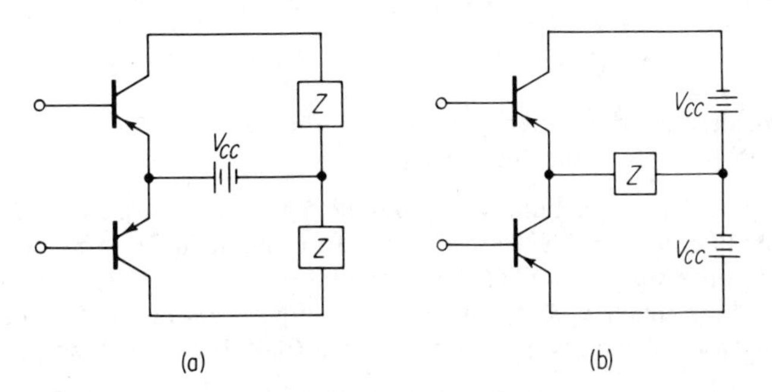

Fig. 4-34. Basic elements of (a) conventional Class B push-pull circuit and (b) bridge circuit. The values of Z and V_{CC} refer to the design equations developed in this chapter.

EXERCISES

1.

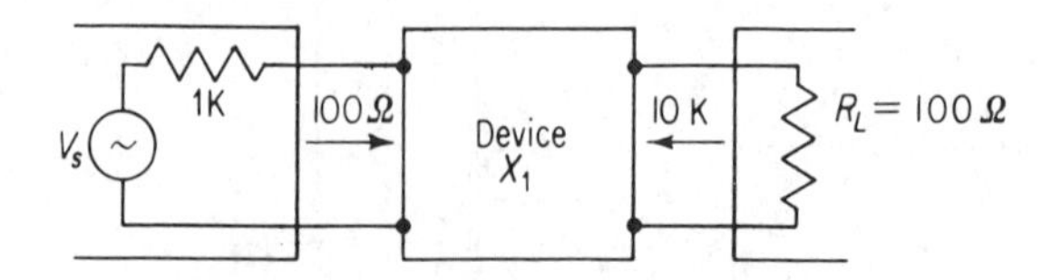

In the foregoing system, given: transducer gain = 40 db and $V_S = 0.1$ volt, find:

(a) P_{AS} = power available from the source
(b) P_L = power into the load
(c) A_P = power gain
(d) A_A = available gain

2. In the drawing of Exercise 1, determine the turns ratio required if transformers are used to match the input and output impedances to the source and load, respectively. (Assume unity coupling.)

3. Use the 2N223, operating at 4.5 volts and 2 ma. The only transformer available for TR2 has a ratio of 20K to 1K. Any desired ratio is available for TR1. What

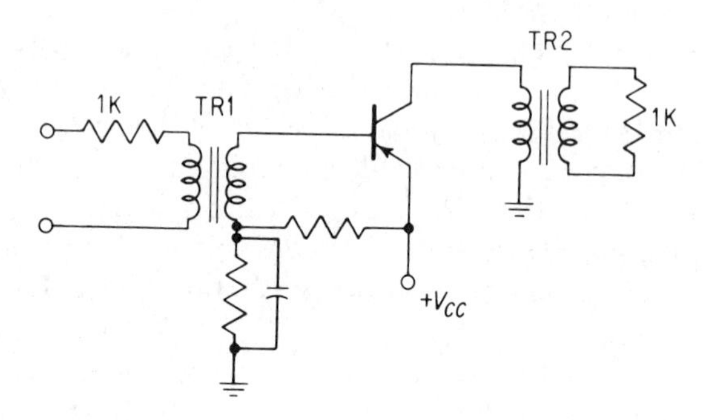

is the maximum input power that will allow the transistor to operate in the linear region? What is the transducer gain?

4 (a) What is the maximum power output that may be expected in a Class A amplifier using a 2N386 power transistor mounted on an infinite heat sink and operating at 10°C? (The maximum rating of the 2N386 is 12.5 watts.)
(b) By how much would the maximum power output of the amplifier be increased if two 2N386s were used in a Class B push-pull amplifier?

5. What is the maximum supply voltage that may be used in the Class A amplifier of part (a) of the accompanying figure? Part (b)? The Class B push-pull amplifier of part (c)?

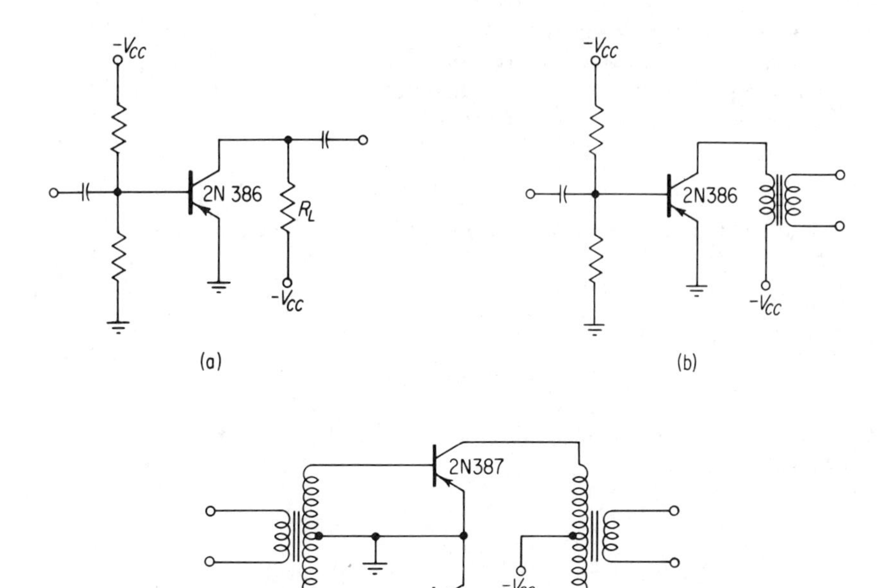

6. Design a Class B amplifier to meet the following requirements:

Power output = ½ watt
Supply voltage = 15 volts
Driver stage impedance = 5000 Ω
Speaker impedance = 3.2 Ω
Operating temperature = 25°C ambient
Transformer efficiency = 80%

(Pick one of the following transistor types: 2N207, 2N1790, 2N225, 2N227, 2N386.)

7. Compare the current gain of a particular transistor operating common emitter and common base into a low resistance load.

8. Write the transistor specifications for a transformer-coupled Class A amplifier to operate under the following conditions (neglect drop in bias and stabilization resistors):

Maximum power output = 3 watts (within 30%, 2.4–3.6 watts)
Operating temperature = 50°C ambient
Power available from
driver transistor = 2 mw
Supply voltage = 12 volts
Transformer efficiencies = 80% each

Assume T_{max} of transistor is 100°C.
Assume $R_{in} = 20\ \Omega$ and input is matched.
Hint: This problem requires information contained in previous chapters. The common-emitter current gain of a transistor is

$$A_i = \frac{h_{fe}}{1 + R_L h_{oe}}$$

In large-signal amplifiers, R_L is usually small and the expression is approximately:

$$A_i \approx h_{fe}$$

The power gain, then, when the input is matched, is

$$P_G = \frac{h_{fe}^2 R_L}{R_g}$$

Give the following specifications:
Maximum ratings

BV_{CES}
P_D at 25°C
Thermal resistance
h_{fe}
$V_{CE} =$ _____, $I_C =$ _____ (min and max)

REFERENCES

1. H. E. Tompkins, *Transistors as Circuit Elements.* Moore School of Electrical Engineering, Phila., Pa., 1956.
2. R. F. Shea, *Transistor Circuit Engineering.* New York: John Wiley & Sons, Inc., 1957.
3. M. P. Ristenbatt and R. L. Riddle, *Transistor Physics and Circuits, 2nd Ed.* Englewood Cliffs, N.J.: Prentice-Hall, Inc., 1966.
4. "Transistor Class B Push-Pull Stages," *Mullard Tech. Communic.*, 3, No. **24** (May, 1957).
5. D. DeWitt and A. L. Rossoff, *Transistor Electronics.* New York: McGraw-Hill Book Company, 1957.
6. R. A. Greiner, *Semiconductor Devices and Applications.* New York: McGraw-Hill Book Company, 1961.
7. M. V. Joyce and K. K. Clarke, *Transistor Circuit Analysis.* Reading, Mass.: Addison-Wesley Publishing Company, Inc., 1962.
8. J. Millman, *Vacuum Tube and Semiconductor Electronics.* New York: McGraw-Hill Book Company, 1958.
9. J. F. Pierce, *Transistor Circuit Theory and Design.* New York: Charles E. Merrill Books, Inc., 1963.

10. P. E. GRAY and C. L. SEARLE, *Electronic Principles—Physics, Models and Circuits*. New York: John Wiley & Sons, Inc., 1969.

11. E. J. ANGELO, JR., *Electronics: BJT's, FET's and Microcircuits*. New York: McGraw-Hill Book Company, 1969.

12. L. G. COWLES, *Transistor Circuits and Applications*. Englewood Cliffs, N.J.: Prentice-Hall, Inc., 1968.

13. C. BELOVE, H. SCHACHTER and D. L. SCHILLING, *Digital and Analog Systems, Circuits and Devices: An Introduction*. New York: McGraw-Hill Book Company, 1973.

14. *Integrated Electronic Systems*. Westinghouse Defense and Space Center. Englewood Cliffs, N.J.: Prentice-Hall, Inc., 1970.

5

Tuned Amplifiers

INTRODUCTION

In Chapter 4 we investigated untuned low-frequency amplifiers. In this chapter we shall look at tuned amplifiers. Although the basic design of tuned circuits is the same for both high- and low-frequency operation, the over-all stage design is different: hence, we divide the discussion of tuned amplifiers into two sections: high-frequency and low-frequency.

The terms *high frequency* and *low frequency* are relative and refer in a very general sense to whether the transistor is operating at a frequency where it is reasonably stable under matched or nearly matched conditions.

First, we look at the theoretical relationships which determine power gain, stability, and neutralization; then, we examine the practical considerations involved in using these devices in high- and low-frequency tuned amplifiers.

Theoretical Power Gain—Frequency Relations

Let us analyze the theoretical relationships between power gain and frequency in tuned transistor amplifiers. By using equivalent circuits, we may deduce certain figures of merit which describe the performance of the transistor in terms of well-known device parameters. Since, in most cases, the common-emitter connection is used for tuned-amplifier applications, we confine most of our analysis to this particular circuit connection.

The power gain may be calculated directly by use of the hybrid-pi equivalent circuit or it may be calculated by use of a "black box" representation of the transistor as outlined in Chap. 2. We shall actually employ a combination of the two approaches by calculating the power gain in terms of the y

parameters and then determining these y parameters from the equivalent circuit. Of course the y parameters could be measured and the gain calculated from these measurements. This also provides a means of checking the accuracy of the equivalent circuit.

The general circuit to be analyzed is shown in Fig. 5-1(a). This circuit is represented as a black box in Fig. 5-1(b). The problem is to calculate the transducer power gain, P_G, as a function of frequency in terms of the known parmeters y_{11}, y_{12}, y_{21}, and y_{22}. Unfortunately, we cannot determine the power gain at any arbitrary frequency and think the problem solved, since the power gain is infinite over a certain frequency range, provided that the source and load admittance have the proper values. In other words, over this critical frequency range (which will be discussed in the low-frequency section of this chapter) the transistor amplifier will be unstable for certain input and output terminations. The power gain for arbitrary source and load admittances cannot be calculated simply, because of this inherent instability which is caused by $r_{b'}$ and the feedback components (r_c and C_{TC}) in the transistor.

For an amplifier in which no attempt is made to cancel out this inherent feedback (such as the amplifier of Fig. 5-1), the matched transducer power gain may be computed for those frequencies for which the transistor is unconditionally stable. By *unconditionally stable*, we mean that the circuit will not oscillate for any real source or load admittance. Under these conditions the matched power gain follows the relationship shown in Fig. 5-2.

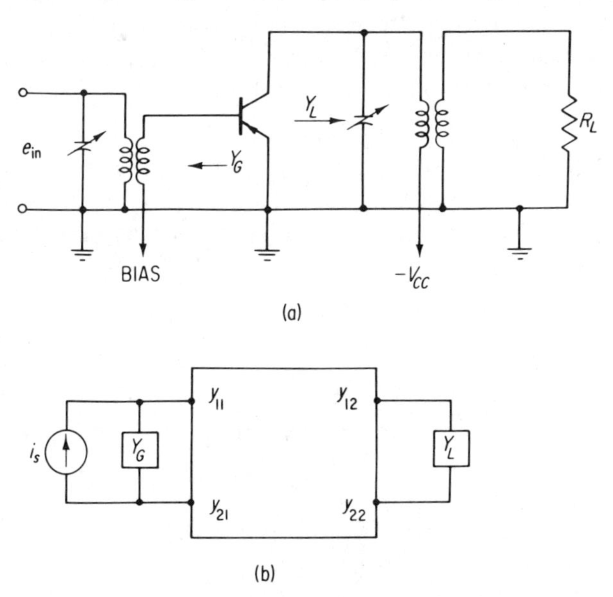

Fig. 5-1. (a) Typical single-tuned amplifier circuit and (b) black-box representation. Lossless transformers are assumed.

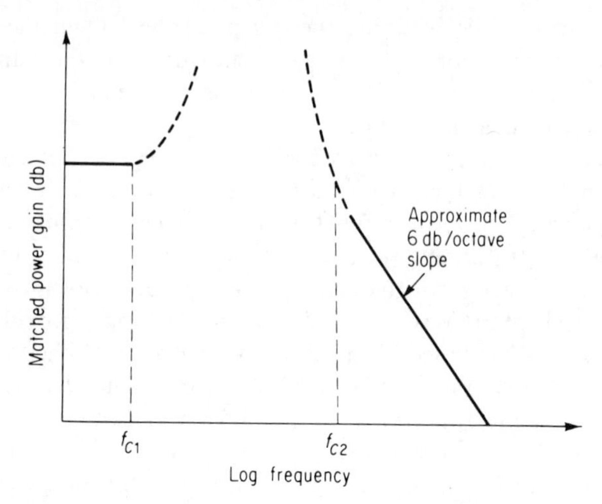

Fig. 5-2. Matched power gain vs frequency for a transistor amplifier with internal feedback. The common-emitter connection was used.

Notice that the gain goes to infinity in the frequency range between f_{C1} and f_{C2}. At low frequencies (below f_{C1}) the gain is relatively independent of frequency. At high frequencies (somewhat above f_{C2}) the gain decreases at approximately 6 db/octave.

The transducer power gain will be determined first for the high-frequency region, assuming perfect transformers. (Lossy transformers will be discussed later in this chapter.) At these high frequencies the transistor is unconditionally stable and transducer power gain is (1)*:

$$P_G = \frac{4Re(Y_G)Re(Y_L)|y_{21}|^2}{[(Y_G + y_{11})(Y_L + y_{22}) - y_{12}y_{21}]^2} \tag{5-1}$$

The common-emitter y parameters for the hybrid-pi circuit have been calculated (2) and are reproduced in expressions 5-2–5-5. The following assumptions were made for these calculations:

$$X_{C_{TC}} \ll r_c$$

$r_{c'}$ and $r_{c'e}$ are negligible.

$$y_{11} = \frac{1 + j\beta_0 \dfrac{f}{f_T}}{r_{b'e} + r_{b'}\left(1 + j\beta_0 \dfrac{f}{f_T}\right)} \tag{5-2}$$

$$y_{12} = \frac{-j\omega C_{TC} r_{b'e}}{r_{b'e} + r_{b'}\left(1 + j\beta_0 \dfrac{f}{f_T}\right)} \tag{5-3}$$

*Numbers in parentheses refer to material listed at the end of the chapter.

$$y_{21} = \frac{r_{b'e}(g_m - j\omega C_{TC})}{r_{b'e} + r_{b'}\left(1 + j\beta_0 \frac{f}{f_T}\right)} \tag{5-4}$$

$$y_{22} = j\omega C_{TC}\left[1 + \frac{g_m r_{b'e} r_{b'}}{r_{b'e} + r_{b'}\left(1 + j\beta_0 \frac{f}{f_T}\right)}\right] \tag{5-5}$$

The maximum transducer power gain is obtained with conjugate matching at input and output. Using this condition and the y parameters of Eqs. 5-2–5-5, calculation of the maximum unneutralized transducer power gain yields expression 5-6, where the following assumptions were made for this calculation:

$$\beta_0 r_e \approx r_{b'e}$$

$$r_{b'e} \gg r_{b'}$$

$$f \ll \frac{1}{2\pi r_e C_{TC}}$$

$$\left(\frac{r_e}{r_{b'}}\right)^3 \ll \frac{C_{b'e}}{C_{TC}}$$

Thus, we obtain

$$P_G = \left(\frac{f_{\max}}{f}\right)^2 \tag{5-6}$$

where

$$f_{\max} \approx \sqrt{\frac{f_T}{8\pi r_{b'} C_{TC}}} \tag{5-7}$$

This equation for power gain shows that the gain decreases at a rate of 6 db/octave. If we know the matched unneutralized gain at any high frequency on the 6 db/octave slope, we can calculate it at any other frequency. The frequency $f_{\max}$ is called the *maximum frequency of oscillation*, since it is the frequency at which the transistor exhibits unity power gain and, therefore, is the frequency above which the transistor will not oscillate. The gain equation (5-6) is valid down to a critical frequency (2) $f_{C2} = r_e f_T / 2r_{b'}$. Below this frequency, the amplifier is only conditionally stable and may oscillate for some source and load terminations. Design in this critical frequency region will be discussed later.

Unilateralization and Neutralization

In practice, the general tuned amplifier-circuit of Fig. 5-1 is often used in conjunction with a feedback network. This network serves the purpose of canceling the feedback inherent in the transistor to a point where feedback between output and input is negligible. This not only makes the circuit stable and free of oscillation but also decreases tuning interaction between input and output. If one can choose an appropriate feedback network which precisely cancels the internal trasistor feedback, we say the amplifier is *unilateralized.* In practical applications the feedback is only *partially* canceled and we term this *neutralization*.

We shall now examine methods of reducing the over-all feedback in the tuned-amplifier circuit of Fig. 5-1. A feedback network is placed around the amplifier as shown in Fig. 5-3(a). Shown in Fig. 5-3(b) is a four-terminal black-box representation of the "new" amplifier consisting of the original amplifier and the feedback network.

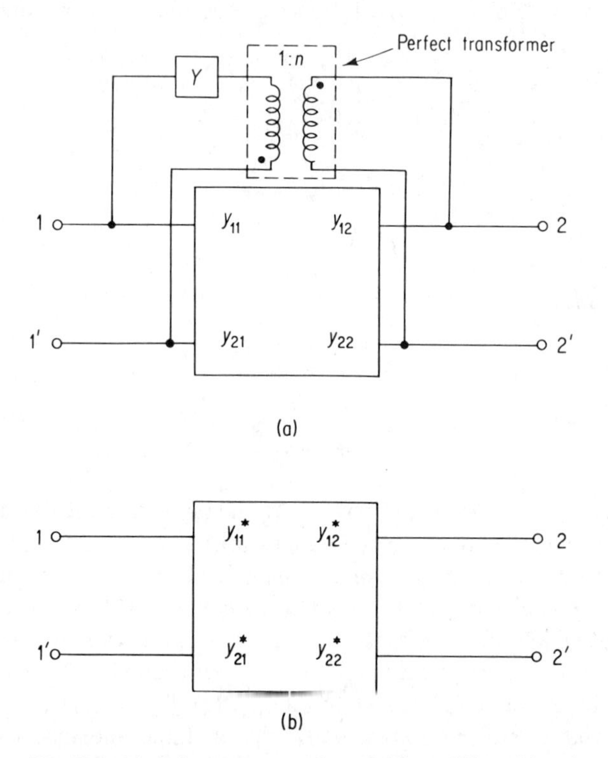

Fig. 5-3. Black-box representation of (a) amplifier with unilateralization network and (b) resulting four-terminal network with new y parameters.

The y parameters of the new amplifier [Fig. 5-3(b)] are

$$y_{11}^* = y_{11} + Y \tag{5-8}$$

$$y_{12}^* = y_{12} + \frac{Y}{n} \tag{5-9}$$

$$y_{21}^* = y_{21} + \frac{Y}{n} \tag{5-10}$$

$$y_{22}^* = y_{22} + \frac{Y}{n^2} \tag{5-11}$$

The condition that the over-all amplifier have no feedback (that is, be unilateral) is that

$$y_{12}^* = 0 = y_{12} + \frac{Y}{n} \tag{5-12}$$

or, therefore, that

$$Y = -ny_{12} \tag{5-13}$$

In this case the maximum unilateralized transducer power gain is

$$P_{G_{MA}} = \frac{|y_{21}^*|^2}{4Re(y_{11}^*)Re(y_{22}^*)} \tag{5-14}$$

Notice that, if we have lossless networks, this represents the maximum gain available from the transistor, and also is similar to the condition described in Chap. 4 for maximum available gain A_{MA}. It is common practice to refer to the low-frequency value of this gain as MAG, although the term is proper and sometimes used at other specified frequencies. Using the y parameters given in Eqs. 5-8–5-11 and 5-2–5-5, the gain in terms of the hybrid circuit parameters is given by

$$P_{G_{MA}} = \left(\frac{f_{\max}}{f}\right)^2 \frac{1 + (r_{b'}f/r_e f_T)^2}{[(r_{b'}/r_e) + (1/n_{\text{opt}})](f/f_T)^2} \tag{5-15}$$

This equation was derived assuming

$$\beta_0 r_e \approx r_{b'e}$$

$$r_{b'} \ll r_{b'e}$$

$$r_e \ll \beta_0 r_{b'} \left(\frac{f}{f_T}\right)^2$$

Also, n_{opt} is the optimum turns ratio (for maximum gain) and is given by

$$n_{\text{opt}} = \frac{1}{\sqrt{2\pi f_T C_{TC} r_{b'}}} \tag{5-16}$$

For typical high-frequency transistors the optimum turns ratio is in the range $1 \leq n_{\text{opt}} \leq 5$. In practice, the existence of an optimum turns ratio is not important because the sensitivity of the gain to variations in the turns ratio is small.

This last equation for unilateralized, matched transducer power gain shows that the gain does not necessarily vary at a 6 db/octave rate unless

$$\frac{r_{b'}f}{r_e f_T} \gg 1 \tag{5-17}$$

and

$$\frac{r_{b'}}{r_e} \gg \frac{1}{n_{\text{opt}}} \tag{5-18}$$

Then, the maximum unilateral gain is identical to the maximum unneutralized gain. We should also note that the unilateral amplifier is unconditionally stable at all frequencies. Since it is impossible in practice to realize exactly the admittance, y_{12}, and a perfect transformer, no truly unilateral circuit can be constructed. They can be approached by careful circuit design.

In practical designs, unilateralization is not required. In most cases, it is possible to obtain very good cancellation of internal feedback at high frequencies by use of capacitive neutralization. An analysis of the typical values of r_c and C_{TC} indicates that $X_{C_{TC}} \ll r_c$ at high frequencies. Thus the feedback is primarily capacitive and may be effectively neutralized with a single capacitor. This form of neutralization is widely used in practice. In this case the feedback admittance, Y, in the circuit of Fig. 5-3 is a capacitance.

The matched unilateral transducer power gain is essentially independent of frequency at low and medium frequencies. Since it is not practical to unilateralize the transistor amplifier, we must be content with operating with some form of neutralization and/or mismatch the load and source admittances in order to provide stability. At low and medium frequencies, neutralization networks that provide a high degree of stability are hard to implement. Hence, mismatching at input and output is utilized as a means for making the unneutralized amplifier stable. These techniques are discussed in detail later in the chapter.

With the foregoing theory as background, let us now look at commonly used tuned-amplifier design techniques and the associated problems. We shall concern ourselves with the high-frequency region, well above f_{C2}. In this region the common-emitter connection is unconditionally stable, without neutralization.

HIGH-FREQUENCY TUNED AMPLIFIERS

Approximate Method for Predicting Tuned-Amplifier Gain

From our previous theory, it is apparent that the variation of unilateralized power gain with frequency follows the general shape of the curve shown in Fig. 5-4. Note, too, that without unilateralization, the transistor is inherently unstable throughout much of the flat region, whenever the load and source resistances approach the matched conditions. In modern high-frequency diffused-base transistors, the MAG in the flat region is in the order of 50–60 db or higher. To realize this requires extremely complex unilateralization networks and great care in adjustment. In general, the MAG in the flat region is much higher than can be utilized in practice.

In order to provide a more useful approach to the design problem, it is usually accurate to consider the maximum useful gain from a high-frequency

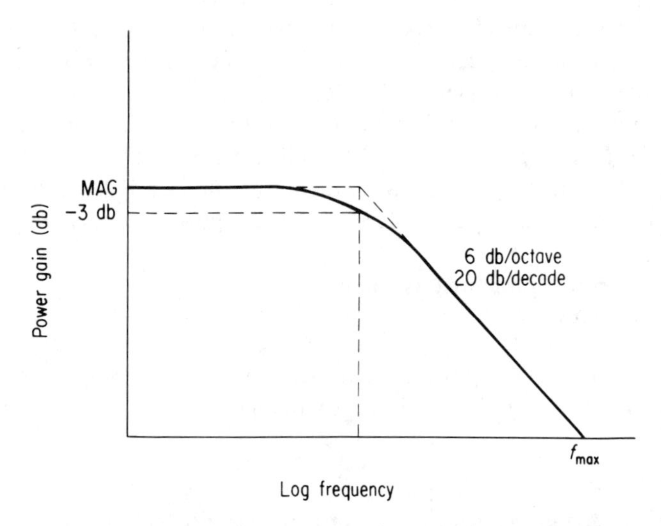

Fig. 5-4. Unilateralized power gain vs frequency for most transistors.

transistor as about 40 db. This gain may be realized in practice by mismatching on the input and output of the transistor. If we assume 40 db as the maximum useful gain, then Fig. 5-4 may be redrawn as shown in Fig. 5-5. Thus, once f_{max} is known, it is possible to estimate the power gain at any frequency, as shown in Fig. 5-5. Note that 6 db/octave is equal to 20 db/decade; therefore, if 40 db is assumed as the useful gain, the knee of the curve occurs at $f_{max}/100$. This provides an easy way of estimating the location of

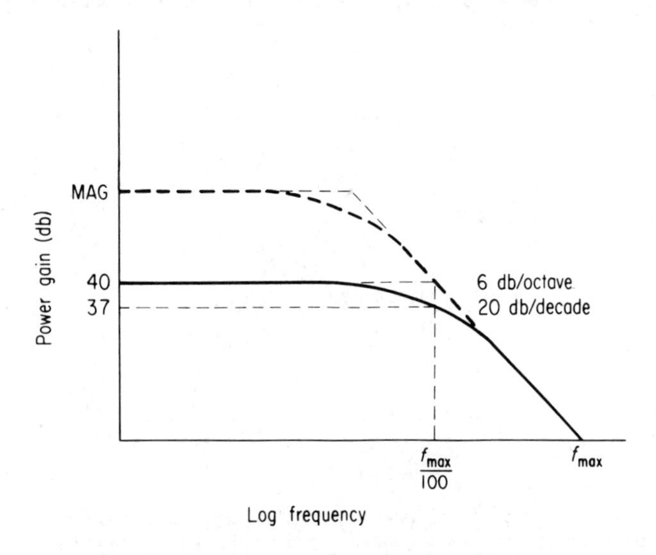

Fig. 5-5. Useful transistor power gain vs frequency.

the knee of power gain versus frequency curve. For example, if f_{max} is 1 GHz (1000 MHz), we would expect 6 db of gain at 500 MHz, 12 db at 250 MHz, etc., and the gain to be essentially independent of frequency below 10 MHz.

Let us now consider the factors affecting the design of tuned amplifiers in the 6 db/octave region. This analysis will usually be useful at frequencies above $f_{max}/50$. In this frequency range the transformers are generally designed to match the source and load to the input and output impedance, respectively.

Common-Emitter, Common-Base, and Common-Collector

In the 6 db/octave region the unilateralized power gain is the same in all three connections and may be approximated from

$$P_G \approx \left(\frac{f_{max}}{f}\right)^2 \tag{5-19}$$

However, the unneutralized power gain is not the same. Figure 5-6 illustrates how the common-base and common-emitter unneutralized gain vary in the high-frequency region as compared to the unilateralized gain. Notice that the unneutralized common-emitter connection has the least gain. This may be understood by remembering that in the common-base connection the intrinsic feedback is from collector to emitter (that is, in-phase or regenerative feedback). The converse is true in the common-emitter connection. Therefore, in the high-frequency region the common-base unneutralized amplifier is regenerative and the unneutralized common-emitter amplifier is degenerative. This extra common-base gain is not obtained with-

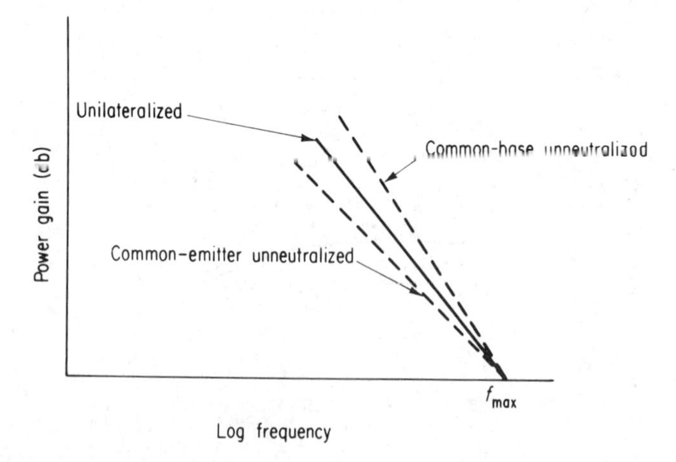

Fig. 5-6. Power gain vs frequency showing eomparison of common-base and common-emitter unneutralized amplifiers with the unilateralized case.

out some sacrifices. Experimental results show that the unneutralized common-base connection will have greater variation in gain and bandwidth from transistor to transistor than the unneutralized common-emitter circuit, and both will be worse than the neutralized connection.

The unneutralized common-base connection is sometimes used at very high frequencies for extra gain in inexpensive applications, to provide some extra gain and save the neutralization expense, and also in very wide band applications. In most other applications, however, the neutralized common-emitter connection is recommended: Neutralization provides much better interchangeability and much less tuning interaction; hence we direct our design approach toward the neutralized common-emitter connection.

The common-collector circuit is usually not used in tuned-amplifier applications and will not be considered here. It is usually used with resistive loads in impedance-matching applications.

TRANSISTOR IMPEDANCES

In Chap. 2 expressions were developed for gain and impedances in terms of h, y, or z parameters. Although the performance of a transistor may be predicted, if the h, y, or z parameters are known at the operating frequencies, it becomes more convenient to use the equivalent circuit approach at high frequencies. The simplest and most useful equivalent circuit for most designs is the hybrid-pi equivalent circuit shown in Chap. 2. This equivalent circuit has the desirable characteristic that none of the elements is frequency dependent. Thus, once all the elements of the hybrid-pi circuit are known, it is possible to calculate the transistor performance at any frequency.

The concept of f_{max} is particularly well suited for tuned-amplifier designs. Therefore, in high-frequency, tuned-amplifier design, it would seem desirable to specify impedances in a manner most useful to the circuit designer who is using the f_{max} concept, and yet still be consistent with the equivalent circuit.

In tuned amplifiers, the tuned circuit normally serves two purposes. First, it provides the desired bandpass; second, it provides impedance matching. In the high-frequency region (on the 6 db/octave slope) it is common practice to match the transistor for maximum power gain. It then becomes necessary to know what impedances to match. A convenient way of expressing the input and output impedances is by use of the short-circuited parallel-equivalent input and output resistance and reactance. The short-circuit parameters represent the parallel combination of R and C (or L) seen looking either into the input terminal with the output shorted (y_{11}), or into the output terminals with the input shorted (y_{22}). When displayed as a function of frequency and operating point, they provide a very close approximation to the impedances under neutralized conditions. Most designs use neutralization at high frequency, and even for the unneutralized case, they provide a good starting point for the design. Curves of the short-circuit input and output impedance of a typical high-frequency transistor are shown in Fig. 5-7.

Note that the parallel equivalent R and C shown in Fig. 5-7 are the impedances which represent y_{11} and y_{22}, except that they are expressed in R

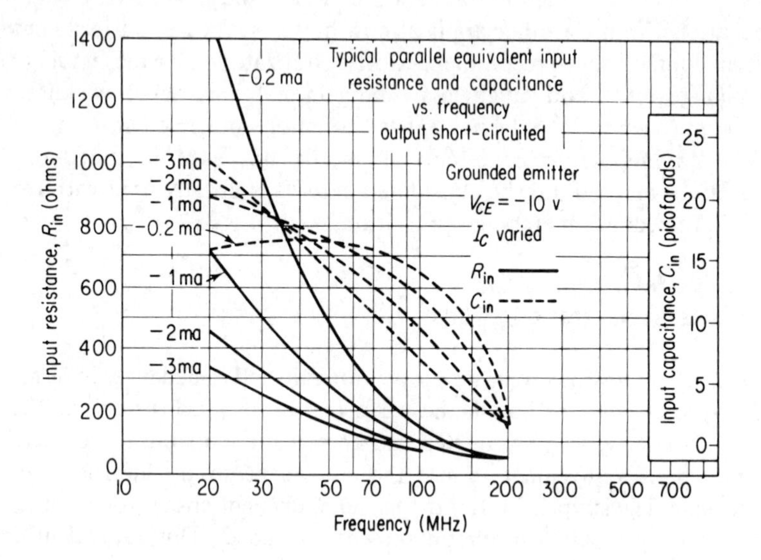

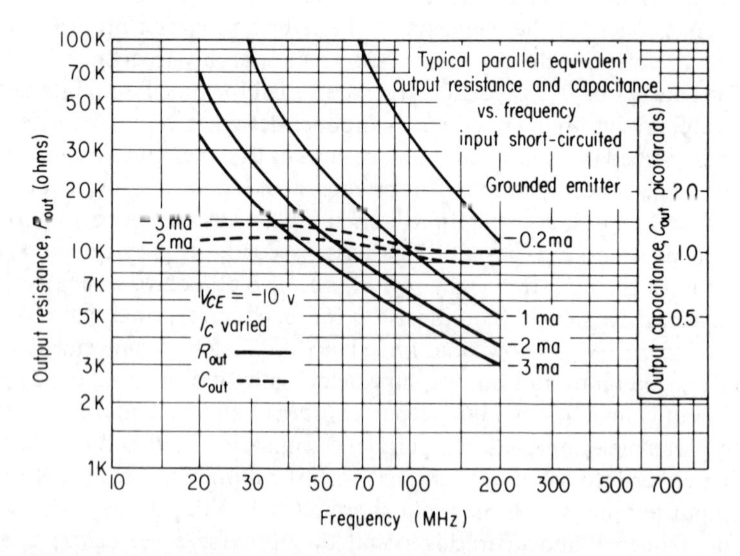

Fig. 5-7. Typical parallel equivalent input and output resistance and capacitance as a function of frequency for the 2N2742.

and C (resistance and capacitance) instead of G and B (conductance and elastance). Manufacturers may present these data in either form. Many manufacturers also include y_{12} and y_{21} curves to enable a complete design even without neutralization. The equivalent-parallel representation is useful because the capacitances are shown, such that they may easily be included as part of the tuning capacitance. Similarly, the resistances are the values which must be matched to provide maximum gain. With this information, plus f_{max}, which predicts the power gain under matched conditions, sufficient transistor information is available to complete the amplifier design with the exception of the neutralization network. We now need to decide upon the desired characteristics of the interstage network.

TUNED-TRANSISTOR AMPLIFIER INTERSTAGE DESIGN

In designing tuned-transistor interstage networks at high frequencies there are three major considerations. They are impedance matching, bandwidth, and neutralization.

Impedance Matching

Unlike most tube amplifier applications, transistors are used as power amplifiers rather than voltage amplifiers. Therefore it is imperative that impedance-matching devices be used to assure maximum power gain because of the wide difference between the input and output impedances of a transistor.

The most common form of impedance-matching device is the transformer. Figure 5-8 illustrates the usual forms of impedance-matching transformers. R_R is the resistance seen looking into the ideal transformer when it is tuned for resonance.

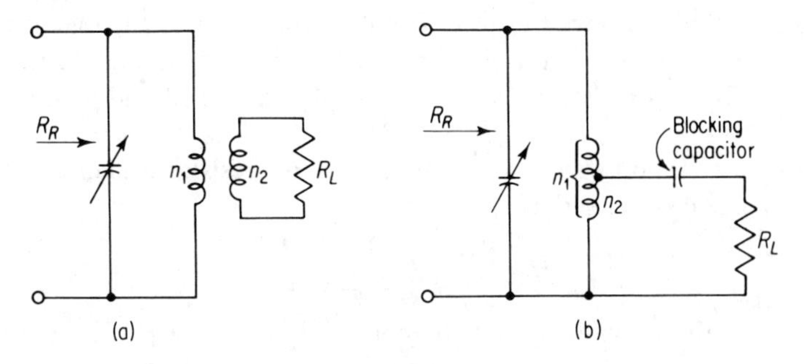

Fig. 5-8. Common impedance-matching transformer connections.

For a lossless transformer with unity coupling ($k = 1$), the transformer matches impedance according to the relationship:

$$R_R = \left(\frac{n_1}{n_2}\right)^2 R_L \tag{5-20}$$

That is, the resistance reflected to the primary of a perfect transformer, whose secondary is terminated in R_L, is equal to R_L times the square of the ratio of primary to secondary turns. The value of the reflected resistance in an actual transformer is also dependent on the coefficient of coupling (k) of the transformer. In many transformers used for high-frequency applications k is sufficiently close to unity so that the foregoing relationship is reasonably accurate for calculating the reflected resistance.

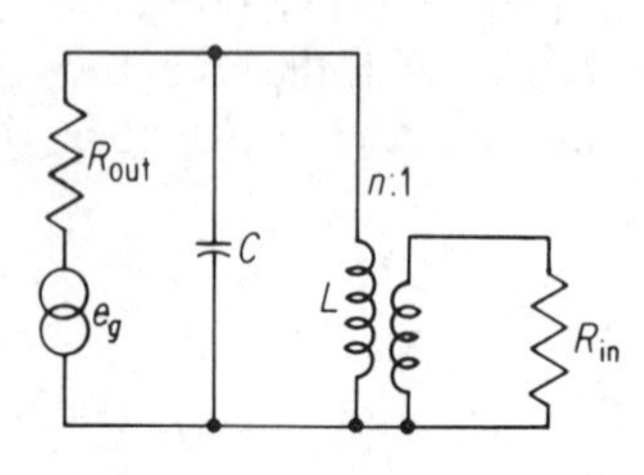

Fig. 5-9. Basic interstage network for tuned transistor amplifier, where R_{out} is the output resistance of the transistor, and R_{in} is the input resistance of the next stage.

In designing interstage networks at high frequencies using ideal transformers, the calculations are straightforward. Consider the network shown in Fig. 5-9 to have lossless reactances and unity coupling. In order to obtain maximum power transfer, n must be chosen to reflect a resistance equal to R_{out} so that

$$R_{out} = n^2 R_{in} \tag{5-21}$$

and

$$n = \sqrt{\frac{R_{out}}{R_{in}}} \tag{5-22}$$

One cannot always consider the ideal case, since transformers usually have a loss resistance associated with them. This is represented by R_P in Fig. 5-10. A simple calculation at resonance gives:

$$R_R = \frac{R_{out} R_P}{R_{out} + R_P} \tag{5-23}$$

Thus, for maximum power transfer from the transistor to the load, the load should be made equal to the parallel combination of the output resistance (R_{out}) and the loss resistance R_P, if R_P and R_{out} are fixed and assumed constant.

Another common way of representing this loss resistance is in terms of Q. The Q of a coil is defined as

$$Q_u = \frac{R_P}{X_L} = \frac{R_P}{2\pi f L} = \frac{R_P}{\omega L} \tag{5-24}$$

where Q_u is the unloaded Q of the coil, and R_P is the parallel-shunt loss resistance.

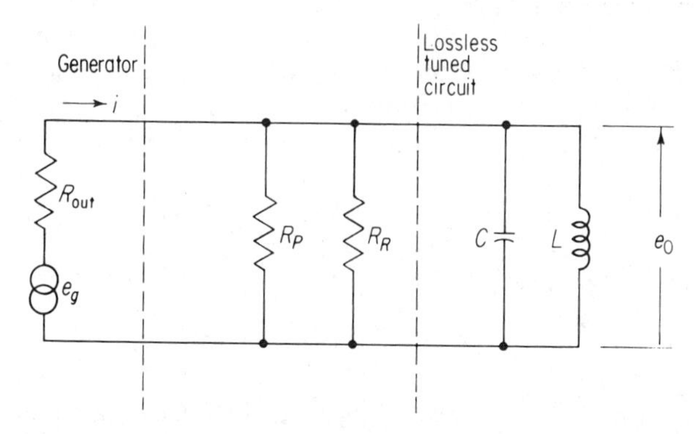

Fig. 5-10. Representation of Lossy tuned circuit where R_P is the parallel loss resistance of the transformer and R_R is the reflected load resistance = n^2R_{in}.

When the coil is tuned to parallel resonance with a lossless capacitor then Eq. 5-24 may be written as

$$Q_u = \frac{R_P}{X_L} = \frac{R_P}{X_C} = 2\pi fCR_P \tag{5-25}$$

If the capacitor is not lossless, R_P must be the total parallel-loss resistance of both the coil and capacitor. In most applications, however, the capacitor losses are negligible. At resonance, the reactances are equal or

$$2\pi f_0 L = \frac{1}{2\pi f_0 C} \tag{5-26}$$

Therefore,

$$f_0 = \frac{1}{2\pi\sqrt{LC}} \tag{5-27}$$

or

$$\omega_0 = \frac{1}{\sqrt{LC}} \tag{5-28}$$

where ω_0 is the resonant frequency.

Combining Eqs. 5-28 and 5-24 when $\omega = \omega_0$ gives

$$Q_u = R_P\sqrt{\frac{C}{L}} \tag{5-29}$$

It can be seen that for a given R_P, which is usually fixed by physical considerations, such as core material, wire size, and spacing, the only way to vary Q_u at a given frequency is by changing the C/L ratio. Hence, the C/L ratio is such an important consideration that many designers simply refer to the C/L, or more commonly the L/C ratio, in transformer designs.

Bandwidth and Insertion Loss

The *bandwidth* of a single-tuned circuit is defined as the frequency separation between the -3 db power points of the gain versus frequency curve. The bandwidth of a single-tuned circuit may be expressed as

$$BW = \frac{f_0}{Q_l} \tag{5-30}$$

where Q_l, the loaded Q is given by

$$Q_l = \frac{R_T}{\omega L} \tag{5-31}$$

Note that for a given bandwidth, Q_l is fixed. Furthermore, in many applications both Q_l and Q_u are fixed because of design considerations. The insertion loss is defined to be the matched interstage loss of the transformer and can conveniently be evaluated after the bandwidth has been determined. If we assume that $Q_u \gg Q_l$, which is rather generally true, then we can show that the value of R_R which minimizes this is R_{out}. Under these circumstances we find that the insertion loss, A_T, is given by:

$$A_T = \left(1 - \frac{Q_l}{Q_u}\right)^2 \tag{5-32}$$

If R_R is mismatched from this value we get additional losses as shown in Fig. 5-11. This must be added to the losses shown in Fig. 5-12, which plots the equation above.

If the maximum gain of the transistor is given as MAG, then the power gain A_P is given by

$$A_P(\text{db}) = \text{MAG(db)} + A_T(\text{db}) \tag{5-33}$$

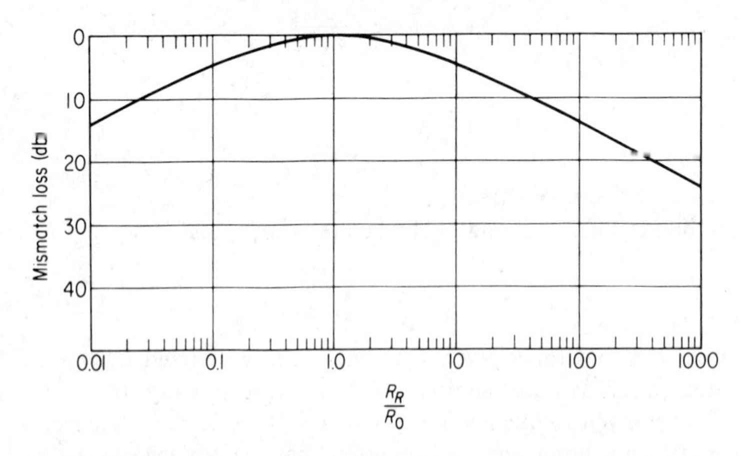

Fig. 5-11. Mismatch loss as a function of the mismatch between the transistor and reflected load or source.

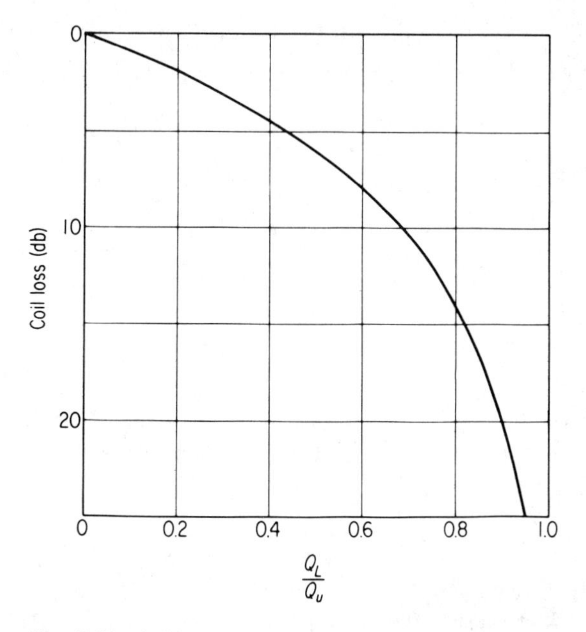

Fig. 5-12. Coil loss as a function of loaded to unloaded Q.

where all three gains are expressed in db. Knowing the transistor gain and the transformer insertion loss, it is possible to determine the stage gain from the input of one transistor to the input of the next. Note that A_T is always negative. Notice that a 2:1 mismatch results in only 0.5 db loss in gain but that Q_l/Q_u must be kept under 0.1 to keep the coil loss under 1 db.

Let us now investigate what happens to the bandwidth of the simple tuned circuit shown in Fig. 5-10. The 3 db bandwidth (BW) of the stage is given by

$$BW = \frac{f_0}{Q_l} \tag{5-34}$$

As noted earlier, the capacitor can be considered lossless in many applications. In this case, at resonance

$$Q_l = \frac{R_T}{\omega_0 L} = 2\pi f_0 R_T C \tag{5-35}$$

and

$$BW = \frac{1}{2\pi R_T C} \tag{5-36}$$

where R_T is the equivalent resistance of R_R, R_P and R_{out} in parallel. From Eq. 5-36 it can be seen that by changing the capacitor in the tuned circuit (the L/C ratio) it is possible to change the bandwidth over wide ranges, for a given R_T. Let us now look at the variation in R_T. If the circuits are narrow band, Q_u will be high and, therefore, R_P will be high (that is, $R_P \gg R_{out}$; $R_P \gg R_R$ for many applications). Under these conditions R_T becomes

$$R_T = \frac{R_{out} R_R}{R_{out} + R_R} \qquad \text{(neglecting } R_P\text{)} \tag{5-37}$$

Under matched conditions $R_{out} = R_R$ and then,

$$R_T = \frac{R_R}{2} = \frac{R_{out}}{2} \tag{5-38}$$

Thus, in many practical matched circuits, the bandwidth is given by

$$BW = \frac{1}{2\pi \frac{R_{out}}{2} C} = \frac{1}{\pi R_{out} C} \tag{5-39}$$

Then, for practical, *matched* transformer-coupled transistor amplifiers, where the output of one transistor drives the input of the next, the maximum bandwidth can be calculated as:

$$BW_{max} = \frac{1}{\pi R_{out}\left(C + \frac{C_i R_i}{R_{out}}\right)} \tag{5-40}$$

where C_i and R_i represent the input parameters of the following transistor.

This is often called the *natural bandwidth* of the transistor. Note that it is completely dependent on transistor parameters and is, therefore, a transistor parameter itself. The natural bandwidth of a transistor is useful because it gives the circuit designer knowledge of the maximum interstage bandwidth which he can achieve without adding substantially to the insertion loss of the transformer. Another way of expressing this parameter is that it is theoretically possible to realize the gain predicted by f_{max} at this bandwidth. The minimum bandwidth is limited by the practical value of Q_u which can be obtained. Figure 5-13 shows a sketch of center frequency gain versus bandwidth for matched transistor amplifiers. Stagger-tuning techniques, which effectively provide higher-gain bandwidth products in tube amplifiers, will not extend the gain-bandwidth product in transistor amplifiers. Although gain and bandwidth can be traded in a stagger-tuned transistor amplifier, the gain will always be less than that obtainable in a synchronously tuned stage.

Bandwidth Reduction

It may be shown that the over-all bandwidth of a number of identical stages is less than the single-stage bandwidth. This reduction has been derived in other textbooks (4) and is included herein for reference.

For single-tuned *identical stages* the bandwidth reduction factor is given by expression 5-41 and shown plotted in Fig. 5-14.

$$BW_n = BW\sqrt{2^{1/n} - 1} \tag{5-41}$$

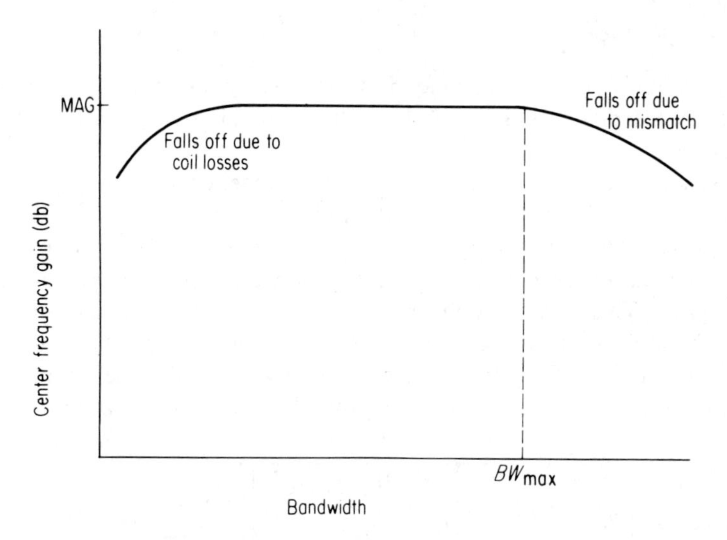

Fig. 5-13. Center frequency gain as a function of interstage bandwidth with transistor matched to load.

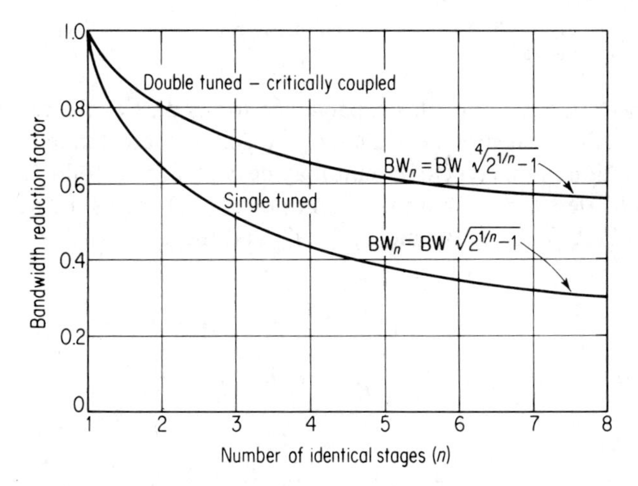

Fig. 5-14. Bandwidth reduction factor vs number of identical cascaded stages.

The double-tuned critically coupled reduction factor is given by expression 5-42 and also plotted in Fig. 5-14 for *identical stages.*

$$BW_n = BW\sqrt[4]{2^{1/n} - 1} \qquad (5\text{-}42)$$

where BW = bandwidth of one tuned circuit
n = number of identical cascaded tuned circuits.

Matching Networks

In addition to the matching networks which use transformers, other networks are often used. Several of these are discussed in the following paragraphs.

Instead of tapping the inductance, the capacitor may be tapped. This technique is used extensively in auto radios and some mobile communications applications. It has the advantage that the manufacturer, by simply changing a capacitor, can change the impedance match after the transformer is designed. This is usually faster and more economical than changing the tap or number of turns on the transformer. Such a capacitance divider is shown in Fig. 5-15. Sometimes the divider network is also the tuning capacitance for the circuit; sometimes, as shown here, it is only part of the tuning

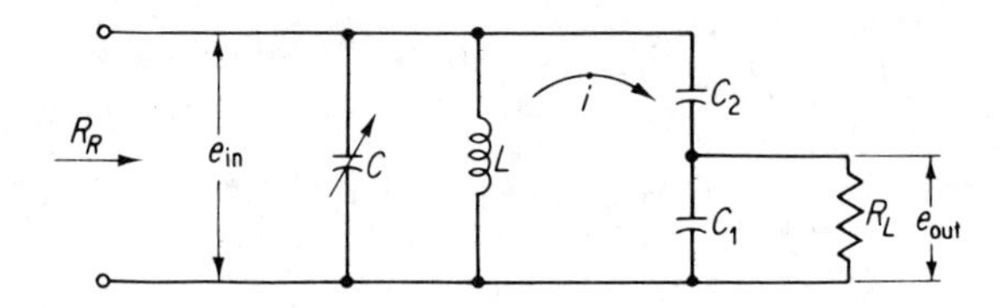

Fig. 5-15. Tuned circuit using capacitance divider matching network.

capacitance. In any event, this capacitance forms part of the capacitance in the tuned circuit; hence the values of capacitance are limited. Impedance transformation in this type of circuit may be explained as follows:

For a lossless tuned circuit the input power (P_{in}) equals the output power to the load (P_{out}). Also, assume $R_L \gg X_{C_1}$ and then,

$$P_{in} = \frac{e_{in}^2}{R_R} = P_{out} = \frac{e_{out}^2}{R_L} \tag{5-43}$$

and

$$R_R = \frac{e_{in}^2 R_L}{e_{out}^2} \tag{5-44}$$

But

$$e_{out} = iX_{C_1} \qquad (if\ R_L \gg X_{C_1}) \tag{5-45}$$

and

$$e_{in} = i(X_{C_1} + X_{C_2}), \qquad (if\ R_L \gg X_{C_1}) \tag{5-46}$$

Combining the preceding:

$$R_R = R_L\left(\frac{C_1 + C_2}{C_2}\right)^2 \tag{5-47}$$

Thus, if the assumption that $R_L \gg X_{C_1}$ is met, the capacitance divider will transform impedances as shown by Eq. 5-47. (*Note that this equation does not contain a frequency term.*)

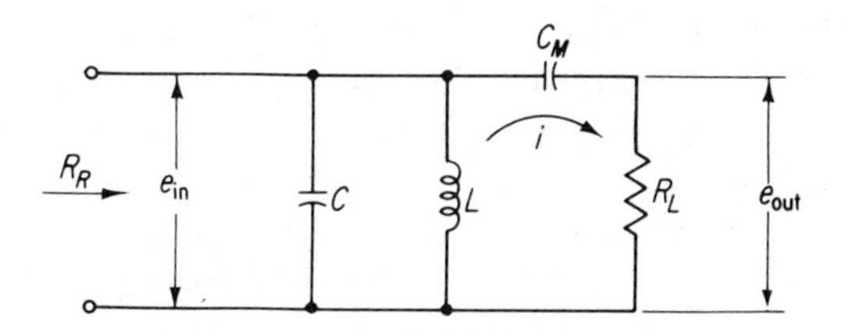

Fig. 5-16. Series matching capacitor as used in simple high-frequency circuit.

Figure 5-16 shows another form of impedance matching that is often used at very high frequencies. This circuit may be analyzed in a manner similar to that shown for Fig. 5-15. Consider again perfect reactances. Then,

$$P_{in} = \frac{e_{in}^2}{R_R} = P_{out} = \frac{e_{out}^2}{R_L} = i^2 R_L \tag{5-48}$$

Assume $X_{CM} \gg R_L$ (C_M is small) and then,

$$e_{in} \approx iX_{CM} \tag{5-49}$$

Therefore,

$$\frac{X_{CM}^2}{R_R} = R_L \tag{5-50}$$

and

$$R_R = \frac{X_{CM}^2}{R_L} = \frac{1}{\omega^2 C_M^2 R_L} \tag{5-51}$$

This method of matching is often used at high frequencies; in examining a high-frequency circuit, one should be careful to determine whether a series capacitor is simply a coupling or blocking capacitor, or whether it is also a matching capacitor. Usually, careful examination will indicate the capacitor's function, because if it is used for matching it is usually small (on the order of 1 to 20 pf). Of course, at lower frequencies it may be larger. The series-matching capacitor provides a simple means of matching a load to a tuned circuit. The range of impedance transformation is limited, however, and in actual circuits care is needed to make certain that complete matching occurs. Often, a circuit will appear to be matched when in fact the series capacitor has been adjusted through the point where it provides the closest impedance to proper match.

NEUTRALIZATION

As noted earlier, there are always internal and external feedback paths between the transistor elements. This feedback is represented in the hybrid-pi equivalent circuit as a parallel resistor (r_c) and capacitor (C_{TC}). In present-day, low-level, high-frequency transistors, the value of the capacitance is usually

0.5–5 pf and the resistor is of the order of 1–10 megohms or more. At frequencies in the 6 db/octave region, the reactance of the feedback capacitance is much smaller than the resistive component, and as such, contributes to the major portion of the feedback.

The feedback network in the transistor results in three undesirable effects: first, the feedback network is a path between the input and output tank circuits and will cause tuning interaction. Second, as a result of the tuned input and output stages, there is at some frequency and load condition sufficient phase shift from the input to output circuit and then back through the feedback network to cause a total phase shift of 360°. That is, the feedback signal is in phase with the input signal. If, under these conditions, the loop gain of the system from input to output and back to input is equal to, or greater than, unity, the amplifier will oscillate. Notice that this condition arises as a result of the reactive input and output circuits. It is particularly apparent at, and below, the knee of the power gain versus frequency curve where the transistor gain is very high. Third, and more important at the higher frequencies, the signal coupled from the collector to the base is out of phase with the input signal, resulting in degeneration and loss of gain.

Neutralization is used to compensate for the transistor feedback and, therefore, reduces these undesirable effects.

In practical common-emitter high-frequency amplifiers, neutralization is usually accomplished by feeding a signal out of phase with the collector signal back through another capacitor to the base, such that the signal through the transistor capacitance and the external neutralizing capacitance cancel each other. A number of methods for accomplishing this are shown in Fig. 5-17. In the circuits of (a) and (b) the transformer provides the phase inversion of the collector signal. Provided that the transformer has unity coupling, the value of C_N will be approximately

$$C_N \approx nC_{fb} \tag{5-52}$$

where C_{fb} = total feedback capacitance
n = transformer turns ratio.

In circuits (c) and (d), the small capacitor (C_P) provides the phase shift, and the value of C_N will be determined by the ratio of the impedance of the tank circuit to the impedance of the phase-shifting capacitor. Note that C_P is not a bypass capacitor and must be small enough to have some impedance at the operating frequency.

In transistor amplifiers, unlike tubes, an additional problem exists in neutralizing. The modified hybrid-pi circuit of Fig. 5-18 shows that the capacitive feedback from collector to base is actually made up of at least two parts. The normal intrinsic capacitance of the collector base junction (C_{TC}) forms a path between collector and intrinsic base. In addition to this capacitance, the case capacitance and other stray lead and socket capacitances form a direct path from collector to the external base connection. Note that

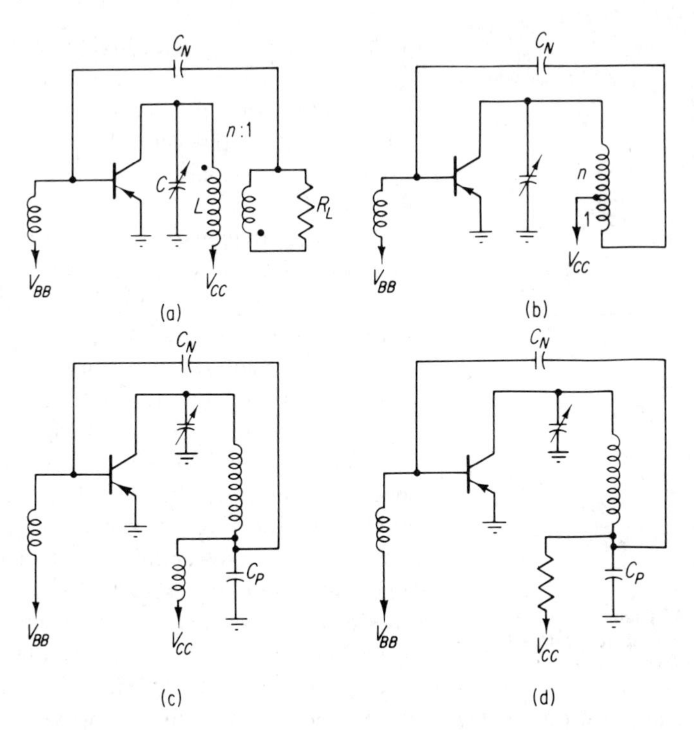

Fig. 5-17. Several typical configurations for neutralizing common-emitter amplifiers.

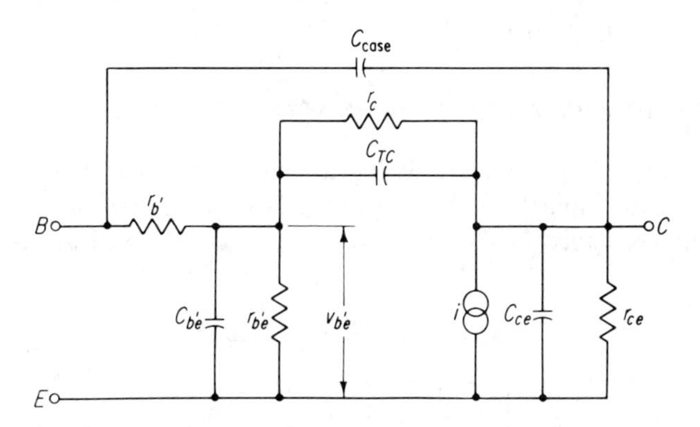

Fig. 5-18. Hybrid-π circuit showing both intrinsic feedback C_{TC} and extrinsic feedback C_{case}, where:

$$r_{b'e} = \beta_0 r_e = \beta_0 \frac{kT}{qI_E} \qquad i = -g_m v_{b'e} = \frac{qI_E}{kT} v_{b'e}$$

C_{TC} and C_{case} are separated at the input by $r_{b'}$. Herein lies the problem. It is impossible completely to neutralize the stage with a simple capacitor because of $r_{b'}$. However, in many high-frequency transistors $C_{TC} \ll C_{case}$ and $r_{b'}$ is small. Thus, at high frequencies simple capacitance neutralization is ade-

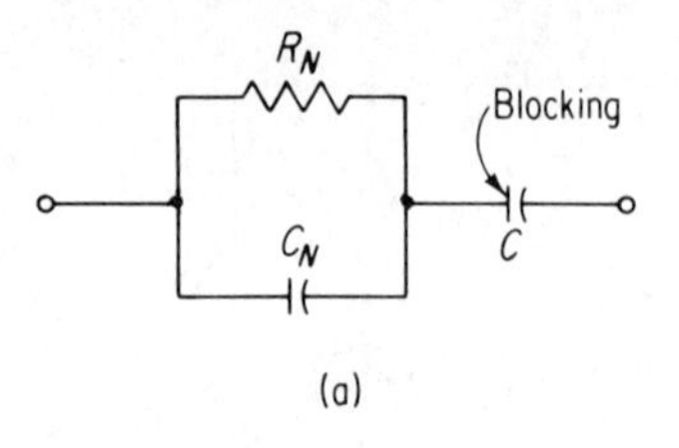

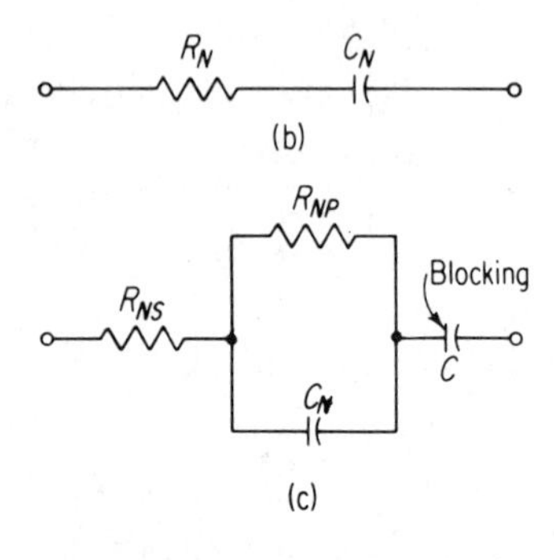

Fig. 5-19. Complex neutralization networks used to approach unilateralization.

quate. In transistors with high $r_{b'}$ and where $C_{TC} \gg C_{case}$, it is often difficult to neutralize the circuit adequately, and the designer must either accept partial neutralization or resort to the more complex neutralization circuits discussed later.

At lower frequencies, and in more exacting applications, it is often desirable to duplicate more closely the feedback characteristics of the transistor (including $r_{b'}$) in designing the external neutralization network. Under these conditions, the feedback circuit may be similar to those shown in Fig. 5-19. R_N and C_N represent the neutralizing components. R_{NS} and R_{NP} refer to the series and parallel neutralizing elements. The blocking capacitors (C) may or may not be included, depending upon the circuit. These types of circuits can be designed to approach the unilateralized case closely, but they are seldom used because of the complexity of adjustment. In most applications which require more than a simple capacitance neutralization, it is better to obtain stability by mismatch rather than by complex neutralization networks. Stabilization by mismatching will be discussed subsequently in the low-frequency portion of this chapter.

CHECKING AND ADJUSTING NEUTRALIZATION

The usual procedure for neutralizing a single stage is to connect a signal generator to the input (as shown in Fig. 5-20) through a resistor simulating the output resistance of the previous transistor. An oscilloscope or other

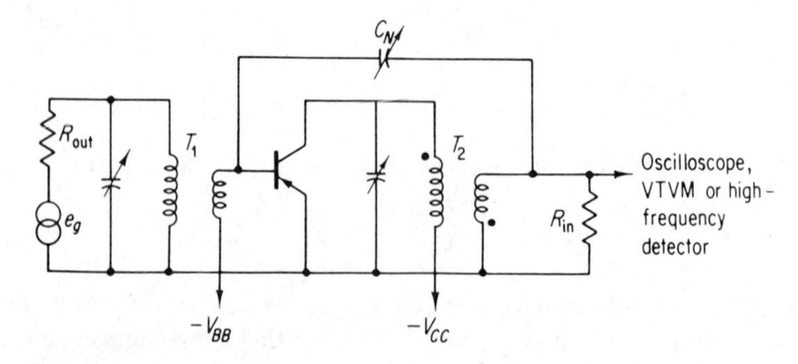

Fig. 5-20. Location of signal generator and indicating device during initial tune-up prior to adjusting C_N (ac circuit only).

indicating device is placed at the output across a resistance simulating the input resistance of the next stage One then adjusts the signal input for a convenient output which does not overdrive the transistor, and tunes transformers T_1 and T_2 for maximum output. The generator is then connected to the output terminals through the resistance simulating the input resistance of the following stage. The indicating device is connected to the base of the transistor as shown in Fig. 5-21. The resistance (R_{out}), simulating the output of the previous stage, should be connected in parallel with the input transformer as shown in Fig. 5-21.*

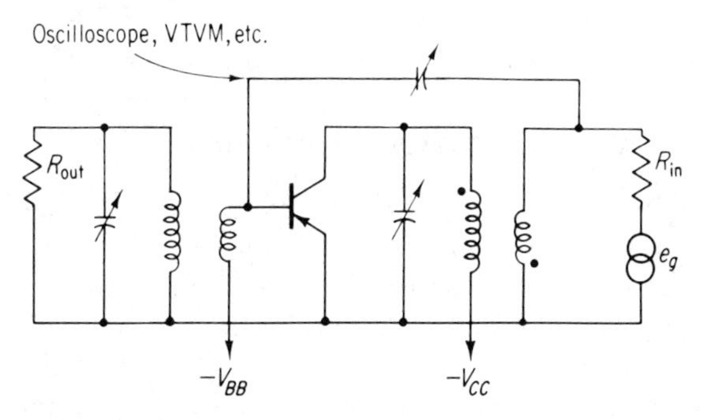

Fig. 5-21. Location of signal generator and indicating device while adjusting C_N (ac circuit only).

The oscilloscope or other indicating device should be chosen to have an input impedance and capacitance such that it does not load the transistor or tuned circuits. For many applications, a good high-frequency oscilloscope is acceptable as an indicator.

Once the circuit is connected as shown in Fig. 5-21, the signal generator voltage is increased until a convenient indication is seen on the oscilloscope. C_N should then be adjusted for minimum signal on the oscilloscope. The signal generator output should be further increased and C_N adjusted again for minimum indication. In a well-designed amplifier, proper adjustment of C_N will result in a very small indication on the oscilloscope on its most sensitive range and with the generator at maximum voltage (1 volt). A good rule of thumb is that the attenuation in this direction, when neutralized, should be at least 10 db more than the gain in the normal connection.

In some amplifiers, particularly where the output circuit is tuned by adjusting the coil rather than the capacitor, it is often necessary to repeat the

*For very high-frequency amplifiers it is usually easier to use a modulated input signal and a high-frequency detector for the indicating device. In this case, a 50 Ω detector is often used, and the input and output are matched to the detector. Then the detector and generator can be directly interchanged when changing from the input to the output. In well-neutralized amplifiers, it is often necessary to add additional gain to the system as the neutralized condition is approached.

tuning and neutralization procedure several times, because the coupling between the primary and secondary changes with tuning and the neutralization network balance is upset. For this reason, it is usually more convenient to use a variable capacitance, rather than a variable inductance, as the tuning element in the output circuit of an amplifier in which the coil provides the phase inversion for neutralization.

As previously noted, neutralization is used to minimize tuning interaction, particularly in multistage IF amplifiers. Conversely, one of the best means of checking how well a stage is neutralized is to check the tuning interaction. This can be done by tuning the stage in the normal fashion. If either the input or output circuits are now detuned, little or no adjustment should be required to obtain the peak response in the other tank circuit.

DESIGN OF NEUTRALIZATION NETWORKS

Although the design equations described earlier in this chapter provide the theoretical means for analyzing neutralization networks, it is often difficult to design a neutralized amplifier because the neutralization voltage is very dependent upon the coupling coefficient of the transformer, or upon the capacitance divider in the tuned circuit; hence it is desirable to know the total effective feedback capacitance of the transistor. When this value is known, it is possible to replace the transistor with this value of capacitance between collector and base. The neutralization capacitor is then adjusted to neutralize exactly this simulated collector-base capacitance as described earlier. The variable neutralization capacitor is then measured and replaced with a fixed-value capacitor. Conversely, a good way to measure the actual value of feedback capacitance is to neutralize the transistor in the operating circuit. When the transistor is fully neutralized, the transistor is removed from the circuit and a variable capacitor inserted between the collector and base pins of the socket. This capacitor is adjusted until the circuit is again neutralized. The capacitor may now be removed and measured. Its value will represent the equivalent collector-base feedback capacitance of the transistor.

The preceding techniques are particularly useful in a multistage IF amplifier where it is very difficult to neutralize several stages at once. They allow the neutralization of each stage individually and provide a means of determining the neutralization capacitance for a number of identical stages.

Comparison of Common-Base and Common-Emitter Neutralization

The neutralization comments made up to this point apply to the common-emitter connection, where the input-to-output feedback (collector to base) is degenerate at high frequencies. To compensate for this, the signal supplied by C_N is in phase with the input and is, therefore, regenerative feedback.

In the common-base connection, the opposite is true. The internal feedback (collector-emitter) is regenerative. Thus neutralization provides degenerate feedback in the common-base connection at high frequencies. Therefore, in circuit design one must take care to provide the proper phase relationships for neutralizing. A typical neutralized common-base stage is shown in Fig. 5-22.

Note that the phase reversal is obtained as in a common-emitter amplifier. In this case, however, the neutralization signal is fed back to the emitter and not to the base.

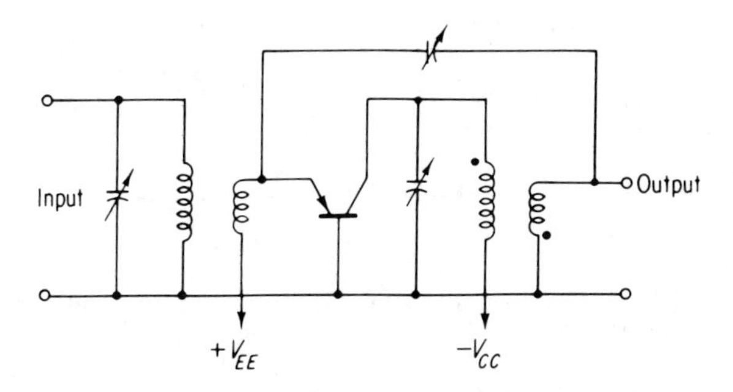

Fig. 5-22. Typical ac circuit of a neutralized common-base amplifier.

LOW-FREQUENCY TUNED AMPLIFIERS

In the first section of this chapter, we discussed the design of high-frequency tuned amplifiers. High frequency and low frequency are relative terms and are related in this discussion to the power gain versus frequency curve shown in Figs. 5-2 and 5-4. A *high-frequency tuned amplifier* is one designed to operate in the frequency range where the power gain curve has a negative slope (approximately −6 db/octave). *Low-frequency tuned amplifiers* are considered to be those designed to operate in the flat portion of the gain-frequency curve. In practice, most of these amplifiers are IF amplifiers, but they may be of any type. Unless complex neutralization networks are used, most present-day transistors are inherently unstable throughout most of the flat region of this curve when operated under matched conditions. Because of the difficulties in constructing and adjusting adequate neutralization networks, mismatch techniques are used to achieve stability.

Stability Considerations

Figure 5-23(a) shows the hybrid-pi equivalent circuit of a common-emitter transistor with its load and source connected. Since $r_{b'e} \gg r_{b'}$, the equivalent circuit at low frequencies, neglecting feedback, may be approximated by the circuit of Fig. 5-23(b).

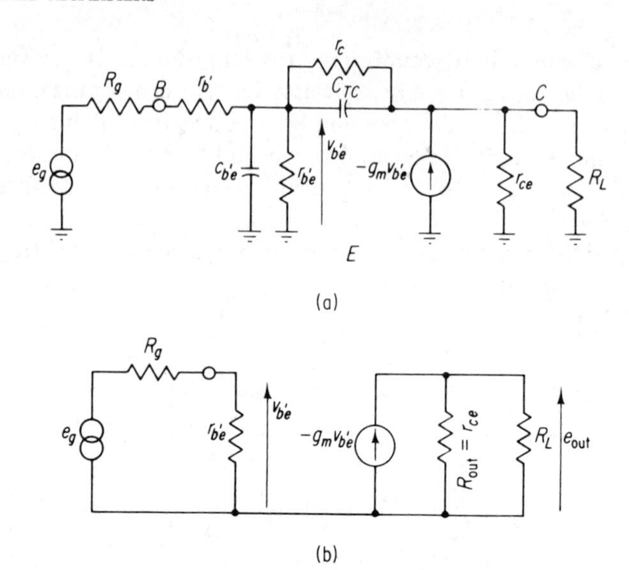

Fig. 5-23. Hybrid-π equivalent circuit showing (a) load and source resistance and (b) low-frequency model.

The power gain of the system may be readily calculated as

$$P_G = \frac{P_{out}}{\text{MAP}} = \frac{\dfrac{e_{out}^2}{R_L}}{\dfrac{e_g^2}{4R_g}} = 4g_m^2 \frac{R_g R_{in}^2}{(R_g + R_{in})^2} \cdot \frac{R_L R_{out}^2}{(R_L + R_{out})^2} \tag{5-53}$$

The maximum available power gain occurs when the transistor input and output resistance is matched to its load and source resistance ($R_g = R_{in}$ and $R_L = R_{out}$) and is given by

$$\text{MAG} = \frac{g_m^2 R_{out} R_{in}}{4} \tag{5-54}$$

The MAG was calculated assuming no feedback was present. At the frequencies normally used for IF amplifiers, the feedback in most present-day transistors intended for this application is primarily capacitive because of the collector-base junction capacitance and the package capacitance. At low frequencies, r_c may also begin to contribute to the feedback. If the equivalent circuit of Fig. 5-23(a) is assumed to be a black box and a capacitor feedback network is added as shown in Fig. 5-24, the input resistance at frequencies below resonance will exhibit a negative component due to the capacitive feedback. It can be shown (5) that the value of negative resistance will be a minimum at a frequency below resonance where the total input and output circuit phase shift is $+90°$. This resistance is given by

$$-R = \frac{2}{A_V \omega C} \tag{5-55}$$

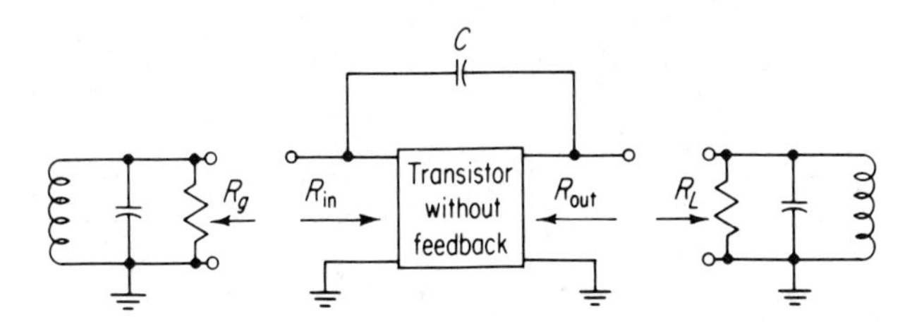

Fig. 5-24. Transistor amplifier represented as a black box without feedback and an external capacitive feedback.

where A_V = amplifier voltage gain at resonance. Oscillations may occur when this negative resistance is equal to or exceeds the positive resistance across the transistor input terminals, or when

$$-R = R_1 \tag{5-56}$$

where

$$R_1 = \frac{R_g R_{\text{in}}}{R_g + R_{\text{in}}} \tag{5-57}$$

The voltage gain of the amplifier is

$$A_V = g_m R_2 \tag{5-58}$$

where R_2 is the effective resistance at the amplifier output terminals and is given by

$$R_2 = \frac{R_{\text{out}} R_L}{R_{\text{out}} + R_L} \tag{5-59}$$

Equation 5-55 may be rewritten as

$$-R = \frac{2}{g_m R_2 \omega C} \tag{5-60}$$

The condition, then, for oscillation is

$$-R = R_1 = \frac{2}{g_m R_2 \omega C} \tag{5-61}$$

Solving for C gives

$$C_{\text{crit}} = \frac{2}{\omega g_m R_1 R_2} \tag{5-62}$$

which is the critical value of feedback capacitance which will just produce oscillation.

In order to obtain a better "feel" for the value of this critical capacitance, let us solve this equation for a typical modern high-frequency transistor. We shall assume values for R_1 and R_2 to be those under matched conditions. Then,

$$R_1 = \frac{R_{\text{in}}}{2}; \qquad R_2 = \frac{R_{\text{out}}}{2}$$

Let us assume $R_{in} = 1000\ \Omega$

$$R_{out} = 100{,}000\ \Omega$$
$$f = 455\ \text{KHz}$$
$$I_E = 2\ \text{ma}$$
$$\frac{1}{g_m} \approx \frac{26}{I_E} = 13\ \Omega$$

Then

$$C_{crit} = \frac{(2)(13)}{(2\pi)(455 \times 10^3)\left(\frac{1 \times 10^3}{2}\right)\left(\frac{1 \times 10^5}{2}\right)}$$

$$C_{crit} = 355 \times 10^{-15} = 0.355\ \text{pf}$$

This value of capacitance is much smaller than one would expect in a transistor (C_{TC}), and so some other way of stabilizing the amplifier must be found. It was stated earlier that neutralizing high-gain, low-frequency tuned amplifiers for stability is difficult so that another method of stabilization must be used as brought out by this example.

An examination of Eq. 5-62 shows that if the value of R_1 and/or R_2 are made smaller, C_{crit} will increase. This means that if the input or output (or both) terminals of the transistor are mismatched on the low side, an unneutralized amplifier may be made stable with its existing feedback capacitance.

In a low-frequency unneutralized tuned amplifier, C_{TC} will be the feedback component. Therefore, Eq. 5-62 can be rearranged to be

$$R_1 R_2 = \frac{2}{\omega g_m C_{TC}} \tag{5-63}$$

This equation states that if the product of $R_1 R_2$ is greater than $2/\omega g_m C_{TC}$, the amplifier will oscillate. Thus, a minimum mismatch limit has been established.

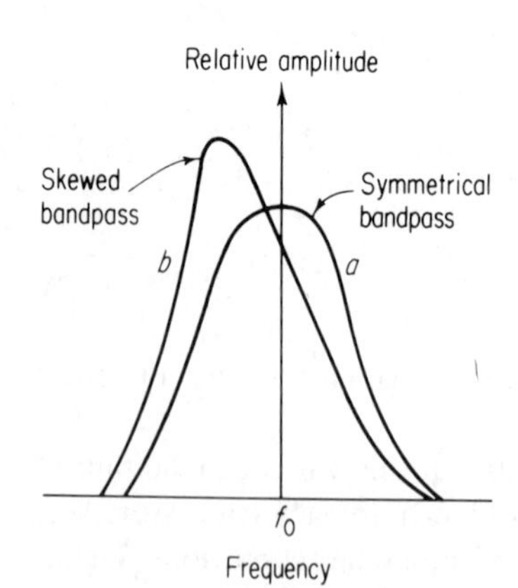

Fig. 5-25. Sketch illustrating the effect of skewing on the bandpass.

Thus far, stability has been discussed only in terms of oscillation or nonoscillation. There are other effects of instability, such as tuning interaction and skewing of the bandpass characteristics. *Tuning interaction* is the effect the output circuit has upon the input circuit, and vice versa, and is particularly noticeable when tuning an amplifier. As the input circuit is tuned to resonance, the output circuit becomes detuned, etc.

Figure 5-25 shows how the bandpass characteristics become skewed with feedback capacitance. Curve *a* is symmetrical with no feedback. Curve *b* shows how the characteristic becomes asymmetrical as feedback is introduced. The increase in gain is also shown. Tuning interaction and skewed bandpass characteristics are reduced by designing for C_{TC} to be much less than C_{crit}.

A stability factor, λ, may be defined as the ratio of the actual feedback capacitance to the amount of capacitance required to produce oscillation:

$$\lambda = \frac{C_{TC}}{C_{\text{crit}}} \tag{5-64}$$

When $\lambda = 0$, no feedback exists, and the amplifier is perfectly stable with a symmetrical bandpass; when $\lambda = 1$, sufficient feedback exists to cause oscillation.

Equation 5-63 can now be modified to include the stability factor as

$$R_1 R_2 = \frac{2\lambda}{\omega g_m C_{TC}} \tag{5-65}$$

A stability factor of 0.4 will give reasonable stability and is often used in amplifier design.

It has been shown that the power gain of the amplifier is modified by a multiplying factor when feedback exists and that it is dependent upon the stability factor (6). In the presence of feedback, Eq. 5-53 is modified to be

$$P_G = 4g_m^2 \frac{R_g R_{\text{in}}^2}{(R_g + R_{\text{in}})^2} \frac{R_L R_{\text{out}}^2}{(R_L + R_{\text{out}})^2} [1 + 2\lambda^2] \tag{5-66}$$

The value of C_{crit} given in Eq. 5-62 is for a single-stage amplifier. If several stages are connected in cascade, C_{crit} will be reduced as shown in Table 1.

TABLE 5-1 Stability Reduction Factor for Cascaded Stages

Number of Stages	Reduction Factor
2	2
3	2.61
4	3

The $R_1 R_2$ product must be changed accordingly.

SINGLE-STAGE DESIGN CONSIDERATIONS

Using the information just obtained for stability considerations, a design procedure can now be outlined for a single-stage, low-frequency amplifier. Knowing the transistor parameters, a number of load and source combinations can be found for a given stability factor. In fact, an infinite number of

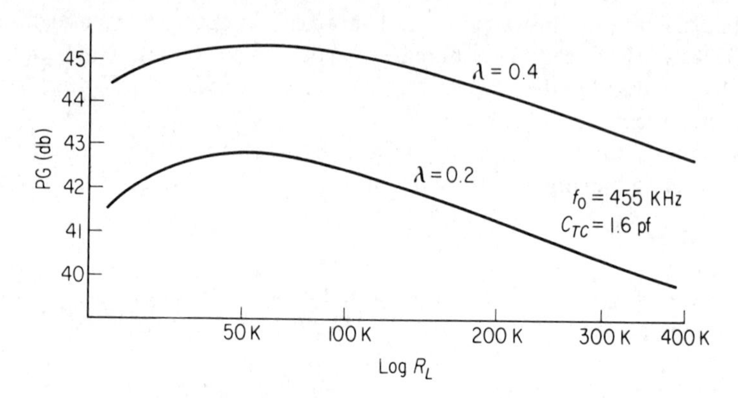

Fig. 5-26. Typical 2N1790 power gain vs load resistance for two stability factors.

load and source combinations will give the same stability factor, but one of these combinations will give the maximum transducer power gain. Figure 5-26 is a plot of power gain versus load resistance for a typical 2N1790 transistor giving stability factors of 0.2 and 0.4. It can be seen from this curve that the peak is very broad and is a maximum near a load resistance of 50 kohms. From Eq. 5-65, R_1 can be found for a given stability factor, from which R_g can then be found. Table 5-2 shows the values of R_g required to give $\lambda = 0.2$ and 0.4.

TABLE 5-2 Combinations of R_L and R_g for a Constant-Stability Factor (2N1790)

R_L	$R_g(\Omega)$	
$K(\Omega)$	$\lambda = 0.4$	$\lambda = 0.2$
400	21.8	11
200	34	17
100	58.5	27
50	112	55
40	134	64
20	278	133
10	620	276
5	1580	600
2.5	—	1600

$f = 455$ KHz; $I_c = 1$ ma

In choosing a value for R_L, remember that the higher the value of R_L, the greater the transformer loss. In the last IF amplifier stage, one must also be careful to choose the value of R_L and the operating point of the transistor such that the amplifier will be capable of delivering sufficient power to the detector circuit.

Once the operating point, R_L, and R_g, have been chosen, the design is then a matter of transformer specification. The design of single-tuned transformers will be discussed first.

SINGLE-TUNED TRANSFORMER DESIGN

Figure 5-27 is a circuit of a typical transformer for the input to a low-frequency tuned amplifier showing the loading resistances. R_p is the shunt loss resistance of the transformer and is given by Eq. 5-25. The output resistance of the preceding stage is R_{out}, R_L is the total reflected resistance presented to the preceding stage, R_{in} is the input resistance of the amplifier stage, and R_g is the source resistance presented to the transistor. Another term, R_T, is important in the transformer design. This is the total resistance reflected across the primary of the transformer and is obtained as

$$\frac{1}{R_T} = \frac{1}{R_P} + \left(\frac{n_2}{n_1}\right)^2 \frac{1}{R_{out}} + \left(\frac{n_3}{n_1}\right)^2 \frac{1}{R_{in}} \tag{5-67}$$

This resistance determines the bandwidth of the transformer, as was shown in Eq. 5-36.

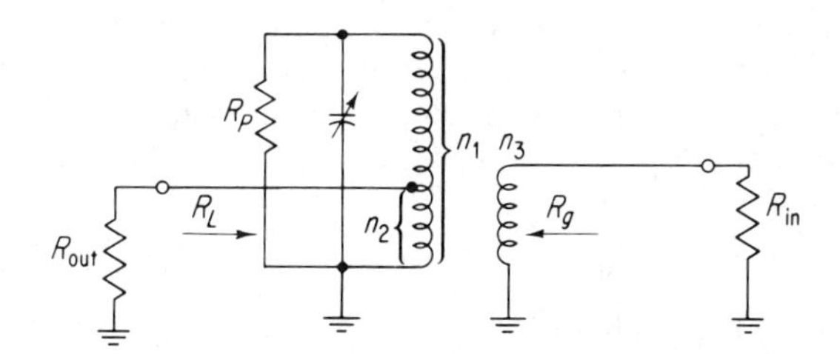

Fig. 5-27. Typical low-frequency interstage transformer showing loading and loss resistances.

The primary inductance of the transformer is variable, but designers often use standard values, such as 1 mh in 455 KHz IF amplifier transformers. Size and economics usually dictate the unloaded Q. The bandwidth requirements are also specified recalling the bandwidth reduction factor.

The turns ratios must be determined in order to provide the desired source resistance and bandwidth with the given input and output resistance of the transistor and the output resistance of the preceding stage. There are two unknowns, namely, n_1/n_2 and n_2/n_3. Two simultaneous equations are required to determine these unknowns. From the bandwidth requirement, R_T can be found as shown in Eq. 5-67. The loss resistance R_P can be found from Eq. 5-25.

The desired value of R_g may be determined in a similar manner as

$$\frac{1}{R_g} = \left(\frac{n_2}{n_3}\right)^2 \frac{1}{R_{\text{out}}} + \left(\frac{n_1}{n_3}\right)^2 \frac{1}{R_P} \tag{5-68}$$

Solving Eqs. 5-67 and 5-68 gives the turns ratio of n_1/n_2 and n_2/n_3.

The output transformer is designed in the same manner as the input transformer, except that R_L rather than R_g is specified.

Design Example

Use of the design procedure is illustrated in the following design example. It is desired to design an amplifier for 455-KHz operation with a 10-KHz bandwidth operating at 1 ma of collector current. The amplifier is to be driven from a 20 KΩ source and operate into a 1.5 KΩ load; the transistor is to be a 2N1790. A stability factor of 0.4 is desired. The primary inductance of the transformer is to be 1 mh, and Q_u is 100.

The 2N1790 typically has the following parameters at 455 KHz for a collector current of 1 ma.

$$\begin{aligned} r_{c'e} &= 500\ \text{K}\Omega \\ r_{b'e} &= 2.5\ \text{K}\Omega \\ g_m &= 37\ \text{mmhos} \\ C_{TC} &= 1.6\ \text{pf} \end{aligned}$$

The curves of Fig. 5-26 were plotted for these conditions, and it can be seen that maximum power gain occurs when R_L is approximately 50 KΩ. Little is lost, however, if the load is 20 KΩ. The insertion loss in the output transformer will be considerably less with this lower value of load. Using a value of 20 KΩ for R_L, R_g can be calculated from Eq. 5-65 and is 262 Ω. The parallel shunt loss resistance is

$$R_P = Q_u X_L = 100(2\pi \times 455 \times 10^3 \times 1 \times 10^{-3}) = 286\ \text{K}\Omega$$

Considering the bandwidth reduction factor discussed earlier, the bandwidth of the individual transformers must be

$$\frac{10\ \text{KHz}}{0.643} = 15.5\ \text{KHz}$$

The factor 0.643 is the bandwidth reduction factor for two identical transformers. Then,

$$Q_l = \frac{f_0}{BW} = \frac{455 \times 10^3}{15.5 \times 10^3} = 29.4$$

and

$$R_T = Q_l X_L = 84\ \text{K}\Omega$$

The input transformer will be loaded as shown in Fig. 5-28.

The simultaneous equations may now be set up to solve for the turns ratios, and we obtain:

$$\frac{n_1}{n_2} = 2.6 \quad \text{and} \quad \frac{n_1}{n_3} = 19$$

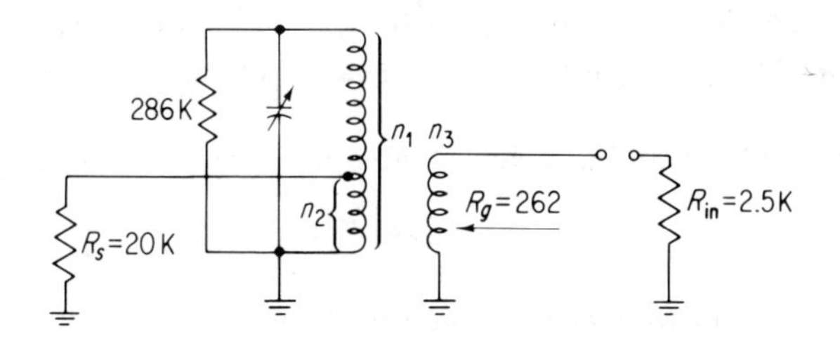

Fig. 5-28. Input transformer loading for design example.

The output transformer is designed in a similar manner except that $R_S = R_{out} =$ 500 KΩ

DOUBLE-TUNED TRANSFORMERS

Often, better selectivity than can be obtained in single-tuned transformers is desired, and double-tuned circuits are used. The design of double-tuned transformers differs considerably from that of single-tuned transformers because they can no longer be considered ideal with their loss resistances represented separately.

In practical design, double-tuned circuits are usually coupled by transformer action or by capacitance. The degree of coupling is a very important parameter in designing for proper impedances and bandwidth. A coefficient of coupling, k, has been defined as (7)

$$k = \frac{M}{\sqrt{L_1 L_2}} \tag{5-69}$$

where M = mutual inductance
L_1, L_2 = inductances of the two coupled circuits.

For the case where the two tuned circuits are capacitively coupled,

$$k \approx -\frac{C_m}{\sqrt{C_1 C_2}} \tag{5-70}$$

where C_1, C_2 = tuning capacitances of the two tuned circuits
C_m = coupling capacitance.

In a double-tuned circuit, maximum energy is transferred when

$$k = \frac{1}{Q} \tag{5-71}$$

This degree of coupling has been defined as the *critical coupling*.

For maximum selectivity, the loaded Q of each individual tuned transformer of a series of cascaded transformers should be equal. For maximum flatness of the response curve within the passband the transformers should be critically coupled. Assuming that these two conditions exist, the overall 3 db bandwidth of n double-tuned transformers in cascade is given by Eq. 5-42.

The bandwidth of an individual double-tuned transformer is given by (4)

$$BW = \frac{\sqrt{2} f_0}{Q_{LU}} \tag{5-72}$$

where f_0 = resonant frequency of the transformer, and
Q_{LU} = loaded uncoupled Q of the transformer.

Combining Eqs. 5-42 and 5-72 and solving for Q_{LU} gives

$$Q_{LU} = + \frac{\sqrt{2} f_0 \sqrt[4]{2^{1/n} - 1}}{BW} \tag{5-73}$$

Because alignment is difficult when the transformers are critically coupled, the usual practice is to make the transformers slightly undercoupled, thereby causing a slight peak in the response curve. The equation for Q_{LU} is still assumed to hold. Figure 5-29 shows how the bandpass characteristics are altered by overcoupling and undercoupling.

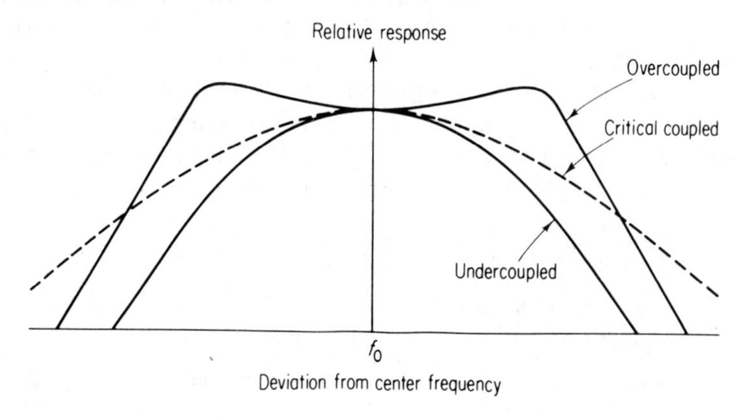

Fig. 5-29. Typical double-tuned transformer response curve showing effects of the degree of coupling.

MULTIPLE-STAGE DESIGN

In a multiple-stage design an additional requirement is placed upon the interstage transformer. The input and output transformer of a single-stage amplifier presents a specific load at only one pair of terminals for stability. In an interstage design the transformer must present a specific resistance at both its input and output terminals. The required resistance will be lower at both terminals than that of the load connected to the terminals. This dictates that the loss of the transformer will be required to contribute a large portion of the loading. This is necessary because the very high output impedance of a transistor cannot reflect a low impedance to the following transistor and, at the same time, reflect the second transistor's input impedance to load the first transistor. In other words, it is impossible to mismatch on the low side on both primary and secondary without the transformer contributing most of the loading. When a condition of this type exists, it can be seen that the ratio of the unloaded Q to the loaded Q will be small; therefore, considerable power gain must be sacrificed for the desired stability. With present-day transistors, however, the useful power gain is more than sufficient for most applications.

It was noted in the single-stage design that there are any number of combinations of load and source resistances in an amplifier for a given stability. Since the interstage transformer in a multiple-stage design presents a large amount of loss, there likewise are an infinite number of combinations which will provide this loading effect. The equations for the optimum load and source resistance in a multiple-stage amplifier have been developed. Since they require solving a fifth-order equation, the usual approach is empirical. The gain is usually optimized for the lower limit of h_{fe} and the stability factor is established using the high limit of h_{fe}.

AGC

Reverse and Forward AGC

The Automatic Gain Control (AGC) of amplifiers is extremely important in almost all communication circuit applications because of the widely varying signal strengths normally encountered in use. It is, therefore, important that the gain of communications transistors be capable of being controlled over a wide range.

The gain of tuned transistor amplifiers may be controlled by two methods. The first of these, known as Reverse AGC, is accomplished in a manner similar to the gain control of tube amplifiers. Although the mechanism of the AGC action is somewhat different, the gain may also be decreased by reducing the bias (collector current). Figure 5-30(a) shows a plot of gain versus collector current for a typical high-frequency amplifier. This curve is obtained by matching and tuning the transistor at the normal maximum gain operating point (2 ma in this case). The base bias is changed, reducing the collector current and, therefore, the gain. No tuning or matching adjustments are made throughout the change in current. When gain is plotted in decibels versus the log of collector current, Reverse AGC action results in a nearly linear reduction in gain from about 1.0 ma collector current. All presently manufactured high-frequency transistors are capable of being "AGC'd" by reducing the collector current. The AGC range of most high-frequency

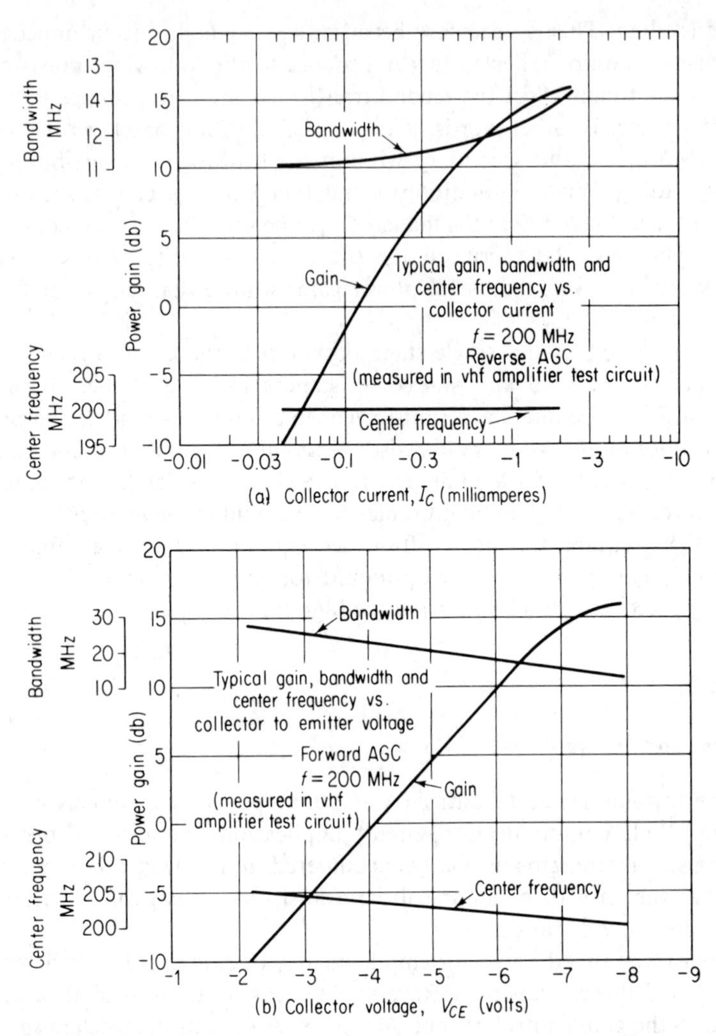

Fig. 5-30. Typical reverse and forward AGC curves for the 2N1742. (a) Reverse AGC; (b) forward AGC. Total series resistance is 1600 ohms.

transistors is from maximum gain to about −10 to −20 db at very low currents.

The second method of transistor AGC action is termed Forward AGC and depends upon our capacity to make the power gain of some transistors decrease, with decreasing collector-to-emitter voltage, over a wide collector-to-emitter voltage range. The gain of all high-frequency transistors decreases with collector voltage, but most do not show significant decrease in gain until the saturation region is approached. Figure 5-30(b) shows a sketch of the gain of a typical amplifier transistor versus collector-to-emitter voltage. The transistor was tuned and matched at normal voltage (8 volts); then the base bias was changed to increase the collector current. By using either the emitter resistor, or this plus an additional bypassed collector resistor, the collector-to-emitter voltage can be varied over the full supply voltage range with only a few milliamperes change in I_C. Again, no additional tuning or matching is done throughout the voltage range.

The mechanism of AGC has been described in detail by Simmons (8). We note that Forward AGC depends on changing the electrical base width. Therefore, one would not expect it to work in the frequency range where gain is not dependent on the electrical base width. Experimental evidence substantiates this theory, and as a general rule Forward AGC will not work below the knee of the power gain–frequency curve. In fact, it is usually not practical below frequencies of $f_{max}/50$. Reverse AGC, on the other hand, works at all frequencies, even audio frequencies, although impedance mismatch becomes an ever-increasing factor in Reverse AGC action as the frequency is lowered.

Forward AGC of Mesa and Planar Devices

Another type of Forward AGC action has been discussed by Weber and Grimes (9) and is presently used with some mesa and planar transistors. This type of Forward AGC occurs because there is an optimum current density in the transistor which produces maximum gain. If the emitter of the transistor is made very small, this optimum current density is reached at low currents (1–5 ma). If the current is increased beyond this point the gain drops. Thus, by applying a forward bias to this type of transistor, Forward AGC action occurs. Gain reduction results from falloff of current gain at higher currents and the mismatch introduced at high currents. Some gain reduction also results because the current gain also falls with voltage, particularly at low voltages (below 2 volts).

This type of AGC is fairly independent of collector voltage and has an advantage over the Forward AGC described earlier, in that it is possible to get this AGC action without changing the collector-emitter voltage. Thus, the problems caused by the transistor saturation in conventional Forward AGC are minimized. More current change, however, is required for the same gain control range. This may cause greater changes in impedances resulting in widening of the bandpass under full AGC.

OVERLOAD, FREQUENCY, AND BANDWIDTH CHANGES WITH AGC

Figure 5-30(a) and 5-30(b) showed typical variation of gain, bandwidth, and center frequency for a 2N1742 operating under both Forward and Reverse AGC action.

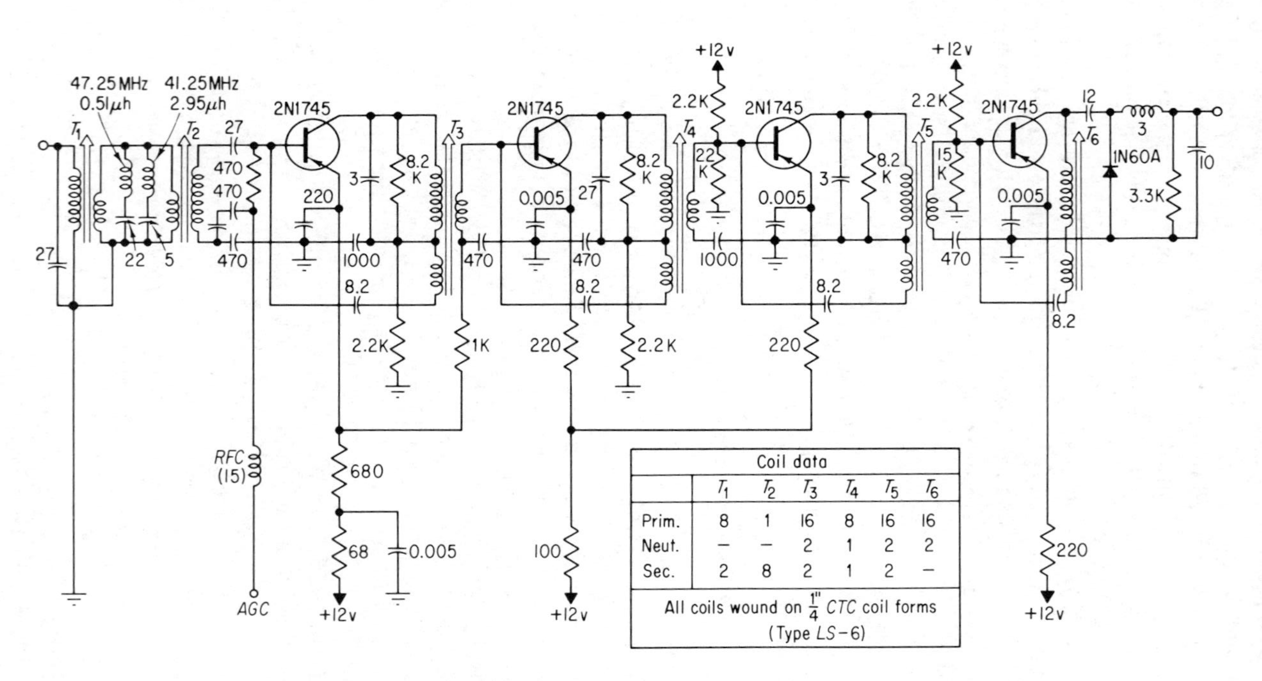

Coil data						
	T_1	T_2	T_3	T_4	T_5	T_6
Prim.	8	1	16	8	16	16
Neut.	–	–	2	1	2	2
Sec.	2	8	2	1	2	–

All coils wound on $\frac{1''}{4}$ *CTC* coil forms (Type *LS*-6)

Fig. 5-31. Four-stage 45 MHz IF amplifier using both Forward and Reverse AGC.

Notice that Forward AGC has a much greater change in bandwidth with AGC, because of the very low output impedance as the transistor nears saturation and/or as the current increases, but the input impedance shift is usually less with Forward AGC. The center frequency shift is also somewhat greater with Forward AGC. It is also apparent that the shifts of bandwidth and center frequency are opposite for Forward and Reverse AGC. This has led some circuit designers to use both types of AGC action in a multistage circuit. By using one or more stages of each type of AGC it is possible to obtain compensating effects to provide a nearly constant bandpass with AGC. An example of an IF strip using both types of AGC action is shown in Fig. 5-31.

In this circuit, the AGC voltage is applied to the base of the first transistor. The base of the second transistor is connected to the emitter of the first stage. Neglecting V_{BE} of the first stage, it may be seen that the same AGC voltage appears on the base of the second transistor. Thus, if the first stage is Forward AGC'd (a negative going voltage with increasing signal), the second stage will also be Forward AGC'd. The third emitter is connected to the second emitter. This will cause the current in the third stage to decrease as it increases in the first two stages. The result is Reverse AGC action in the third stage.

Although Forward AGC has some disadvantages as was shown in Fig. 5-30(b), it is favored in many circuit designs because of its far superior overload performance. This is shown in Fig. 5-32.

With Reverse AGC, the *overload level* (defined as the input power above which the amplifier is no longer linear) actually reduces as we reduce the

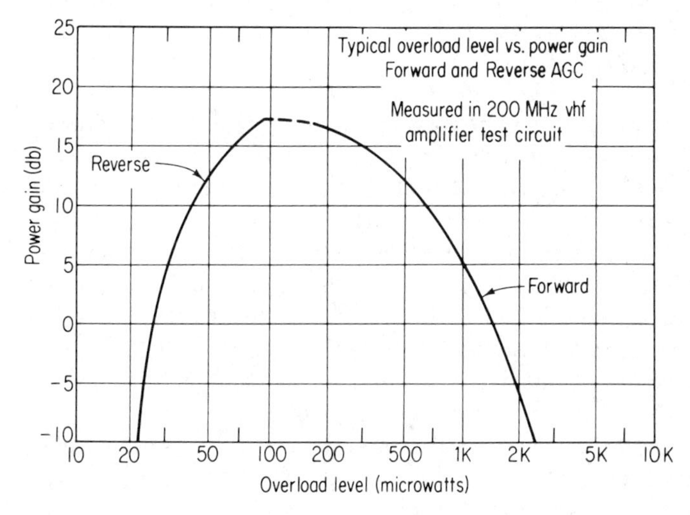

Fig. 5-32. Typical overload curves comparing Forward and Reverse AGC for the 2N1742.

gain by Reverse AGC. Thus, as Reverse AGC action is applied we are able to handle less and less signal. This is contrary to what is desirable because the signal is increasing when the transistor is being AGC'd. Forward AGC has the desired characteristic, and under full AGC it handles several hundred times greater input signals than Reverse AGC without distortion.

Forward AGC provides better overload and cross-modulation characteristics than Reverse AGC, but note that, even with Forward AGC, these characteristics are generally worse than with tubes. Herein lies a major problem in applying transistors to communications equipment.

A word of caution is necessary to all designers of transistorized AGC circuits. Because transistors are usually dc stabilized with emitter resistance, the base voltage is often several volts off ground. This means that the AGC voltage must be referenced on a dc voltage equal to the base voltage at the normal operating point. Care must be taken in an AGC design to make certain that the proper reference voltages are available. Figure 5-33 shows a television IF amplifier circuit where this problem is solved in the video driver stage by using a high positive voltage to obtain approximately +10 volts reference voltage across the 4.7 KΩ collector resistor.

Wide Band Amplifiers

In designing wide band tuned amplifiers, where the bandwidth is in the order of 30 per cent of the center frequency or greater, it is usually impossible to neutralize. Furthermore, because of the small loads in these designs, the feedback is usually not serious. In these applications it is more desirable to use the common-base stage to take advantage of its large natural bandwidth.

Narrow Band Amplifiers

Two techniques may be used to design narrow band amplifiers. The first involves "tapping" the transistor down on the tank circuit as was shown in Fig. 5-27. This causes the output resistance of the transistor to be stepped up across the entire tank circuit by a factor of $[(n_1 - n_2)/n_2]^2\ R_{out}$. Thus, for a given Q_u, the Q_l can be made higher because the load resistance and output resistance are reflected across the entire tank circuit. The resulting total resistance across the tank is higher than would exist if the collector were connected directly to the top of the tank circuit.

Another technique which will narrow the bandwidth is making the stage regenerative so that the transistor provides either a very high or a negative load or source impedance. This may be done by over-neutralizing or by designing the stage to be regenerative. Adjustment is critical, however; one should exercise care whenever this is used.

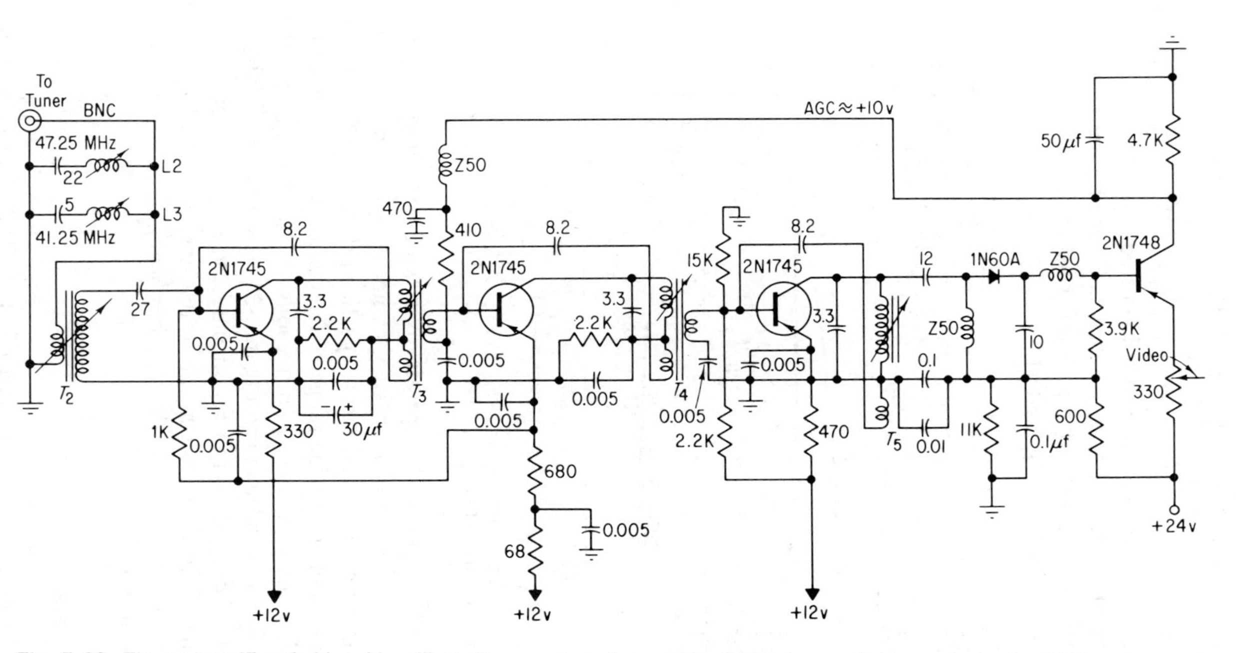

Fig. 5-33. Three-stage IF and video driver illustrating one technique of developing large reference voltages for AGC system.

PERFORMANCE WITH TEMPERATURE

Figure 5-34 shows experimental results of the variation of gain, noise figure, and bandwidth with temperature in a high-frequency amplifier. For these measurements, the amplifier was matched and tuned at room temperature and then only the transistor temperature was varied. No further tuning adjustments were made during the test.

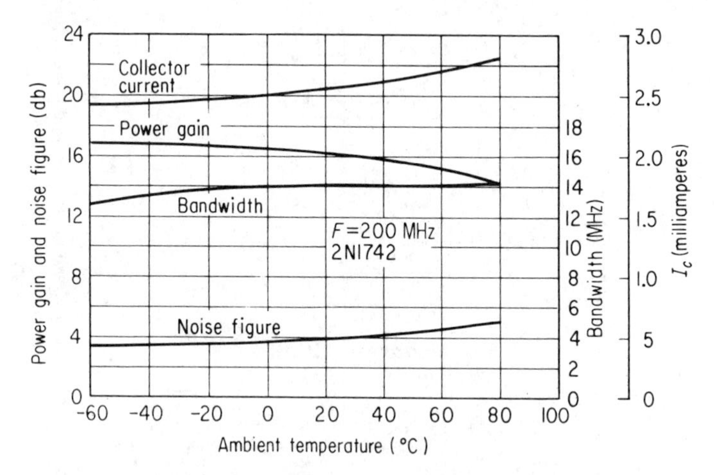

Fig. 5-34. Variation of power gain, noise figure, bandwidth, and collector current with temperature.

In designing amplifiers to operate over wide temperature ranges, the designer must be careful to temperature-compensate not only the transistor, but also the passive circuit elements, particularly L and C. Passive components are available in a wide range of temperature coefficients. One must ascertain which circuit elements are responsible for performance variation in order to compensate adequately. Figure 5-35 illustrates the pitfalls awaiting the unsuspecting designer who uses uncompensated passive elements which were not designed for wide temperature variations. Note the large difference in gain which occurs when the entire circuit is heated, as compared to heating the transistor alone.

Performance Versus Life

Figure 5-36 shows results of some life data. Note that gain, bandwidth, and noise figure are virtually independent of life, within measurement accuracy. As a result of these measurements, there is good reason to believe that the parameters which affect high-frequency performance of the transistor do not materially change with life. It can further be concluded that if the

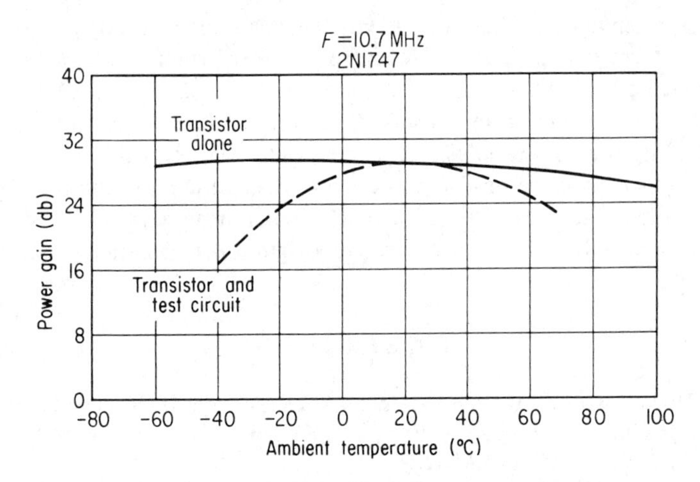

Fig. 5-35. Power gain vs temperature showing effects of using uncompensated passive circuit elements.

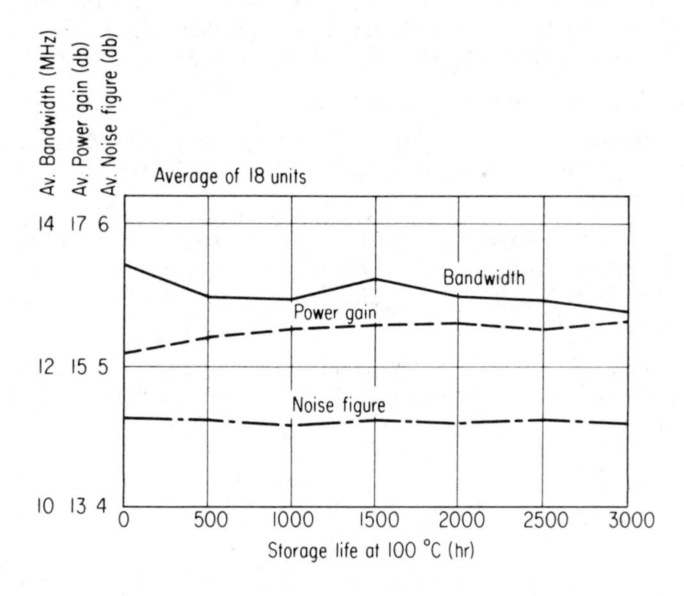

Fig. 5-36. Life-test data indicating virtually constant gain, bandwidth, and noise figure with life.

static characteristics of the transistor, that is, I_{CBO}, β_0, etc., do not change, then the noise figure, gain, and impedances of a transistor operating in the typical high-frequency circuit will not change. If the static characteristics do change, then these may affect the operating point and cause the transistor bias, impedance levels, etc., to change, thereby causing a small change in the actual performance of the circuit. On the basis of experimental evidence, however, these changes would not affect the performance of the transistor in a typical circuit until such time as the transistor had reached what would normally be considered an end-of-life condition. It can, therefore, be concluded that normal static life test measurements are valid for the high-frequency performance of a transistor in a practical circuit and that, as long as the static characteristics do not reach their normal end-of-life condition, the transistor will perform at or near its original performance in the circuit.

EXERCISES

1. A 2N107 transistor is used in a *CB* class A RF amplifier. Typical parameters are: $h_{ib} = 32$ ohms; $h_{rb} = 3 \times 10^{-4}$; $h_{fb} = 0.95$; $h_{ob} = 10^{-6}$ mhos; $C_{ob} = 40$ pf; $f_\alpha = 0.6$ MHz.
A single-tuned load consists of a 1-mh coil in parallel with a 100-pf capacitor. The resonance Q of the load is 200.
(a) Determine the voltage gain of the stage at the resonant frequency.
(b) Determine the gain 10 KHz above and below resonance.
(c) Does the f_α of the transistor restrict the response greatly?

2. Repeat Exercise 1 for the *CE* connection.

3. In a single-tuned direct-coupled amplifier stage using a 2N107 transistor that is tuned to 1200 KHz it is found that the bandwidth is 20 KHz. Determine the Q of the circuit.

4. A direct-coupled single-tuned amplifier has a bandwidth of 180 KHz and a resonant shunt impedance of 50 kilohms. What must be the value of a shorting resistance across the tank if the gain is to be constant within 10 per cent over the 180-KHz band?

5. The circuit of a cascade bandpass amplifier is shown in the accompanying diagram, where z_1 and z_2 are parallel resonant circuits tuned to the same frequency.

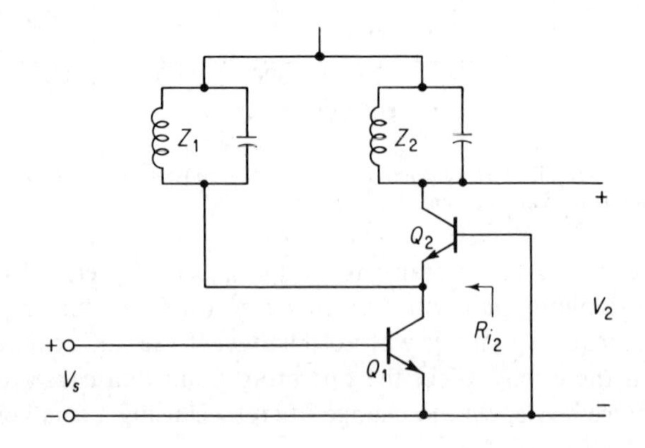

(a) Find an expression for R_{i_2}
(b) Deduce an expression for the over-all gain of the amplifier.
(c) How does it compare with that of a single *CE* state with a parallel resonant load?

6. Determine the proper design for the winding of an i-f transformer with $L_1 = L_2$ and each winding tuned to resonance by a capacitance of 100 pf. The secondary voltage is not to fall below 0.90 of the peak value in a 10-KHz band, centered at 455 KHz. Find k, L_1, L_2, Q_1, Q_2, and the secondary voltage, with 1 volt, 455 KHz to the primary. Assume critical coupling.

7. A single-tuned direct-coupled amplifier is to operate at high frequency. Using the hybrid-pi equivalent circuit of the transistor:
(a) Deduce an expression for the voltage gain of the stage.
(b) Find an expression for the main-bandwidth product.
(c) An alloy transistor has the following hybrid-pi parameters:

$r_{b'} = 80$ ohms $\quad C_{b'e} = 1600$ pf $\quad g_m = 30$ millimhos
$r_{b'e} = 1200$ ohms $\quad C_{b'c} = 10$ pf
$r_{ce} = 100$ kilohms $\quad r_{b'c} = 3$ megohms

Calculate the gain, gain-bandwidth product, assuming an effective $R =$ 5 kilohms and a shunt impedance of 1 kilohm.

REFERENCES

1. W. W. GARTNER, "Maximum Available Power Gain of Linear Four-Poles," *IRE Trans. Circuit Theory*, **5** (December, 1958), 375–76.

2. J. W. HALLIGAN, "Prediction of Transistor Amplifier Performance at High Frequency," Application Lab Report 652, Philco Corp., April, 1960.

3. A. J. COTE, JR., "Evaluation of Transistor Neutralization Networks," *IRE Trans. Circuit Theory*, **5** (June, 1958), 95–103.

4. S. SEELY, *Electron Tube Circuits*. New York: McGraw-Hill Book Company, 1956.

5. D. D. HOLMES and T. O. STANLEY, "Stability Considerations in Transistor Intermediate-Frequency Amplifiers," *Transistors* I, RCA Laboratories, Princeton, N.J., March, 1956, pp. 403–406.

6. J. A. EKISS, "Calculation of Optimum Load and Source Resistances for MADT IF stages (455 kc)," Application Lab Report 545, Philco Corp., Lansdale Division.

7. *Reference Data for Radio Engineers*, 4th ed. New York: American Book-Stratford Press, 1958, pp. 236–40.

8. C. D. SIMMONS, "A Low-Cost, High-Gain TV IF Transistor," Paper presented at the Radio Fall Meeting, IRE, EIA, Syracuse, N.Y., Oct. 29, 1958.

9. R. WEBER and R. GRIMES, "Forward and Reverse AGC Characteristics of Germanium Mesa Transistors," Paper presented at the Radio Fall Meeting, IRE, EIA, Syracuse, N.Y., November, 1960.

10. L. WELDON, "Designing AGC for Transistorized Receivers," *Electronic Design*, Sept. 13; Oct. 11, 1962.

11. P. H. VANANROOY, "Some Design Aspects of Automatic Gain Control for Television Receivers," Paper presented at Chicago Spring Conference, IRE, July, 1960.

12. "Automatic Gain Control for Television Receivers," *RCA Rev.* **IX**, No. 3, September, 1948.

13. "Automatic Volume Control as a Feedback Problem," *Proc. IRE*, **36**, No. 4, April, 1948.

14. E. G. NIELSON, "Behavior of Noise Figure in Junction Transistor," *Proc. IRE* (July, 1957), 957–63.

15. C. R. GRAY and T. C. SOWERS, "Variation of High-Frequency Power Gain and Noise Figure versus Temperature," Paper presented at Radio Fall Meeting, IRE, EIA, Syracuse, N.Y., Oct. 31, 1961.

16. C. R. GRAY and C. D. SIMMONS, "Variation of Transistor High-Frequency Noise Figure and Gain with Life," Paper presented at Radio Fall Meeting, IRE, EIA, Syracuse, N.Y., Nov. 1, 1960.

17. D. O. PEDERSON, J. J. STUDER and J. R. WHINNERY, *Introduction to Electronic Systems, Circuits and Devices*. New York: McGraw-Hill Book Company, 1966.

18. H. E. STEWART, *Engineering Electronics*. Boston, Mass.: Allyn & Bacon, Inc., 1969.

19. C. BELOVE, H. SCHACHTER and D. L. SCHILLING, *Digital and Analog Systems, Circuits and Devices: An Introduction*. New York: McGraw-Hill Book Company, 1973.

6

Video Amplifiers

INTRODUCTION

A *video amplifier* may be defined as a high-fidelity low pass or bandpass amplifier. If the amplifier is dc coupled it is a low pass amplifier; if it is ac coupled there is a low-frequency as well as a high-frequency rolloff so that it must be defined as a bandpass amplifier. The video amplifier may be likened to a super high-fidelity audio amplifier. In an audio amplifier, the frequency range of interest is on the order of three decades. In a video amplifier, the frequency range may extend to five or six decades or more.

One of the most common uses of video amplifiers is in television receivers where it is required to reproduce signals over the frequency spectrum of approximately 30 hertz to 4 MHz. In TV transmitters, where a large number of stages are cascaded, as in a camera chain, the bandwidth requirements are much greater than those of a single stage, on the order of 2 Hz to 8 MHz. Video amplifiers are also used in radar display systems. The function of the video amplifier in this application is essentially the same as that in the TV receiver. Another area of use is in the oscilloscope for both horizontal and vertical amplifiers. In these applications the frequency response may run from dc to 60 MHz. Video amplifiers are also used in vacuum tube voltmeters where the frequency response requirements may run from 2 Hz to 6 MHz or more.

PERFORMANCE CRITERIA

The salient characteristics of the video amplifier for most applications are the uniformity of the amplitude versus frequency characteristic, and the uniformity of the phase versus frequency characteristic. If true high fidelity performance is to be the criterion, the variation in amplitude versus frequency must be no greater than a few tenths of a decibel and the variation in time delay must be no greater than a few thousandths of a microsecond (at the high-frequency end) over the entire band of frequencies of interest.

Let us look at why these criteria are important in a video amplifier. In any linear system, with no feedback, the resultant amplitude response is equal to the product of the individual amplitude responses of the components of that system, and also the equivalent phase response of the system is the sum of the individual phase responses of the components of the system. Also, the output of a transmission system is equal to the product of the input and the transmission characteristics of the system under discussion. Figure 6-1 is an idealized amplitude and phase versus frequency characteristic of a video transmission system. If we assume the input to the system is a "square wave" (containing frequency components f_i, $3f_i$, $5f_i$, . . .) and then calculate or measure the output from the system, we shall find that the output waveform is an exact reproduction of the input waveform, except that the scale factor has been changed. This holds true as long as the maximum frequency components of significant amplitude in the input waveform are equal to or less than f_1, Fig. 6-1. Figure 6-2 is the assumed input square wave. This may be broken up into component parts by the use of a Fourier expansion. We have shown the fundamental and the third, fifth, and seventh harmonics as the components in the input waveform. When this waveform is multiplied by the transmission characteristics of the amplifier under discussion, it may be seen that the relative amplitudes of the harmonics are unaltered, and that the phase of each

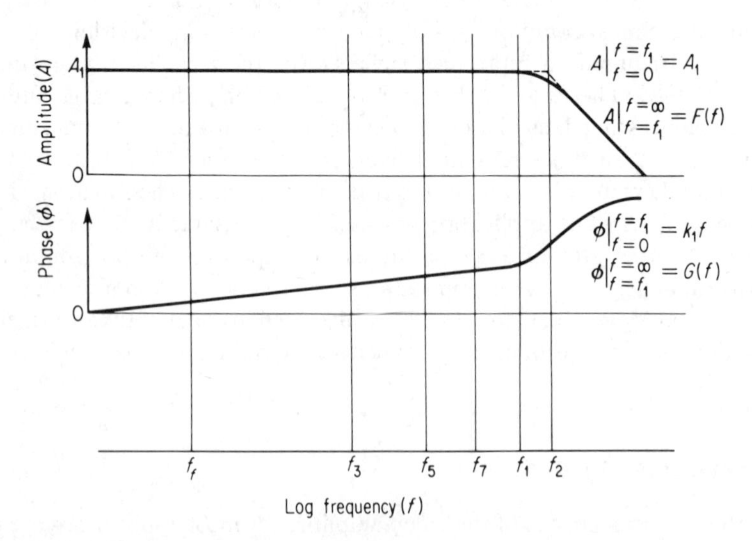

Fig. 6-1. Ideal amplitude-phase response of video amplifier.

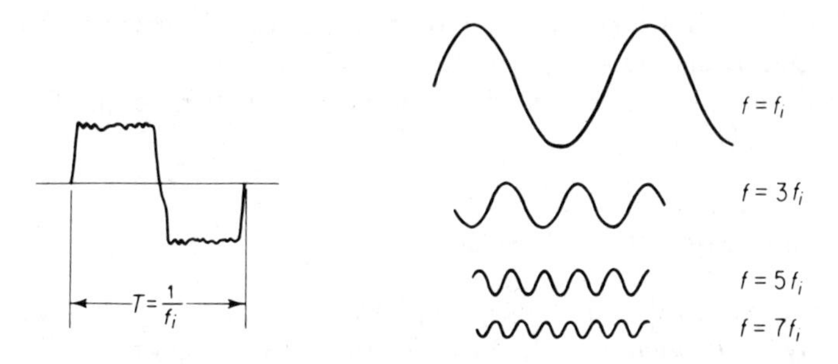

Fig. 6-2. Square wave at frequency f_i which is assumed to be made up of signals with frequency components of f_i, $3f_i$, $5f_i$, and $7f_i$.

component is altered by a constant times the frequency, so that the net over-all result is a faithful reproduction of the input waveform. It may be seen from Figs. 6-1 and 6-2 that, if any of the frequency components of the input waveform extend to frequencies above f_1, for example f_2, distortion will result. The criteria of constant amplitude response and linear phase response become self-evident for faithful reproduction of the input signal.

CASCADED STAGES

If several video stages are cascaded, the degree of perfection of the individual stages must be much greater than if a single stage were to be used. Again referring to Fig. 6-1, if there are two identical stages, and if f_2 is the frequency at which the amplitude response of each of the stages is down 3 db, the overall amplitude characteristic, which is the product of the individual amplitude responses, will be down 6 db. If n uncompensated video amplifiers are cascaded, it was shown earlier that the bandwidth reduction at the high-frequency end of the band is given by

$$BW_n = \sqrt{2^{1/n} - 1}\, BW_1 \tag{6-1}$$

where n = number of identical stages
BW_n = 3-db bandwidth of n cascaded stages
BW_1 = 3-db bandwidth of each stage.

This assumes that the rolloff, at the high end of the band, follows a 6-db per octave slope. It is interesting to note in passing that synchronously single-tuned amplifiers obey the same law, provided that they too roll off at the same rate. If the rate of rolloff is different, the bandwidth shrinkage factor will also be different. If, for example, the amplifiers were perfectly flat to the cut-off frequency and dropped instantaneously to zero, there would be no bandwidth shrinking, no matter how many stages were cascaded.

It has been shown (1)* that the over-all bandwidth of several stages may be predicted if the individual bandwidths of the stages are known. This was done for the case of two- and three-cascaded amplifiers, each of a different natural bandwidth.

CHARACTERISTICS OF RC's

In dealing with video amplifiers, and with transistors in general, we are constantly confronted with the behavior of series and parallel *RC* networks. It might be of interest to review the characteristics of an *RC* circuit under several different conditions. First, let us consider a series connection of a resistance, capacitance, and a generator. We shall assume that the generator is a very high-impedance generator so that it produces a constant current independent of frequency, Fig. 6-3(a). We then look at the voltage across the resistance and the capacitance as a function of frequency (this is the condition in an ac coupled amplifier and is very important at relatively low frequencies).

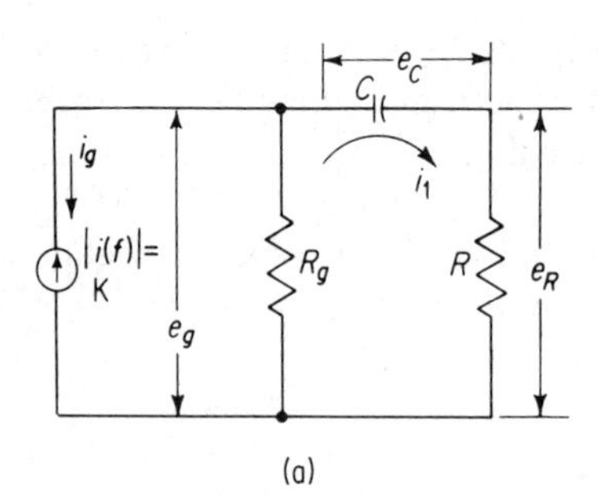

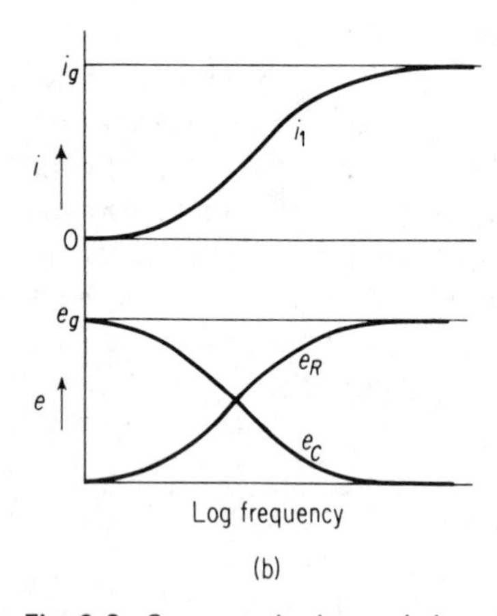

Fig. 6-3. Current and voltage relationships in a simple series RC circuit driven from a current source.

Figure 6-3(b) shows the voltages around the circuit as well as the current through the circuit. As the frequency increases, the current (i_1) approaches the value of i_g. If the capacitor is very large, this will happen at a relatively low frequency but the shape of the curves will be the same. The frequency at which the voltage across the capacitor is equal to the voltage across the resistor is defined as the bandwidth, or the 3-db bandwidth of the circuit. At this point the phase shift is 45°. If we are considering a coupling network for video frequencies, this 3-db frequency must occur at a very low frequency or the phase shift will distort any low-frequency signals. To move this crossover frequency to a lower frequency it is necessary to increase either *R* or the value of *C*. If *R* is fixed, there is no alternative to increasing *C*.

*Numbers in parentheses refer to material listed at the end of the chapter.

Figure 6-4(a) is the same series circuit but now we are driving from a voltage source, where the voltage across the generator is a constant, independent of frequency, with its associated series generator impedance, R_g. In Fig. 6-4(b) we have plotted the voltage and current characteristics that result from this network as the frequency is varied. It may be seen that the curves are very similar to those of Fig. 6-3(b), but the limits are somewhat different. The series generator impedance limits the maximum current that may flow in the circuit. Figures 6-5(a) and (b) show the same characteristics for a parallel RC and a current source driving the combination. Again, we find a crossover frequency which is still defined as the rolloff or 3-db frequency. In this case we find the voltage across the combination decreasing with frequency. This occurs because the capacitor is shunting the high-frequency currents around the resistance.

Figures 6-6(a) and (b) show the parallel RC circuit driven from a voltage source, again with R_g the series generator impedance. This is the condition that occurs at the input to a transistor where R would correspond to $r_{b'e}$ (of the hybrid-pi equivalent circuit). The C shown in the diagram would correspond to the $C_{b'e}$ of the same equivalent circuit, and R_g may be thought of as $r_{b'}$ plus the series resistance of the driving generator.

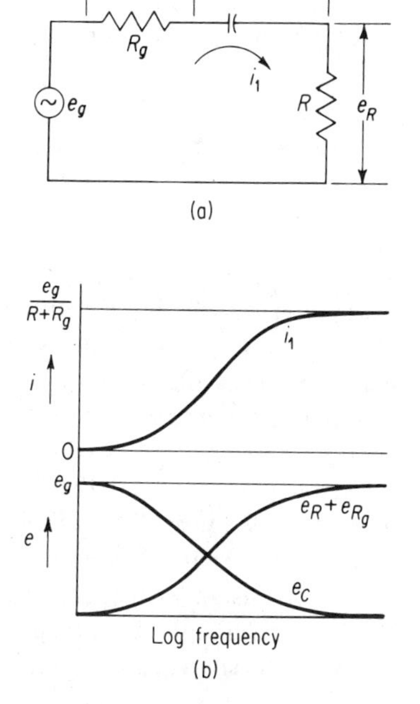

Fig. 6-4. Current and voltage relationships in a simple series RC network driven from a voltage source.

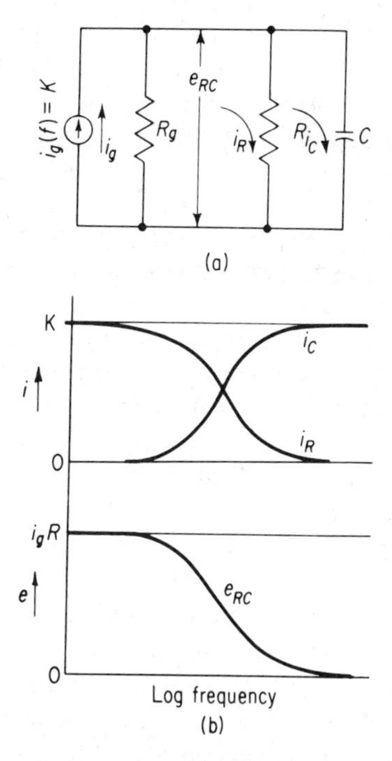

Fig. 6-5. Current and voltage relationships in a simple parallel RC network driven from a voltage source.

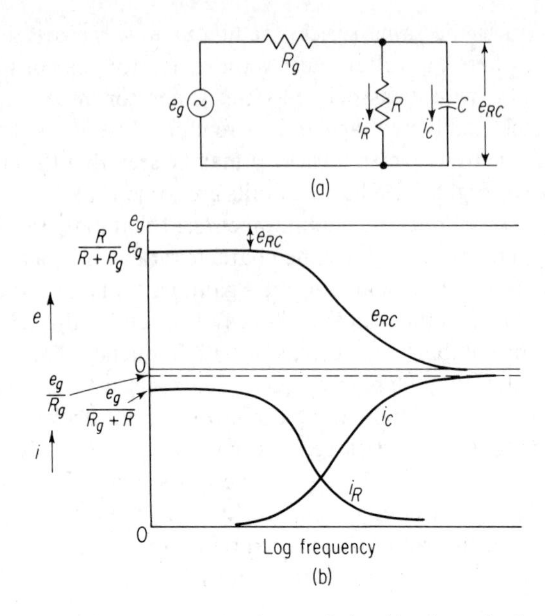

Fig. 6-6. Current and voltage relationships in a simple parallel RC network driven from a voltage source.

It may be seen from Fig. 6-6(b) that the voltage across the RC combination decreases as the frequency increases.

TRANSISTOR VIDEO AMPLIFIER CHARACTERISTICS—GENERAL

Since we are dealing with wide bandwidths, it is almost impossible to use transformer coupling between stages in a cascaded video amplifier because of the bandwidth limitations of transformers. Therefore, almost all video amplifiers are either *RC* or dc coupled. Thus, in general, the transistors are not matched. That is, mismatch occurs at both input and output of the amplifier.

In Chap. 4 expressions were developed which indicated that the input resistance of an amplifier is a function of the load resistance. Differences in the three possible connections were also discussed and the common-base connection was shown to have the lowest input resistance, the common-emitter somewhat higher, and the common-collector the highest. The output resistance as a function of the source resistance was also shown. The common-collector connection was shown to have the lowest output resistance, the common-emitter a higher value, and the common-base the highest output resistance, as a function of source resistance. The current gains of the three connections were also shown as a function of R_L. The common-collector connection had the highest current gain, the common-emitter a slightly lesser value, and the common-base the least current gain of the three connections.

Since all these degrees of freedom exist, the design of a video amplifier must be based on the specific application that the amplifier is to serve. One cannot state generally that there is a best way to design a video amplifier without knowing the full requirements. Some of the things that must be known are the allowable input impedance, the magnitude of the load impedance, the over-all gain and bandwidth required, as well as the usual requirements as to temperature variation, supply voltages available, and the other transistor variables.

GAIN BANDWIDTH

Let us now look at the gain bandwidth product of the three connections, since in essence this is the real criterion of the transistor in video amplifier service. We shall look at the gain bandwidth product under the conditions of short-circuit output, high-impedance output, low-impedance input, and high-impedance input. We should then be able to make an intelligent guess as to what type of connection best suits our needs for a given application.

Common-Base

Let us first consider the common-base connection. If we assume a short-circuit output and a high-impedance input, we can see that we are measuring the parameter $i_c/i_e = \alpha$ as a function of frequency. If we defined the bandwidth of the stage as the frequency at which the current gain is down 3 db, we find that, by definition, this frequency is the alpha cut-off frequency (f_α). As pointed out in Chap. 2, alpha cutoff is usually about twice f_T for diffused base transistors. Figure 6-7 shows a T equivalent common-base high-frequency

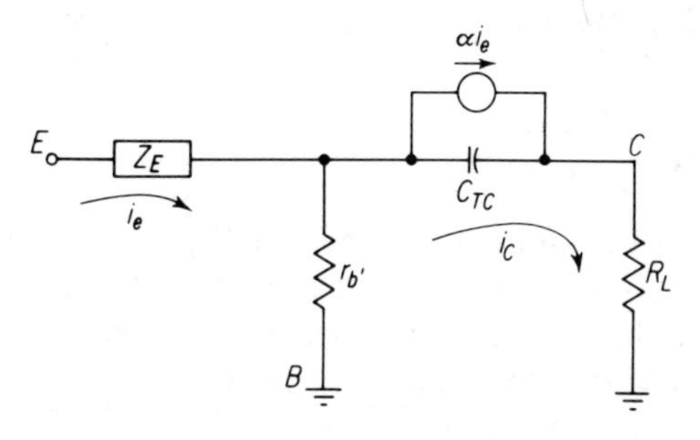

Fig. 6-7. High-frequency T-equivalent common-base circuit with load R_L.

circuit. The equation which describes the voltage drops around the loop for a constant emitter current is

$$-i_e r_{b'} + i_c\left(r_{b'} + \frac{1}{j\omega C_{TC}} + R_L\right) - \frac{\alpha i_e}{j\omega C_{TC}} = 0 \tag{6-2}$$

Rearranging, we obtain

$$i_e\left(r_{b'} + \frac{\alpha}{j\omega C_{TC}}\right) = i_c\left(\frac{1}{j\omega C_{TC}} + R_L + r_{b'}\right) \tag{6-3}$$

and so

$$A_{ib} = \frac{i_c}{i_e} = \frac{\alpha + j\omega C_{TC} r_{b'}}{1 + j\omega C_{TC}(R_L + r_{b'})} \tag{6-4}$$

This is the current gain for the grounded-base connection.

Normally, in video amplifiers, the phase shift and amplitude of $\alpha + j\omega C_{TC} r_{b'}$ can be ignored with respect to $1 + j\omega C_{TC}(R_L + r_{b'})$. Alternatively, we may state that if the 3-db bandwidth of the video amplifier is f_K then $f_K \ll f_\alpha$ and $f_K \ll 1/2\pi C_{TC} r_{b'}$. With these assumptions, then, we may define a cut-off frequency

$$f_{co} = \frac{1}{2\pi C_{TC}(R_L + r_{b'})} \tag{6-5}$$

Note that the bandwidth reduces 6 db per octave with increasing $R_L + r_{b'}$. A rough evaluation may now be made of the bandwidth of the grounded-base stage as a function of $R_L + r_{b'}$. For very small loads and $r_{b'} \to 0$, the bandwidth is approximately f_α, the alpha cut-off frequency. As $R_L + r_{b'}$ increases to a value of $(1/2\pi C_{TC} f_\alpha)$, the bandwidth will be reduced 3 db. From that point on, the bandwidth reduces at the rate of 6 db per octave with increasing load resistance.

We have shown in Chap. 4 that the input impedance of the grounded-base amplifier increases with frequency. Therefore, if the stage is being driven from a voltage source, the input current will fall as the input impedance rises. Thus the stage gain will be reduced as the frequency increases, or, to look at it another way, a decrease in bandwidth will result. Therefore, *if maximum gain-bandwidth is to be preserved, the grounded-base stage should be driven from a current source.*

Common-Collector

By rearranging the equivalent circuit of Fig. 6-7, we may examine the properties of the common-collector circuit. This is shown in Fig. 6-8 where it has been assumed that the input impedance is large compared with $r_{b'}$.

The common-collector current gain is given by:

$$A_{ic} = \frac{-i_e}{i_b} = \frac{1}{(1-\alpha) + j\omega C_{TC} R_L} \tag{6-6}$$

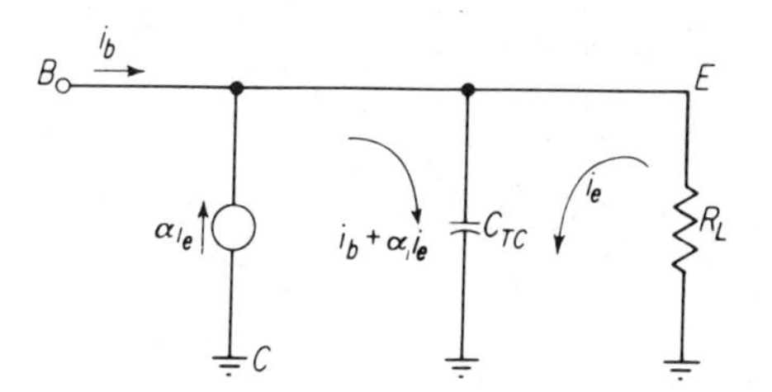

Fig. 6-8. Common-collector equivalent for high frequencies.

or, in somewhat more standard form,

$$A_{ic} = \left(\frac{1}{1-\alpha}\right)\left(\frac{1}{1 + \dfrac{j\omega C_{TC} R_L}{1-\alpha}}\right) \tag{6-7}$$

Let us analyze how the current gain varies with load resistance. If α is near unity, then the first part of Eq. 6-7 is approximately beta. If we also assume R_L is very small, then the bandwidth will be approximately the beta cut-off frequency. If on the other hand R_L gets large, so that the first term is negligible, we see that R_L determines the bandwidth of the circuit. A plot of the bandwidth as a function of R_L is shown in Fig. 6-9. We have shown in Chap. 4 that the input impedance of the common-collector stage is a function of the current gain and the load resistance. Since this is true, the input impedance will follow the current gain curve for a given load resistance, and since the gain falls with frequency, the input impedance will fall. It is, therefore, *necessary to drive this type of amplifier, a common-collector connection, from a voltage source if the maximum gain-bandwidth is to be achieved.*

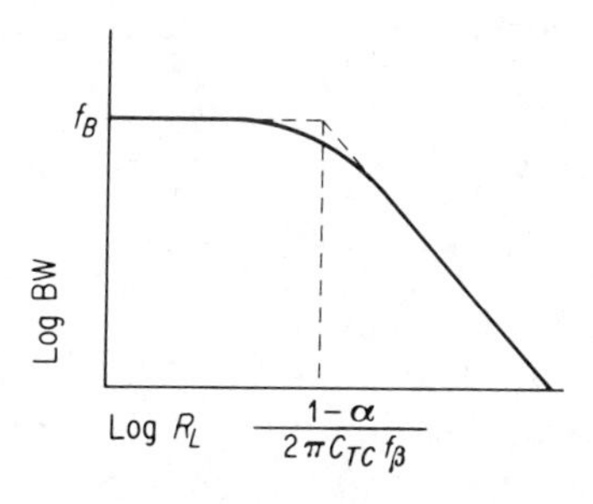

Fig. 6-9. Bandwidth as a function of load resistance for a common-collector video amplifier.

Common-Emitter

The third connection, the grounded-emitter connection, is probably the configuration most used for video amplifier service. In general, this connection yields the greatest gain, and is of the most utility in a cascaded amplifier where more than two stages are required. Starting with the hybrid-pi equivalent circuit—Fig. 6-10(a)—the output impedance of the stage may be calculated with the input open-circuited.

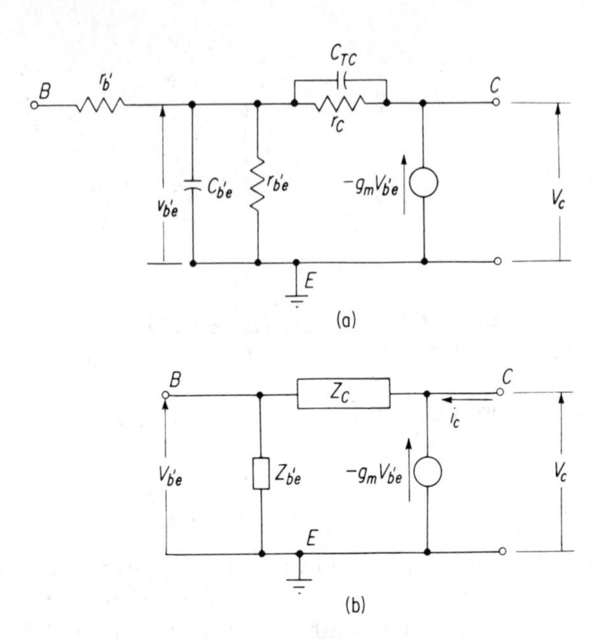

Fig. 6-10. Hybrid-π equivalent circuit.

With reference to Fig. 6-10(b), which is Fig. 6-10(a) for the case of open-circuited input, we can write, using compressed notation,

$$Z_{(\text{out})} = \frac{v_c}{i_c} = \frac{(i_c - g_m v_{b'e})(Z_{b'e} + Z_e)}{i_c}$$

and

$$v_{b'e} = Z_{b'e}(i_c - g_m v_{b'e}) \tag{6-8}$$

Here we have put

$$Z_{b'e} = \frac{r_{b'e}}{1 + j\omega r_{b'e} C_{b'e}} = \frac{r_{b'e}}{1 + \dfrac{j\omega}{\omega_B}}$$

$$\omega_B = \frac{1}{r_{b'e} C_{b'e}} \tag{6-9}$$

and

$$Z_c = \frac{r_c}{1 + j\omega r_c C_{TC}} = \frac{r_c}{1 + \dfrac{j\omega}{\omega_c}} \tag{6-10}$$

where

$$\omega_c = \frac{1}{r_c C_{TC}}$$

Putting all this in we obtain, finally:

$$Z_{(\text{out})} = \frac{Z_{b'e} + Z_c}{1 + g_m Z_{b'e}} = \frac{\dfrac{r_{b'e}}{1 + j\omega/\omega_B} + \dfrac{r_c}{1 + j\omega/\omega_c}}{1 + \beta_0/(1 + j\omega/\omega_B)} \tag{6-11}$$

where

$$\beta_0 \equiv g_m r_{b'e}$$

In most practical cases we can assume without significant loss of accuracy that

$$\frac{r_c}{1 + \dfrac{j\omega}{\omega_c}} \gg \frac{r_{b'e}}{1 + \dfrac{j\omega}{\omega_B}} \tag{6-12}$$

since $r_c \gg r_{b'e}$, and for most frequencies of interest $\omega < \omega_c, \omega_B$. If we use this result in Eq. 6-11 and separate this into its real and imaginary parts, we then have, for the parallel resistive and capacitive components,

$$\begin{aligned} R_{P(\text{out})} &\approx r_c(1-\alpha)\left(\frac{1 + (\omega/\omega_B)^2}{1 + \dfrac{\omega^2}{\omega_B\omega_c}}\right) \\ C_{P(\text{out})} &\approx \frac{C_{TC}}{1-\alpha}\left(\frac{1 + \omega^2(1-\alpha)/\omega_B^2}{1 + \left(\dfrac{\omega}{\omega_B}\right)^2}\right) \end{aligned} \tag{6-13}$$

If the input is now short-circuited, parallel components of the output impedance, computed in the same manner as previously, will be modified to be:

$$\begin{aligned} R_{P(\text{out})} &\approx \frac{r_c(1-\alpha)}{K}\left(\frac{1 + (K\omega/\omega_{B'})^2}{1 + \dfrac{K^2\omega^2}{\omega_B\omega_c}}\right) \\ C_{P(\text{out})} &\approx \frac{KC_{TC}}{1-\alpha}\left(\frac{1 + K\omega^2(1-\alpha)/\omega_B^2}{1 + \left(\dfrac{K\omega}{\omega_{B'}}\right)^2}\right) \end{aligned} \tag{6-14}$$

where

$$K = \frac{r_{b'}}{r_{b'} + r_{b'e}}$$

and

$$\omega_{B'} = \omega_B\left(\frac{r_{b'} + r_{b'e}}{r_{b'}}\right) = \frac{\omega_B}{K} \tag{6-15}$$

The output impedances in terms of parallel resistance and capacitance as calculated from Eqs. 6-13 and 6-14 are plotted in Figs. 6-11 and 6-12 as a function of frequency. To calculate the input impedance with the output short-circuited, the equivalent circuit of Fig. 6-13 is used. The loop equations are

$$\begin{aligned} i_c &= -g_m v_{b'e} \\ v_{b'e} &\approx -i_b Z_{b'e} \end{aligned} \tag{6-16}$$

where we have again assumed $Z_c \gg Z_{b'e}$, which is usually the case in most transistors.

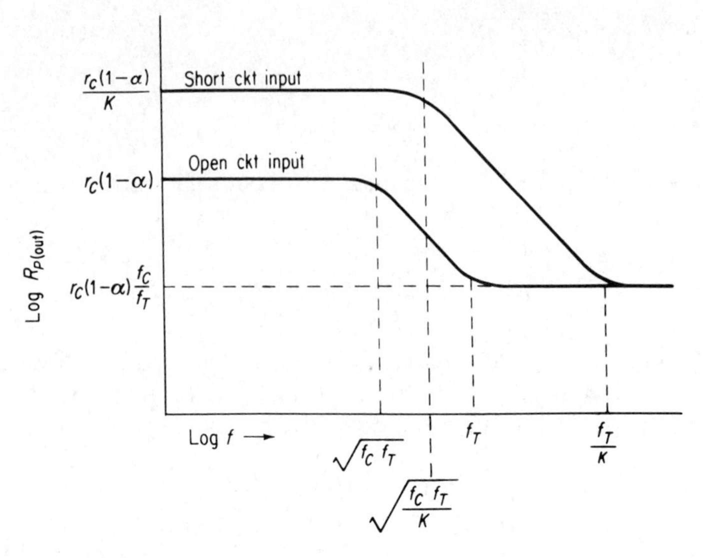

Fig. 6-11. Common-emitter output resistance vs frequency for open- and short-circuited input.

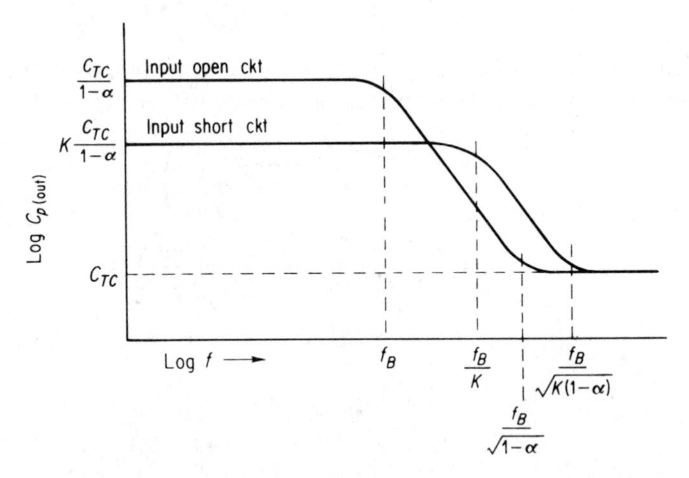

Fig. 6-12. Common-emitter output capacitance vs frequency for open- and short-circuited input.

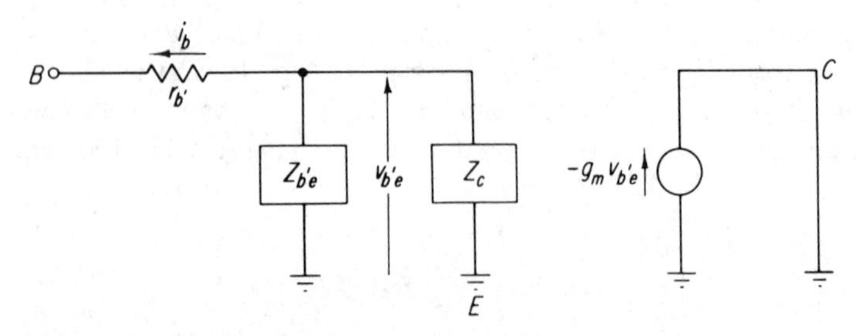

Fig. 6-13. Common-emitter equivalent circuit for determining input impedance with output shorted.

Putting these results together, we obtain

$$i_c \approx g_m i_b \frac{r_{b'e}}{1 + j\omega C_{b'e} r_{b'e}} = \frac{\beta_0 i_b}{1 + \dfrac{j\omega}{\omega_B}} \tag{6-17}$$

We thus have finally:

$$A_{ie} = \frac{i_c}{i_b} \approx \frac{\beta_0}{1 + \dfrac{j\omega}{\omega_B}} \tag{6-18}$$

From Eq. 6-18 we see that f_B corresponds to the frequency at which beta would be down 3 db from β_0. It is, therefore, called the *beta cut-off frequency*.

If the load impedance Z_L is not small compared to the output impedance of the grounded-emitter stage, there is an equivalent Miller effect coupling impedance caused by the coupling (C_{TC}) from output to input. This results in a current being fed back to the input circuit from the output circuit. This current is a function of the gain of the stage, so that the higher R_L is, the greater the feedback current. If the load impedance is a pure resistance, the reflected impedance is capacitive so that the effect is a capacitance shunting $C_{b'e}$ and $r_{b'e}$. The value of this reflected capacitance is $(1 + g_m R_L)C_{TC}$. This added capacitance reduces the bandwidth. The gain-bandwidth product, with a current source drive, and with a load impedance not close to a short circuit is given by

$$(A_{ie})(BW) = \frac{g_m}{2\pi[C_{b'e} + C_{TC}(1 + g_m R_L)]} \tag{6-19}$$

$$\approx \frac{1}{2\pi r_e C_{b'e}\left[1 + \dfrac{C_{TC}(1 + g_m R_L)}{C_{b'e}}\right]} \tag{6-20}$$

In more compact notation then,

$$(A_{ie})(BW) \approx \frac{f_T}{1 + \dfrac{\omega_T}{\omega_{CL}}} \tag{6-21}$$

where

$$f_T \equiv \frac{1}{2\pi r_e C_{b'e}} = \frac{\beta_0 \omega_B}{2\pi} = \beta_0 f_B$$

$$\omega_T = 2\pi f_T, \qquad \omega_{CL} = \frac{1}{C_{TC} R_L}$$

and

$$\beta_0 = \frac{r_{b'e}}{r_e} = g_m r_{b'e}$$

It may be seen from Eq. 6-20 that if the load resistance R_L increases, the gain-bandwidth product is reduced. Computation of gain and bandwidth may be made, for a given load resistance, from Eq. 6-21. Note that the trade of gain for bandwidth is not on a one-to-one basis.

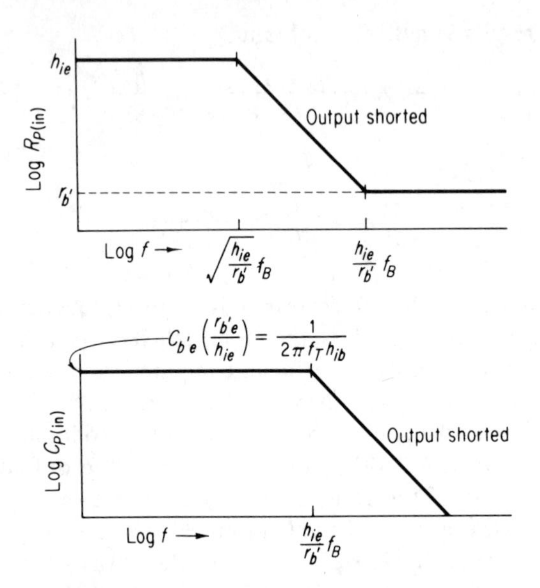

Fig. 6-14. Common-emitter parallel equivalent input resistance and capacitance vs frequency with output shorted.

The resistive and capacitive components of the short-circuit input impedance characteristics are given in Eqs. 6-22 and 6-23, which follow, and asymptotically plotted as a function of frequency in Fig. 6-14.

$$R_{P(\text{in})} = h_{ie}\frac{1 + \dfrac{\omega r_{b'}}{h_{ie}\omega_B}}{1 + \left(\dfrac{\omega r_{b'}}{h_{ie}\omega_B}\right)^2} \tag{6-22}$$

$$C_{P(\text{in})} = \left(\frac{C_{b'e}r_{b'e}}{h_{ie}}\right)\left[\frac{1}{1 + \left(\dfrac{\omega r_{b'}}{h_{ie}\omega_B}\right)^2}\right] \tag{6-23}$$

Note that with the output of the grounded-emitter stage short-circuited or operating into a very low impedance, and with the input driven from a current source, that the gain becomes essentially β_0 and the bandwidth is essentially f_T/β_0. If the bandwidth is calculated for the condition of a finite source resistance, with the output short-circuited, a bandwidth shrinkage occurs which is dependent on R_g, the source resistance. This is

$$BW = \frac{f_T}{\beta_0}\left(\frac{r_{b'e} + r_{b'} + R_g}{r_{b'} + R_g}\right) \tag{6-24}$$

It should be emphasized in passing that Eq. 6-18 expresses the variation of beta as a function of frequency. If the magnitude of beta is plotted as a function of frequency the curve will be as shown in Fig. 6-15. This shows two very

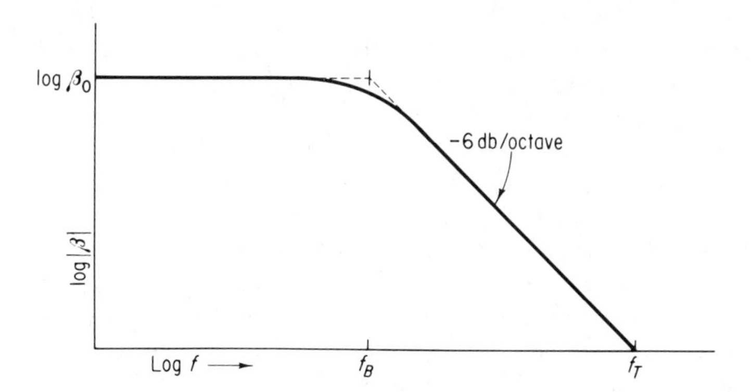

Fig. 6-15. Common-emitter current gain as a function of frequency.

important and often mentioned frequencies. The frequency at which beta is 3 db down from its low-frequency value of β_0 is termed the *beta cut-off frequency* (f_β or f_{hfe}). The frequency at which the magnitude of beta is equal to unity (or 0 db if plotted on a decibel scale) is f_T, which is defined as the gain-bandwidth product. The slope back toward f_β from f_T is a constant at the rate of -6 db per octave up to a value of β_0. To measure f_T, it is only necessary to measure beta at any frequency along this line, and since a point on the line is known and the slope is known, the line may be extrapolated to where it intersects the beta-equal-to-unity line. This by definition is f_T and is a relatively simple measurement to make.

CASCADED STAGE CONSIDERATIONS

If a video amplifier is to be constructed of several cascaded stages, the question arises as to what type of connection to use for the individual stages. When looking at this problem, it is difficult to say how the amplifier should be constructed without knowing the requirements of the input and output circuits. Let us make some general assumptions which are often met in practice. Let us assume that the amplifier will be driven from a voltage source. Let us also assume that we desire a voltage gain throughout the amplifier and that the load resistance is low compared to the output impedance of a grounded-emitter amplifier (but not approaching a short circuit).

We may divide the amplifier into three convenient sections. The last section, which is driving a load of fairly high impedance (compared to a short circuit), the preceding section which is driving the last transistor and in all probability, therefore, a fairly low impedance, and the input transistor. That is, the input impedance of the last transistor is the load for the preceding transistor and this is true until we get back to the input transistor. The input transistor drives the following one. It is driven from a relatively low source resistance as we have postulated. Therefore, we have to determine what combination of transistor connections would best suit our purpose. Since the output stage must deliver a fairly high voltage, it may be classified as a voltage amplifier. Since a transistor used in the common-collector configuration produces no voltage amplification, we can eliminate this connection. If we consider the common-base circuit, we find that it is capable of producing voltage gain since the input impedance is much lower than the load impedance. Also, we have determined that the source impedance for the common-base connection should be a current source, that is, a high-impedance source, if maximum bandwidth is to be preserved. It would, therefore, seem that the common-base stage would make an ideal output stage. For intermediate stages, we would probably use the grounded-emitter configuration, since it has current gain, whereas the grounded-base connection does not. The condition for maximum bandwidth of the common-collector stage is a low-impedance source. Thus it probably would not be used in intermediate stages. The input stage would probably be a common-emitter stage too, again, because it has greater gain than the other connection. If a high-input impedance were desired, the common-collector stage would be used.

TABLE 6-1

Circuit	2-stage Voltage Amplification	2-stage Half-power Bandwidth	Amplifier Bandwidth Product/Stage
CE–CE	$\left(\frac{\alpha_0}{1-\alpha_0}\right)^2$	$0.64 f_T(1-\alpha_0)$	$\alpha_0 f_T$
CB–CB	$(\alpha_0)^2$	$0.64 f_\alpha$	$\alpha_0 f_\alpha$
CC–CC	1		
CB–CC	$\frac{\alpha_0}{1-\alpha_0}$	$f_T(1-\alpha_0)$	$1.56 f_T\sqrt{\alpha_0(1-\alpha_0)}$
CB–CE	$\frac{\alpha_0^2}{1-\alpha_0}$	$f_T(1-\alpha_0)$	$1.56 f_T\sqrt{\alpha_0(1-\alpha_0)}$
CE–CC	$\frac{K\alpha_0}{(1-\alpha_0)^2}$	$0.64 f_T(1-\alpha_0)$	$f_T\sqrt{K\alpha_0}$

Note: $K < 1$.

Angell (2) has compiled a table of six combinations of two-stage amplifiers with the important characteristics for the different combinations listed. This is shown in Table 6-1. It should be pointed out that these stages were to have been part of another iterative amplifier. That is, when driven from, and driving into, like circuits as part of an infinite cascade.

COMPENSATING VIDEO AMPLIFIERS

Some techniques are applicable to transistor amplifiers (as in vacuum tube amplifiers) for extending the frequency range of the uncompensated amplifier. These are known as compensating or peaking circuits. They may be used at both high and low ends of the frequency band. It is found that at the low end of the frequency band, the frequency-limiting components are the coupling and decoupling capacitors. These may be eliminated if one resorts to a dc amplifier, but the problems involved in the dc amplifier are many and intricate. The usual solution to the problem of low-frequency compensation is to make the coupling capacitors as large as possible. Some degree of compensation can be achieved by the technique shown in Fig. 6-16. It may be seen that at some low frequency there will be a phase shift and a loss of signal across the capacitor C. A means of compensating this loss, or extending the low-frequency amplification of Q_1, is to choose values of C_f and R_f such that as the drop across C becomes greater, the load for Q_1 increases. That is, C_f becomes less of a bypass. The design of this network is accomplished by setting the ratio of impedance z_1 to z_2 to be equal to a constant K(3). This will insure that the current i_1 will be independent of frequency as in a compensated attenuator (ignoring the variation of input impedance of the transistor). We thus have

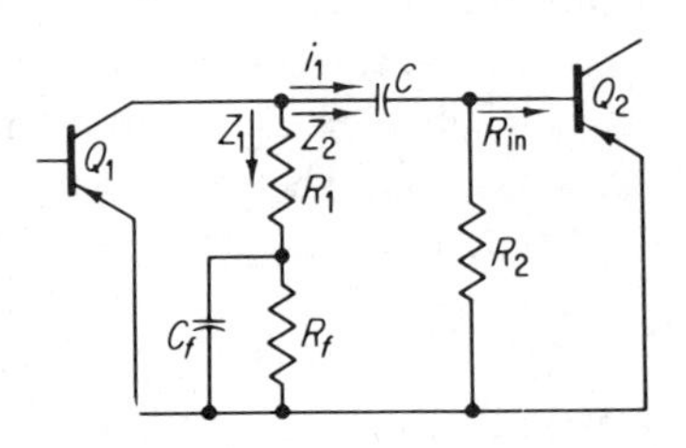

Fig. 6-16. AC circuit for low-frequency compensation of video amplifier.

$$\frac{z_1}{z_2} = K \tag{6-25}$$

where

$$z_1 = R_1 + \frac{R_f}{1 + j\omega C_f R_f} \tag{6-26}$$

$$z_2 = \frac{1}{j\omega C} + \frac{R_2 R_{\text{in}}}{R_2 + R_{\text{in}}} \tag{6-27}$$

By equating real and imaginary parts in Eqs. 6-26 and 6-27, two design equations are obtained and are

$$R_1 + \frac{R_f}{1 + \omega^2 C_f^2 R_f^2} = \frac{K R_2 R_{\text{in}}}{R_2 + R_{\text{in}}} \tag{6-28}$$

and

$$\omega^2 C C_f R_f^2 = K(1 + \omega^2 C_f^2 R_f^2) \tag{6-29}$$

The high-frequency end of the bandpass of the network may be extended by the addition of inductances either in series or in shunt with the capacitances of the circuit. The effect is somewhat different from that in the case of the vacuum tube amplifier, since there is absent there a parasitic resistance which cannot be removed in the case of transistors. That is, it is not possible to connect one end of the compensating inductance directly to one end of the capacitance that one is trying to tune out. It must also be determined whether increasing the current can be achieved by adding a resonant element. If a current source is supplying current to a load, then trying to tune the load may do no good, since the total current flowing from the generator is already flowing into the load. If the load on a transistor is a capacitive circuit, as for example the output of the final transistor in the chain which is driving a load such as a cathode ray tube, the insertion of inductance will effect some good. Also, if the input to a transistor amplifier, driven from a voltage source, exhibits a decreasing input impedance with frequency, anything that would increase the current flow into this impedance would tend to maintain the gain at a higher value. If the input impedance is capacitive, an inductance will serve this function. Therefore, in a common-emitter stage driven from a voltage source, an inductance in series with the input will help extend the frequency response.

Several feedback techniques are also used to extend the high-frequency performance. The simplest form of this technique is a partially bypassed emitter resistor in the common-emitter connection. At low frequencies the bypass across the emitter resistor is not a complete bypass so that low-frequency degeneration is purposely introduced. One may see that, as the frequency increases, the degeneration decreases so that, if the correct value of R and C is selected, a broadening of the frequency response is possible at the sacrifice of low-frequency gain. Figure 6-17(a) shows this circuit and Fig. 6-17(b) shows the effect of this network on the gain and frequency response. Figure 6-17(b) shows that initially, with no degeneration, the gain was A_1, and the bandwidth was f_1, but with the application of degeneration, the gain at low frequencies has gone down to A_2, but at the same time we have extended the bandwidth to f_2. In such a network, R is usually 10–100 Ω, and C has a value of up to 50 pf. Other techniques involving feedback are applicable, and feedback around several stages may be used more effectively than around a single stage (Fig. 6-18).

Distributed amplifiers are often used for producing extremely wide band video amplification. It has been found that distributed techniques are more practical for bandwidths in excess of about 75 MHz. This is done by constructing a transmission line or delay line in such a fashion that the transistor inputs can be tapped along this delay line; then, the outputs of the transistors connected to taps along a similar delay line. As a wave propagates down the input delay line, the outputs of the amplifiers add up in phase along the output delay line, so that gain over a very wide band of frequencies is obtained. The gain per transistor is low but the product of the gains of all the transistors can be very high, and in fact stages can be added to get any desired

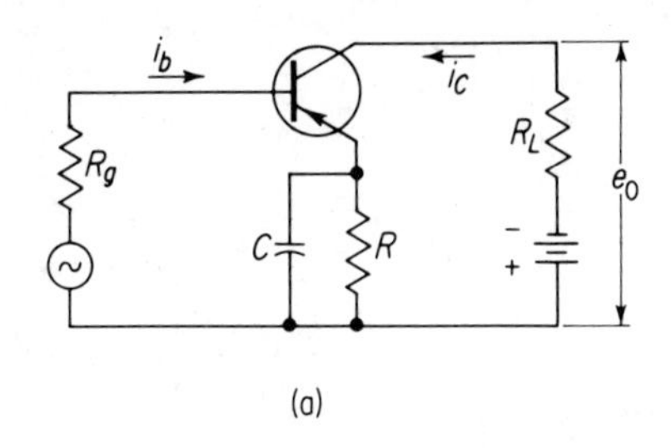

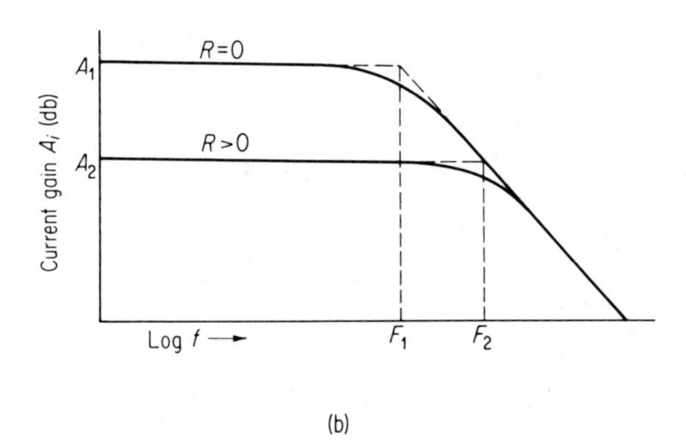

Fig. 6-17. (a) Video amplifier using an emitter peaking network and (b) plot of current gain vs frequency to show change in gain and bandwidth with and without peaking.

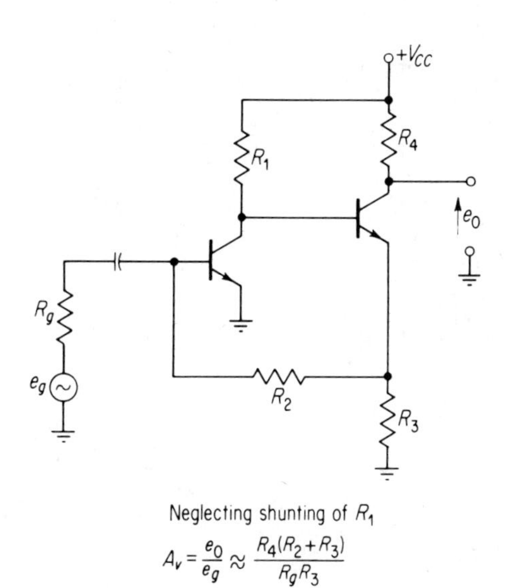

Fig. 6-18. Two-stage direct coupled video amplifier with feedback. For reasonable values of beta (> 30) gain is only dependent on resistor ratios.

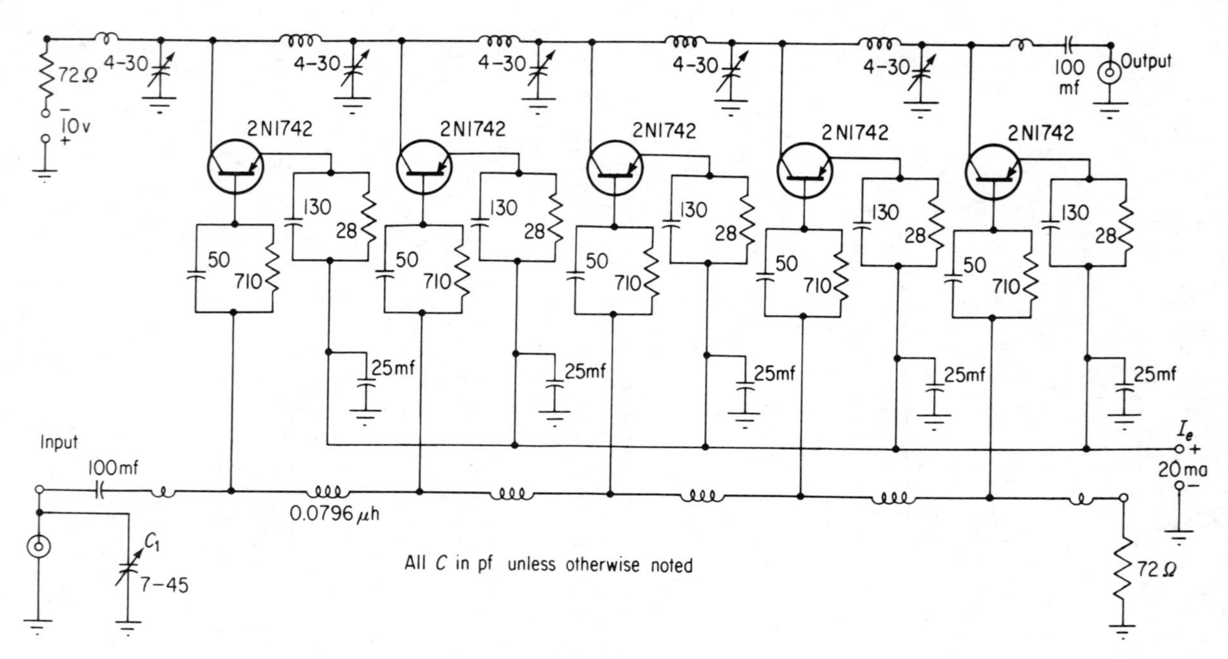

Fig. 6-19. Five-stage distributed amplifier with 72Ω input and output impedance.

degree of gain. The linearity is usually good at high signal levels because of the small signal swing per stage. The design and construction of such an amplifier is extremely complicated, and padding the variable elements of the delay line is a very time-consuming and expensive procedure. A schematic of such a distributed amplifier is shown in Fig. 6-19 (4).

Response Compensation

Another type of compensation may be utilized which might be called *response compensation*. This is illustrated schematically in Fig. 6-20 where two stages are cascaded—stage B and stage C with the approximate asymptotic responses indicated. These individual responses would have a product, shown as response A. This approach may be used if, for instance, amplifier B was the first amplifier in a chain, and it may have very desirable low-frequency noise characteristics, but lack high-frequency response. By properly choosing the response of stage C and matching the asymptotic curves, one may obtain an over-all response of good uniformity with the desirable low noise characteristic of amplifier B.

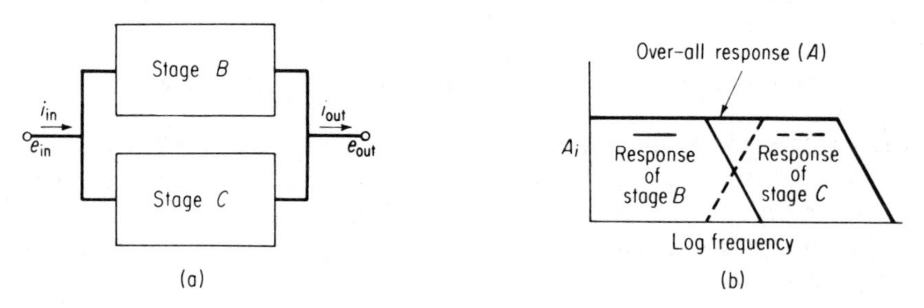

Fig. 6-20. Paralleled video amplifiers. (a) with different responses; (b) idealized response curves of each stage and the over-all amplifier.

Video Polarity

A point not mentioned before but which may be very important in a video amplifier is the polarity of the output signal. If we are talking about an amplifier for television use, for instance, a definite polarity indicates either black or white signal. In some cases it may be necessary to add an inverter amplifier stage just for the purpose of inverting polarity. The common-collector and common-base configurations do not invert the input signal. The common-emitter stage does.

Voltage Swing

It sometimes happens that the output voltage requirement of a video amplifier is greater than the permissible voltage swing of a single transistor. In a case of this sort, the circuit shown in Fig. 6-21 has been used with good

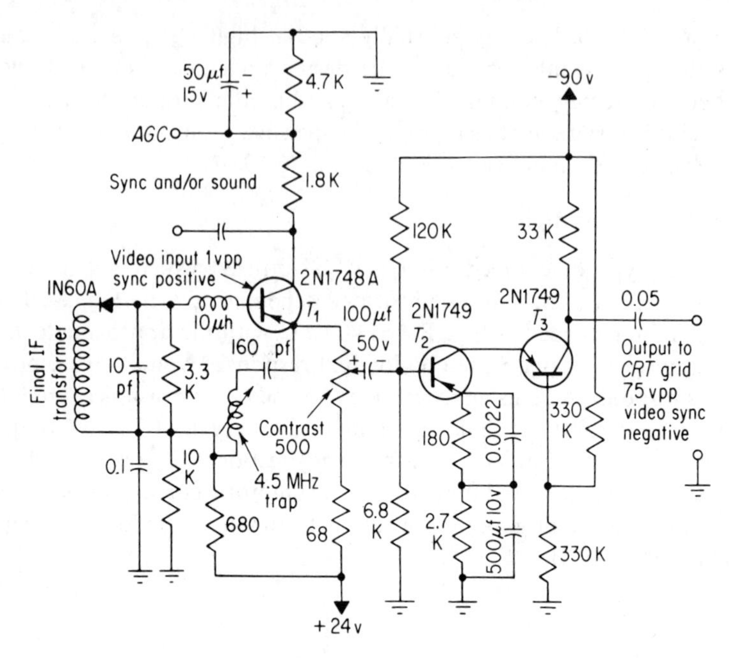

Fig. 6-21. Typical voltage doubler video amplifier for television applications.

success. It is referred to as a *voltage doubler*. It might also be called a *cascade amplifier*. It is in essence a grounded-emitter stage driving a grounded-base amplifier stage. The voltage is now divided between the two amplifiers in series, so that the maximum allowable collector voltage is not exceeded on either stage.

NOISE IN TRANSISTOR VIDEO AMPLIFIERS

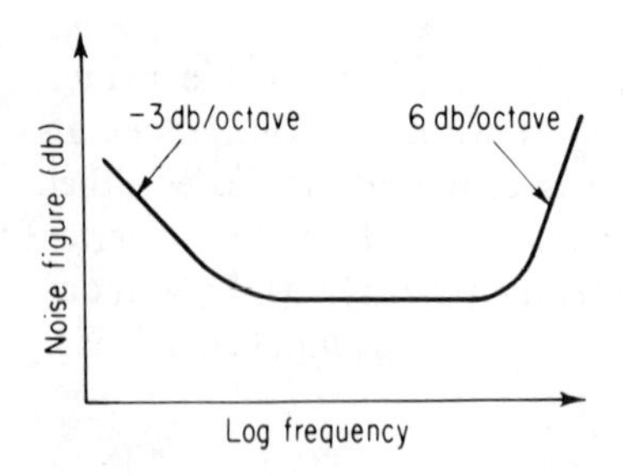

Fig. 6-22. Noise figure vs frequency for a transistor. The noise figure in the mid-frequency range is about 3 db for most transistors under matched conditions.

For completeness, a word should be said about noise figure in the video range. The general characteristic of the noise figure behavior in transistors as a function of frequency is shown in Fig. 6-22. This subject is discussed in detail in Chap. 12. The designer of transistor video amplifiers should, however, be aware that the noise figure of transistors (NF) increases at a 3 db/octave rate in the low-frequency region. This can cause serious problems in an amplifier designed for very low-frequency operation (a few hundred cycles or less). The low-frequency knee of the NF versus frequency curve is often in the region of 1–10 KHz. Thus, the noise figure at 100 Hertz is often 10–20 db.

In general, the high-frequency noise is much less of a problem because if the transistor has the frequency response required to produce the desired video bandpass, the noise figure will usually be adequate throughout the useful high-frequency portion of the bandpass.

DC AMPLIFIERS

Direct-coupled amplifiers are commonly used for amplification of small dc signals or for signals of very low frequency. In addition, pulse amplifiers are often directly coupled to minimize low-frequency distortion. The major problem in designing any amplifier of this type then is to minimize the drift in the dc operating characteristics of the device. That is, stabilizing the quiescent operating point.

From the earlier discussion, we have seen that the major problem in designing stable biasing circuits was the considerable temperature variation in all of the parameters of the transistor. We saw that we had to use emitter or series feedback, for example, in order to keep the dc quiescent operating point, gain, etc., from varying as I_{CBO}, the collector cut-off current, increased. For a dc amplifier, we cannot bypass the emitter resistor, and therefore must use other compensating techniques. These were mentioned briefly in Chap. 3, and included using diodes, thermistors, or even other transistors.

One design technique which uses transistors is that of the differential amplifier. Basically, this consists of two transistors, one connected as a common-collector amplifier and the other as a common-emitter amplifier, connected together through a common-emitter resistor as shown in Fig. 6-23. Of course, this can also be used as a difference amplifier over a wide frequency range, as a low-level amplifier, or for many other applications (5, 7). The analysis is somewhat tedious and is found in many standard works. Suffice it to say that the essential result is that the imbalance of one transistor's operating point is correspondingly matched by that in the other, assuming identical transistors and initial balance, so that no output results owing to temperature, power supply, or resistor variation. In order to use this connection to stabilize an amplifier, for example, it is used as the first stage of the dc amplifier in a

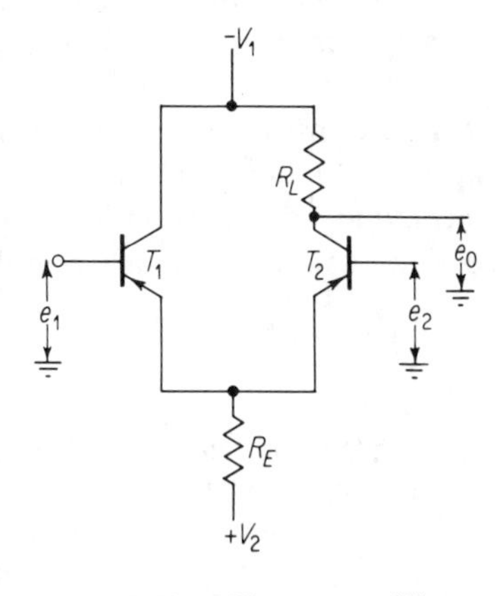

Fig. 6-23. Difference amplifier.

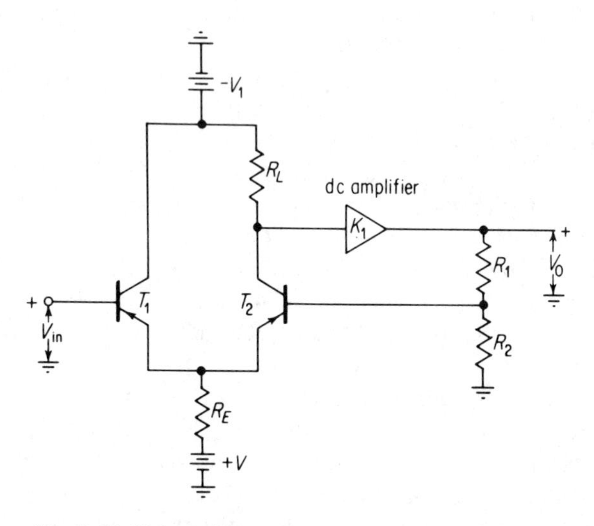

Fig. 6-24. DC amplifier stabilized by a differential amplifier.

manner indicated in Fig. 6-24. This technique can be used to stabilize an amplifier to better than 5 millivolts drift (5).

Another technique, used primarily to stabilize high-gain amplifiers, such as those used in the operational amplifiers of an analog computer, is to use ac signals, derived from a chopper, to stabilize the amplifier by feedback. That is, for example, to sample the drift voltage, to chop it into a pulsating dc by using a transistor chopper, then amplify it, using conventional amplifiers, and apply it to the input stage as in a conventional feedback amplifier. This technique is discussed in the chapter on choppers where examples of analog or operational amplifiers are also described.

EXERCISES

1. Calculate the maximum β_0 and minimum f_T required for a common-emitter, uncompensated video amplifier driven from a current source and driving a 50 Ω resistive load to provide a bandwidth between 8 and 12 MHz. (Assume that $C_{TC} = 0$.)

2. It has been suggested that a series feedback network, consisting of an inductance and a resistance in series, provides a potential frequency compensation technique for use in video amplifiers. The ac circuit proposed is shown in the accompanying diagram.

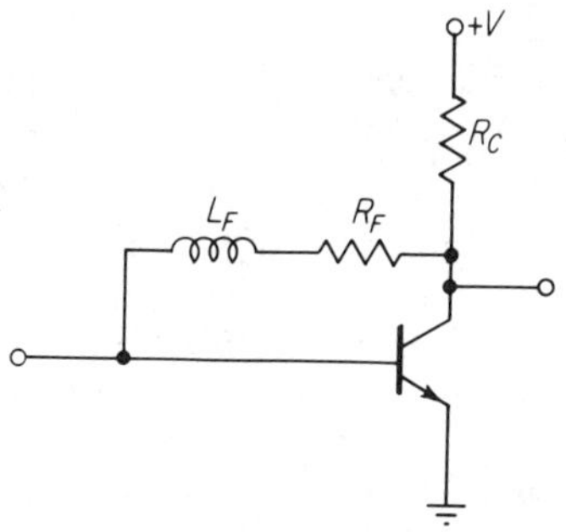

Derive a general expression for the voltage gain of such an amplifier stage, assuming the stage is driven from a voltage source and is driving a high impedance load ($R_L \longrightarrow \infty$). To simplify the analysis use the unilateral equivalent circuit derived using the Miller effect transformation. Assume also that $r_{b'}$ is zero. The circuit to be analyzed is

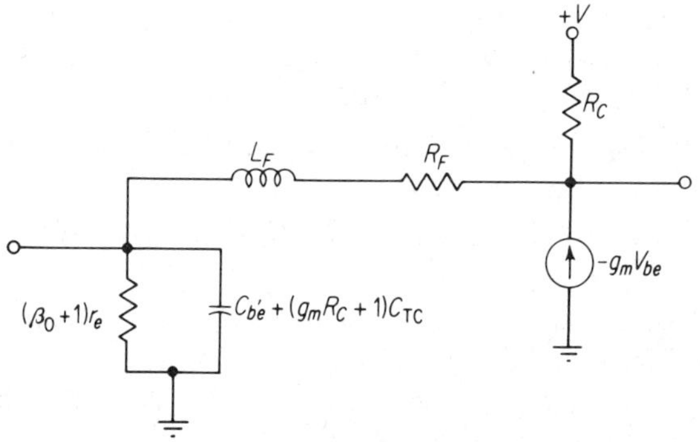

What conclusions may you draw with respect to this compensation technique?

3. A video amplifier is used to provide the gain between a 600 Ω source and a 5000 Ω load. Coupling capacitors will be used to isolate the amplifier from the transducer and load. The lower cutoff frequency for voltage gain is to be at 75 Hz. For the amplifier $R_{in} = 1.5$ KΩ and $R_0 = 10$ KΩ. Find the size of the capacitance required. Repeat the above if the frequency cutoff is now

(a) 180 Hz

(b) 20 Hz

4. Design a video amplifier to meet the following specifications for use in line amplifier service as part of a TV studio camera chain: A line amplifier is inserted in a coaxial line to raise the level of the incoming signal. Since it may be inserted in any position in any line, it should be designed to terminate and drive the same impedance line. It should not invert the signal (input and output in phase). The required amplifier specifications are

Voltage gain = 100 minimum
Low-frequency 3-db point = 2 Hz
High-frequency 3-db point = 7 MHz
Source and load impedance = 75 Ω

The design should require a minimum number of stages. Do not use any frequency compensation techniques. Determine what transistor characteristics are required to meet the performance. Do not be concerned with the dc bias circuitry.

REFERENCES

1. F. E. TERMAN, *Electronic and Radio Engineering*. New York: McGraw-Hill Book Company, 1953.
2. J. B. ANGELL, "High Frequency and Video Amplification," *Handbook of Semiconductor Electronics*. New York: McGraw-Hill Book Company, 1956.
3. J. MILLMAN and H. TAUB, *Pulse and Digital Circuits*. New York: McGraw-Hill Book Company, 1965.
4. M. V. JOYCE and K. K. CLARKE, *Transistor Circuit Analysis*. Reading, Mass.: Addison-Wesley Publishing Co., 1962.
5. R. F. SHEA, *Transistor Circuit Engineering*. New York: John Wiley & Sons, Inc., 1957.
6. J. R. MILLER *et al.*, *Communications Handbook*. Dallas, Texas: Texas Instruments, Inc., 1965.
7. J. MILLMAN, *Vacuum Tube and Semiconductor Electronics*. New York: McGraw-Hill Book Company, 1958.
8. P. M. CHIRLIAN, *Analysis and Design of Electronic Circuits*. New York: McGraw-Hill Book Company, 1965.
9. J. J. CORNING, *Transistor Circuits Analysis and Design*. Englewood Cliffs, N.J.: Prentice-Hall, Inc., 1965.
10. J. D. RYDER, *Engineering Electronics*. New York: McGraw-Hill Book Company, 1967.
11. P. M. CHIRLIAN, *Integrated and Active Network Analysis and Synthesis*. Englewood Cliffs, N.J.: Prentice Hall, Inc., 1967.
12. B. L. COCHRUN, *Transistor Circuit Engineering*. New York: The Macmillan Company, 1967.

7

DC Characteristics and Low-Level Switching Circuits (Choppers)

INTRODUCTION

In a large number of applications, the transistor performs an electrical function very similar to that of the simple single-pole, single-throw switch. Such applications are generally grouped under the broad classification of *transistor switching circuits*. In many of these switching applications, the transistor closely approximates the electrical behavior of an electromechanical relay, in that a small current supplied to the device can control the opening and closing of a pair of electrical contacts. The transistor switch possesses many advantages over the electromechanical switch, the most notable being extremely fast operation, small size, and freedom from mechanical problems, such as contact pitting and wear, contact "bounce," shock, and vibration.*

Of the three standard connections of the transistor, the common-emitter configuration is by far the most popular in switching circuits. The common-emitter connection is the almost universal choice in switching circuits because this configuration provides both voltage gain and current gain. Hence, the output of a common-emitter switch can drive several succeeding transistor switches, an extremely desirable property in electronic computer circuitry. Of the remaining configurations, the common-collector connection is occasionally useful because it provides current gain and possesses essentially unity voltage gain. The common-base connection is almost never used in switching circuitry. The transistor switching theory which follows is, therefore, limited to the common-emitter connection.

Since the transistor switch is to perform an electrical function very similar to that of the mechanical single-pole, single-throw switch, we com-

*The MOSFET is also used in such circuits and will be discussed in a later chapter.

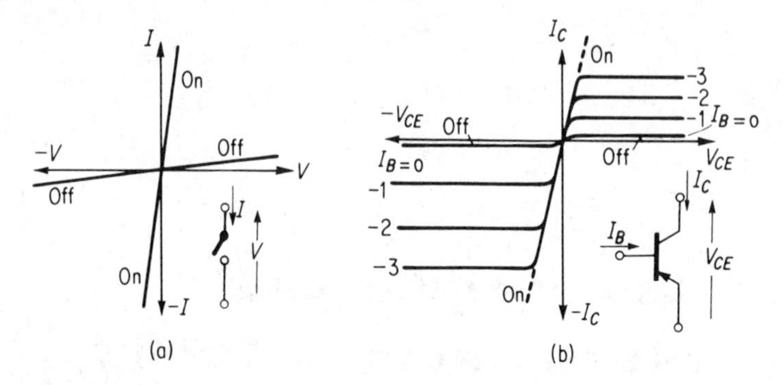

Fig. 7-1. Output characteristics (a) of a mechanical switch; (b) of a transistor in the common-emitter connection.

pare the two devices: Figure 7-1 shows (a) the output characteristics of a mechanical switch, (b) the output characteristics of a transistor in the common-emitter connection.

It is immediately obvious that the two devices are alike in that each can operate in the first and third quadrants of the *V-I* characteristics. The deviation of the mechanical switch output characteristics from the axes has been greatly exaggerated, however, in order to show the similarity. In the mechanical switch, the ON resistance is very nearly zero, and the OFF resistance is very large. The degree to which the analogy is accurate is thus dependent on the degree to which the transistor approximates an open or short circuit in its OFF and ON states, respectively.

For the moment, let us confine our attention to operation of the transistor switch in the third quadrant only. This region corresponds to "normal" operation of the PNP transistor shown in Fig. 7-1(b). For convenience, we may symmetrically invert the transistor output characteristics, as shown in Fig. 7-2. Also, the regions of interest in which the transistor output characteristics somewhat simulate those of the mechanical switch have been indicated. Region I corresponds to the OFF condition of the mechanical switch and is called *cutoff*. Region II is the linear active region of operation of the transistor for small-signal applications. Region III corresponds to the ON condition of the mechanical switch and is called the *saturation region*, for reasons which will now be explained.

Consider first the cut-off region of Fig. 7-2. Since any negative base current serves to turn the transistor on, the boundary of this region is taken as the static output characteristic when $I_B = 0$, or when we open-circuit the base. It is readily seen that the transistor switch has some reverse current flowing, even at relatively low voltage. The path through which this current flows may be represented by a resistance shunting the switch terminals which is called the *leakage* resistance. For some transistors, it may be desirable to operate

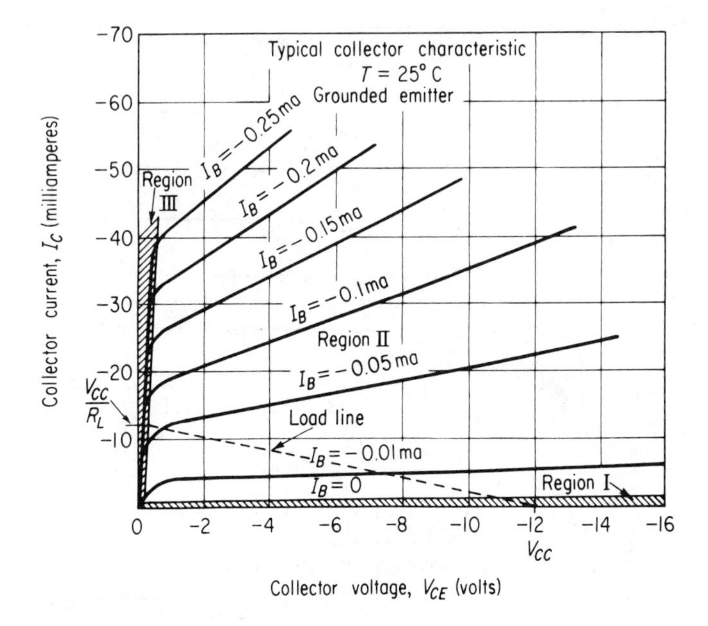

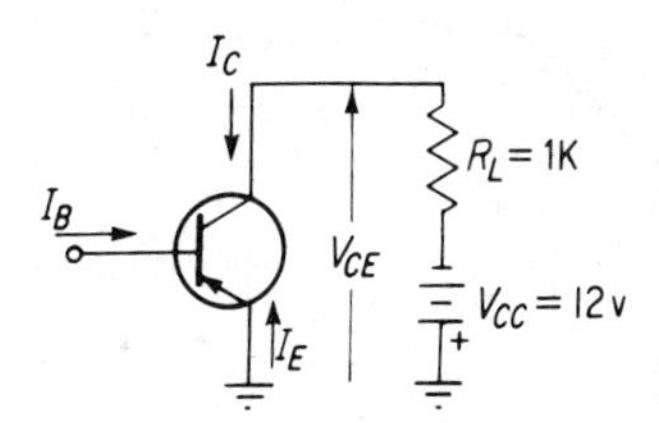

Fig. 7-2. Common-emitter output characteristics showing three regions of operation.

well into the cut-off region in order to obtain a very high leakage resistance. This desirable condition may require a reverse bias on the base of the transistor to hold it in cutoff.

Figure 7-3(a) shows an enlarged plot of Region III, and Fig. 7-3(b) shows the input characteristic of the transistor. This input characteristic is used to establish the boundary of Region III. For proper switching action, it is desirable that the collector-to-emitter voltage be as low as possible. For the circuit accompanying Fig. 7-2, a loadline may be constructed on the output characteristics. If the device is driven by a base current of —0.15 ma, the collector current is very nearly equal to $-V_{CC}/R_L$. If the base current is further increased to —0.25 ma, however, there is essentially no increase in collector current, and the transistor is said to be *saturated.* The increase in

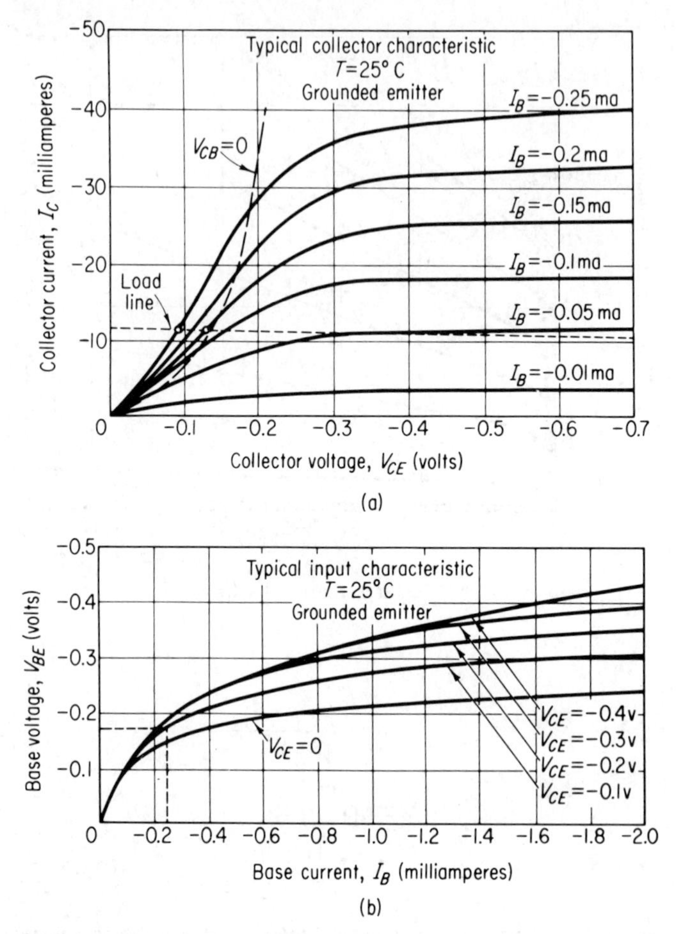

Fig. 7-3. Saturation and input characteristics of the common-emitter transistor.

base current does produce a reduction in the collector-to-emitter voltage, so that operation well into the saturation region is desirable from a dc power dissipation viewpoint.

From the input characteristic of Fig. 7-3(b), we see that operation of the device at a base current of -0.25 ma requires a base-to-emitter voltage of approximately -0.18 volt, but the collector-to-emitter voltage at this operating condition is only -0.1 volt. This occurs because the device is being overdriven, and the collector junction has become forward biased. The saturating voltage $V_{CE\,\text{sat}}$ is the algebraic sum of the individual junction voltages. Since the collector junction is forward biased, $V_{CE\,\text{sat}}$ is equal to the difference in magnitude between V_{BE} and V_{BC}.

The *edge of saturation*, or the boundary between Region II and Region III, is defined by superimposing, on the output characteristics, a locus of points where $V_{CB} = 0$, or $V_{CE} = V_{BE}$. Referring to Fig. 7-3(b), it is seen that, at any given base current, $V_{BE} < V_{CE}$ along the uppermost curve of the input characteristic. To obtain the boundary of the saturation region, choose a value of V_{CE} and determine the value of I_B for which $V_{BE} = V_{CE}$ in Fig. 7-3(b). Each point of the curve for the boundary of Region III is then transferred to the output characteristics.

The slope of the output characteristics at very low voltages has the units of conductance. The inverse of the slope is, therefore, the saturation resistance, r_{sat}, of the transistor, and may be as large as several hundred ohms. In our analogy with the mechanical switch, then, the transistor may be represented by a perfect switch in series with a saturation resistance and a parallel shunt leakage resistance. To a good approximation, the values of these equivalent resistances may be obtained from the slope of the output characteristics at low voltage and at zero base current.

THE EBERS AND MOLL EQUATIONS

The equivalent circuit of a transistor switch evolved in the preceding section provides no qualitative information about the use of the transistor as a switch, or the influence of external circuit elements. A much more detailed understanding of the transistor switch may be obtained from the Ebers and Moll equations mentioned in an earlier chapter. For convenience, they will be restated here. There are two equivalent forms of the Ebers and Moll equations: the symmetrical voltage equations:

$$I_E = I_{ECS}(e^{qV_{EB}/kT} - 1) - \alpha_I I_{CES}(e^{qV_{CB}/kT} - 1) \tag{7-1}$$

$$I_C = -\alpha_N I_{ECS}(e^{qV_{EB}/kT} - 1) + I_{CES}(e^{qV_{CB}/kT} - 1) \tag{7-2}$$

and the symmetrical current equations:

$$I_E = \frac{I_{EBO}}{1 - \alpha_N \alpha_I}(e^{qV_{EB}/kT} - 1) - \alpha_I I_C, \qquad \left(\frac{-V_{CB}}{kT} \gg 1\right) \tag{7-3}$$

$$I_C = \frac{I_{CBO}}{1 - \alpha_N \alpha_I}(e^{qV_{CB}/kT} - 1) - \alpha_N I_E, \qquad \left(\frac{-V_{EB}}{kT} \gg 1\right) \tag{7-4}$$

In addition, for most low-level and medium-level switching transistors,

$$\alpha_N I_{EBO} = \alpha_I I_{CBO} \tag{7-5}$$

The various quantities were defined in Chap. 1. The equivalence of the two forms is dependent on the two identities:

$$I_{ECS} = \frac{I_{EBO}}{1 - \alpha_N \alpha_I} \tag{7-6}$$

$$I_{CES} = \frac{I_{CBO}}{1 - \alpha_N \alpha_I} \tag{7-7}$$

The derivation of these equations by Ebers and Moll is quite rigorous and is independent of the physical geometry of the transistor (1).* A less rigorous derivation is described here. We assume that the transistor may be physically separated into two linear transistors in parallel. As we saw in Fig. 7-1(b), the transistor is operable in both the first and third quadrants of the current-voltage output characteristics. For convenience, we choose our parallel transistors so that one transistor (the "normal" transistor) will be a linear, unilateral device in the third quadrant, and the other transistor (the "inverted" transistor) will be a linear, unilateral device in the first quadrant. These assumptions are made for a PNP transistor only, but the method is generally valid for NPN transistors.

Since our fictitious transistors are linear and unilateral, they may be represented by the simple equivalent circuits shown in Fig. 7-4. Combining

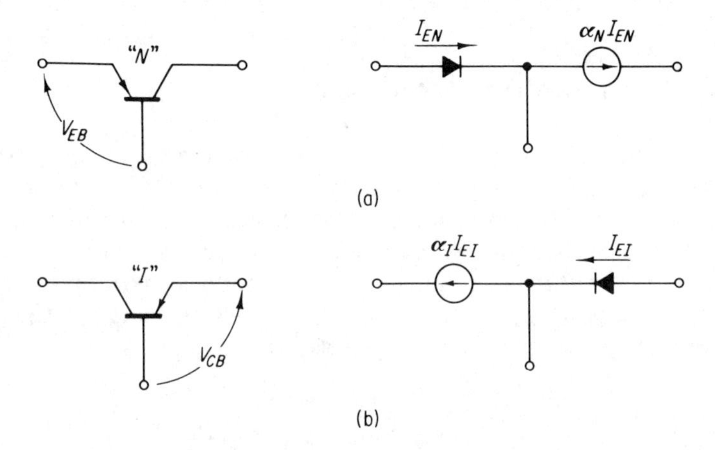

Fig. 7-4. (a) Equivalent circuit of "normal" transistor; (b) equivalent circuit of "inverted" transistor.

the two equivalent circuits into a composite representation for the physical transistor, we have an equivalent circuit which represents the real transistor over its entire range of operation. This circuit is shown in Fig. 7-5. Expressions for the emitter diode currents I_{EN} and I_{EI} are readily obtained from the discussions of Chap. 1 as

$$I_{EN} = I_{SN}(e^{qV_{EB}/kT} - 1) \tag{7-8}$$

$$I_{EI} = I_{SI}(e^{qV_{CB}/kT} - 1) \tag{7-9}$$

Note that in the circuit of Fig. 7-5, if $V_{CB} = 0$, then $\alpha_I I_{EI} = 0$, and $I_E = I_{EN}$. The emitter diode saturation current I_{SN} is, therefore, the reverse emitter

*Numbers in parentheses refer to material listed at the end of the chapter.

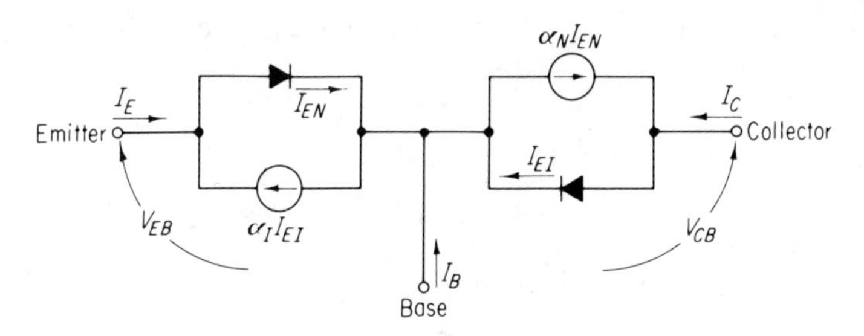

Fig. 7-5. Transistor equivalent circuit formed from equivalent circuits for the normal and inverted transistors.

leakage current of the real transistor with base shorted to collector, or $I_{SN} = I_{ECS}$. Similarly, $I_{SI} = I_{CES}$.

From this relatively simple equivalent circuit, the symmetrical voltage form of the Ebers and Moll equations may be written by inspection and give expressions 7-1 and 7-2. The symmetrical current form of the equations may also be obtained from the equivalent circuit of Fig. 7-5.

Let us consider next the application of these equations to the output characteristics of Fig. 7-2. We shall redefine the three regions of interest in terms of the terminal voltages of the transistor as follows:

Region I (cutoff): emitter and collector junctions reverse-biased, or

$$V_{CB} < 0, \qquad V_{EB} < 0$$

Region II (active): emitter forward-biased, collector reverse-biased, or

$$V_{CB} < 0, \qquad V_{EB} > 0$$

Region III (saturation): both junctions forward-biased, or

$$V_{CB} > 0, \qquad V_{EB} > 0.$$

In Region I, if $V_{CB} \lesssim -4(kT/q)$ and $V_{EB} \lesssim -4(kT/q)$, then

$$I_{C\,\text{off}} \approx (\alpha_N I_{ECS} - I_{CES}) = -\frac{(1-\alpha_I)I_{CBO}}{1-\alpha_N\alpha_I} \tag{7-10}$$

Thus, for a reverse emitter bias of at least 4 kT/q (approximately 100 mv at room temperature), the OFF collector current is slightly less than I_{CBO}. If $V_{EB} = 0$, the OFF collector current is, by definition, $-I_{CES}$, or $-I_{CBO}/(1 - \alpha_N\alpha_I)$. The cutoff collector current, then, has a value between $-I_{CES}$ and approximately $-I_{CBO}$.

A particular case of interest in the active region (Region II) is when $V_{EB} \approx +4(kT/q)$, which is of the order of $V_{CE\,\text{sat}}$. Under this condition, $(qV_{EB}/kT - 1) \approx 50$, and the collector current is approximately

$$I'_C \approx -(1 + 50\alpha_I)I_{CES} \tag{7-11}$$

In Region III, both junctions are forward biased and the currents in the circuit are essentially determined by external circuit elements. It is convenient to solve Eqs. 7-1 and 7-2 for the junction voltages. From

$$V_{EB} = \frac{kT}{q} \ln\left[1 + \frac{I_E + \alpha_I I_C}{I_{EBO}}\right] \tag{7-12}$$

and from

$$V_{CB} = \frac{kT}{q} \ln\left[1 + \frac{I_C + \alpha_N I_E}{I_{CBO}}\right] \tag{7-13}$$

we can obtain the saturation voltage as

$$V_{CE\,\text{sat}} = \frac{kT}{q} \ln\left[\frac{\alpha_I(I_C + \alpha_N I_E + I_{CBO})}{\alpha_N(I_E + \alpha_I I_C + I_{EBO})}\right] \tag{7-14}$$

where we have used $V_{CE} \equiv V_{CB} - V_{EB}$. Since we are interested only in the common-emitter configuration, it is highly desirable to write Eq. 7-14 in terms of base current. From Fig. 7-2, $I_C + I_E + I_B = 0$. Also, $I_B = (1 - \alpha_N)I_E$. Making these substitutions into Eq. 7-14, the common-emitter saturation voltage becomes

$$V_{CE\,\text{sat}} = \frac{kT}{q} \ln\left[\frac{\alpha_I\left(1 - \dfrac{I_C}{\beta_N I_B}\right)}{1 + \dfrac{I_C}{I_B}(1 - \alpha_I)}\right] \tag{7-15}$$

where, as before, $\beta_N = \alpha_N/(1 - \alpha_N)$. In order to drive the transistor into saturation, $\beta_N I_B > I_C$. Let us define a *circuit* current gain of $I_C/I_B = \beta_C$, where β_C is the "circuit beta." Note that β_C is always less than β_N. Then,

$$V_{CE\,\text{sat}} = \frac{kT}{q} \ln\left[\frac{\alpha_I\left(1 - \dfrac{\beta_C}{\beta_N}\right)}{1 + \beta_C(1 - \alpha_I)}\right] \tag{7-16}$$

As mentioned in the earlier sections of this chapter, the saturation resistance of the transistor should be as low as possible if the ON transistor is to simulate a short circuit. To obtain the dynamic saturation resistance, we may differentiate Eq. 7-16 with respect to I_C to obtain

$$\frac{\partial V_{CE\,\text{sat}}}{\partial I_C} = r_{\text{sat}} = -\frac{kT}{q}\left[\frac{1}{\beta_N\left(I_B - \dfrac{I_C}{\beta_N}\right)} + \frac{1}{\beta_I\left(\dfrac{I_B}{\alpha_I} + \dfrac{I_C}{\beta_N}\right)}\right] \tag{7-17}$$

In Eq. 7-17, the quantity I_C/β_N is the base current required to just saturate the transistor at a collector current of I_C. Equation 7-17 shows that the harder the transistor is overdriven, the lower is the saturation resistance.

TRANSISTOR CHOPPERS

The chopper is a special category of transistor switching circuit. Choppers are used primarily in applications where the input is a low-level, very low-frequency (essentially dc) signal. Using a chopper and a subsequent ac amplifier eliminates need for a stable dc amplifier. Let us look at the special transistor requirements for this application.

An expression for the saturation voltage of a common-emitter switch, derived from the Ebers and Moll equations, is given by expression 7-15. For the transistor switching circuit shown in Fig. 7-6(a), the collector current is near zero, since there is no collector supply voltage. Under these conditions, Eq. 7-15 reduces to

$$V_{CE} = \frac{kT}{q} \ln (\alpha_I) = -\frac{kT}{q} \ln \left(\frac{1}{\alpha_I}\right) \tag{7-18}$$

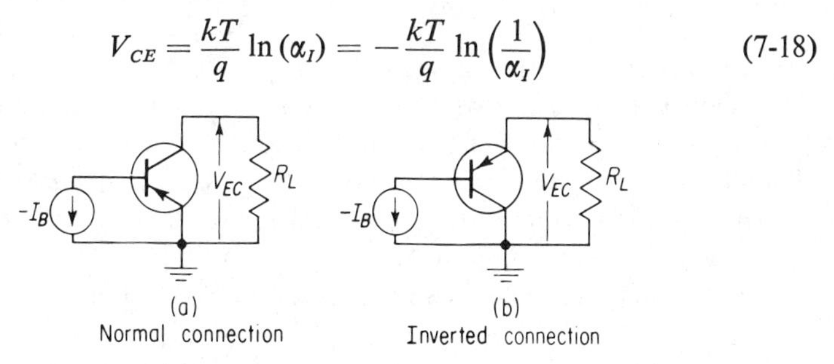

Fig. 7-6. Normal and inverted transistor connections. The collector-emitter voltage is a function of the dc β of the transistor.

Equation 7-18 shows that for α_I very nearly equal to unity, the saturation voltage is near zero, a very desirable condition; for most transistors, however, α_I is considerably less than unity. But α_N is normally very near unity, and since the saturation voltage of a normally connected transistor switch is expressed in terms of α_I, there may be some merit in using the transistor in the inverted connection, as shown in Fig. 7-6(b). Returning to expressions 7-12 and 7-13 for the individual junction voltages, we may develop an expression for the inverted saturation voltage, V_{EC}, and for $I_E = 0$, as:

$$V_{EC} = -\frac{kT}{q} \ln \frac{1}{\alpha_N} \tag{7-19}$$

Now α_N is very nearly unity, so that we may expand the logarithmic term and the inverted saturation voltage at zero emitter current is thus approximately:

$$V_{EC} \approx -\frac{kT}{q}\left(\frac{1-\alpha_N}{\alpha_N}\right) = -\frac{kT}{q\beta_N} \tag{7-20}$$

The quantity kT/q is approximately 26 mv at room temperature, so that the inverted saturation voltage, or "offset voltage," is of the order of a millivolt or less. The saturation resistance for very low emitter currents is found by differentiating Eq. 7-20 with respect to I_E to obtain

$$r_{\text{sat (inverted)}} = \left.\frac{\partial V_{EC}}{\partial I_E}\right|_{I_E=0} = -\frac{kT}{qI_B}\left[\frac{1}{\beta_I} + \frac{\alpha_N}{\beta_N}\right] \tag{7-21}$$

For the values of β_I and β_N normally encountered, r_{sat} is of the order of tens of ohms, or less. The inverted transistor switch is thus an almost perfect low-level switch, having a low ON resistance and a very low ON voltage across its terminals.

In the OFF condition, with both junctions reverse biased by at least 100 mv, the Ebers and Moll equations may be solved for the inverted "offset current," I_{OS}

$$I_{OS} = \frac{-(1 - \alpha_N)I_{EBO}}{1 - \alpha_N \alpha_I} \tag{7-22}$$

which shows that I_{OS} is always less than I_{EBO}. For good silicon transistors, the OFF leakage resistance may be as large as several thousand megohms.

From the preceding paragraphs, we can now construct the output characteristics of the low-level inverted transistor switch. At zero load current, the offset voltage $V_{EC(O)}$ is $kT/q\beta_N$, and the saturation resistance is approximately $-kT/q\beta_I I_B$. In the OFF condition, the offset current I_{OS} is $(1 - \alpha_N)I_{EBO}/(1 - \alpha_N\alpha_I)$. These points are shown in Fig. 7-7.

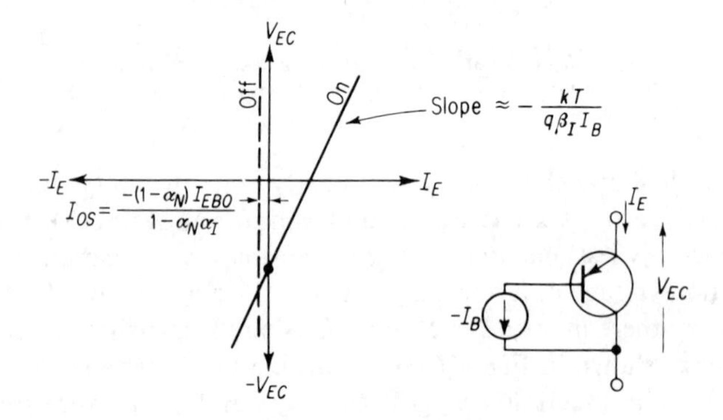

Fig. 7-7. The variation of offset voltage with emitter current in the inverted connection.

When the transistor is operated in this low-level region, either in the normal or inverted connection, the switching circuit is often referred to as a *transistor chopper;* the name refers to the ability of the circuit to "chop" dc and very low-frequency signals so that they may be amplified by conventional ac amplifiers.

In deriving the low-level parameters of the transistor switch, we assumed that the load current was zero. Examination of Eq. 7-16 shows that the conditions of zero load current are satisfied if $I_L \ll \beta_I I_B$. Practically, this requires only that the transistor be held well into the saturation region.

The Ebers and Moll equations used to describe the behavior of the transistor switch are strictly true only for the intrinsic device. In general, the extrinsic elements, such as spreading resistances, may be neglected in medium-level switching circuits. At the very low signal levels for which the transistor chopper is intended to be used, the effects of the extrinsic resistances may not be neglected. Figure 7-8 shows the complete equivalent circuit of a low-level inverted transistor switch. Generally, r_{cl} and r_{el} are negligibly large. The extrinsic base resistance also may be neglected since it does not appear between the output terminals of the switch. A reasonably accurate equivalent circuit for the transistor chopper is therefore shown in Fig. 7-9. The only

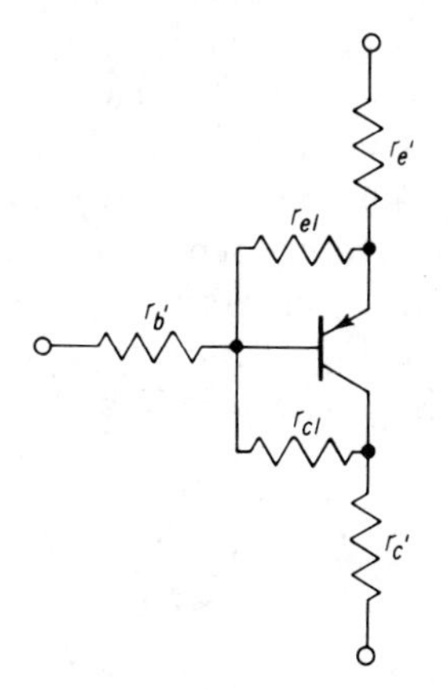

Fig. 7-8. Equivalent circuit for the chopper transistor showing parasitic shunt and series resistances.

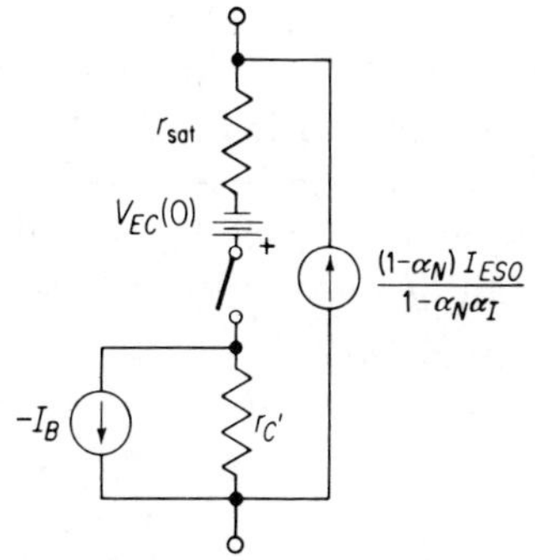

Fig. 7-9. Equivalent circuit for the chopper transistor valid in the ON and OFF states.

extrinsic resistance of any significance in the inverted connection is r_C', since the relatively large base current produces a voltage across this resistance. Because of this resistance, the offset voltage is a function of base current, as shown in Fig. 7-10. The "hump" at low base current occurs because of the current dependence of β_N at very low currents. At high base currents, the curve has a slope equal to r_C'. The extrinsic offset voltage is composed of two

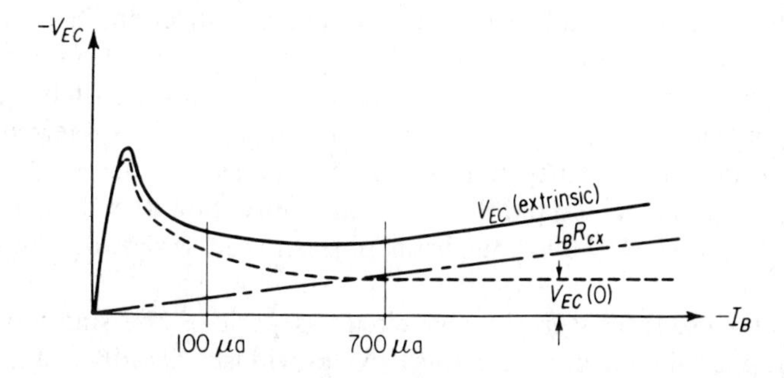

Fig. 7-10. The variation of offset voltage with base current.

components; this results in a nonlinear variation with temperature. The intrinsic offset voltage $V_{EC(O)}$ is a function of temperature, both explicitly (kT/q) and implicitly $(1/\beta_N)$. The extrinsic collector resistance r'_C is also a function of temperature. The net effect of the interactions between these quantities is shown in Fig. 7-11. At relatively high base currents (I_{B3}) where $I_B r'_C$ is the predominant term in the expression for the extrinsic offset voltage, V_{EC} increases linearly with increasing temperature. At very low base currents (I_{B1}), V_{EC} decreases with increasing temperature at low temperatures, because β_N is increasing. Near room temperature, however, the kT/q factor begins to increase faster than β_N, and V_{EC} begins to increase. At moderate base currents (I_{B2}), of the order of 100–500 μa, V_{EC} is essentially constant with temperature, because of the interaction of the various temperature-dependent quantities. Note that the range of base currents for which V_{EC} is essentially constant is also the range for which V_{EC} is minimum in Fig. 7-10. Operation of the chopper with I_B in the order of 100–500 μa is thus very desirable.

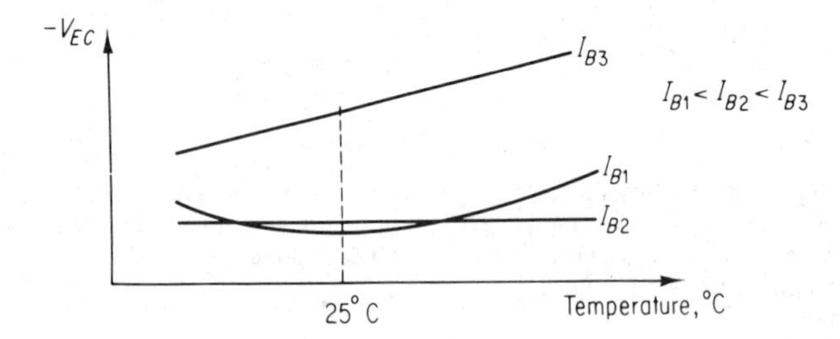

Fig. 7-11. The variation of offset voltage with temperature. The base current may be adjusted to yield a zero temperature coefficient of offset voltage.

As an example of the use of a transistor switch as a chopper, consider the simple modulator circuit shown in Fig. 7-12. When the base voltage is positive, the transistor is OFF and the output is essentially equal to the input signal. When the base drive voltage is negative, the transistor is saturated and the output signal is virtually zero. The chopper circuit "chops" signal source, the transistor switch, and the external load. In one connection, the output signal is taken directly across the terminals of the transistor switch so that the chopper appears in shunt with the output terminals. This is referred to as a *shunt chopper*. In the other connection, the transistor appears as a series switch between the source and the load. This connection, therefore, is a series chopper. Figure 7-13 shows this distinction between series and shunt choppers.

The simplest transistor chopper circuit is perhaps the single transistor shunt chopper shown in Fig. 7-13(a). As previously described, the steady-

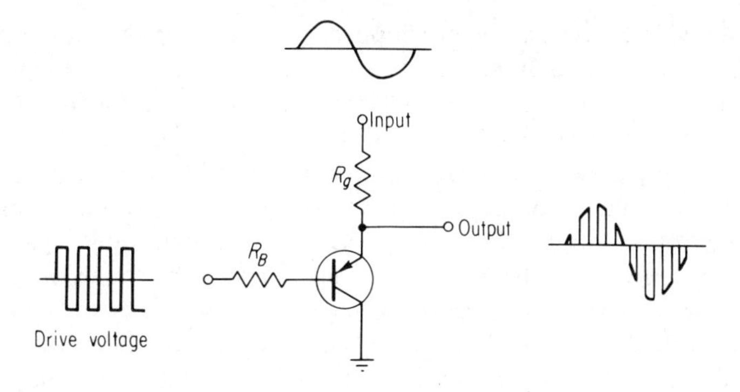

Fig. 7-12. A simple modulator circuit.

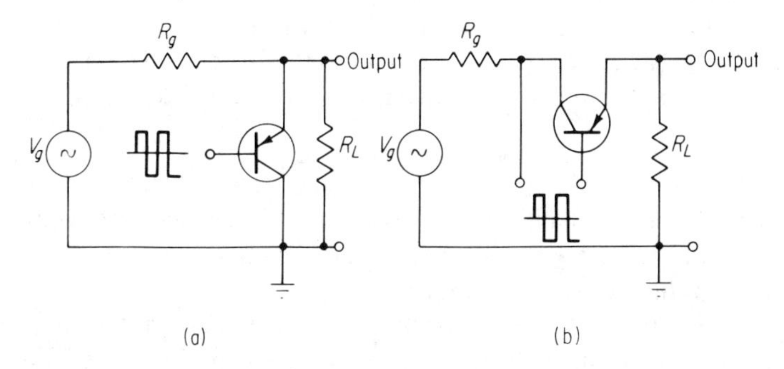

Fig. 7-13. (a) Shunt and (b) series choppers.

state conditions existing in the circuit may be characterized by an equivalent circuit which approximates the behavior of the transistor switch. Since, ideally, the steady-state output voltage appearing at the transistor terminals is zero when the transistor is conducting, the transistor is used in the inverted connection so as to approximate a true short circuit more closely. Note, however, that the range of signal currents which can be accommodated by the shunt chopper of Fig. 7-13(a) is determined primarily by the magnitude of the base drive voltage which is available. The ON base drive must be sufficiently large to hold the inverted transistor heavily in saturation while conducting the load current to ground, and the OFF base voltage must be sufficiently large to maintain a reverse bias on the emitter junction with the maximum positive amplitude of input signal present at the emitter terminal.

A source of error is introduced at the output terminals if the offset voltage of the shunt chopper varies significantly with ambient temperature variations. However, the factors which contribute to the net offset voltage produce a somewhat compensating temperature effect, for some particular value of base drive current. Since this same value of base current also produces the minimum offset voltage, the transistor chopper should be operated at this optimum value. Because the base drive current is restricted to this value, a high degree of saturation can be maintained only by externally limiting the maximum value of the signal current which flows through the ON transistor.

The maximum value of input signal which can be chopped is primarily limited by the magnitude of base drive voltage that is available. For very low-level chopper applications, however, the minimum input signal which may be chopped is of interest. For the single transistor shunt chopper, the minimum input signal which can be recognized is obviously of the order of the offset voltage, V_{EC}. For input signals of less than several millivolts, then, the single transistor chopper is unsuitable. A novel circuit (2) that may be used to chop extremely small input signals is shown in Fig. 7-14. This circuit consists of two single transistor choppers connected as a back-to-back pair, with the base drive circuit isolated from ground. In this back-to-back connection, the offset voltages of the pair of transistors tend to cancel, so that the net voltage at the output terminals with both transistors ON is

$$V_O = V_{EC_1} - V_{EC_2} \tag{7-23}$$

By proper selection of matched pairs of transistors, the net offset voltage may be minimized. Transistor pairs are available which are matched over a wide temperature range to within ± 25 μvolts.

The offset current of such a back-to-back matched pair is equal to the offset current of a single transistor chopper. In the circuit of Fig. 7-14, however, no turn-off bias need be applied to the base of the transistors in order to block input signals of either polarity. In the OFF state, with no reverse bias applied, the circuit shown in Fig. 7-15 will not conduct in either direction because the emitter diodes are in an opposing direction. The circuit is thus capable of chopping input voltages of considerable magnitude without requiring a high reverse base drive.

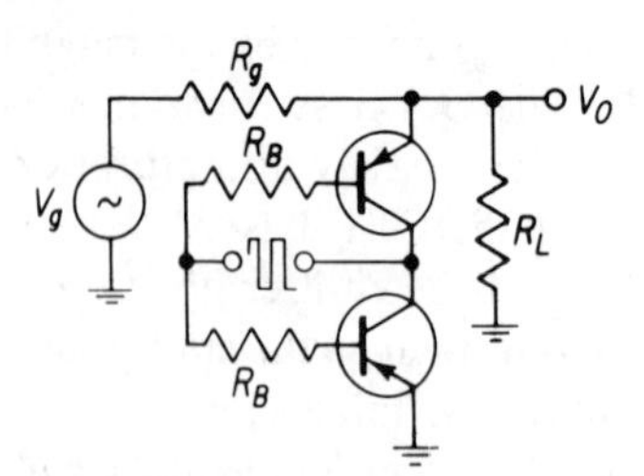

Fig. 7-14. A back-to-back transistor pair used as a shunt chopper. The individual offset voltages cancel, reducing the net offset voltage across the pair.

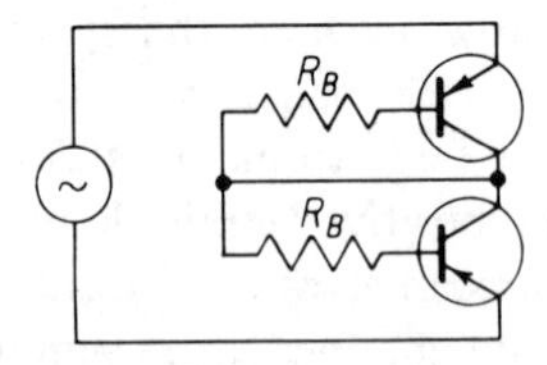

Fig. 7-15. The back-to-back pair blocks signals of either polarity.

THE ANALOG SWITCH

The back-to-back matched-pair chopper circuit just described has many advantages over the single-transistor chopper. In addition to its usefulness as a series or shunt chopper for dc and very low-frequency modulation, however, the back-to-back matched pair of chopper transistors is extremely useful in what may be loosely described as *analog switching circuits.* Such circuitry is also known by a variety of names, such as *sampling*, *time division multiplexing*, or *commutating* circuits.

In many fields, it is often desirable to combine the continuous analog data output from a number of sources in some manner so that these data may be transmitted via a single communications channel. As an example, consider the system shown in Fig. 7-16. As the multiposition switch S_1 sweeps counterclockwise, it samples each input signal instantaneously. The combined output of switch S_1 is then suitable for transmission over a single wire to some remote location. At the receiving terminal, a similar multiposition switch is synchronously swept in a clockwise direction such that it distributes the combined signals to the proper output terminal. For proper operation of the system, a synchronizing signal must also be transmitted over the system. The switches S_1 and S_2 are essentially "analog switches." Figure 7-17 shows a method of implementing the system of Fig. 7-16 using the back-to-back matched pair of chopper transistors. Each input is sampled in a sequential manner by sequentially turning on the corresponding matched pair of transistors. The combined output then appears as a pulse train across a common load resistor, R_L. In order to maintain a high degree of accuracy in representing the various input signals, each input must be sampled at a rate considerably higher than the highest frequency of interest in the input signals.* The combined output will, therefore, consist of pulses having a

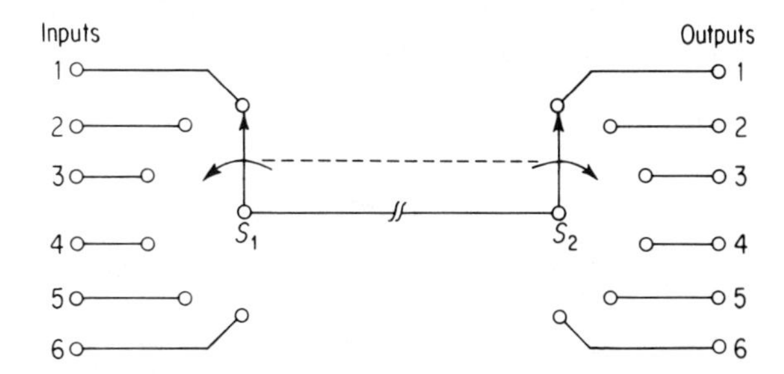

Fig. 7-16. Schematic representation of an analog switching network.

*With the addition of low-pass filters in reconstruction, one needs only to sample in excess of twice the highest-frequency component in order to recover the signal completely.

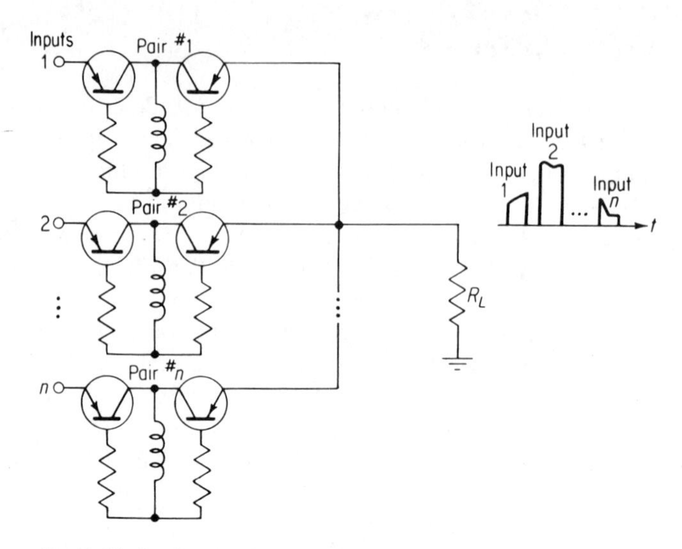

Fig. 7-17. Implementation of the analog switching network with back-to-back chopper pairs.

relatively short duration and fast rise and fall times. Now consider the effect of the number of system inputs on the output. In Fig. 7-17, when transistor pair no. 1 is ON, pairs 2 through n are OFF. Each pair of transistors has some output capacitance which appears in parallel with the load resistor, R_L. Although the individual output capacitances are low (of the order of 2–3 pf), the combined capacitance shunting the load resistor may be sufficiently large to affect the system response time. In such cases, the maximum number of analog switch outputs which can be connected to the load resistor is fixed. If more than this number of input terminals are required, a pyramiding technique, such as the one shown in Fig. 7-18 must be used. Here, the analog switches have been shown as ideal switches for convenience. This technique is applicable to any number of system inputs.

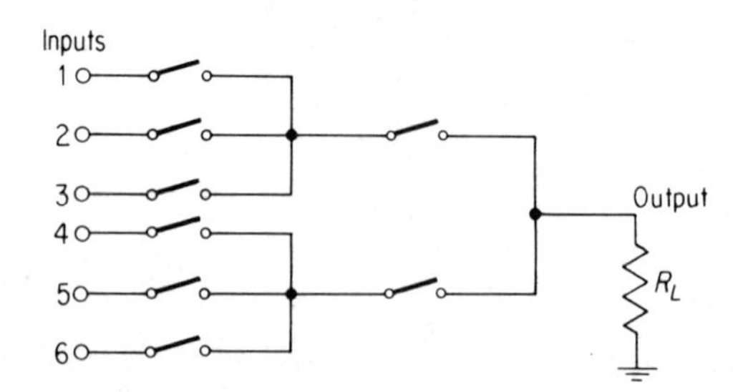

Fig. 7-18. Pyramiding of the analog switches is used to reduce the effect of capacitance.

CHOPPER TRANSIENT PERFORMANCE

In all the foregoing discussion, we have been dealing with steady-state conditions in the transistor switch. In addition to the steady-state offset voltage and offset current, the output signal of a simple transistor chopper contains electrical transients which occur when the switch changes from the ON to the OFF state, and vice versa. These transients arise from charge storage in the intrinsic transistor itself and from capacitive feedthrough from the base drive circuit to the output terminals.

Consider the single transistor shunt chopper circuit shown in Fig. 7-19. In order to turn the transistor ON, sufficient charge must be stored in the base region of the device to saturate it. The charge may be considered to be stored in an internal capacitance. Also, the various circuit capacitances shown in Fig. 7-19 are charged to their steady-state value in the ON condition. When the base drive voltage changes abruptly from $-V_{BB}$ to $+V_{BB}$, as shown, the voltages across the various capacitances cannot change instantaneously, and the output voltage tends to follow the rising base voltage. Initially, the transistor is ON, and thus a relatively low resistance path exists in the circuit. As soon, however, as sufficient charge exists to allow the emitter diode to reach zero bias, the diode begins to recover to its high-resistance OFF state, so that the remaining stored charge in the transistor and in the various circuit capacitances discharges relatively slowly. The turn-off transient, then, is a sharply rising positive "spike" with a relatively long decay time, as shown in Fig. 7-19.

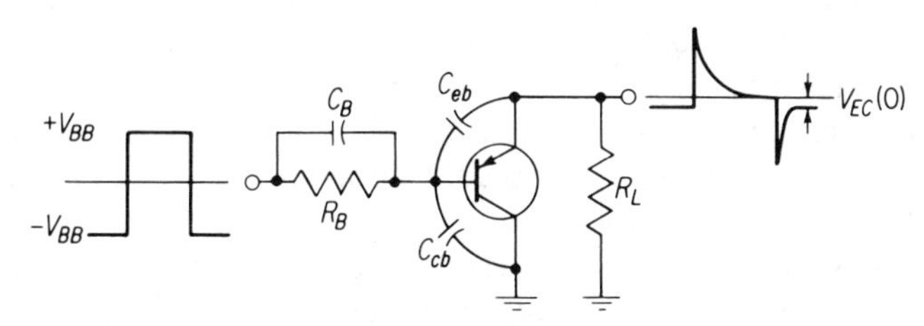

Fig. 7-19. Transient spikes occur during the switching intervals.

In the steady-state OFF condition, the emitter and collector junction capacitances are fully charged to the reverse base turn-off voltage, $+V_{BB}$, whereas any stray circuit capacitance across the base resistor, such as C_B, is uncharged. When the base drive voltage changes abruptly from $+V_{BB}$ to $-V_{BB}$, a very sharp negative spike appears at the output terminals, because of the capacitive coupling between input and output. As soon as the junction capacitances are discharged to zero volts, however, the transistor begins to turn ON, and presents a very low resistance to any further charging currents. The turn-on transient thus decays very rapidly to zero.

In general, the amplitude of both the turn-on and turn-off transients is directly proportional to load resistance, whereas the decay time of both transients is independent of the load resistance. If the load resistance becomes too large, the effects of stray load capacitance become apparent. The amplitude of both transients is inversely proportional to load capacitance. The turn-off transient decay time of both transients is independent of the load resistance. If the load resistance becomes too large, however, the effects of stray load capacitance become apparent. The amplitude of both transients is inversely proportional to load capacitance. The turn-off transient decay time is greatly increased by load capacitance, so that this condition is to be avoided.

The amplitude of both transients increases with increased base drive voltage (at constant turn-on base current). The decay time of both transients is independent of base drive voltage, but, the method in which the drive voltage is introduced at the base of the transistor has a significant effect. This is perhaps best shown by example. Figure 7-20 depicts four simple methods of connecting the drive to the base load. In Fig. 7-20(a), a ± 10 volt square wave drive is supplied to the base through a 47 kilohm resistor. Both the leading and trailing edges of the square wave are capacitively coupled to the output terminals and produce "spikes" of the order of several volts. Shunting a diode clamp from base to collector of the transistor, as shown in Fig. 7-20(b), limits the magnitude of the turn-off bias to several tenths of a volt and results in a significant reduction in both the amplitude and duration of the transients. Since the turn-off bias is not sufficient to maintain the OFF condition in the presence of any high positive input signals, this circuit is restricted to use with low-level input signals. If this is the case, still further reduction in the transients can be obtained by returning the base resistor to ground during the OFF period of the chopping cycle, that is, by supplying only a negative turn-on pulse to the base, as shown in Fig. 7-20(c). Note that, in this circuit, although the amplitude of the transients is of comparable magnitude, the turn-off transient is of considerably longer duration. If the amplitude of this transient is of more interest than the duration, a further reduction by an order of magnitude in the spike amplitude can be achieved by the circuit of Fig. 7-20(d), with very little sacrifice in the transient duration. Here, the negative drive pulses are supplied to the base of the transistor through a diode in series with a current-limiting resistor. As previously mentioned, the insertion of a small capacitor from the emitter-to-collector terminals of the transistor results in a trade of transient duration for amplitude. The insertion of a fast recovery diode in series with the base resistor, as in Fig. 7-20(d), results in a much more efficient trade of spike duration for spike amplitude. The diode eliminates the feedthrough of the positive-going edge of the base drive voltage, so that the turn-off transient is due almost entirely to the exit of the charge stored internally in the transistor.

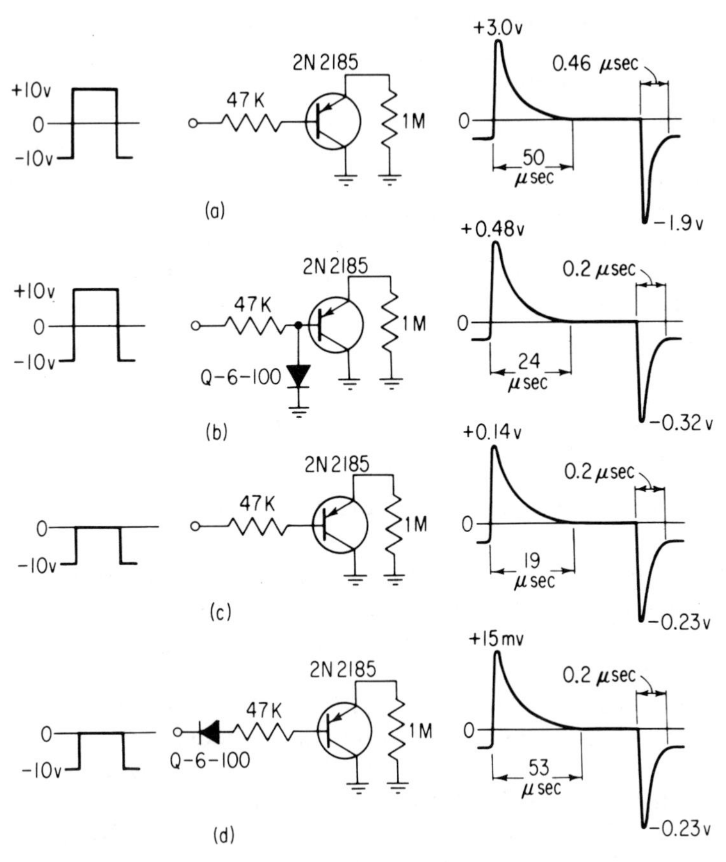

Fig. 7-20. Typical switching transients occurring in a shunt chopper. The 2N2185 is a silicon alloy transistor.

The chopper transients of the back-to-back pair are very similar to those of the single-transistor chopper. Since the matched pair requires an isolated base drive, usually supplied through a transformer, the analysis of these transients is considerably more complicated.

CHOPPER-STABILIZED AMPLIFIERS

DC amplification was discussed briefly in Chap. 4. In that chapter it was pointed out that it is possible to design and construct dc amplifiers with equivalent input drifts in the order of 5 μv/°C. This represents a sufficient level of stability for many applications, but often represents a limitation where a relatively high stable gain is required, or where extremely small dc signals must be amplified. If an amplifier has an equivalent input drift of

5 μv/°C, then a temperature change of 100°C has the same effect on the output as an input signal of 500 μv.

Generally, ac amplifiers can be designed to provide a more stable gain than dc amplifiers because of the relative lack of bias stability problems in the ac amplifier and the low dc stage gain that can be used (negative dc feedback). This situation is used to advantage in a chopper-stabilized amplifier. A transistor chopper is used to "chop" or modulate a dc signal, converting it to an ac signal. The ac signal is then amplified to the desired level and the original dc signal recovered from it by demodulating the ac signal. The block diagram for such an amplifying system is shown in Fig. 7-21.

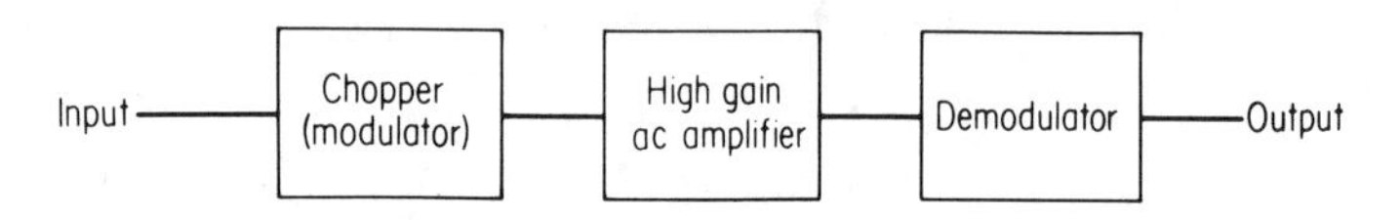

Fig. 7-21. Chopper stabilized dc amplifier system.

The chopper could be any one of the types referred to in previous sections of this chapter. The back-to-back chopper pair is generally preferred because of the lower offset voltage that can be obtained by using matched transistors. The amplifier design can employ conventional techniques (see Chap. 3). The only unique requirement of this amplifier is that it employs a differential output stage. This is required to drive the demodulator circuit which is shown in simplified form in Fig. 7-22. A square wave switching input, e_c, which is either in phase or 180° out of phase with the switching voltage for the chopper, is used to switch Q_1 and Q_2 ON and OFF alternately. The chopped and amplified differential input signal is demodulated through this switching action of Q_1 and Q_2 and the amplified dc is recovered as the output. With such an amplifying system it is possible to obtain extremely high gains (10^6–10^8) and very low drift rates (0.5 μv/°C). This is far better performance than can be obtained with dc amplifiers.

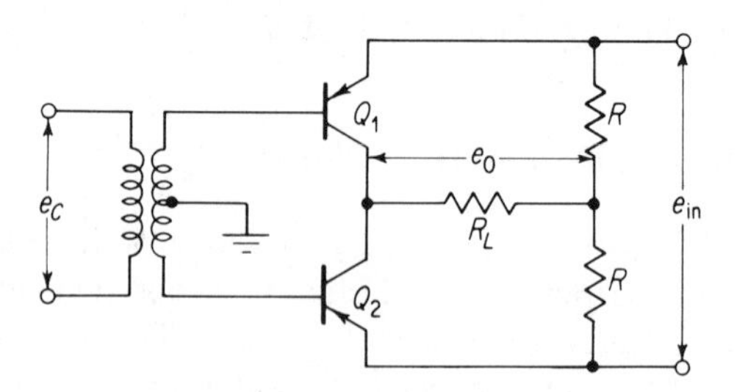

Fig. 7-22. A simple demodulator circuit.

AN APPLICATION OF HIGH-GAIN AMPLIFIERS

High-gain amplifiers have a variety of uses but perhaps the most widely known application is in operational amplifiers. An *operational amplifier* is an amplifying device which performs a mathematical operation, such as multiplication, addition, or integration.

Suppose we have an amplifier with an extremely high voltage gain $-K$, an infinite input impedance, and zero output impedance. How can this device be used as an operational amplifier?

Let us analyze the circuit shown in Fig. 7-23 and find the voltage gain of the circuit $G = e_0/e_1$. From Fig. 7-23 we can write two circuit equations

$$\frac{e_1 - e_1'}{R_1} = -\frac{e_0 - e_1'}{R_2} \tag{7-24}$$

and

$$e_0 = -\frac{1}{K}e_1' \tag{7-25}$$

If we eliminate e_1' from these two equations and solve for $G = e_0/e_1$, we obtain

$$G = \frac{-\dfrac{R_2}{R_1}}{1 - \dfrac{1}{K}\left(\dfrac{R_2}{R_1} - 1\right)} \tag{7-26}$$

Now suppose K is so large that

$$\frac{1}{K}\left(\frac{R_2}{R_1} - 1\right) \ll 1$$

Then, we have

$$G = \frac{e_0}{e_1} \approx -\frac{R_2}{R_1} \tag{7-27}$$

In other words, the output voltage is equal to the input voltage multiplied by a constant. This circuit can be used for multiplication (or division).

Look at the circuit of Fig. 7-24. Here a capacitor replaces R_2. One can readily show that this circuit performs the integration operation. Proceeding

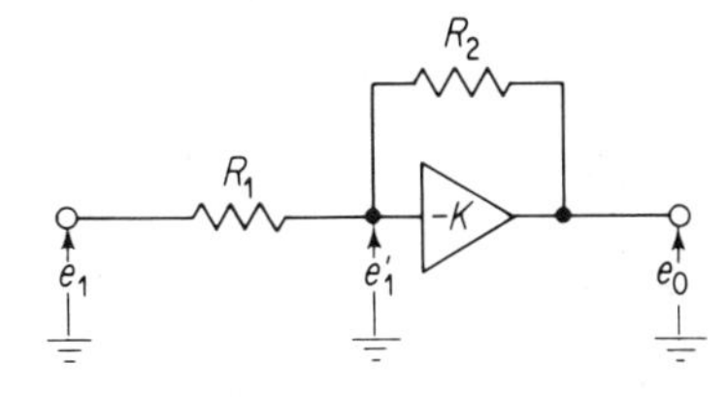

Fig. 7-23. A simple application of an operational amplifier.

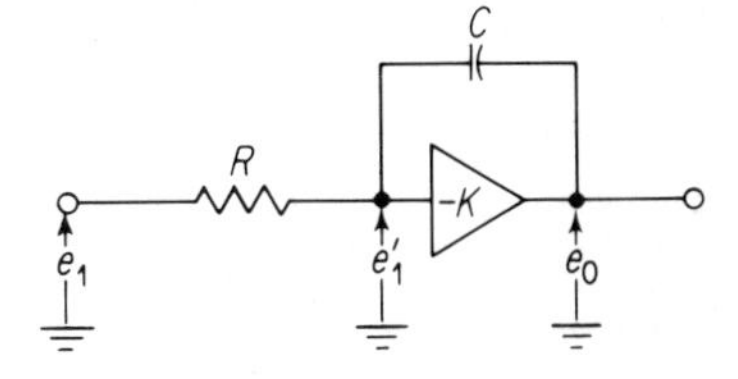

Fig. 7-24. An operational amplifier circuit which performs integration.

as in the prior example (using transform notation)

$$\frac{e_1 - e_1'}{R} = -sC(e_0 - e_1') \tag{7-28}$$

$$e_1' = -\frac{1}{K}e_0 \tag{7-29}$$

Letting $K \longrightarrow \infty$, then

$$e_0 = -\frac{1}{RCs}e_1 \tag{7-30}$$

Mathematically this is equivalent to the integration of e_1 to obtain

$$e_0 = \frac{1}{RC}\int_t e_1\,dt \tag{7-31}$$

Additional mathematical operations may be performed by proper selection of feedback components.

EXERCISES

1. Given the symmetrical voltage forms of the Ebers and Moll equations 7-1 and 7-2, and the equivalent circuit of Fig. 7-5 show that expressions 7-3 and 7-4 hold.
2. From the Ebers and Moll equations derive an expression in terms of I_B for the emitter current at which the intrinsic offset voltage of an inverted transistor switch equals zero.
3. An ideal transistor (no extrinsic elements) has the following characteristics at 25°C:

$$\beta_N = 50$$
$$\beta_I = 5$$
$$I_{EBO} = 10^{-9}\text{ amp}$$

   Construct the inverted common-emitter output characteristics at 25°C showing all pertinent regions and identifying all pertinent points. For a fixed-base current $I_B = 10^{-3}$ amp, what is the offset voltage for emitter currents of $I_E = 1\ \mu$a, 10 μa, and 100 μa?
4. An operational amplifier configuration is connected as shown. What mathematical operation does this circuit perform? What is the output voltage if the input voltage is a ramp function? Assume $K \longrightarrow \infty$.

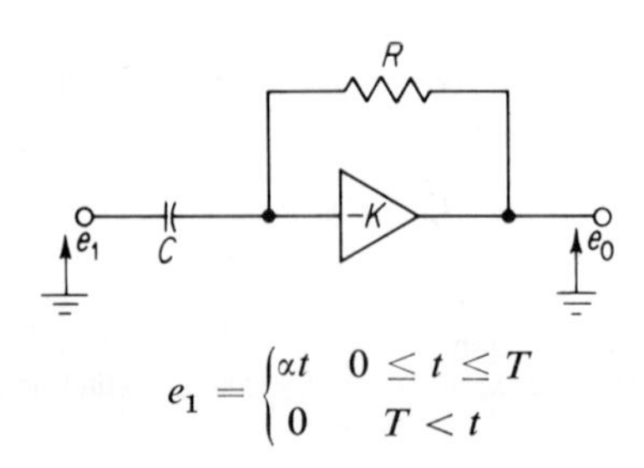

$$e_1 = \begin{cases} \alpha t & 0 \leq t \leq T \\ 0 & T < t \end{cases}$$

REFERENCES

1. J. B. EBERS and J. L. MOLL, "Large Signal Behavior of Junction Transistors," *Proc. IRE*, **42** (December, 1954), 1761–78.

2. R. L. BRIGHT, "Junction Transistors Used As Switches," Paper 55-156, *Trans. AIEE*, March, 1955.

3. A. K. SUSSKIND, *Notes on Analog-Digital Conversion Techniques.* New York: Technology Press of MIT and John Wiley & Sons, Inc., 1957.

4. C. BELOVE, H. SCHACHTER and D. L. SCHILLING, *Digital and Analog Systems, Circuits and Devices: An Introduction.* New York: McGraw-Hill Book Company, 1973.

5. S. SEELY, *Electronic Circuits.* New York: Holt, Rinehart & Winston, Inc., 1968.

6. H. E. STEWART, *Engineering Electronics.* Boston, Mass.: Allyn & Bacon, Inc., 1969.

7. F. C. FITCHEN, *Electronic Integrated Circuits and Systems.* New York: D. Van Nostrand Reinhold Company, 1970.

8. *Integrated Electronic Systems*, Westinghouse Defense and Space Center. Englewood Cliffs, N.J.: Prentice-Hall, Inc., 1970.

8

Operational Amplifiers

The previous chapter briefly introduced the operational amplifier. Here, because of its great importance in modern circuit design, we expand on that discussion.

The operational amplifier or op amp is an amplifier that exhibits very high ac and dc gain. Generally these amplifiers have a high input impedance and a low output impedance. This circuit derives its name from the traditional application of performing mathematical operations such as integration and summing. While the analog computer always required high quality, high precision, and high cost op amps, recent advances in microcircuit and integrated circuit techniques have reduced the size and cost of op amps to the point that a greater variety of practical applications is now feasible.

Op amps are now used in the fields of servomechanisms and automatic control, active network synthesis, instrumentation, and others. Signal conditioning, digital to analog conversion, regulation, and waveshaping are only a few of the many functions now performed quite effectively by the op amp. Because of the growing importance of this component in electronics and related areas, it is appropriate to consider the theory and applications of the op amp in some detail.

THE IDEAL OP AMP

A truly ideal op amp would exhibit infinite input impedance, zero output impedance, infinite voltage gain, and infinite bandwidth. Since these specifications are unattainable we will define the ideal op amp as a circuit having very high input impedance (no loading of preceding stages), negligibly low output impedance (compared to the load), very high voltage gain, and a bandwidth that exceeds that required by the particular application. We will further restrict our discussion to op amps having both inverting and noninverting inputs since this type of op amp is presently very popular. Figure 8-1 shows the schematic representation of an ideal differential op amp.

In practice it is easy to obtain op amps with gains ranging from 10^4 to 10^7, open loop bandwidths ranging from a few hertz to several kilohertz, input impedances in the megohm range, and output impedances in the ohm range. The asymptotic falloff of gain with frequency is generally 6 db/octave or

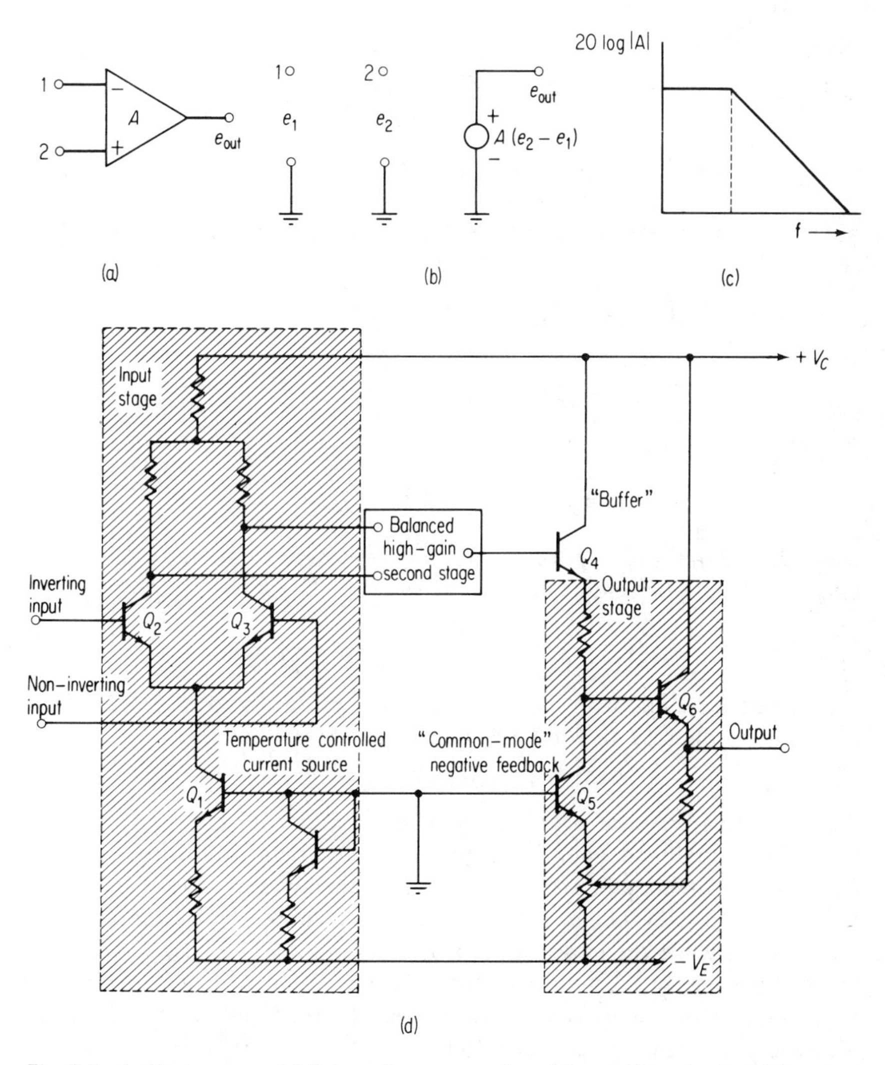

Fig. 8-1. An ideal op amp: (a) Schematic representation; (b) equivalent circuit; (c) frequency response; (d) circuit equivalent in integrated form.

occasionally 12 db/octave. While the orders of magnitude of output impedance and bandwidth may seem to invalidate the concept of the ideal op amp, such is not the case. The bulk of applications require the use of negative feedback and/or frequency compensation to extend the available bandwidth and decrease the output impedance of the amplifier in any event. We shall now consider some applications of the op amp.

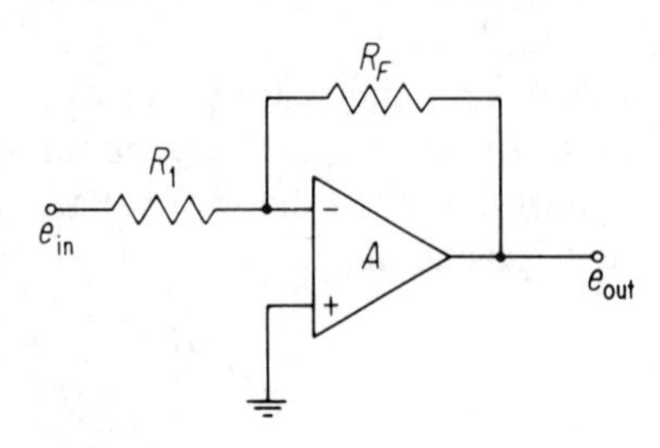

Fig. 8-2. An inverting amplifier.

Figure 8-2 shows the op amp being used as an inverting amplifier. If the open loop gain is given by

$$A = \frac{A_{MB}}{1 + j\omega/\omega_2}, \tag{8-1}$$

where A_{MB} is the midband gain, then the gain with feedback can be found to be:

$$G = \left(\frac{-A}{1 + AF}\right)\left(\frac{R_F}{R_1 + R_F}\right) \tag{8-2}$$

$$= \left(\frac{-A_{MB}}{1 + A_{MB}F}\right)\left(\frac{R_F}{R_1 + R_F}\right)\left(\frac{1}{1 + j[\omega/\omega_2(1 + A_{MB}F)]}\right) \tag{8-3}$$

In this case the feedback factor is

$$F = \frac{R_1}{R_1 + R_F} \tag{8-4}$$

If we assume that the return ratio $A_{MB}F$ is very large compared to unity, the closed loop gain becomes

$$G = \frac{-R_F}{R_1}\left(\frac{1}{1 + j[\omega/\omega_2(1 + A_{MB}F)]}\right) \tag{8-5}$$

The closed loop midband gain is determined by the ratio of the feedback resistance to the resistor R_1. The amplifier bandwidth is

$$BW = \omega_2(1 + A_{MB}F) \tag{8-6}$$

a considerable increase over the open loop bandwidth ω_2. For values of closed loop midband gain considerably greater than unity, insuring that $R_F \gg R_1$, midband gain and bandwidth are directly exchanged. The closed loop midband gain will be related to the open loop gain by the factor $1/(1 + A_{MB}F)$ while the closed loop bandwidth will be $(1 + A_{MB}F)$ times the open loop value. In effect the overall gain–bandwidth product of the amplifier is constant so long as R_F is much larger than R_1. We should further note that we have restricted our considerations thus far to amplifiers exhibiting a 6 db/octave rolloff with frequency.

As an example, let us consider the case where $R_F/R_1 = 1000$, $A_{MB} = 10^5$, and $\omega_2 = 2\pi \times 10^3$ rad/sec. When feedback is applied the closed loop gain is expressed quite accurately by

$$G = \frac{-10^3}{1 + j[\omega/2\pi \times 10^5]}$$

The midband gain is equal to -1000 while the bandwidth is $2\pi \times 10^5$ rad/sec.

The impedance seen by a voltage source is very nearly equal to R_1 for the inverting amplifier. In the midband region the loading effect of R_F on the amplifier input can be represented by the reflected impedance $R_F/(1 + A_{MB})$. The input impedance to the circuit is $R_1 + R_F/(1 + A_{MB}) \approx R_1$ for practical values of A_{MB}. It is noteworthy that the effective impedance between the amplifier input and ground is approximately zero. The voltage at this point is consequently extremely small in most instances and in fact can be considered a virtual ground for purposes of calculation.

It now becomes a simple matter to calculate the current through R_1 and R_F. All current through R_1 must also flow through R_F since this path presents an approximate short circuit path and the amplifier itself has negligibly high input impedance. The current through both R_1 and R_F is

$$i_1 = \frac{e_{\text{in}}}{R_1} \tag{8-7}$$

Loading on the source is determined by R_1 in the case of the inverting amplifier. A noninverting stage is shown in Fig. 8-3.
While this stage does not invert the input signal, negative feedback is still being applied since the output is fed back to the inverting input. The output voltage is proportional to the difference between e_{in} and e_m; thus

$$e_{\text{out}} = A(e_{\text{in}} - e_m). \tag{8-8}$$

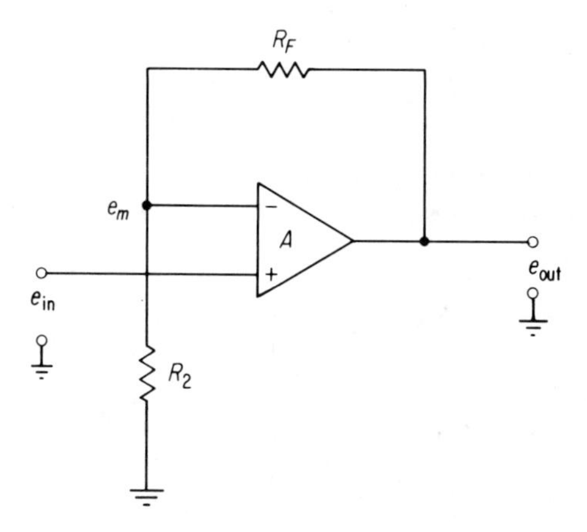

Fig. 8-3. A noninverting amplifier.

The voltage e_m is related to the output voltage by

$$e_m = Fe_{\text{out}}$$

where $F = R_2/(R_F + R_2)$. The closed loop gain is found to be

$$G = \frac{e_{\text{out}}}{e_{\text{in}}}\left(\frac{A}{1 + AF}\right) \tag{8-9}$$

If $A = A_{MB}/(1 + j\omega/\omega_2)$ the gain G can be expressed as

$$G = \left(\frac{A_{MB}}{1 + A_{MB}F}\right)\left(\frac{1}{1 + j\omega/\omega_2(1 + A_{MB}F)}\right) \tag{8-10}$$

The bandwidth with feedback is equal to that for the inverting amplifier and represents a considerable improvement over the open loop bandwidth. If the return ratio $A_{MB}F$ is very high compared to unity, the midband closed loop gain is

$$G_{MB} = \frac{1}{F} = \frac{R_F + R_2}{R_2} = 1 + \frac{R_F}{R_2} \tag{8-11}$$

For the numerous applications requiring a relatively high gain, the ratio R_F/R_2 approximates the midband gain. This circuit allows gain and bandwidth to be exchanged directly for all values of gain including very low values. The gain–bandwidth product is constant with variation of G_{MB}.

The input impedance is a function of the input impedance of the op amp. This point will be discussed in more detail later.

NONIDEAL EFFECTS IN THE OP AMP

Some of the practical effects that account for departures from ideal operation include finite input impedance, nonzero output impedance, dc drift or offset, noise, instability, and greater than 6 db/octave asymptotic falloff with frequency.

Impedance Effects

A suitable equivalent circuit of the op amp with sufficient accuracy for almost any application is shown in Fig. 8-4. The resistors r_1 and r_2 are

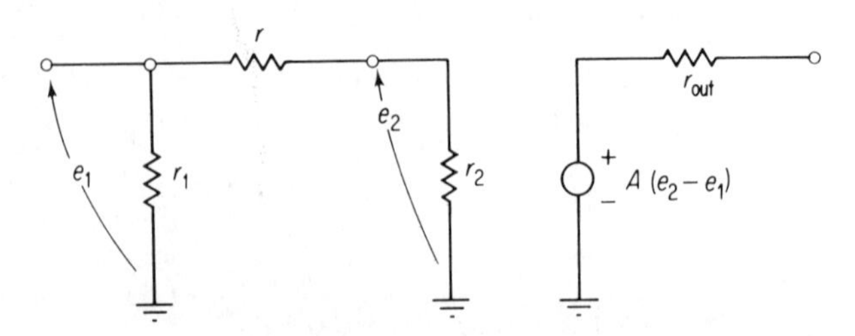

Fig. 8-4. Equivalent circuit for the op amp.

generally in the range of 10 MΩ to 1000 MΩ for transistor or integrated circuit differential stages. The resistance r, appearing between the input terminals, is typically between 10 KΩ and 1 MΩ while r_{out} may range from a few ohms to a few kilohms. For most op amps, r_1 and r_2 will be approximately equal and are specified in terms of the common mode impedance r_{cm}. This impedance corresponds to the input impedance of the amplifier when the input terminals are connected together. Thus, in terms of r_1 and r_2,

$$r_{cm} = \frac{r_1 r_2}{r_1 + r_2} \tag{8-12}$$

The values r_1 and r_2 can be estimated to be twice the common mode impedance.

Some manufacturers do not bother to list a value for r_{cm} indicating that perhaps r_1 and r_2 are insignificant. This is true to a large extent since r will normally have a greater influence on input impedance in most configurations. In the case that feedback is used to increase input impedance considerably, r_1 or r_2 may become more significant. The equivalent circuit of Fig. 8-5 is then sufficient for all but the preceding case.

When the op amp is used as an inverting amplifier as shown in Fig. 8-3, the effect of r is negligible. The impedance r actually appears in parallel with the reflected impedance which has a value of $R_F/(1 + A_{MB}) \approx 0$. The negative terminal appears as a virtual ground, consequently the parallel path to ground presented by r is ignored. For the noninverting amplifier the input impedance is influenced by r. The usual approximations of zero output impedance for either inverting or noninverting amplifiers, zero impedance from the inverting terminal to ground for the inverting amplifier, and infinite impedance from the noninverting terminal to ground for the noninverting amplifier are generally quite reasonable.

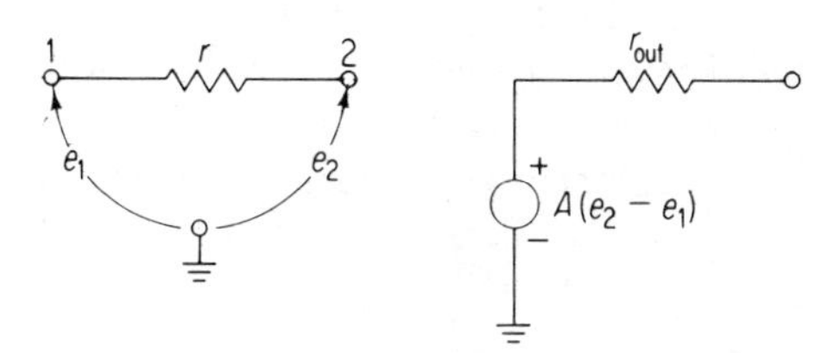

Fig. 8-5. A practical equivalent circuit for the op amp.

Drift and Offset

Drift and offset represent two problems that have great practical significance. Ideally, zero input to the op amp would result in zero output. The actual open loop amplifier, however, will almost always show an output signal with no input present. Changes in voltage drops across junctions, reverse bias leakage current changes, and changes in device current gains in early stages with temperature or aging result in this nonzero output voltage. Relatively small changes in these variables in the first few stages are greatly amplified before reaching the output. At a given temperature, application of a very small input signal will offset these device effects and zero the output voltage. This applied signal is defined as the offset signal. The offset signal can be applied as a current or voltage. The voltage (applied through zero source impedance) necessary to zero the output is called the *offset voltage* while the *offset current* is that current (applied from a true current source with infinite impedance) necessary to zero the output.

Unfortunately, the offset signal is a function of temperature and aging since the effects leading to a nonzero output are functions of these variables. Thus, temperature changes cause the output voltage to vary from its zero value. The change in offset signal required to zero the output as temperature varies is normally called *drift*. Note that both offset and drift are referenced to the input of the amplifier rather than the output. This removes the necessity of considering gain and allows one to meaningfully compare the drift and offset of amplifiers with different values of gain.

In order to assess the effects of offset on performance of the op amp, the equivalent circuit of Fig. 8-6 has been proposed. The sources e_{os} and i_{os} represent the effects within the amplifier that lead to offset voltage and current. If an input signal of e_{in} is applied, the apparent voltage at the input consists of e_{in} plus terms directly proportional to e_{os} and i_{os}. Since R_F will generally be equal to R_1 (unity gain) or larger in an inverting amplifier, the contribution to v_s from e_{os} will be a factor of one to two times the offset voltage. The contribution to v_s from i_{os} is proportional to R_1. This can lead to problems if a high input impedance to the amplifier is required. For this case R_1 must be large, leading to a large error term involving i_{os}. In a specific application a bias voltage can be applied to cancel the errors due to e_{os} and i_{os}. However, as temperature changes, these values drift, leading to finite errors.

As an example of the relative importance of the drift values, let us consider a stage with drift coefficients of 25 μV/°C and 1n A/°C at room temperature. If R_1 is selected to be 1 kΩ, the drift in v_s will be 25 μV/°C +

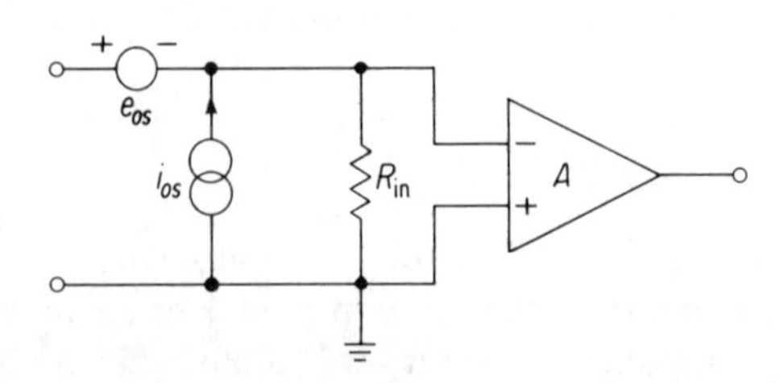

Fig. 8-6. Equivalent circuit reflecting offset effects.

$(\ln A/°C)(1\ k\Omega) = 26\ \mu V/°C$. The drift in v_s is determined mainly by drift in e_{os}. If R_1 is 100 kΩ, the drift in v_s is $25\ \mu V/°C + (\ln A/°C)(100\ k\Omega) = 125\ \mu V/°C$ resulting mainly from drift in i_{os}. It is often true that some compromise between high input impedance and low drift must be made for the inverting amplifier.

The noninverting amplifier results in errors that are of the same order of magnitude as the inverting case. In particular, the error in v_s due to i_{os} is again given by $R_1 i_{os}$. However, the noninverting stage does not require that R_1 be large to achieve high input impedance. Low values of R_1 can be used in this case leading to low drift without substantially decreasing the high input impedance.

When differential stages are used, the resistances presented to both input terminals should be equal to cancel drift effects. The circuit of Fig. 8-7 shows one configuration that compensates somewhat for current drift effects if R_C is chosen to equal $R_1 R_F/(R_1 + R_F)$. The capacitor is used to restore midband gain just as an emitter bypass capacitor is used.

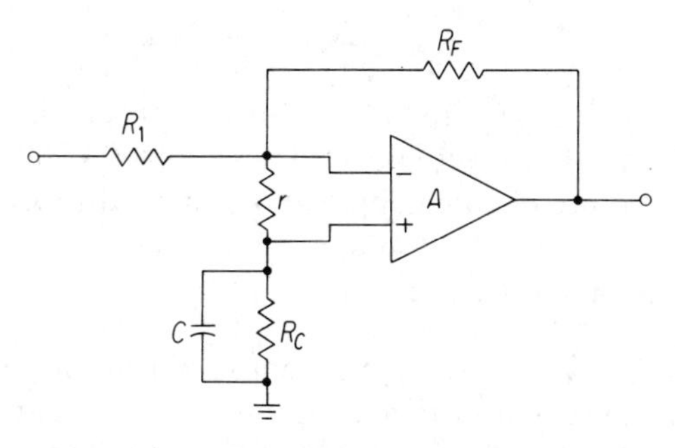

Fig. 8-7. Inverting amplfier with compensation.

When the op amp is used in a high gain configuration to amplify small signals, it is often necessary to adjust the offset to zero at some typical operating temperature. The configuration of Fig. 8-8 shows one method of zeroing the inverting stage. We note that since the apparent impedance from the inverting terminal to ground is very small, loading of R_B is negligible for typical values. There are of course many applications wherein it is unnecessary to consider offset effects. Circuits with small overall gains generally require little attention to offset.

Common Mode Rejection

When we apply the op amp in a differential configuration, we assume that the open loop gain from the negative input to output is equal in magnitude to the gain from the other input to output. If this were the case, a very large signal could be applied simultaneously to both inputs with no output signal

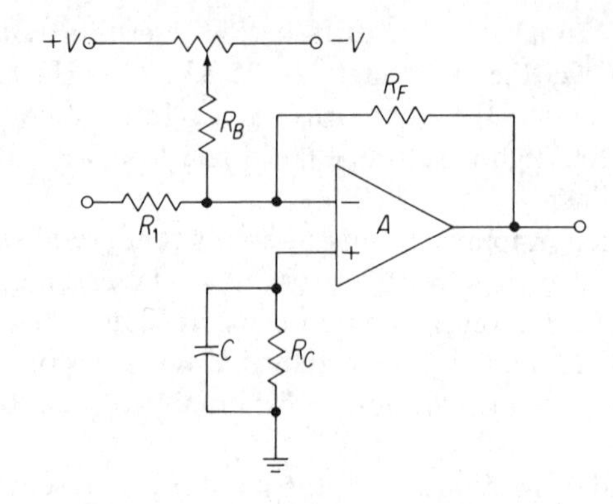

Fig. 8-8. Bias circuit for offset adjustment.

resulting. When a signal is applied simultaneously to the actual op amp, a small output signal results. A comparison of the resulting signal to the input signal is done in terms of the common mode rejection ratio

$$CM_{\text{rej}} = 20 \log \frac{e_{\text{in}}}{e_{\text{out}}} \tag{8-13}$$

It is assumed that e_{in} is applied to both inputs in this equation. Typically this figure is 60 to 100 db. This figure is of interest in precise difference circuits, and also gives a measure of common noise or temperature drift rejection.

Frequency Response and Compensation

One of the most obvious departures from ideal behavior exhibited by the practical op amp is the tendency to oscillate unless some sort of compensation is used. In fact, it is typical of an uncompensated amplifier to oscillate with no external feedback present. It is informative to investigate the reasons leading to this oscillatory tendency.

We have assumed in the previous discussions that the op amp has an open loop gain that is flat from dc to some corner frequency ω_2. Above ω_2 it was assumed that the magnitude of the gain fell at the rate of 6 db/octave. In general the actual frequency response of the op amp may appear as shown in Fig. 8-9. The frequency response shown in Fig. 8-9 results from an open loop gain expressed as

$$A = \frac{A_{MB}}{(1 + j\omega/\omega_2)(1 + j\omega/\omega_3)(1 + j\omega/\omega_4)} \tag{8-14}$$

When negative feedback is applied, the gain for the inverting stage is

$$\left(\frac{-A}{1 + AF}\right)\left(\frac{R_F}{R_1 + R_F}\right) \tag{8-15}$$

If the loop gain AF attains a phase shift of 180° before the magnitude drops

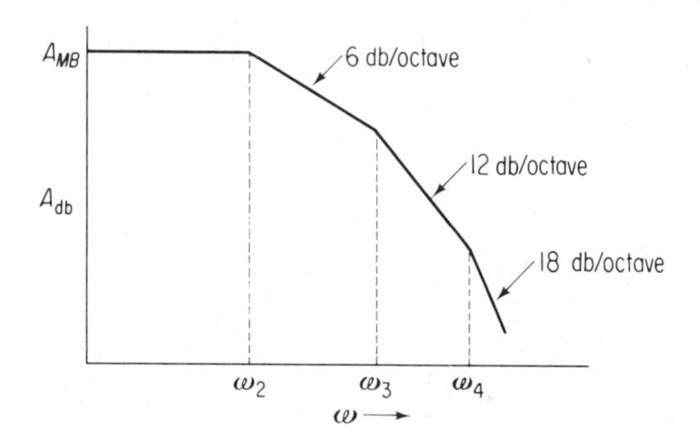

Fig. 8-9. Frequency response of a practical op amp.

below unity, the circuit will satisfy the conditions for oscillations. When it was assumed that the open loop gain could be expressed as

$$A = \frac{A_{MB}}{1 + j\omega/\omega_2} \tag{8-16}$$

there was no stability problem as long as F was real. In this instance, 90° is the maximum possible phase shift. For the gain of Eq. 8-15, even though F may not be complex, the quantity AF can have 180° phase shift. Unless appropriate measures are taken, the practical op amp will oscillate when feedback is applied. In fact, the circuit will generally oscillate when no feedback is applied since stray capacitances link input and output. Some of the popular compensation schemes, designed to stabilize the circuit against oscillations, will now be considered.

Some op amps are compensated by the manufacturer, but most require an additional compensating network. The basic idea is to add a circuit that forces the magnitude of the loop gain to fall below unity at some frequency before the rolloff rate reaches 12 db/octave. This guarantees that AF will never have a phase shift of 180° when the magnitude exceeds unity. At frequencies above the point where $|AF|$ has dropped to unity, the falloff can exceed 12 db/octave without affecting the stability of the circuit.

Compensating networks can be added to the appropriate points of an integrated op amp by means of terminals provided by the manufacturer. The compensating terminal is connected to some node within the amplifier as shown in Fig. 8-10. The squares represent amplifying stages within the op amp. The capacitor C is selected such that a new corner frequency is introduced far below the corner frequency of the uncompensated stage. Figure 8-11 shows the response of both cases.

If the closed loop gain is selected such that the intersection of the open loop and closed loop gains occurs between ω_C and ω_2, operation of the amplifier is stable. However, if the intersection is extended beyond ω_2, instability can take place.

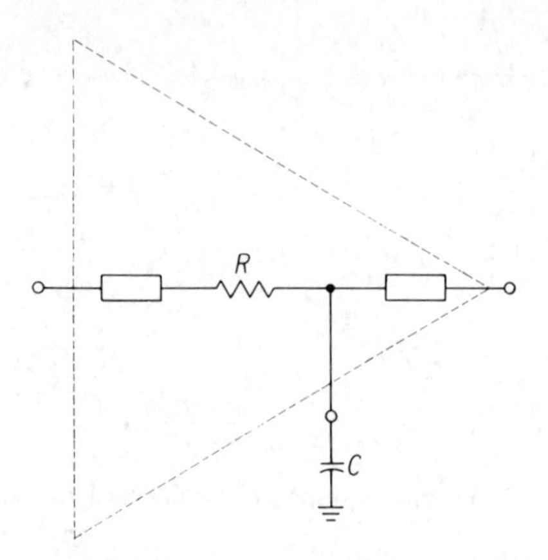

Fig. 8-10. Compensating node within the op amp.

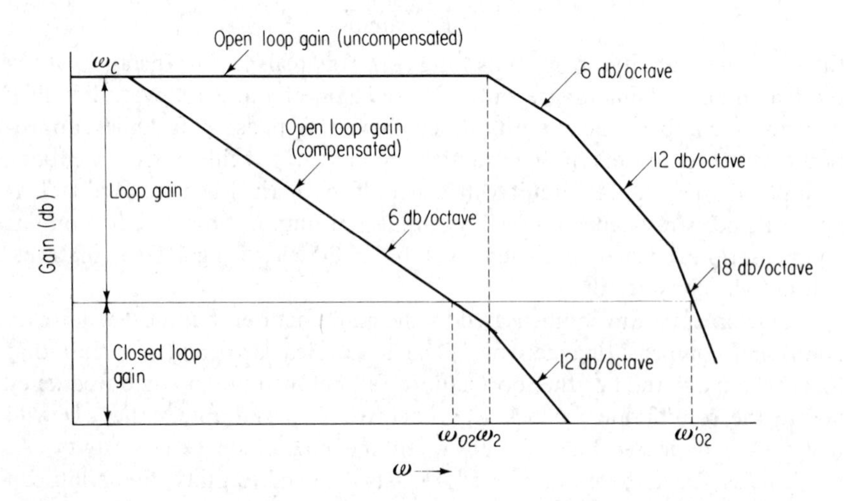

Fig. 8-11. Frequency responses of uncompensated and capacitor compensated amplifiers.

It is possible to extend the region of stable operation by the lag compensator of Fig. 8-12. Since closed loop gain is usually accurately approximated by

$$G = \frac{1}{F}$$

the point in frequency where $|AF| = 1$ is also the point where $|A| = 1/|F|$. This point is referred to as the *crossover frequency*. If we assume that $|F|$ is constant, Fig. 8-11 clearly locates the critical points of the uncompensated and compensated amplifiers. The closed loop gain has a magnitude of $1/|F|$ and the intersection of this line with the open loop gain (uncompensated) occurs at ω'_{02}. The falloff of the open loop gain at this point is 12 db/octave

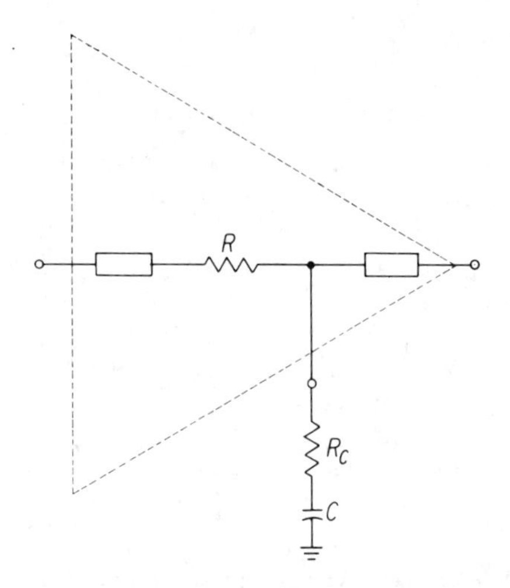

Fig. 8-12. Compensation using RC network.

and a phase shift of 180° is possible. In practice, oscillations are almost certain for this case. With compensation added, the new corner frequency of the open loop gain occurs at ω_{02} and is much lower than the corner frequency of the uncompensated amplifier. In this instance, the intersection of the open loop and closed loop gains occurs at ω_{02} and the falloff is only 6 db/octave at this point. At this point in frequency, even though $|AF| = 1$, oscillations cannot occur because the phase shift of AF will be less than 90°.

The frequency ω_C is determined by

$$\omega_C = \frac{1}{CR_{eq}} \tag{8-17}$$

where R_{eq} is the equivalent resistance presented to the compensating capacitor. Manufacturers generally supply graphical information relating several values of C to corresponding corner frequencies. The uncompensated and compensated cases are shown in Fig. 8-13 where

$$\omega_{c2} = \frac{1}{R_c C}$$

and (8-18)

$$\omega_{c1} = \frac{1}{(R + R_c)C}$$

where we have chosen $\omega_{c2} = \omega_2$. Note that since the asymptotic falloff reaches 12 db/octave at ω_3, stable operation can be expected for crossover frequencies nearing ω_3. Again most manufacturers will graphically relate values of C, R_C, and ω_{C1} to aid in the proper selection of the compensating network.

It should be noted that if the manufacturer builds the compensating network into the op amp, flexibility is sacrificed. For many low frequency

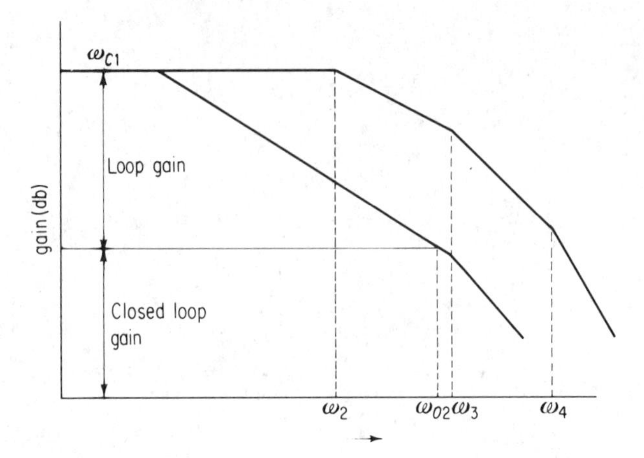

Fig. 8-13. Frequency responses of uncompensated and RC compensated amplifiers.

applications this flexibility is entirely unnecessary. However, as high frequency work is undertaken, access to a compensating terminal may prove to be very useful.

Rather than compensating the open loop gain of the amplifier, additional elements can be used in the feedback loop to achieve stable operation. A capacitance placed in parallel with the feedback resistance of the inverting stage will stabilize the amplifier if appropriately selected. In this case, the feedback transfer function F increases at high frequencies as A decreases in magnitude. If the feedback resistance is selected such that $1/R_F C = \omega_F$ is approximately equal to the corner frequency of the uncompensated stage, more stable operation will result.

Operational amplifiers are becoming extremely important in the fields of circuit and system design. The op amp reduces considerably the number of discrete element amplifiers that are now used. In addition op amps are used as summers, integrators, or digital to analog converters. They are used in active filter design and precision sweep circuitry.

LINEAR CIRCUIT APPLICATIONS OF OPERATIONAL AMPLIFIERS

Differential dc Amplifiers

The amplifiers to be discussed in this section are most descriptively known as *differential dc amplifiers*, denoting the fact that they amplify the difference between two signals, and that the inputs are directly coupled. Other common terms used for this basic type of amplifier are: *transducer amplifier, bridge amplifier, data amplifier, instrumentation amplifier, difference amplifier*, and *error amplifier*. Such amplifiers are easily realized through the use of one or more operational amplifiers with linear feedback. Inputs are typically from transducers, which convert a physical parameter and its variations to electric signals. Examples of such transducers include thermocouples and straingage bridges. Several types of such differential dc amplifiers, of varying complexity and performance characteristics, are discussed in the following paragraphs.

Differential dc Amplifiers Using One Operational Amplifier

The circuit of Fig. 8-14 has the virtue of simplicity in that it uses only one operational amplifier and four matched resistors. The presence of a common–mode voltage e_{cm} and a differential voltage $(e_1 - e_2)$ is characteristic of most transducers. The common-mode voltage may represent a dc level, as in a bridge, or noise pickup. If an ideal operational amplifier is assumed, and if $R_2/R_1 = R_4/R_3$, we obtain:

$$e_0 = \frac{R_2}{R_1}(e_2 - e_1) \qquad \text{(8-19)}$$

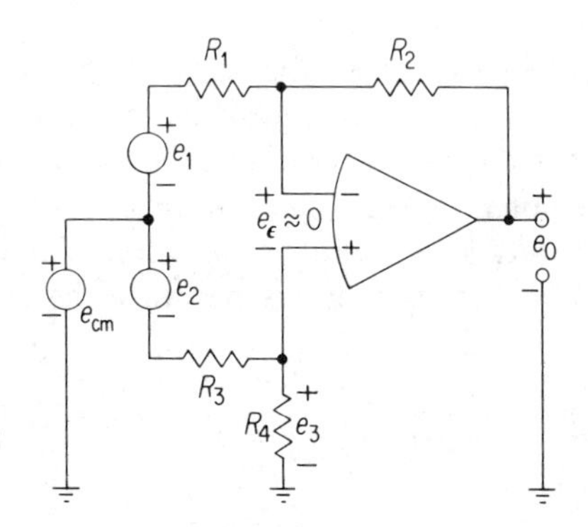

Fig. 8-14. Simple differential amplifier. A zero source impedance.

The resistor ratios, R_2/R_1 and R_4/R_3, must be carefully matched in order to insure the rejection of common–mode signals. The value of these resistor ratios sets the gain for differential signals. These equations illustrate the performance of the circuit when dealing with zero source impedances and nonzero common-mode signals. For zero source impedance, the gain is determined solely by the feedback resistors and, if these resistors are matched in pairs as indicated, common-mode signals are rejected completely. Actually, of course, the operational amplifier has been assumed ideal in having infinite input impedance, infinite gain, and infinite common–mode rejection. If these factors are given real values and their effects evaluated, it will be found that the finite input impedance of the operational amplifier and its inherent finite common-mode rejection will place limits on the overall common-mode rejection of the closed-loop differential amplifier. The finite open-loop gain will limit the gain accuracy of the overall circuit.

The principal limitations of this circuit are its low input impedance and the difficulty of varying the gain. The input impedance, of course, is determined by the feedback and input resistors. If these resistors are made large in order to increase the input impedance, the dc errors due to bias currents will be proportionately increased, thus placing an upper limit on the feasible values of input impedance. The gain of the differential amplifier can be changed only by varying the ratios of the feedback resistors. Because of the necessity for maintaining the equality of the resistive ratios, it is quite difficult to vary the gain continuously. Gain steps can be achieved if the common-mode rejection is carefully adjusted at each gain setting.

The differential amplifier circuit of Fig. 8-15 is a similar type of circuit, with the added feature of a gain vernier that allows the gain to be continuously varied without affecting the common-mode rejection of the circuit. The output voltage is

$$e_0 = 2\left(1 + \frac{1}{K}\right)\frac{R_2}{R_1}(e_2 - e_1) \tag{8-20}$$

Note, however, that this circuit requires four matched resistors of value R_2 and two matched resistors of value R_1. The gain is an inverse function of the setting of the vernier potentiometer and, as such, is highly nonlinear. The potentiometer can, however, provide approximate linearity over limited ranges. The circuit still suffers from the limitations of low input impedance. The dc offset errors are much the same as those for the circuit of Fig. 8-14.

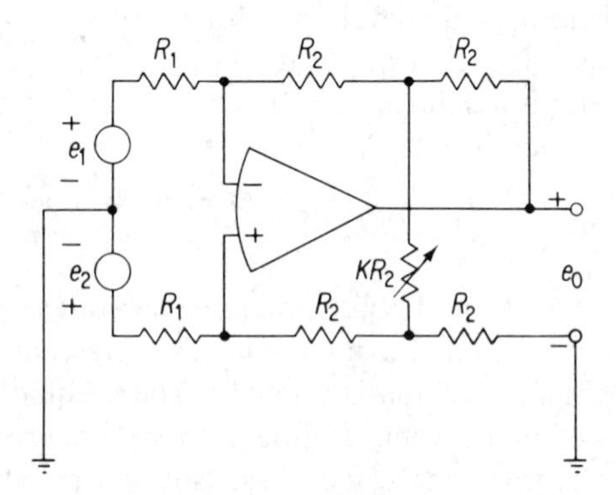

Fig. 8-15. Simple adjustable-gain differential amplifier.

Differential dc Amplifiers Using More Than One Operational Amplifier

The circuit of Fig. 8-16 provides another low-impedance alternative to those of Figs. 8-14 and 8-15. The two amplifiers required operate in the inverting mode and need not have a noninverting capability. Thus they can be chopper-stabilized amplifiers for low drift or may be field-effect transistor (FET) input types, which may have rather poor linearity when used non-invertingly. The output voltage is

$$e_0 = \frac{R_2}{R_1}(e_2 - e_1) \tag{8-21}$$

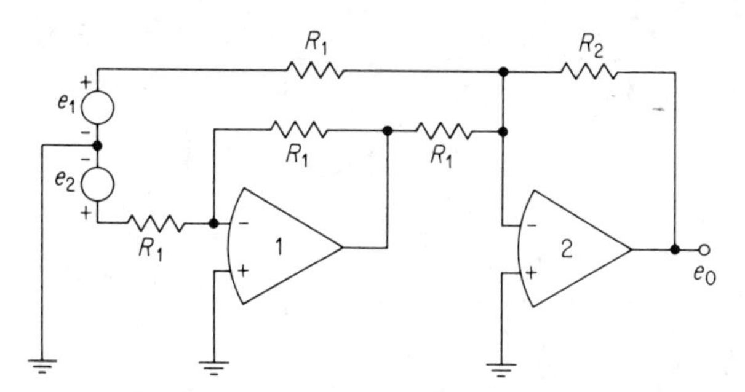

Fig. 8-16. Differential dc amplifier using inverting operational amplifiers.

The gain can be easily varied, in steps or continuously, by changing the value of R_2, without affecting the common-mode rejection properties. Good common-mode rejection requires four closely matched resistors of value R_1. Note that the dc offset error is approximately four times that of a single amplifier.

Since the common-mode rejection of the operational amplifiers is not a factor, the common-mode rejection of the closed-loop amplifier can be trimmed to quite high values by simply allowing a small amount of adjustability of one of the R_1 resistors. The common-mode voltage capability of the circuit is limited only by the output voltage capability of the unity gain inverter. This capability can be increased by making the gain of amplifier 1 less than unity. The gain of amplifier 2 must then be increased accordingly, however, thereby increasing the output offset error.

Another differential dc amplifier circuit using two operational amplifiers is shown in Fig. 8-17. This circuit provides the high input impedance lacking in the circuits discussed up to now. For this circuit,

$$e_0 = \left(1 + \frac{R_4}{R_3}\right)(e_2 - e_1), \qquad \text{if } \frac{R_1}{R_2} = \frac{R_4}{R_3} \tag{8-22}$$

Again, equality of the two resistor ratios is required if the circuit is to reject common-mode signals. The operational amplifiers, since they operate in the noninverting mode, must have good common-mode properties. The input impedance at each terminal of the differential amplifier is simply the common-mode input impedance of the operational amplifiers. This can be quite large (10 MΩ and up) depending on the type of operational amplifier used. For fixed gains, or gain steps, the circuit is quite useful, but it is not feasible for continuously variable gain. Also, since the input voltage of the upper amplifier must be less than $R_1/(R_1 + R_2)$ times the output saturation voltage, the common-mode voltage range is very limited at low values of overall gain. This is not considered a serious limitation, since such amplifiers are usually used at gains of 10 or greater.

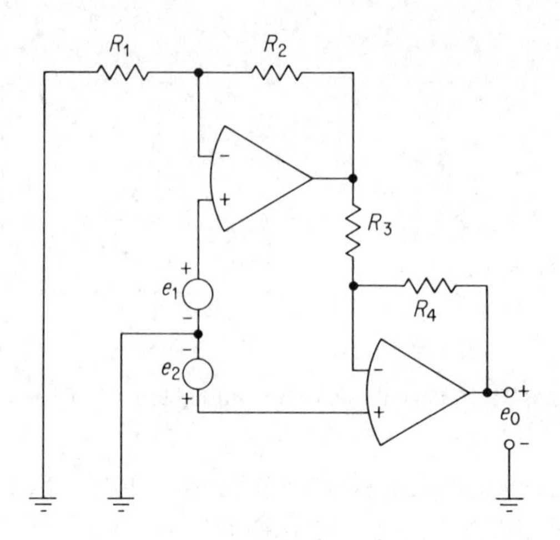

Fig. 8-17. High-input-impedance differential amplifier.

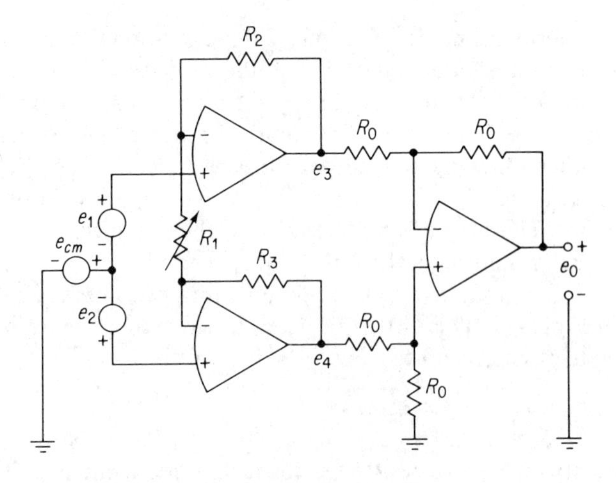

Fig. 8-18. High-input-impedance adjustable-gain differential amplifier.

The differential dc amplifier circuit of Fig. 8-18 overcomes most of the weaknesses of the circuits discussed up to this point. Analysis of the circuit yields, if $R_2 = R_3$,

$$e_0 = \left(1 + \frac{2R_2}{R_1}\right)(e_2 - e_1) \tag{8-23}$$

The two input amplifiers constitute a differential buffer amplifier with a gain of $1 + (2R_2/R_1)$ for differential signals, and unity gain for common-mode signals. The noninverting configuration of these input amplifiers insures high input impedance at both inputs. The gain is easily varied by a single resistor R_1. The effect of mismatch in resistors R_2 and R_3 is simply to create a gain error without affecting the common-mode rejection of the circuit. The resistors R_o of the output amplifier must be accurately matched, or trimmed, to insure the rejection of common-mode signals at this point. This final amplifier acts simply as a differential-input/single-ended-output converter. Feedback impedances in both stages can be relatively low in value to minimize the effects of bias current, since these feedback elements do not affect the input impedance of the differential amplifier. Usually, all of the gain of this differential amplifier is in the input stage, thus insuring that only the offset voltages of these two operational amplifiers are significant in determining the output offset. Since the output voltage offset is proportional to the difference of the voltage offsets of these two amplifiers, it is desirable to use amplifiers whose voltage offsets tend to track with temperature. Such techniques are the basis for some low-drift differential-amplifier modules. The bias currents of these input amplifiers will flow through the impedance of the source, and will thus generate additional offset voltage, which will appear at the output of the differential amplifier amplified by the differential gain factor. The use of amplifiers with FET input stages will greatly reduce this effect.

Analog Integrators

The analog integrator is extremely useful in computing, signal-processing, and signal-generating applications. It uses an operational amplifier in the inverting configuration, as shown in Fig. 8-19. The equations of operation are derived assuming an ideal operational amplifier of gain A. These are

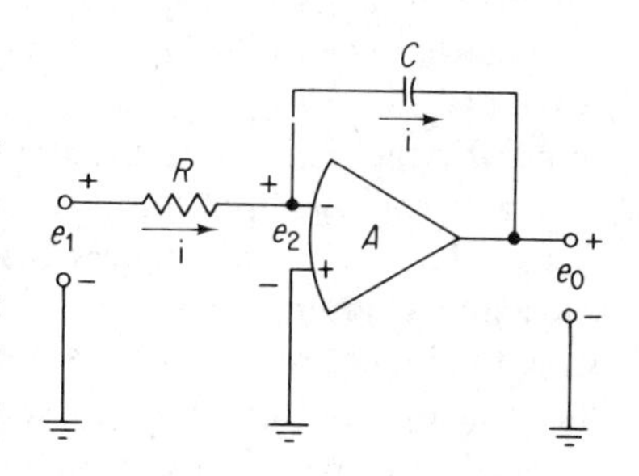

Fig. 8-19. Analog integrator.

$$\left.\begin{aligned} \frac{e_1 - e_2}{R} &= i \\ e_2 - e_o &= \frac{1}{C}\int_0^t i\,dt = \frac{1}{RC}\int_0^t (e_1 - e_2)\,dt \\ e_2 &= -\frac{e_o}{A} \end{aligned}\right\} \tag{8-24}$$

If $A \longrightarrow \infty$, then $e_2 \longrightarrow 0$, and

$$e_o = -\frac{1}{RC}\int e_1\,dt \tag{8-25}$$

As in the inverting amplifier, the summing point is held at a virtual ground by the high gain of the amplifier and its feedback network. Since no current flows into the input terminal of the operational amplifier, all of the input current, $i = e_1/R_1$, is forced to flow into the feedback capacitor, causing a charge voltage to appear across this element. Because one end of the capacitor is tied to the virtual ground point, the output voltage of the amplifier equals the capacitor-charging voltage. The overall integrator circuit has the low output impedance normally associated with a feedback amplifier.

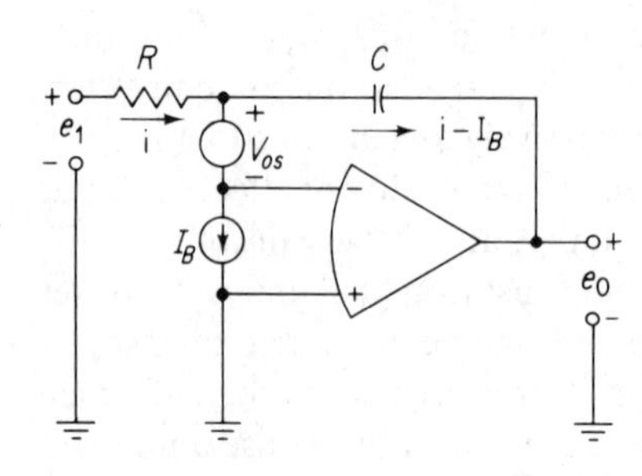

Fig. 8-20. Effects of offset voltage and bias current in integrator circuit.

The dc offset and bias current of the analog integrator are taken into account in the more realistic model of Fig. 8-20. Because these dc errors exist, the output of the integrator now consists of two components: the integrated signal term and an error term

$$e_o = -\frac{1}{RC}\int e_1\,dt + \frac{1}{RC}\int V_{os}\,dt + \frac{1}{C}\int I_B\,dt + V_{os} \qquad (8\text{-}26)$$

The error term itself is made up of a component due to the input offset voltage and another due to the input bias current. The integral of the dc offset voltage results in a ramp voltage, a linearly increasing term whose polarity is determined by the polarity of the input offset voltage. In addition to this ramp-voltage error, the input offset voltage creates an output offset voltage equal to it in value. The bias current flows almost entirely through the feedback capacitor, charging it in ramp fashion, similar to the ramp voltage resulting from the input offset voltage. These two ramp-voltage errors will continue to increase until the amplifier reaches its saturation voltage or some limit set by external circuitry. These error components usually set the upper limit on feasible length of integration time. The error component caused by bias current can be minimized by increasing the capacitance of the feedback element. This can be done only by decreasing the value of the input resistor, if a specific value of the RC time constant is to be achieved. A lower limit usually exists on R because of current limitations and loading of the input signal source.

The effects of bias current can be reduced by inserting a resistance R between the noninverting input of the amplifier and ground. This equalizes the resistances at the two inputs and changes the effects of bias current to that of offset (difference) current. Thus, in the equation for output voltage, the bias current I_B should be replaced by the offset current I_{OS} if the compensating resistor is used. The error ramp due to voltage offset is fixed by the chosen value of RC time constant.

To realize the performance possibilities of an operational amplifier as an integrator, a feedback capacitor must be selected with a dielectric leakage current that is less than the bias current of the amplifier. Polystyrene and Teflon are usually the best choices for the ultimate in long-term integrating accuracy. If shorter integration times are required, the requirements on capacitor quality can accordingly be relaxed. Mylar capacitors may then prove satisfactory, as will silver-mica types if small values of capacitance, corresponding to high-speed integration, are to be used.

The choice of the type of amplifier is also governed by the length of computing time and the desired accuracy. Chopper-stabilized amplifiers are usually used for long-term integrators because of their superior long-term dc stability. FET amplifiers are used for medium-length integration because of their low bias current. Amplifiers with bipolar transistor input stages may be used in very short-term integration, such as in signal generation (sweep generation, triangle waves, etc.). The resulting integrator response function is

$$\frac{E_0}{E_1}(s) \approx \frac{-A_0}{\left(\frac{\tau_0}{A_0}s + 1\right)(A_0RCs + 1)} \tag{8-27}$$

Line-Driving Amplifiers

One of the primary areas of application for the operational amplifier is that of buffering between a signal source and the desired load. Usually the signal source is very limited in power, has relatively high internal impedance, and is of a low level. The load is relatively low in impedance (possibly capacitive) and requires high-level signals. Thus the amplifier must provide impedance buffering, signal scaling, and power gain. Needless to say, it must be stable under the desired conditions of loading and feedback, and must have sufficient gain and bandwidth to insure accurate response to input signals. A typical example of such an application is the line-driving amplifier.

When data signals must be transmitted over long signal lines from a remote measuring station, the line-driving amplifier is usually required. Figure 8-21 illustrates a simulated load of this type. The capacitance is that of a shielded cable and may be as little as a few picofarads or as much as several

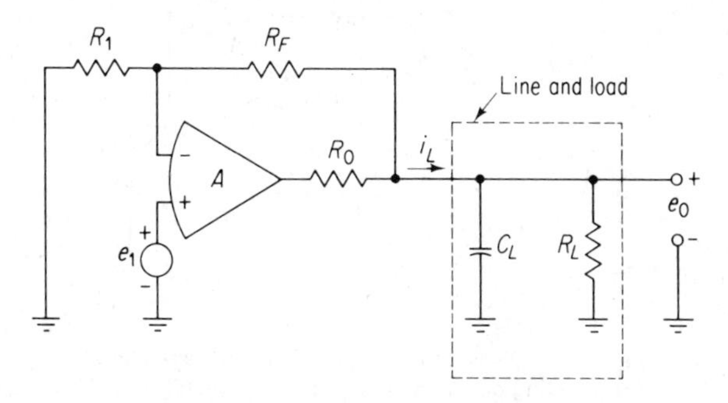

Fig. 8-21. Line-driving amplifier.

microfarads. As an example, the amplifier must be capable of supplying 63 mA to the capacitive load if $C_L = 10{,}000$ pF and the output voltage is a 10-volt sine wave at 100 KHz.

AC-Coupled Feedback Amplifiers

Although the operational amplifier is designed to amplify dc signals, it has a rather broad frequency response and is consequently quite useful for strictly ac signals. The feedback network can be tailored for exactly the desired passband. One of the simplest ac amplifiers is that shown in Fig. 8-22(a) where the closed-loop gain is given by

$$\frac{E_0}{E_1}(s) = -\left(\frac{R_F}{R_1}\right)\frac{s}{s + (1/R_1C_1)} \tag{8-28}$$

The dc gain is zero, whereas the high-frequency gain approaches $-R_F/R_1$. The lower cutoff frequency is

$$f_c = \frac{1}{2\pi R_1 C_1} \tag{8-29}$$

The dc output offset voltage E_{os} is equal to the dc input offset voltage plus the dc offset voltage generated by the input bias current flowing through R_F.

$$E_{\text{os}} = V_{\text{os}} \times 1.0 + I_{B1}R_F \tag{8-30}$$

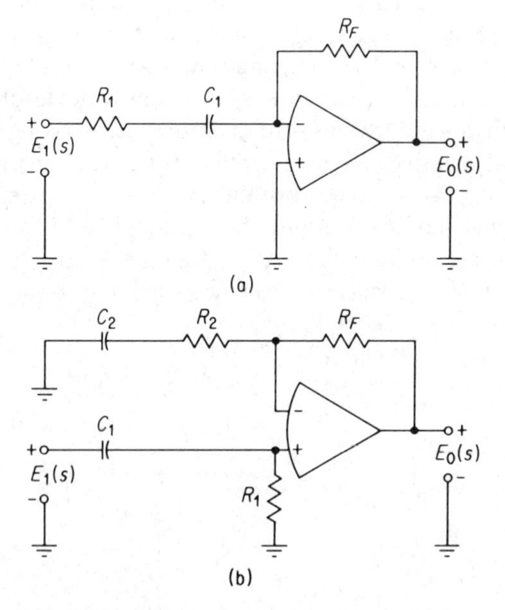

Fig. 8-22. Typical ac-coupled feedback amplifiers; (a) inverting circuit; (b) noninverting circuit.

A noninverting ac amplifier is shown in Fig. 8-22(b). The response is given by

$$\frac{E_0}{E_1}(s) = \frac{s}{s + (1/R_1C_1)}\left[\frac{(R_2 + R_F)C_2s + 1}{R_2C_2s + 1}\right] \tag{8-31}$$

Both circuits of Fig. 8-22 have relatively low input impedance above the cutoff frequency, determined by the resistors denoted R_1 in both cases.

The circuit of Fig. 8-23 is an ac amplifier whose input impedance is "bootstrapped" to a high value. Resistor R_2 provides a decoupling for dc input signals. However, for high-frequency signals the voltage across R_2 becomes very small. Consequently, very little current flows through R_2, and the effective input impedance is very high. Differential ac amplifiers are also easily realized through the use of operational amplifiers.

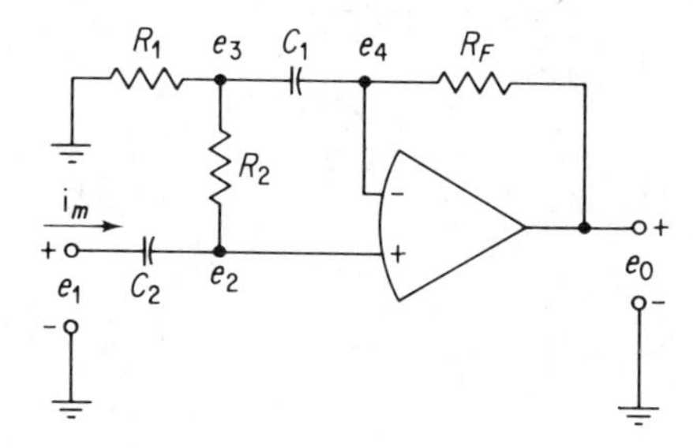

Fig. 8-23. Bootstrapped ac amplifier.

Voltage-to-Current Converters

In applications such as coil-driving and transmission of signals over long lines, it is sometimes desirable to convert a voltage to an output current. With operational amplifiers this is quite easily done. Several realizations of the voltage-to-current converter will be examined in this section.

The simplest V-to-I converters are those for floating loads. The circuits of Fig. 8-24 are the prime examples of this type. Figure 8-24(a) illustrates a simple inverting circuit. The input current is given by $i_1 = e_1/R_1$, since R_1 is terminated at the virtual ground of the summing junction. This same current flows through the feedback load impedance Z_L in the feedback loop. The current i_1 is independent of the value of Z_L. Both the signal source and the operational amplifier must be capable of supplying the desired amount of load current. The circuit of Fig. 8-24(b) operates in the noninverting mode and, hence, presents a high impedance to the driving source. The current is again given by the equation $i_1 = e_1/R_1$ and, again, i_1 is the load current. Very little current, however, is required from the signal source because of the high input impedance of the noninverting amplifier.

For loads grounded on one side there are also circuits that give V-to-I conversion. The single-amplifier circuit of Fig. 8-25 acts as a current source

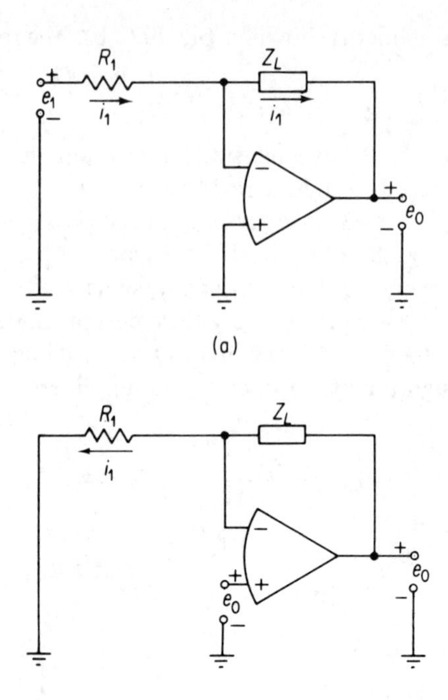

Fig. 8-24. V-to-I converters floating loads: (a) Inverting amplifier type; (b) noninverting amplifier type.

controlled by e_1; i.e.,

$$i_L = -\frac{e_1}{R_2} \quad \text{if} \quad \frac{R_3}{R_2} = \frac{R_F}{R_1} \tag{8-32}$$

If these ratios of resistances are matched, the circuit will function as a true source of current with very high internal impedance. A mismatch of the ratios will be seen as a decreased internal impedance of the current source. Fluctuations in effective load impedance will then cause fluctuations of the output current. The operational amplifier for the circuit of Fig. 8-25 must have an output voltage range sufficient to provide the maximum load voltage plus the voltage drop across R_3. Normally, R_1 and R_2 will be chosen to draw small currents and R_F and R_3 will be made small to minimize voltage drops.

Reference-Voltage Sources and Regulators

Because of its high input impedance and easily adjustable gain, the operational amplifier may be used as a reference–voltage source with very low output impedance and substantial output–current capability. Two circuits for

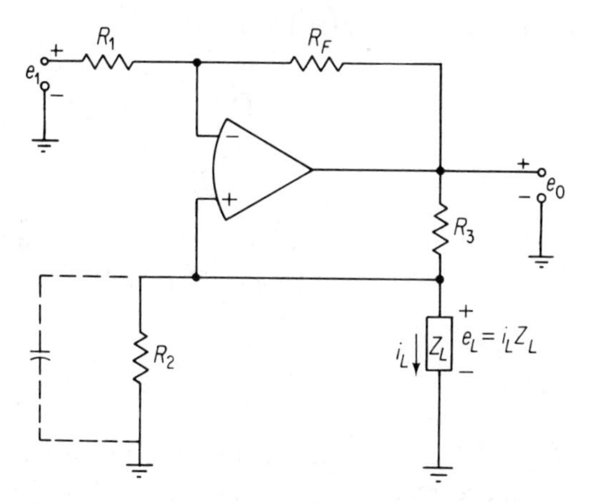

Fig. 8-25. Voltage-to-current converter grounded load.

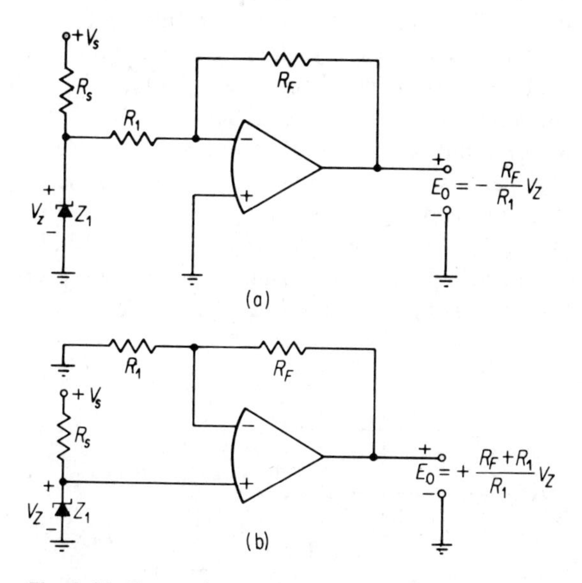

Fig. 8-26. Zener reference sources. (a) Inverting; (b) non-inverting.

use with Zener diodes are shown in Fig. 8-26(a) and (b). The loading conditions on the Zener diodes are constant and the load regulation is

$$\text{Regulation (percent)} = \frac{R_o}{A\beta R_L} \times 100 \tag{8-33}$$

where R_L is the minimum load impedance, and

$$\text{where} \quad \begin{cases} R_{\text{out}} = \dfrac{R_o}{A\beta} \\ \beta = \dfrac{R_1}{R_1 + R_F} \end{cases} \tag{8-34}$$

and R_o is the open-loop output impedance. Regulation with respect to the input voltage V_s depends upon the dynamic resistance of the reference Zener diode Z_1.

Voltage Regulators

Any one of the voltage references described in the preceding section may be considered a voltage regulator, with extremely tight regulation characteristics. Where higher output currents are required, a power booster can be added, inside the feedback loop. However, in speaking of voltage regulators, it is more usual to consider operation from a single source of unregulated dc voltage, rather than the dual supplies tacitly assumed in the reference-voltage circuits. Figure 8-27 shows such a regulator. The amplifier, which normally operates on dual power supplies of opposite polarity, is biased for operation on a single unregulated power supply. The negative supply terminal is grounded and the noninverting input is biased at the Zener voltage. The Zener diode Z_1 operates at constant load current, since the output current is provided by the transistor Q_1. If the amplifier has a minimum (balanced) supply rating of $\pm V_m$, then V_s must be larger than $2V_m$. Similarly, if $\pm V_M$ is the maximum (balanced) supply rating, V_s must not exceed $2V_M$. The amplifier will saturate as the output voltage approaches either supply voltage.

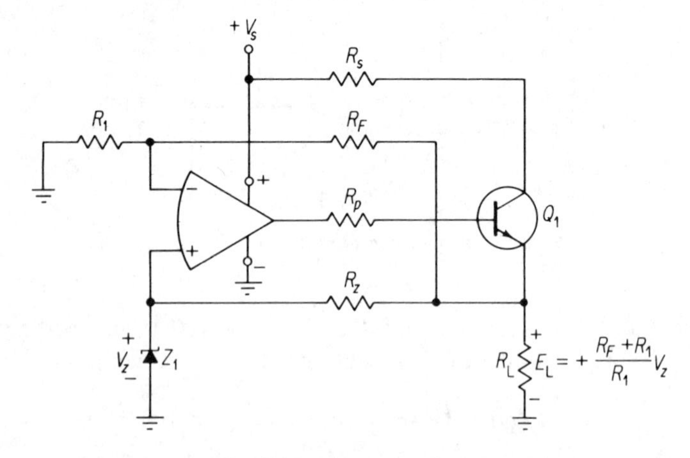

Fig. 8-27. Voltage regulator.

This determines the limit on output, whereas the common–mode voltage range sets the lower limit on Zener voltage.

Although the amplifier may have an internal current limit, the resistor R_p is required to protect against short circuit in this type of regulator. This is because a short circuit to ground is equivalent to a short circuit to the negative supply. This causes a power dissipation equal to twice that of a short circuit to ground when operating on balanced dual supplies. Thus the internal protection may not be sufficient. The value of R_p should be chosen to limit the amplifier short-circuit current to approximately half the internal current-limit value when the output is at positive saturation voltage. The resistor R_s provides current-limiting to protect Q_1.

The load regulation of this type of regulator can exceed 0.01 percent, since the effective output impedance is very low. The line regulation is increased beyond that of the Zener by using the output voltage as excitation for the Zener.

Current Amplifiers

Current amplifiers, or current-to-voltage converters, are realized very simply by the use of operational amplifiers. An ideal current source has infinite output impedance and an output current that is independent of load. Photocells and photomultiplier tubes are basically current sources with an output impedance that is finite but very large. For small load impedances, the output impedance may be considered infinite.

The current-to-voltage converter of Fig. 8-28 presents almost zero load impedance to ground because the inverting input appears as a virtual ground. The input current, however, flows through the feedback resistor, generating an output voltage

$$e_o = -i_s R_F \tag{8-35}$$

The actual input impedance Z_{in} of the current-to-voltage converter, taking

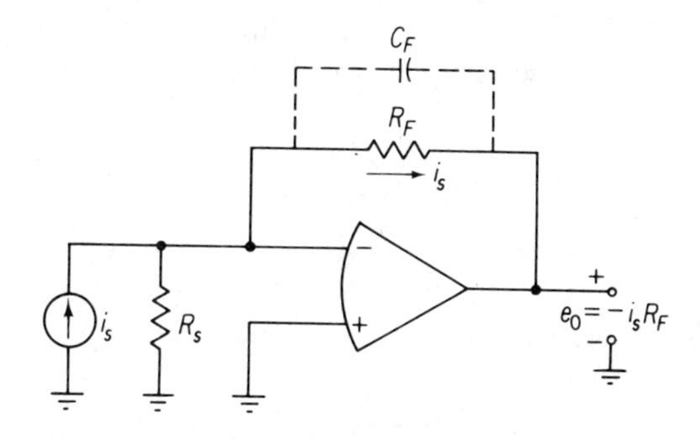

Fig. 8-28. Current amplifier (I-to-V converter).

into account the finite gain A and differential (open-loop) input impedance Z_{id}, is

$$Z_{\text{in}} = \frac{Z_{id}}{1 + \frac{Z_{\text{id}}}{R_F}(1 + A)} \approx \frac{R_F}{1 + A} \tag{8-36}$$

The lower limit on measurement of current input is determined by the bias current of the inverting input. For greatest resolution, FET or varactor bridge amplifiers are usually employed.

The gain of the amplifier for dc offset voltage and noise voltage is given by

$$\frac{R_F + R_S}{R_s} \approx 1.0 \qquad \text{since } R_s \gg R_F \tag{8-37}$$

Although errors resulting from these parameters are very small, current noise can be a factor because of the very large impedances. Since most such measuring circuits are used for very-low-frequency signals, it is usual to parallel R_F with a capacitor C_F to reduce the high-frequency current noise. Output impedance of the current-to-voltage converter is very low because of the nearly 100 per cent feedback.

EXERCISES

1. The connection shown below is referred to as a voltage follower. Discuss why $V_o = V_I$ and comment upon terminal impedance levels.

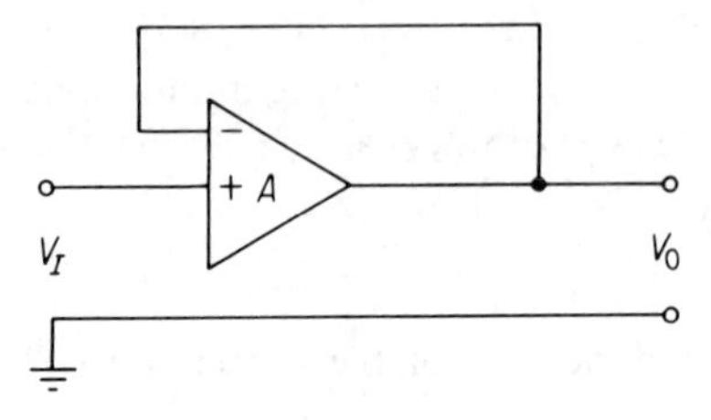

2. The circuit shown is a differential input/output amplifier. Explain why the differential gain is approximately equal to R_2/R_1.

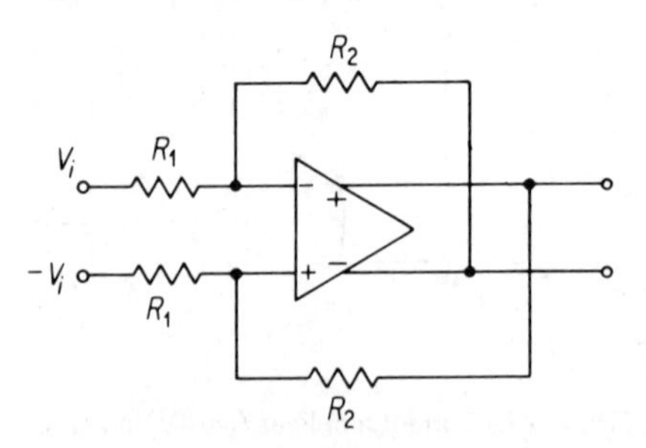

REFERENCES

1. "Applications Manual for Operational Amplifiers," *Philbrick/Nexus Research*, 1965.
2. "Handbook of Operational Amplifier Applications," *Burr-Brown Research Corp.*, Tucson, Ariz., 1961 (out of print).
3. "Handbook and Catalog of Operational Amplifiers," *Burr-Brown Research Corp.*, Tucson, Ariz., 1969 (out of print).
4. G. A. KORN and T. M. KORN, *Electronic Analog and Hybrid Computers*. New York: McGraw-Hill Book Company, 1964.
5. T. MIURA et al., "On computing errors of an integrator," Proc. 2nd AICA Conf., *Presses Academiques Europeanes*, Brussels, 1958.
6. G. TOBEY, "Analog instrumentation," *Instr. Control Syst.*, Jan. 1969.
7. N. D. DIAMANTIDES, "Improved electronic differentiator," *Electronics*, July 27, 1962.
8. G. A. KORN, "Exact design equations for operational amplifiers with four-terminal computing networks," *IRE Trans. Electronic Computers*, vol. EC-11, (Feb. 1962), pp. 82–83.

9

Switching Circuits—Function and Design

INTRODUCTION

In Chap. 7 we discussed the dc characteristics of the transistor switch. In this chapter we continue discussing the switching properties of transistors and consider digital circuit design and implementation. The investigation of the transient times involved in switching the transistor from an OFF state to an ON state, and vice versa, is analogous to determining the time for a set of relay contacts to open or close when the relay coil is activated.

The switching speed of transistors is extremely important in digital circuits as well as in a host of other applications, such as the horizontal output stage in a TV set, dc-to-ac inverters, and power supply regulators. The development in this chapter is primarily aimed at describing the switching properties of transistors in digital circuit applications. The equations developed for switching times are not applicable to the low-level chopper circuits described in Chap. 7. The transient characteristics will not be derived in detail here. However, the Appendix does so using the transistor charge control model.

CONVENTIONAL DEFINITIONS OF DELAY, RISE, STORAGE, AND FALL TIMES

The definitions of the four switching times used in this chapter differ somewhat from those conventionally used. Here we define the times in terms of the edge of conduction and the edge of saturation. We do this in order to make the analysis as clear as possible and the equations as simple as possible.

Conventionally, the times are defined in terms of 10 per cent and 90 per cent of the output waveforms. This is done to simplify measuring the times.

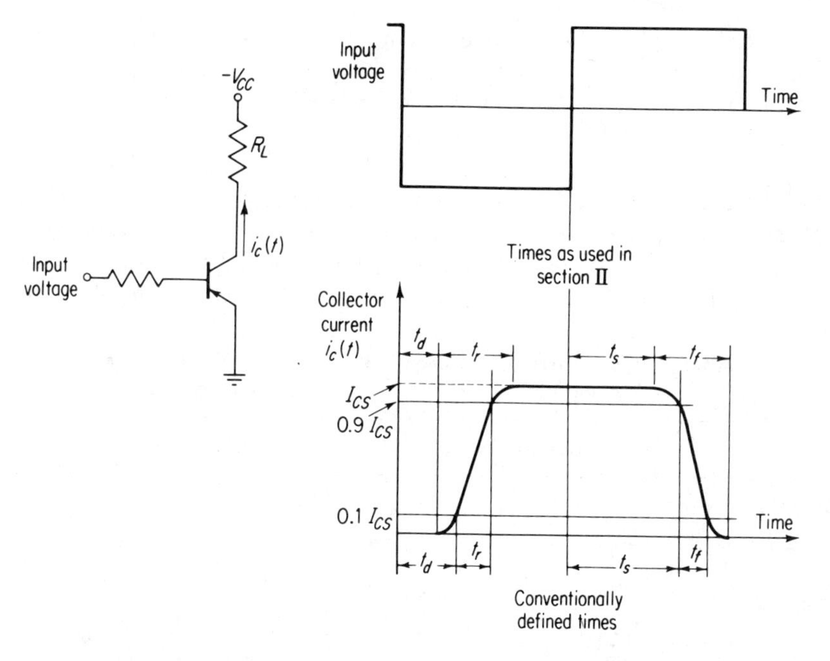

Fig. 9-1. Conventional definitions of switching times.

The definitions used here and the conventional definitions are shown diagrammatically in Fig. 9-1.

Including 10 per cent of the rise time in the delay time, 10 per cent of the fall time in the storage time, and altering the rise and fall time to include only the 10–90 per cent change in the collector current, necessitates slight modifications to the previous equations for switching time. The required modifications follow readily from the previous analysis. The equations for the conventionally defined times for constant current base drive are

Delay time:

$$
\begin{aligned}
t_d = \frac{1}{3\bar{\omega}_T} &+ \frac{\phi_e C_{EO}}{1-n_e}\left[\left(1+\frac{V_{BEI}}{\phi_e}\right)^{1-n_e} - \left(1-\frac{V_i}{\phi_e}\right)^{1-n_e}\right] \\
&+ \frac{\phi_c C_{CO}}{1-n_c}\left[\left(1+\frac{V_{BEI}+V_{CC}}{\phi_c}\right)^{1-n_c} - \left(1+\frac{V_{CC}-V_i}{\phi_c}\right)^{1-n_c}\right] \\
&+ h_{FE}\left[\frac{1}{\bar{\omega}_T} + R_L\bar{C}_{TC}\right]\ln\left(\frac{1}{1-0.1\dfrac{\beta_C}{h_{FE}}}\right)
\end{aligned}
\tag{9-1}
$$

Representation was selected for the frequency response of α in which $\alpha = \alpha_0(1+jf/f_\alpha)$, with f_α being the common-base current gain cut-off frequency. Moll's original analysis assumed the effect of the collector capacitance–load resistance time constant to be negligible.

An extremely clever technique was developed for calculating the storage time. The saturated transistor was split into two unsaturated transistors: a "normal transistor," characterized by the normal parameters α_N and ω_N, and an "inverted transistor," characterized by α_I and ω_I.

A comparison of the common-emitter switching time equations for constant base drive for the Moll model and the charge-control model is shown in Table 9-1. The comparison is made for the case $R_L \longrightarrow 0$. Because of the assumption made for the frequency response of α, the basic time constant in the Moll model is ω_N, instead of ω_T, as in the charge-control model.*

TABLE 9-1 Comparison of Common-emitter Switching Times Calculated from the Moll Model and the Charge-control Model

	Moll Model	Charge-Control Model
Rise time	$t_r = \frac{h_{FE}}{\omega_N} \ln \left(\frac{1}{1 - \frac{\beta_C}{h_{FE}}} \right)$	$t_r = \frac{h_{FE}}{\omega_T} \ln \left(\frac{1}{1 - \frac{\beta_C}{h_{FE}}} \right)$
Fall time	$t_f = \frac{h_{FE}}{\omega_N} \ln \left(1 + \frac{\beta_{CO}}{h_{FE}} \right)$	$t_f = \frac{h_{FE}}{\omega_T} \ln \left(1 + \frac{\beta_{CO}}{h_{FE}} \right)$
Storage time	$t_S = \frac{\omega_N + \omega_I}{\omega_N \omega_I (1 - \alpha_N \alpha_I)} \ln \left(1 + \frac{\beta_{CO}}{\beta_C} \left\{ \frac{1 - \frac{\beta_C}{h_{FE}}}{1 + \frac{\beta_{CO}}{h_{FE}}} \right\} \right)$	$t_S = \frac{Q_{SX}}{I_{BX}} \times \ln \left(1 + \frac{\beta_{CO}}{\beta_C} \left\{ \frac{1 - \frac{\beta_C}{h_{FE}}}{1 + \frac{\beta_{CO}}{h_{FE}}} \right\} \right)$

Quantitatively this is a significant difference in diffused base transistors where $\omega_T \approx 0.5\omega_N$. The charge-control model presents no direct method for relating the storage-time constant to physical device parameters. This is handled nicely in the Moll model. However, the biggest difference between the two models is the conceptual difference; namely, in the equivalent circuit, the transistor responds to a control current, whereas in the charge-control model the transistor responds to a charge, the time integral of a current.

Rise time:

$$t_r = h_{FE}\left[\frac{1}{\bar{\omega}_T} + R_L \bar{C}_{TC}\right] \ln \left(\frac{1 - 0.1\frac{\beta_C}{h_{FE}}}{1 - 0.9\frac{\beta_C}{h_{FE}}} \right) \tag{9-2}$$

*See Appendix for details.

Storage time:

$$t_S = \tau_S \ln\left(1 + \frac{I_{B1}}{I_{B2}}\left\{\frac{1 - \dfrac{\beta_C}{h_{FE}}}{1 + \dfrac{\beta_{CO}}{h_{FE}}}\right\}\right)$$

$$+ h_{FE}\left[\frac{1}{\bar{\omega}_T} + R_L\bar{C}_{TC}\right]\ln\left(1 + 0.1\frac{\beta_{CO}}{h_{FE}}\right) \tag{9-3}$$

Fall time:

$$t_f = h_{FE}\left[\frac{1}{\bar{\omega}_T} + R_L\bar{C}_{TC}\right]\ln\left(\frac{1 + 0.9\dfrac{\beta_{CO}}{h_{FE}}}{1 + 0.1\dfrac{\beta_{CO}}{h_{FE}}}\right) \tag{9-4}$$

The general switching equations can be similarly modified by assigning new limits of integration and adding the appropriate rise and fall time contributions to the delay and storage time intervals.

COMPARISON OF THE CHARGE-CONTROL MODEL WITH MOLL'S EQUIVALENT-CIRCUIT MODEL

Looking at the development of the understanding and characterization of transistor switching times from a historical viewpoint, the charge-control model, as derived in the Appendix, was preceded by several years by an equivalent-circuit model developed by Moll (9).

In this model the active region switching times are calculated using conventional analysis techniques. The equivalent-circuit model chosen by Moll was the T equivalent as described in Chap. 2. Moll chose to neglect C_{TE} in comparison with the diffusion capacitance.

CODING SYSTEMS

In the preceding discussion of switching statics and transients, it was shown that only two levels or states of the transistor are involved: ON and OFF. It was also shown that the transistor is a very good switch because its ON and OFF states are well defined, very stable, and very rapid in transition between states. Before discussing the function and design of transistor switching circuits, it is worthwhile to review just how logic, or information transmission, may be performed by a group of circuits switching ON and OFF. In other words, how information can be conveyed by simple "yes" and "no" instructions and/or answers.

Information theory (1)* shows that any continuously time-varying signal can be represented by n different discrete levels that are limited in subdivision by noise (where *noise* may be defined as the inability to recognize a signal level). Furthermore, these levels may change in a minimum time interval, τ, limited by energy storage in the system (an inverse of system bandwidth). Finally, the sampling theorem states that any bandwidth-limited signal may be completely characterized by $2f_s$ independent samples/second (where f_s is the maximum frequency component in the Fourier expansion of the signal of interest). Thus, if we look at an interval in time T seconds long, one would expect any of $(n)^{T/\tau}$ possible signal combinations. The signal during T could be recreated very accurately by supplying the appropriate level (signal strength) for each interval from 0 to $n - 1$ of length $1/2f_s$ to replace the continuously varying signal. This requires a system capable of n different levels, whereas the transistor switch offers only two possible levels. The solution is to encode these levels in to a timed binary sequence. This is illustrated by the following table of a binary coding for six levels (with a sequence of three bits, one could represent up to eight levels):

TABLE 9-2 Six-Level Binary Coding

0	000
1	001
2	010
3	011
4	100
5	101

A binary number may be written as

$$N_2 = a_n a_{n-1} \ldots a_i \ldots a_2 a_1 a_0 a_{-1} a_{-2} \ldots$$

where

$$a_i \leq (\text{Base} - 1), \text{ Base} = 2 \text{ for binary}$$

and thus

$$a_i = 1 \text{ or } 0.$$

This actually means (expanded into the equivalent number in Base 10)

$$N_{10} = a_n 2^n + a_{n-1} 2^{n-1} + \ldots + a_i 2^i + \ldots + a_2 2^2 + a_1 2^1 + a_0 2^0 + a_{-1} 2^{-1} + \ldots$$

The information content, I, of such a coding process may be evaluated as $n^{T/\tau} = B^I$ or, since $B = 2$ for binary encoding

$$I = \frac{T}{\tau} \log_2(n) \text{ bits (binary digits)} \tag{9-5}$$

*Numbers in parentheses refer to references listed at end of chapter.

The system capacity, or information rate, is

$$C = \frac{I}{T} = \frac{1}{\tau}\log_2(n) \text{ bits/sec.} \tag{9-6}$$

The binary coding system is desirable from two standpoints: (1) it is obviously the simplest method devisable for uniquely describing a signal; (2) requirements on the level generating and recognizing devices are minimal, as only two levels are needed. Another important factor is related to the second reason cited: the fewer the required levels, the greater the over-all system reliability.

Our argument has been limited to information transmission because this is more universally understood than logic operation. For the present, suffice it to say that logic operations are easily performed by simple two-level devices connected in accordance with the rules of switching (Boolean) algebra. The algebra of logic permits us to perform the various functions—addition, subtraction, shifting, etc.—necessary in a computation system.

LOGIC FUNCTIONS

A few of the basic logic functions, or building blocks, are illustrated in Fig. 9-2. The AND gate performs the Boolean Multiply: $C = a \cdot b$—an output c exists *if a and b* are present. The OR gate performs the Boolean Addition: $c = a + b$—an output c exists *if a or b or both* are present. The Inhibit gate, of which the inverter is a special case, performs the Boolean AND-NOT: $d = a \cdot b \cdot \bar{c}$—output *d if a and b but not c* are present. Finally there is the buffer, which may or may not perform inversion or gating, which is used to drive many loads and represents a logic pyramiding function.

These logic functions, their symbols, and their truth tables are illustrated in Fig. 9-2. A *truth table* lists all outputs for all input combinations. For binary switching, there will be $(2)^n$ possible input combinations where n is the number of inputs. In these tables, a "1" is defined as the "true" logic level and a "0" as the "false" logic level.

These functions do not directly apply to the transistor as a logic element, however. A transistor operating in the common-emitter mode is an inverter, so that any gating signal into its base is inverted, and we inherently have NOT-AND or NAND (not to be confused with AND-NOT), NOT-OR or NOR, and NOT gates. This is illustrated in Fig. 9-3 for the NAND and NOR gates.

Let us synthesize a gate using a current-driven (large-series base resistance) common-emitter switch. This is illustrated in Fig. 9-4. Notice that the transistor will not conduct and that the switch will be OFF (output near $+V_{CC}$) when all its inputs are low, near zero. Thus, this is a NAND gate for low inputs. However, the transistor will be conducting (saturated) and the

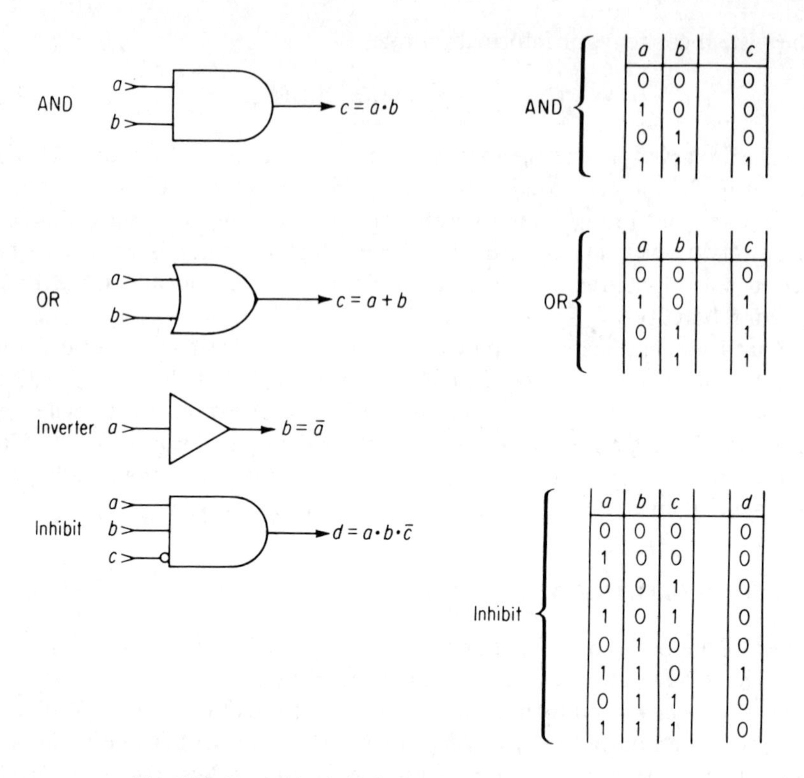

AND

a	b	c
0	0	0
1	0	0
0	1	0
1	1	1

OR

a	b	c
0	0	0
1	0	1
0	1	1
1	1	1

Inhibit

a	b	c	d
0	0	0	0
1	0	0	0
0	0	1	0
1	0	1	0
0	1	0	0
1	1	0	1
0	1	1	0
1	1	1	0

Fig. 9-2. Several basic logic functions and their truth tables.

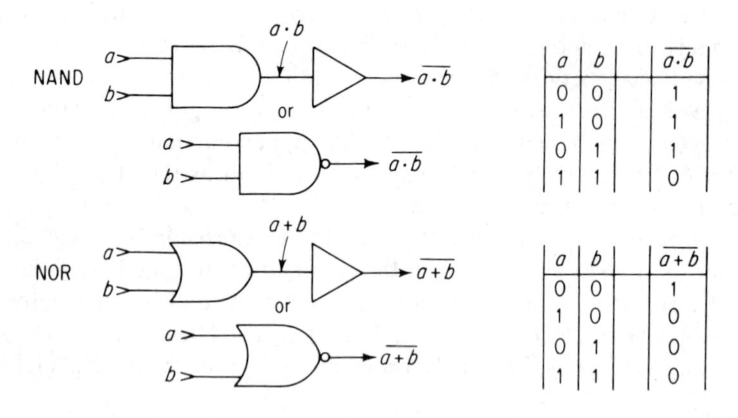

a	b	$\overline{a \cdot b}$
0	0	1
1	0	1
0	1	1
1	1	0

a	b	$\overline{a + b}$
0	0	1
1	0	0
0	1	0
1	1	0

Fig. 9-3. Inverted gating functions.

switch will be ON, output near zero, when one or more inputs are positive (enough to provide base current). This indicates that a choice of "true level" must be made to define what logic function a given gate will perform. The accepted definition is based upon the low output obtained for high inputs.

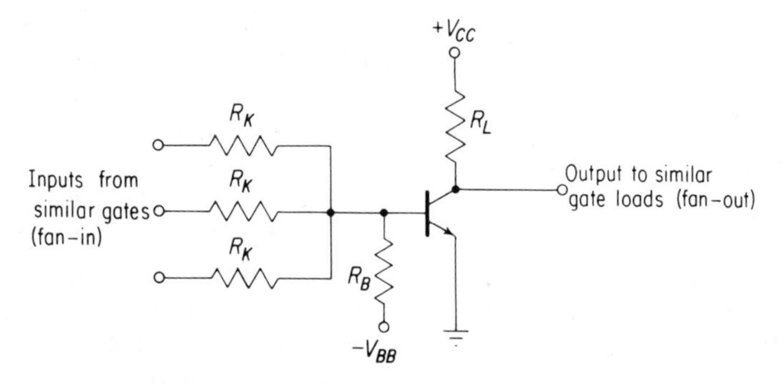

Fig. 9-4. Common-emitter NAND-NOR gate.

Thus, Fig. 9-4 represents a NOR gate as any or all inputs high will cause a low output.

DESIGN METHODS AND CRITERIA

Before discussing the various types of transistor logic and their respective design equations, let us consider the requirements or criteria of a design and the methods of achieving these criteria.

Obviously, the first requirement that a design must satisfy is insuring that a given circuit will function at all. Such designing may be called *nominal design* where all parameters are taken at their nominal (zero tolerance, 25°C) values.

In "absolute worst-case design," all parameters are allowed to assume values (reflecting operating temperature, aging, and tolerance) which will restrict operation or performance in the most devastating manner. Thus, an absolute worst-case design insures that a circuit will function when every restrictive parameter is as bad as it could possibly become. This requires using high and low temperature values and maximum and minimum tolerances simultaneously—a rapid, safe, but not very realistic design approach.

One may instead employ a "marginal worst-case design." Here, parameters are assigned limits on a more realistic basis. For example, if a specific set of parameters is known to possess a positive temperature coefficient, one can estimate the maximum circuit-to-circuit temperature gradient, ΔT_c, using the maximum tolerance at the high temperature, T_H, but using the minimum tolerance at the lower temperature, $T_H - \Delta T_c$, instead of at the lowest temperature, T_L. Note that this approach requires more than one design computation; in fact, it requires $[(T_H - T_L)/\Delta T_c]$ design steps (one general method requires analysis in the -55 to $+25$°C range and another in the $+25$°C to $+125$°C range).

Finally, one may employ a "statistical design" approach where the circuit is designed with a knowledge of the distributions of each parameter. Greater

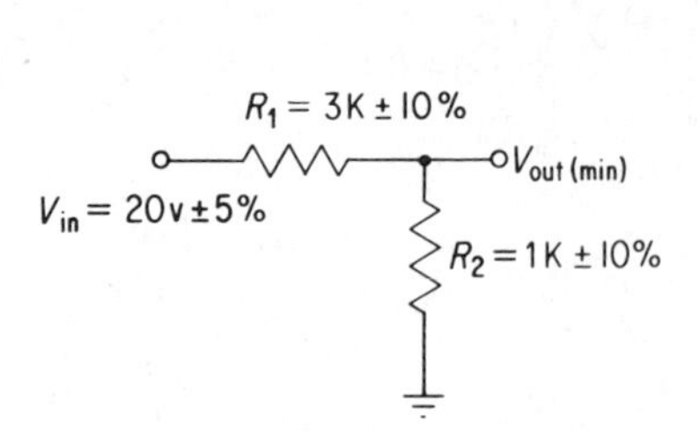

Fig. 9-5. Simple voltage divider for design examples of worst case and statistical design.

performance is possible because all parameters are not worst-case simultaneously but are given statistical "weights" in the design equations.

Let us illustrate each of the aforementioned design procedures with the following design example—the simple voltage divider of Fig. 9-5. We wish to calculate the minimum output voltage for a given input; over-all tolerances (including temperature) are ± 10 per cent for resistances and ± 5 per cent for the input voltage.

Nominal Design:

$$V_{out} = \frac{R_2}{(R_1 + R_2)} V_{in} = \frac{(1\text{ K})(20\text{ v})}{(1\text{ K} + 3\text{ K})} = 5\text{ v}$$

Absolute Worst-case Design:

$$V_{out} \geq \frac{\underline{R_2}}{(\overline{R_1} + \underline{R_2})} \underline{V_{in}}$$

or

$$\underline{V_{out}} = \frac{(1 - 0.1)(1\text{ K})(1 - 0.05)(20\text{ v})}{(1 + 0.1(3\text{ K}) + (1 - 0.1)(1\text{ K})} = 4.07\text{ v}$$

(Note the presence of $\overline{R_1}$ and $\underline{R_2}$, where a bar above a variable means its largest value and one below, its lowest value.)

A more realistic approach would be to use

Marginal Worst-case Design:

$$V_{out} \geq \frac{\underline{R_2}}{(\overline{R_1'} + \underline{R_2})} \underline{V_{in}}$$

where $\overline{R_1'} \leq \overline{R_1}$.

Suppose it were known that the minimum tcr (temperature coefficient of resistance) was 10 per cent for the temperature range of interest. Then $\overline{R_1'} = (1 + 0.1 - 0.1)R_1 = R_1$
and

$$\underline{V_{out}} = \frac{(1 - 0.1)(1\text{ K})(1 - 0.05)(20\text{ v})}{(1 + 0)(3\text{ K}) + (1 - 0.1)(1\text{ K})} = 4.38\text{ v}$$

Statistical Design

The following computations are included only for completeness; it is the result, not the method that is important at the moment. Suffice it to say that a component parameter may be assigned a pessimistic "normal distribution" where the majority of parameter values will be centered around the typical or nominal value. Using statistical techniques, it is possible to obtain a solution to the problem which states that the majority of voltage dividers built will yield greater than a specific minimum output (and less than a specific maximum output).

Assume that the tolerances given on $V_i(\equiv V_{in})$, R_1, and R_2 represent the 3σ points on the respective distributions and that the nominal values are the means of the distributions. Taking the nominal design equation, where $V_o \equiv V_{out}$

$$V_o = \frac{R_2}{R_1 + R_2} V_i$$

the necessary partial derivatives may be obtained:

$$\frac{\partial V_o}{\partial V_i} = \frac{R_2}{R_1 + R_2} = \frac{1}{4}; \quad \frac{\partial V_o}{\partial R_1} = \frac{-R_2 V_i}{(R_1 + R_2)^2} = -\frac{5}{4}\text{ma};$$

$$\frac{\partial V_o}{\partial R_2} = \frac{R_1 V_i}{(R_1 + R_2)^2} = \frac{15}{4}\text{ma}$$

Using the statistical relationship,

$$(3\sigma \,|\, V_o)^2 = \left(\frac{\partial V_o}{\partial V_i}\right)^2 (3\sigma \,|\, V_i)^3 + \left(\frac{\partial V_o}{\partial R_1}\right)^2 (3\sigma \,|\, R_1)^2 + \left(\frac{\partial V_o}{\partial R_2}\right)^2 (3\sigma \,|\, R_2)^2$$

and the three σ points,

$$3\sigma \,|\, V_i = 1 \text{ v}, \quad 3\sigma \,|\, R_1 = 0.3 \text{ K}, \quad 3\sigma \,|\, R_2 = 0.1 \text{ K}$$

results in

$$(3\sigma \,|\, V_o)^2 = \left(\frac{1}{16}\right)(1) + \left(\frac{25}{16}\right)(0.09) + \left(\frac{225}{16}\right)(0.01) = 0.586 \text{ v}$$

The nominal output has already been shown to be 5.0 v. Thus,

$$V_o = V_o(\text{NOM}) \pm 3\sigma \,|\, V_o = 5.0 \pm 0.586 \text{ v} = 5.0 \text{ v} \pm 11.75 \text{ per cent at } 3\sigma$$

At the lower 3σ point, the output is $V_o(-3\sigma) = 4.414$ v. Since 0.135 per cent of the normal distribution lies below the lower 3σ point, only 1350 out of 1,000,000 voltage dividers would have V_{out} less than 4.414 volts.

As many design problems are solved on a worst-case basis, the following guide is offered for obtaining the worst-case values of transistor parameters from the specification given on the manufacturer's data sheets.

PARAMETER VARIATIONS FOR WORST-CASE DESIGN

There are many opinions about what rules of thumb should be used in worst-case design. The following are given only as a guide for most general cases and should not be considered rigid for such designs. Let us consider parameter variations with temperature and aging separately.

Temperature

The largest variations with temperature seem to exist for I_{CBO} which is an exponential function of temperature. I_{CBO} approximately doubles for every 10°C rise in junction temperature.

$V_{BE\ \mathrm{ON}}$ is found to decrease with temperature at a rate of approximately 2 mv per °C rise in junction temperature.

DC current gain, h_{FE}, is lowest at low temperatures, decreasing approximately 30 per cent for junction temperature variations from +25°C to −25°C and approximately 50 per cent to −55°C. This change is dependent on the type of transistor and its construction.

Saturation voltage increases slightly with temperature. The variation is usually less than 20 per cent for junction temperature variation from +25°C to +85°C.

For most transistors, variations in ω_T, C_{TE}, and C_{TC} are so small that they are usually neglected. The storage time constant τ_s is quite temperature dependent. The increase in τ_s is usually about 30 per cent for an increase in junction temperature from 25°C to 55°C.

Aging

Here, of course, the degradation in parameter performance is dependent upon the length of aging. Therefore, it is only plausible to mention which parameters will show the greatest degradation during life. $V_{CE\ \mathrm{SAT}}$ and $V_{BE\ \mathrm{ON}}$ are found to be very stable with life and are usually derated by no more than 10 per cent. However, h_{FE} and I_{CBO} are found to be less stable with life and are usually derated by approximately 20 per cent and 100 per cent, respectively, depending upon the transistor type.

Variations in transient response parameters with life are usually neglected because of the small changes involved.

From the foregoing discussion, it should be apparent that it is necessary to incorporate these transistor parameter variations into the circuit design. In addition, the newer planar devices will operate within these tolerances for several thousand hours at maximum conditions.

FORMS OF TRANSISTOR LOGIC (OPERATION AND DESIGN)

The design engineer is often faced with the problem of selecting the proper transistor for his particular application (2). In the field of "switching," there are various circuits which are used as basic building blocks. These building blocks are then interconnected in some fashion to form the over-all system, perhaps a computer. In this section we shall consider two of the most commonly used building blocks, namely, the gated inverter and the flip-flop. The following is mainly a discussion of the pertinent properties of the transistor which are important and necessary to obtain the optimum performance of these two building blocks in different types of "transistor logic." In this discussion, the inverter and the flip-flop are certainly related since the flip-flop is nothing more than the interconnection of two inverters. Also, the basic *worst-case* design equations are given for the most important of these logic forms.

From a circuit viewpoint, different types of transistor logic are formed essentially by changing the coupling element between transistor inverters. Coupling elements, such as direct connections, resistors, diodes, and resistor-capacitor combinations, are most common. Let us discuss the two basic building blocks for various types of transistor logic and thus derive the desirable qualities of the transistor for these applications along with the respective design equations.

Certainly the simplest coupling element between transistors is a direct connection. The logic thus formed is termed *Direct Coupled Transistor Logic* (DCTL).*

DCTL

The basic inverter and flip-flop are shown in Fig. 9-6 and Fig. 9-7. DCTL is one of the many types of saturated transistor logic; that is, when the transistor is ON it is saturated and thus the voltage from collector to emitter is

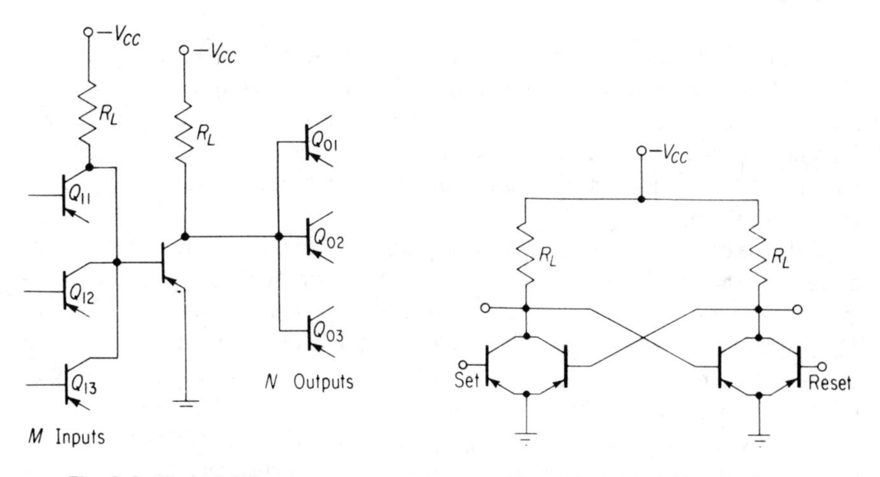

Fig. 9-6. Basic DCTL gate.

Fig. 9-7. Basic DCTL flip-flop.

*While today almost all logic design is done with integrated circuit gates, it is still instructive to consider discrete design since it more clearly indicates interactions and tolerances.

very small. With reference to Fig. 9-6, consider the case where any or all of the input transistors are ON. Then, their output voltage $V_{CE\ SAT}$ will be applied to the next transistor. The base voltage needed to turn a silicon transistor ON is approximately 0.5–0.7 volt and therefore the middle transistor will be OFF. Since this middle transistor is OFF, its collector current will be small, and the collector-to-emitter voltage will try to take on the value $-V_{CC}$. This action will turn the output stages ON. The forward-biased base-emitter diodes of the output stages will clamp the collector voltage of the middle stage to $V_{BE\ ON}$ of the output transistors. It is seen that the collector voltage swing of any given stage is from $V_{CE\ SAT}$ to $V_{BE\ ON}$. A typical range of transistor input characteristics is shown in Fig. 9-8.

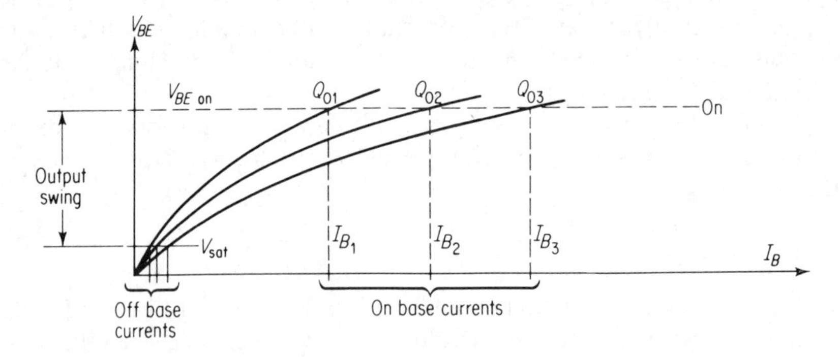

Fig. 9-8. Sketch of the typical input characteristics of three transistors illustrating spread in I_B on base current.

Notice also that the ON base current is different for variations in input characteristics of the output transistors. The lack of unique ON and OFF states, where the OFF base current is dependent upon $V_{CE\ SAT}$ and the ON base current is dependent upon $V_{BE\ ON}$, places severe requirements upon the transistor. Let us discuss the importance of the various dc transistor parameters in this type of logic.

A low saturation voltage $V_{CE\ SAT}$ is desirable, for this limits the collector current in the OFF state. This OFF collector current determines the maximum number of fan-ins. From the foregoing discussion of variations in $V_{BE\ ON}$, we see that this parameter has the most effect in determining the maximum number of fan-outs. Also, notice that, since $V_{BE\ ON}$ decreases with increasing temperature, the output voltage swing will also decrease.

A transistor with moderately high dc current gain, h_{FE}, is desirable so as to insure saturation of the transistor which receives the lowest base current (transistor Q_1 in Fig. 9-8). Of course, a tight specification on the variation of $V_{BE\ ON}$ will lessen the need for a high h_{FE}.

A transistor with low I_{CBO} is required for DCTL because the OFF collector current due to I_{CBO} is multiplied by some fraction of h_{FE}. Since I_{CBO}

increases exponentially with temperature, and since the switched collector current normally employed in DCTL is low, a condition may be reached where the transistor collector voltage will be held at a low value because of the fractional $h_{FE}I_{CBO}R_L$ drop at high temperatures. It is interesting to note that the temperature variation of I_{CBO} and $V_{BE\,\mathrm{ON}}$ limits the maximum reliable operating temperature of DCTL to approximately 55°C for germanium transistors and higher for silicon transistors.

To summarize, a transistor to be used in DCTL should possess the following dc qualities: a low maximum $V_{CE\,\mathrm{SAT}}$, a high minimum $V_{BE\,\mathrm{ON}}$ with a very narrow spread in this parameter, a moderately high h_{FE}, and a low I_{CBO}. Because of the small voltage swings involved, breakdown voltage ratings have very little importance in DCTL.

Since switching speed is an important consideration in most applications, a discussion of the relative importance of the transistor transient response parameters in DCTL is in order.

In most cases the ON base current of the transistor is large; that is, the transistor is driven hard into saturation. Thus, one would expect short rise times. Since the overdrive base current is large and no turn-off current is supplied, one can expect long storage times and moderate fall times. Also, as a consequence of not having a turn-off current, turn-on delay times are quite small. From this, one may well conclude that the transient response parameter of greatest importance is the storage factor τ_s, which should be as low as possible.

Now let us derive the worst-case design equations for DCTL. Consider first the circuit of Fig. 9-6. This circuit may be analyzed as follows: Generally the collector supply voltage (V_{CC}) and the ON collector current (I_{CS}) are defined before design by systems requirements, such as power dissipation and other related considerations. This fixes the load resistor as

$$R_{L\,\mathrm{nominal}} = \frac{V_{CC}}{I_{CS}} \tag{9-7}$$

The minimum current through R_L when the transistor is OFF is

$$I_{RL} = \frac{\underline{V_{CC}} - \overline{V_{BE\,\mathrm{ON}}}}{\overline{R_L}} \tag{9-8}$$

This current must be sufficient to supply the base drive current for the N fan-outs. Only one fan-out stage (Q_{01} of Fig. 9-6 and 9-8) will have $\overline{V_{BE\,\mathrm{ON}}}$ in the worst case and will require only $\underline{I_B}$, whereas the $(N - 1)$ other fan-outs (Q_{02} and Q_{03} of Figs. 9-6 and 9-8) will have $\overline{V_{BE\,\mathrm{ON}}}$ and "hog" $\overline{I_B}$. Therefore,

$$\underline{I_{RL}} \geq \underline{I_B} + (N - 1)\overline{I_B} + \overline{I_{CEX}} \tag{9-9}$$

where

$$\underline{I_B} = \frac{\overline{I_{CS}}}{\underline{h_{FE}}} = \frac{\overline{V_{CC}}}{\underline{R_L}\,\underline{h_{FE}}}, \qquad (\overline{I_B} \text{ from input curves}) \tag{9-10}$$

and $\overline{I_{CEX}}$ is known at the maximum temperature for $V_{CE} = \overline{V_{BE\,ON}}$ and $V_{BE} = \overline{V_{CE\,SAT}}$. The fan-in is given by

$$\underline{V_{CC}} - M\overline{R_L}\,\overline{I_{CEX}} \geq \overline{V_{BE\,ON}} + \underline{I_B}\,\overline{R_L} \tag{9-11}$$

We can describe this circuit with two independent design equations. These are

The OFF equation:

$$\frac{\underline{V_{CC}}\,\overline{V_{BE\,ON}}}{R_L} - \overline{I_{CEX}} \geq \frac{\overline{V_{CC}}}{\underline{R_L}\,\underline{h_{FE}}} + \overline{I_B}(N-1) \tag{9-12}$$

The ON equation:

$$\underline{V_{CC}} - [M\overline{I_{CEX}} + \underline{I_B}]\overline{R_L} \geq \overline{V_{BE\,ON}} \tag{9-13}$$

where the parameters are as previously defined. Both these equations must be satisfied for operation to be insured. The degree to which they are satisfied is the margin of safety in the design.

Another configuration is possible in DCTL. This is the same as Fig. 9-6 but without the inverter and is shown in Fig. 9-9.

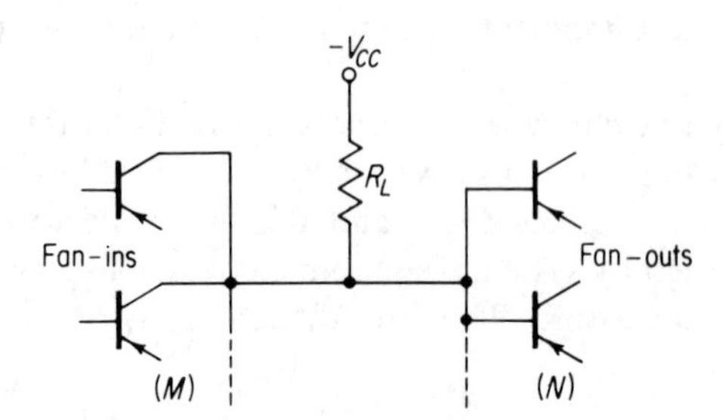

Fig. 9-9. Another DCTL configuration.

In this case, expression 9-8 still holds but Eq. 9-9 now becomes

$$\underline{I_{RL}} \geq \underline{I_B} + (N-1)\overline{I_B} + M\overline{I_{CEX}} \tag{9-14}$$

and expression 9-11 is not needed.

Notice that the I_{CS} of any DCTL fan-in transistor depends on the number of fan-ins that are ON as

$$I_{CS} = \frac{\overline{V_{CC}}}{\underline{R_L}}\left(\frac{1}{m}\right) \tag{9-15}$$

where m is the number of fan-ins that are ON. This puts a severe limitation on the design, as the input characteristics will vary widely with a m-to-one change in I_{CS}.

Notice that equations so written require no signs affixed to the parameter, and absolute values are used. Thus, they are correct as they stand for either PNP or NPN devices.

One of the obvious advantages of this form of logic is its simplicity: only one power supply is needed; only resistors and transistors are required; the power dissipation is very low. This logic is very uneconomical in terms of transistors as they are used for both gating and gain. Also, the small voltage swing, the temperature limitations, and the lack of turn-off current place restrictions on this logic. Furthermore, the tight input characteristic specifications make transistor selection rather difficult.

While these remarks were certainly true in 1967 they are no longer limiting because of the tremendous improvements made in integrated circuit manufacture. It is now the situation that transistors and diodes are cheaper to make on an IC chip than resistors and most capacitances are now made by using reverse biased p-n junctions. The photolithographic processes discussed in a later chapter have so changed logic circuit hardware in the last five years that T^2L (new name for integrated circuit version of DCTL) is now responsible for more than half of all logic design packages in computers. IC chips containing temperature compensated flip flops, multi input gates, inverters etc., are now the major building blocks used in design rather than the old DCTL. The many other types of logic described in the earlier text such as RTL, Resistor Transistor Logic; RCTL, Resistor and Capacitor Transistor Logic; DTL, Diode Transistor Logic; and DDTL, Double Diode Transistor Logic, are still used occasionally but have been developed in other forms such as ECL, Emitter Coupled Logic, or MOS, Metal Oxide Semiconductor FET logic. While we discuss these here briefly for completeness, we delay until later in the text their more complete treatment.

In the next few paragraphs we will consider a few of the more popular logic families.

A. Resistor-Transistor Logic (RTL)

A four-input RTL gate is shown in Fig. 9.10. For positive logic the gate functions as a NOR gate. The RTL family is constructed in relatively simple configurations and is consequently one of the cheapest lines of logic. On the other hand, this family is more inflexible than other popular families. The fan-out specification is usually small with a typical value of four or five for gates. Buffer or current amplifier stages are available to increase the number of parallel inputs that can be driven. The noise margin for RTL is low as also is the switching speed. Because of the disadvantages of RTL, it finds limited application in digital system design.

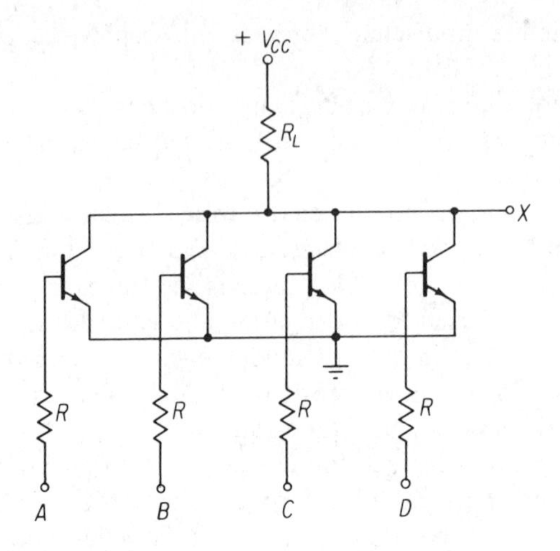

Fig. 9-10. An RTL four input gate.

B. Diode-Transistor Logic (DTL)

The DTL family generally has a better fan-out specification than the RTL family although this figure is still not outstanding. The noise margin is better and a variation of the DTL line can lead to a High-Threshold Logic family (HTL) with a very good noise margin. Switching times are of the same order of magnitude as the RTL circuit.

A typical DTL gate is depicted in Fig. 9.11. This gate functions as a NAND gate for positive logic since all three inputs must be high to cause T_2

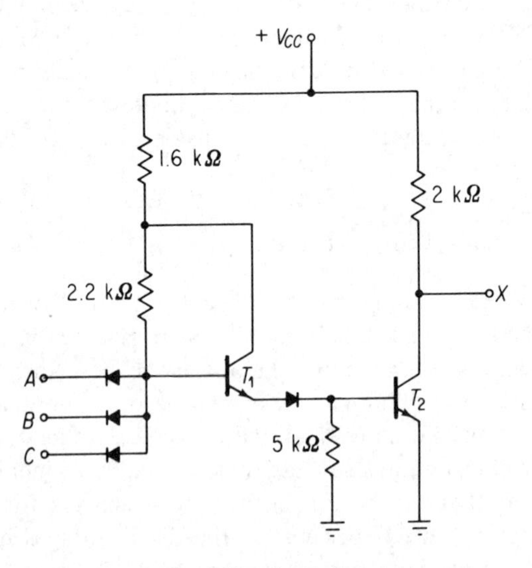

Fig. 9-11. A three input DTL gate.

to turn on. If any input is low the base of T_1 is clamped to a voltage near ground by the forward biased diode. The drop across the emitter-base junction of T_1 plus the drop across the following diode keeps the base of T_2 negative and insures that T_2 is off for this condition. Typical levels for binary 0 and binary 1 are 0.4 v and 2.4 v, respectively.

C. High-Threshold Logic (HTL)

If greater logic level voltage swings are desired along with a greater noise margin, the diode connected to the base of T_2 can be replaced by an avalanche or Zener diode. This HTL circuit is shown in Fig. 9.12.

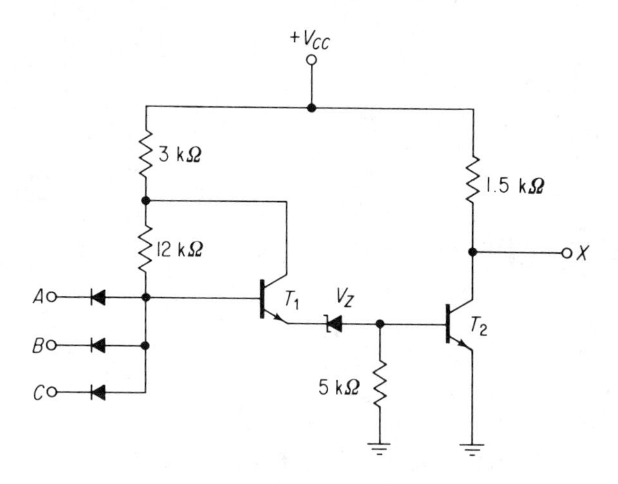

Fig. 9-12. A three input HTL gate.

If V_Z is 6 volts, the smallest input signal must exceed 6 v to turn T_2 on. For a 15 v power supply, the maximum value of a binary 0 voltage is 1.5 v while the minimum value of the binary 1 voltage is 12.5 v. An 11 v transition would be present in this circuit.

D. Transistor-Transistor Logic (TTL)

One of the most popular logic families at the present time is the TTL family (T^2L). Since the late 1960's this line has emerged as the most flexible and continues to be in great demand after the mid-1970's. This family possesses good fan-out figures and relatively high speed switching. The Schottky-clamped TTL lowers switching times even further with propagation delay of gates in the area of 2 nanosecs. The basic TTL gate is shown in Fig. 9.13. The TTL family is based on the multiemitter construction of transistors made possible by integrated circuit technology. The operation of the input

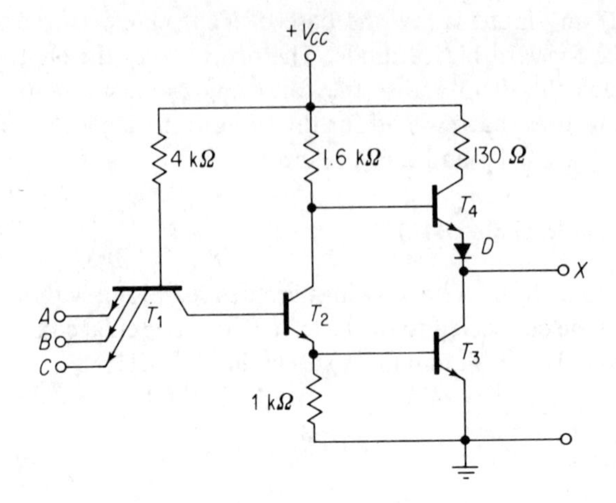

Fig. 9-13. Basic TTL gate.

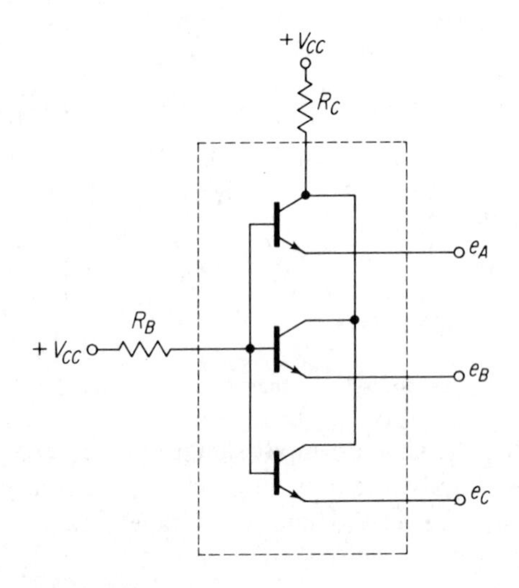

Fig. 9-14. Discrete circuit equivalent to the multiemitter transistor.

transistor can be visualized by the circuit of Fig. 9.14 which shows the bases of the three transistors connected in parallel as also are the collectors while the emitters are separate.

If all emitters are at ground level the transistors will be saturated due to the large base drive. The collector voltage will be only a few tenths of a volt above ground. The base voltage will equal $V_{(BE\ \mathrm{on})}$ which may be 0.5v. If one or two of the emitter voltages are raised, the corresponding transistors will shut off. The transistor with an emitter voltage of zero volts will still be saturated, however, and this will force the base voltage and collector voltage to remain low. If all three emitters are raised to a higher level, the base and collector voltages will tend to follow this signal.

Returning to the basic gate of Fig. 9.13 we see that when the low logic level appears at one or more of the inputs, T_1 will be saturated with a very small voltage appearing at the collector of this stage. Since at least 2 $V_{BE(\text{on})}$ must appear at the base of T_2 in order to turn T_2 and T_3 on, we can conclude that these transistors are off at this time. When T_2 is off, the current through the 1.6 kΩ resistance is diverted into the base of T_4 which then drives the load as an emitter follower.

When all inputs are at the high voltage level, the collector of T_1 attempts to rise to this level. This turns T_2 and T_3 on which clamps the collector of T_1 to a voltage of approximately 2 $V_{BE(\text{on})}$. The base-collector junction of T_1 appears as a forward-biased diode while the base-emitter junctions are reverse-biased diodes in this case. As T_2 turns on, the base voltage of T_4 drops, decreasing the current through the load. The load current tends to decrease even faster than if only T_4 were present due to the fact that T_3 is turning on to divert more current from the load. At the end of the transition T_4 is off with T_2 and T_3 on. For positive logic the circuit behaves as a NAND gate. This arrangement of the output transistors is called a *totem pole.*

E. Emitter-Coupled Logic (ECL)

One of older families is the ECL family. For many years this configuration was unrivalled in high-speed switching applications, but now competes with the Schottky-clamped TTL family in the high-speed area. The good switching characteristics of this family again result from the avoidance of heavy saturation of any transistors within the gate.

Figure 9.15 shows an ECL gate with two separate outputs. For positive logic X is the OR output while Y is the NOR output.

Often the positive supply voltage is taken as zero volts and V_{EE} as -5 v. The diodes and emitter follower T_5 establish a base reference voltage for T_4. When inputs A, B, and C are less than the voltage V_B, T_4 conducts while T_1, T_2, and T_3 are cut off. If any one of the inputs are switched to the 1 level which exceeds V_B, the transistor turns on and pulls the emitter of T_4 positive enough to cut this transistor off. Under this condition output Y goes negative while X goes positive. The relatively large resistor common to the emitters of T_1, T_2, T_3, and T_4 prevents these transistors from saturating. In fact, with nominal logic levels of -1.9 v and -1.1 v, the current through the emitter resistance is approximately equal before and after switching takes place. Thus, only the current path changes as the circuit switches. This type of operation is sometimes called current mode switching. Although the output stages are emitter followers, they conduct reasonable currents for both logic level outputs and, therefore, minimize the asymmetrical output impedance problem.

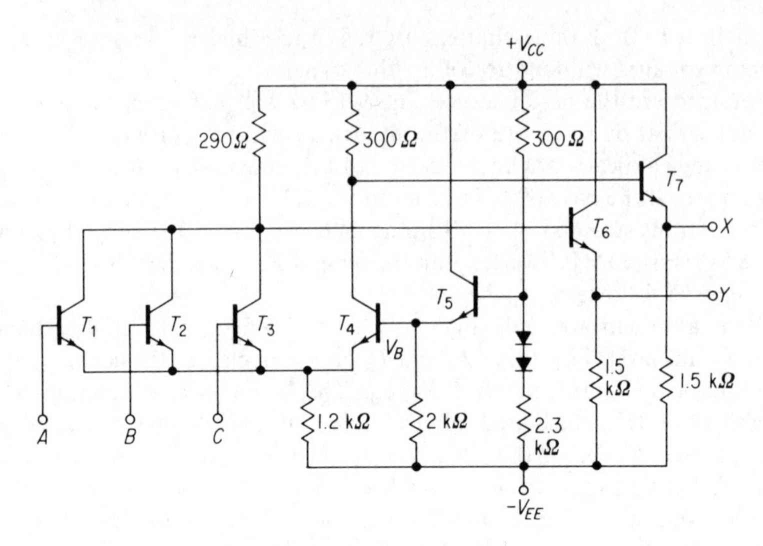

Fig. 9-15. An ECL gate.

The ECL family has a disadvantage of requiring more input power than the TTL line. Furthermore, the great variety of logic circuits realizable with TTL cannot be duplicated with ECL.

F. Complementary-Symmetry MOS (CMOS)

Although p-channel (p-MOS) and n-channel (n-MOS) MOS devices offer advantages in terms of packing density over the CMOS family, the latter has become very popular in recent years. Two major reasons for the growing popularity of CMOS are the extremely low power dissipation of CMOS gates and other logic circuits and the compatibility of this family with TTL logic circuits. While p-MOS and n-MOS logic families continue to grow in importance, we shall here consider only the well-established CMOS family. Because switching speeds are lower than many bipolar logic families, CMOS is used in low to medium frequency applications.

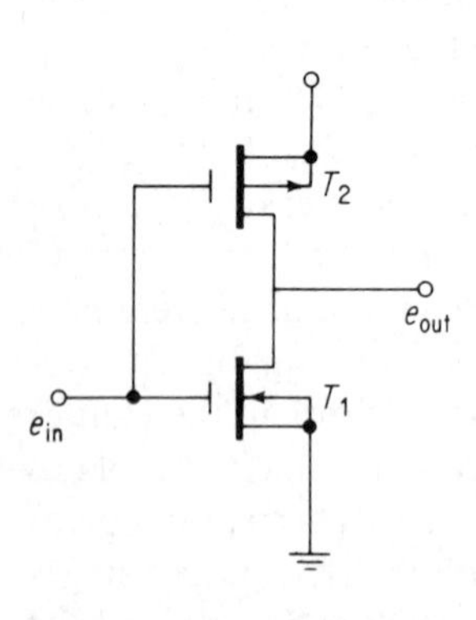

Fig. 9-16. Basic CMOS inverter.

The inverter of Fig. 9.16 is the basic building block for the CMOS gate. Both p-channel and n-channel devices are enhancement type MOS transistors. When the input voltage is near ground potential the n-channel device T_1 is off. The voltage from gate to source of T_2 is approximately $-V_{DD}$ and, therefore, T_2 is on. With T_2 in the high conductance state and T_1 in the low conductance state, the power supply voltage drops across T_1 and appears at the output. When e_{in} increases and approaches V_{DD}, T_2 turns off while T_1 turns on dropping the power supply voltage across T_2 and leading to zero output voltage. The high impedance of the off transistor results in negligible current drain on the power supply. The threshold voltage can be controlled during fabrication and is generally designed so that switching occurs at an input voltage of approximately $V_{DD}/2$.

The output voltage levels of a gate are typically less than 0.01 v for the 0 state and 4.99 v to 5.00 v for the 1 state ($V_{DD} = 5$ v). The required input voltages for this case might be 0.0 v to 1.5 v for a 0 and 3.5 v to 5.0 v for a 1. The noise margins in either state are then approximately 1.5 v.

Figure 9.17 shows a two-input NOR gate using CMOS. If both A and B are held at the 0 logic level, both n-channel devices are off while both p-channel transistors are on. The output is near V_{DD}. If input A moves to the 1 state the upper p-channel device turns off while the corresponding n-channel turns on. This leads to an output voltage near ground. If input B is at the upper logic level while A is at the 0 level, the lower p-channel device shuts off while the corresponding n-channel device turns on again resulting in an output 0. When both A and B equal 1, the output is obviously 0.

A positive logic NAND gate is shown in Fig. 9.18. In this case at least one of the series n-channel devices will be off and one of the parallel p-channel devices will be on if A, B, or both A and B are at the low logic level. The output will then be high. Only when both A and B are raised to

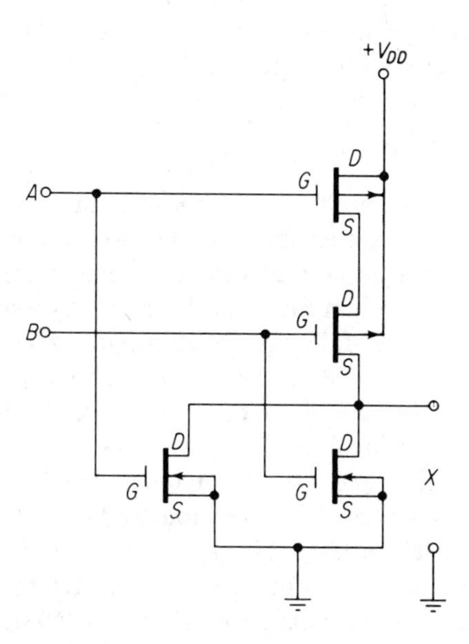

Fig. 9-17. A CMOS NOR for positive logic.

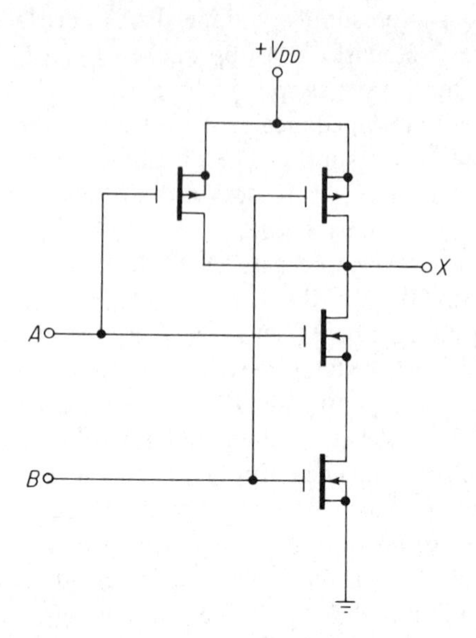

Fig. 9-18. A CMOS NAND gate for positive logic.

the high logic level will X equal the low level. For this input combination, both p-channels are off while both n-channels will be on.

There are several other logic circuits that can be fabricated in the CMOS family. Various types of flip-flops, memories, and shift registers represent only a few of the useful CMOS circuits.

Nonsaturated Logic

Current switching is probably the most popular of the nonsaturated switching logics. As a consequence of the nonsaturated operation, little turn-on delay time and no storage-time problems present themselves. The basic building block, shown in Fig. 9-19, is essentially a split load differential amplifier, with an emitter-follower input driving a common-base output stage. This gives rise to complementary outputs. If only one output is needed, the emitter may be clamped to ground, as shown in Fig. 9-20.

This very fast logic involves a signal swing about an operating point ($-V_{Cq}$) rather than ON and OFF switching. Therefore, no large signal parameter linearity is needed. Since the input and output signals are centered about different dc levels, a level shifting device is needed between stages if identical current switches are to be cascaded. The alternate solution, not always easily realizable, is to cascade PNP and NPN stages, whereby each input operates at the level of the preceding output.

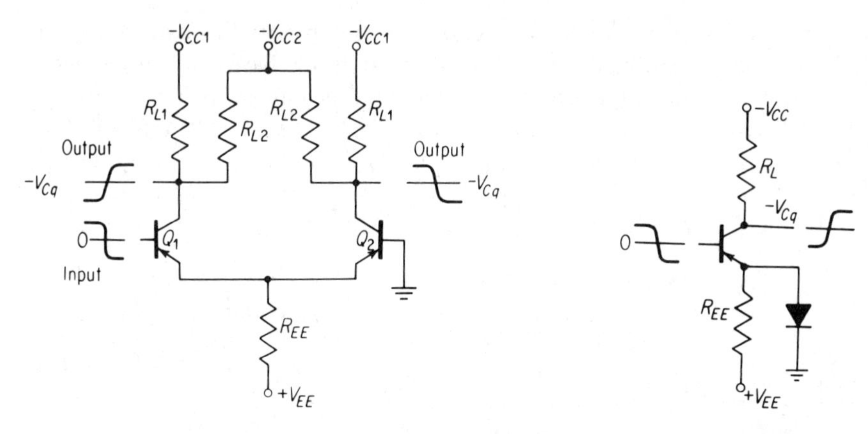

Fig. 9-19. Typical nonsaturating logic circuit.

Fig. 9-20. Typical single output nonsaturating circuit.

The major requirement of a transistor to be used in current switching is capability for large power dissipation and high-gain bandwidth product. Large power dissipations may present a reliability problem. Notice also that the power dissipation of circuit components is high.

Because of the low base impedance of the two stages, a low base spreading resistance, r'_b, is desirable. $V_{BE\ ON}$ is relatively unimportant and $V_{CE\ SAT}$ is of little importance, h_{FE} need only be moderate, and I_{CO} is of little interest.

The operating point is usually chosen at the point of highest f_T, thus insuring small rise and fall times. Rise time is proportional to the magnitude of the input voltage swing, although a limit is imposed upon this by the swing of previous stages. It is sometimes advantageous to use an amplifier-type transistor in the application (amplifier transistors usually have a lower r'_b than switching types).

Table 9-3 gives the circuit values for a current switch using the 2N1204. Also included are the response times as measured at both outputs.

Referring to Fig. 9-19 and the detail of Fig. 9-20, the circuit may be analyzed as follows: V_{EE} and R_{EE} are both large and provide a current source. Consider Q_1 to be an emitter follower. When the base of Q_1 is positive, the emitter potential will be positive with respect to ground. Thus current will flow through Q_2 and pull its collector toward zero (but not

TABLE 9-3 Current Switch Using the 2N1204

$R_{L_1} = 180$	$V_{CC_1} = -4.5$ volts	$I_E = 15$ ma
$R_{L_2} = 1.8\text{K}$	$V_{CC_2} = -9$ volts	$V_{\text{in}} = 0 \pm 0.5$ volt
$R_{EE} = 4.7\text{K}$	$V_{EE} = +70$ volts	$V_{\text{out}} = 4.5 \pm 0.5$
Q_1	$t_r = 12$ nsec	$t_f = 8$ nsec
Q_2	$t_r = 7$ nsec	$t_f = 12$ nsec

saturate it). As Q_1 is turned ON, its base potential goes through zero when both Q_1 and Q_2 will be conducting equally. With the base of Q_1 more negative than the emitter, Q_2 will be cut off and all the current will flow through Q_1's collector. The two collector supplies make it possible to obtain an equal voltage swing around a quiescent point ($-V_{Cq}$). Notice that this circuit will lend itself ideally to voltage comparator applications.

The design equation may be written as follows:

$\underline{|V_{\text{in}}| < |V_{BE\,\text{ON}}|\,(Q_{1\,\text{OFF}})}$

$$V_{CB_1} = \left(\frac{V_{CC_2}R_{L_1} + V_{CC_1}R_{L_2}}{R_{L_1} + R_{L_2}}\right) - I_{CBO}\left(\frac{R_{L_1}R_{L_2}}{R_{L_1} + R_{L_2}}\right) \tag{9-16}$$

$\underline{|V_{\text{in}}| > |V_{BE\,\text{ON}}|\,(Q_{1\,\text{ON}})}$

$$V_{CB_1} = \left(\frac{V_{CC_2}R_{L_1} + V_{CC_1}R_{L_2}}{R_{L_1} + R_{L_2}}\right) - \left(\frac{h_{FE}}{1 + h_{FE}}\right)\left(\frac{R_{L_1}R_{L_2}}{R_{L_1} + R_{L_2}}\right)\left(\frac{V_{EE} - V_{BE\,\text{ON}}}{R_{EE}}\right) \tag{9-17}$$

$\underline{V_{\text{in}} = 0}$

$$V_{CB_2} = V_{CB_1} = V_{Cq} = \left(\frac{V_{CC_2}R_{L_1} + V_{CC_2}R_{L_2}}{R_{L_1} + R_{L_2}}\right) - \left(\frac{h_{FE}}{1 + h_{FE}}\right)\left(\frac{R_{L_1}R_{L_2}}{R_{L_1} + R_{L_2}}\right)\left(\frac{V_{EE} - V_{BE\,\text{ON}}}{2R_{EE}}\right) \tag{9-18}$$

Usually, the design may be made such that

$$|V_{CB\,\text{OFF}} - V_{Cq}| = |V_{Cq} - V_{CB\,\text{ON}}|$$

resulting in equal voltage swings about V_{Cq}. This is then translated to $V'_{Cq} = 0$ for the next current switch input.

The main advantage of current switching is its high speed. Current switching circuits have been made to operate at 250 MHz clock rates (4). Its main disadvantages are high component count, high power consumption, and low noise immunity due to low signal swings.

COMBINATIONAL LOGIC CIRCUITS

In the preceding section, we discussed transistor logic in terms of the coupling or gating elements. We have looked primarily at NAND and NOR gates. Now let us consider a few more complex switching circuits that involve the interconnection of these gates to achieve some desired operation.

Exclusive-OR

We note from our discussion of the OR gate that the output function could be described as an "either-or-both" operation in terms of the inputs. Thus, this gate could be called an *Inclusive*-OR gate. It may often be necessary

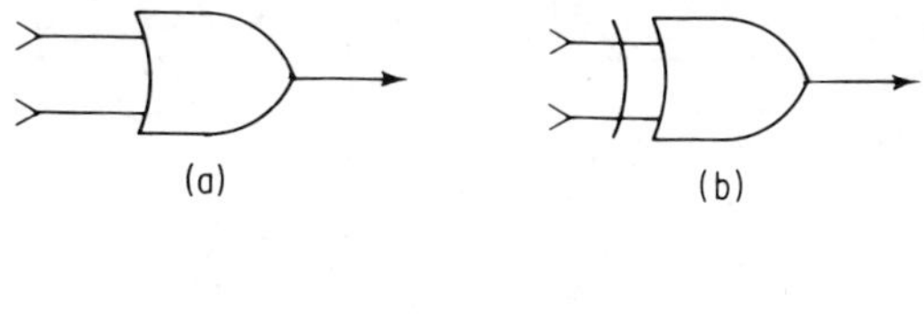

x	y	$z = x + y$	$w = x \oplus y$
0	0	0	0
1	0	1	1
0	1	1	1
1	1	1	0

(c)

Fig. 9-21. (a) Inclusive and (b) exclusive-OR gates and (c) truth table.

to obtain an *Exclusive*-OR gate giving an output describable as "either-but-not-both" in terms of its inputs. The two gate symbols and truth tables are compared in Fig. 9-21.

Just what combination of simple AND and OR gates will perform this Exclusive-OR operation? A procedure (5) is available which permits us to generate the specific simple functions required.

1. List all possible (2^n) input combinations
2. List the desired outputs for each combination of inputs
3. Ascribe OR functions to each combination
4. Choose those (3) that give a "0" output and combine them by an AND function.

In our case of the Exclusive-OR, we proceed as illustrated in Fig. 9-22. It is now a simple matter to provide the function $z = (\bar{x} + \bar{y}) \cdot (x + y)$ as shown in Fig. 9-23. The 2^n input combinations are shown in serial form at x and y to check the result.

We should now mention the principle of duality that exists for Boolean algebra. This is called the De Morgan theorem, and states two important equivalents:

$$\bar{x} + \bar{y} \equiv \overline{x \cdot y}$$
$$\bar{x} \cdot \bar{y} \equiv \overline{x + y}$$

All 2^n possible input combinations:

x	y	Desired output	OR function
0	0	0	$\bar{x} + \bar{y}$
1	0	1	$x + \bar{y}$
0	1	1	$\bar{x} + y$
1	1	0	$x + y$

Function to be generated: $(\bar{x} + \bar{y}) \cdot (x + y)$

Fig. 9-22. Generation of Exclusive-OR.

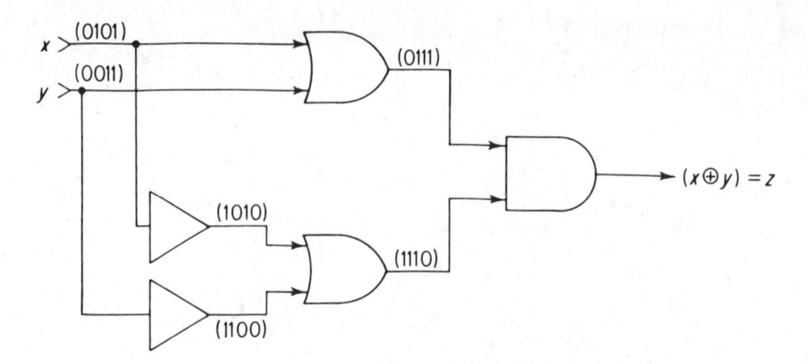

Fig. 9-23. Exclusive-OR gate derived from simple AND and OR gates and inverters.

where x and y may actually be multiple functions themselves. Two additional equivalences are

$$x \cdot \bar{x} \equiv 0$$

$$x + \bar{x} \equiv 1$$

where x may be a multiple function. These identities are useful in transforming and simplifying logic expressions. For example, expanding the Exclusive-OR expression and utilizing the fact that $x \cdot \bar{x} = 0$ gives

$$z = (\bar{x} + \bar{y}) \cdot (x + y) = \bar{x} \cdot x + \bar{x} \cdot y + \bar{y} \cdot x + \bar{y} \cdot y = (\bar{x} \cdot y) + (x \cdot \bar{y})$$

This illustrates that the Exclusive-OR could also be formed as shown in Fig. 9-24. (Compare this with that of Fig. 9-23.)

Now let us build up an RTL Exclusive-OR, using PNP transistors, for the logic form of Fig. 9-23. We know that RTL gives NOR for negative ($-V_{CC}$) levels and NAND for zero ($-V_{CE\,SAT}$) levels; thus we will assign "1" = OFF and "0" = ON implying "negative true" logic. The resultant circuit is shown in Fig. 9-25, where the logic inputs and the resulting logic levels throughout the circuit are shown in serial form. Notice that for this

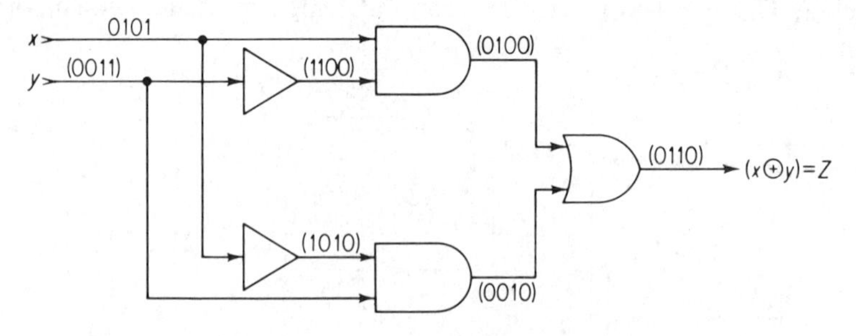

Fig. 9-24. Alternate derived Exclusive-OR.

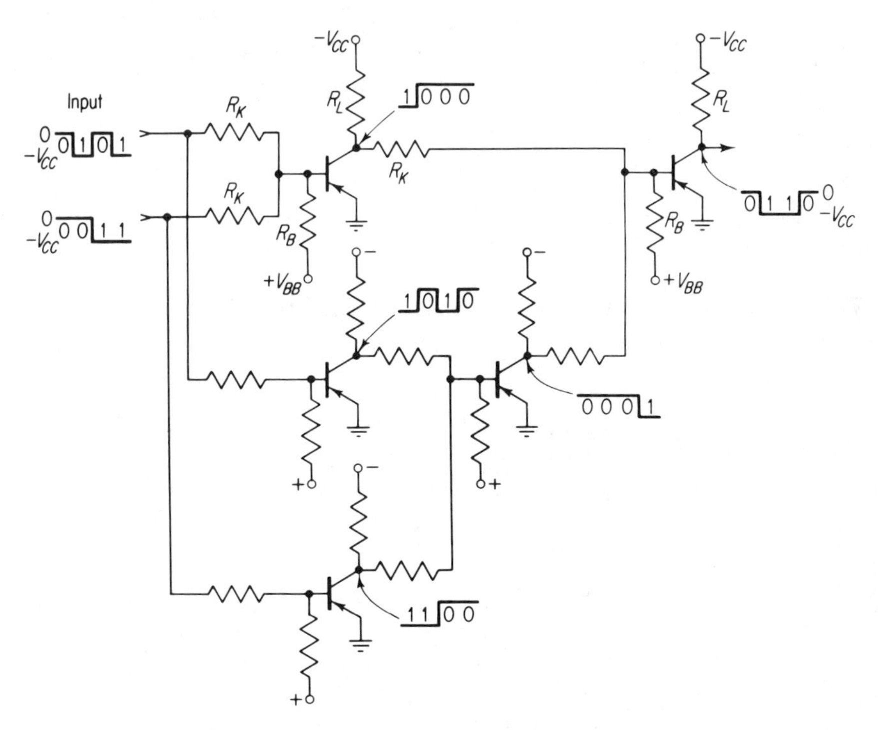

Fig. 9-25. Exclusive-OR RTL circuit using PNP transistors.

circuit, which simply recognizes a difference in input levels, there is no need to have a specific "true" signal. We could just as easily have chosen "zero true" logic. The same circuit would have resulted, but now a "zero" output would indicate unlike input levels. Though it will indicate a difference in inputs, the Exclusive-OR is not capable of telling which input (x or y) is at its true level.

The Exclusive-OR truth table shows that it provides a true output for dissimilar inputs. It will be shown (see next section on the half-adder) that its output actually represents the sum bit resulting from binary addition. The Exclusive-OR is often referred to as a quarter-adder (because it is essentially one quarter of a full adder).

Half-Adder

Another operation often performed in computation systems is the arithmetic addition of binary numbers. The basic addition table for binary addition is given in Table 9-4.

TABLE 9-4 Binary Add Table

Add	Sum	Carry
0 + 0	0	0
1 + 0	1	0
0 + 1	1	0
1 + 1	0	1

To illustrate this, let us perform the following addition in both base 10 and base 2.

Base 10		*Base* 2	
29	=	0 0 1 1 1 0 1	
+11	=	0 0 0 1 0 1 1	
30		0 0 1 0 1 1 0	—S } half-adder
c		c c	—C } solutions
40		0 0 0 0 1 0 0	
		c c	
		0 1 0 0 0 0 0	
		c	
40	=	0 1 0 1 0 0 0	—Complete Solution

The half-adder (one half of a full adder) will provide only a partial solution; it will provide the sum bit, S, and the carry bit, C, but does not have any provision for a previous carry input bit. Our half-adder would take the form shown in Fig. 9-26. Notice that this implementation of the half-adder inherently provides the inverted carry ($\bar{c}$). Thus, the additional AND gate may not be needed if $\bar{c}$ can be utilized.

Full Adder

Although the half-adder provides the basic output needed for binary addition, bit, sum, and carry, it sums only $x + y$, but no previous carry. In other words, provision is needed for a carry input (c_i) in addition to the two numbers (bits) to be added (x_i and y_i). The output of the full adder consists of the required sum and carry. It is now the sum of x_i, y_i, and c_i and the appropriate carry. The full adder generates the following:

$$S_o = (\bar{x} \cdot \bar{y} \cdot c_i) + (\bar{x} \cdot y \cdot \overline{c_i}) + (x \cdot \bar{y} \cdot \overline{c_i}) + (x \cdot y \cdot c_i) \tag{9-19}$$

$$c_o = (\bar{x} \cdot y \cdot c_i) + (x \cdot \bar{y} \cdot c_i) + (x \cdot y \cdot \overline{c_i}) + (x \cdot y \cdot c_i) \tag{9-20}$$

The truth table and symbol are shown in Fig. 9-27.

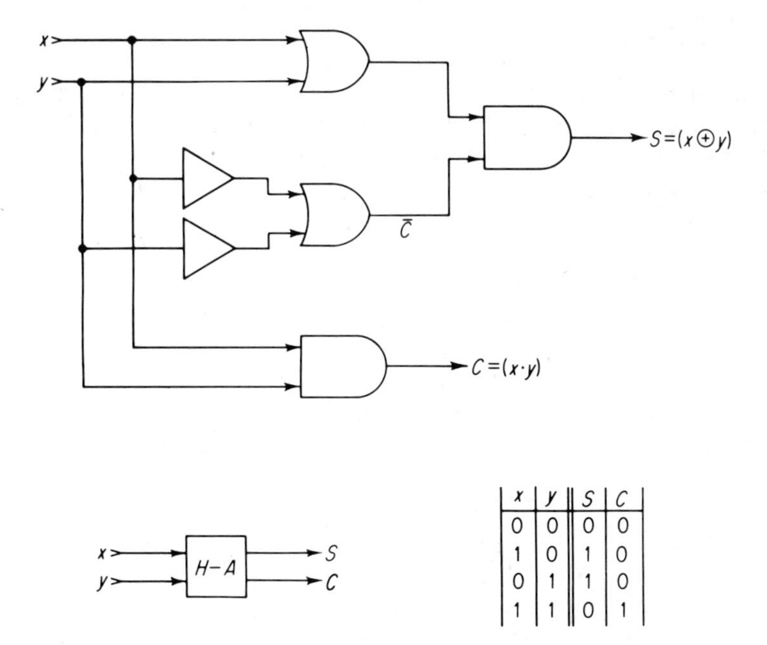

x	y	S	C
0	0	0	0
1	0	1	0
0	1	1	0
1	1	0	1

Fig. 9-26. Typical half-adder and truth table.

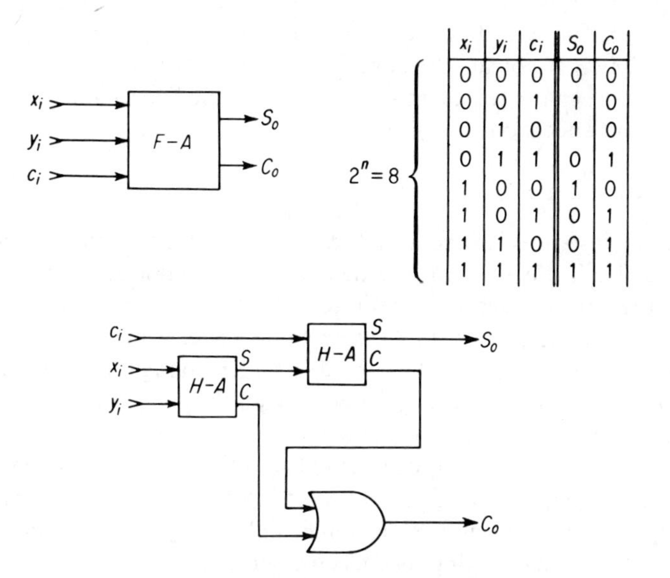

x_i	y_i	c_i	S_o	C_o
0	0	0	0	0
0	0	1	1	0
0	1	0	1	0
0	1	1	0	1
1	0	0	1	0
1	0	1	0	1
1	1	0	0	1
1	1	1	1	1

Fig. 9-27. Full-adder and truth table.

The Emitter Follower

Although it is actually a noninverting gate, the emitter follower usually finds application where low impedance loads (many outputs) are to be driven. This is essentially a common-collector switch—it does not give signal level gain and it does not give inversion but it does transform high to low impedances. A typical *EF* circuit is shown in Fig. 9-28.

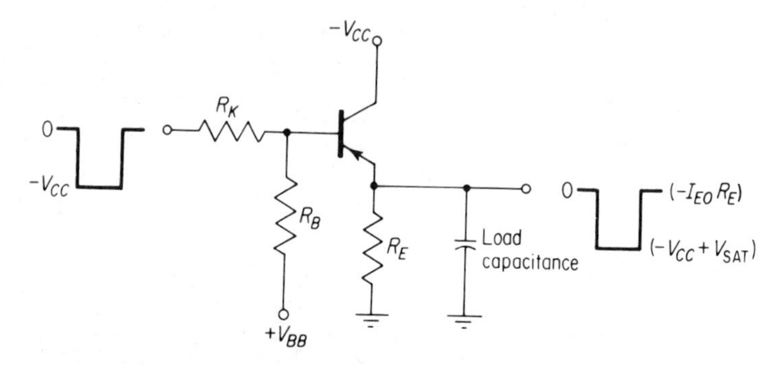

Fig. 9-28. Typical emitter follower circuit.

The emitter follower develops its output voltage across the load utilizing the fast turn-on characteristics of the transistor. The load capacitance will limit the turn-off time owing to its *RC* time constant. This is generally minimized by operating with a sufficiently small R_E, at the expense of increased power dissipation and drive requirements. In some applications, an R_E of less than 200 Ω is not uncommon. When driving transmission lines, R_E is matched to the characteristic impedance, Z_0, of the line to minimize undesirable transient effects due to mismatch.

The Buffer

In many instances, where a high capacitance exists in an otherwise high impedance load, the small R_E needed with the emitter follower cannot be used for the large voltage swing required, and the resulting power dissipation can become prohibitive. In this case a buffer gate, which may or may not perform inversion, is used. One possible buffer configuration is illustrated in Fig. 9-29. Gating may readily be performed before, or at, the base of Q_1. When the input is near zero, Q_1 and Q_3 are OFF, resistor R_2 and diode D hold Q_2 ON, pulling the output near $-V_{CC}$. When the input is negative, Q_2 is OFF because Q_1 is ON and its collector is low. Q_3 is also ON, pulling the output to $V_{CE\,SAT}$. Resistor R_3 is small and merely provides short-circuit protection for Q_2 by limiting its collector current.

The waveforms shown in Fig. 9-30 illustrate the comparison between the inverter, the emitter follower, and the buffer for a high-capacitance load. The

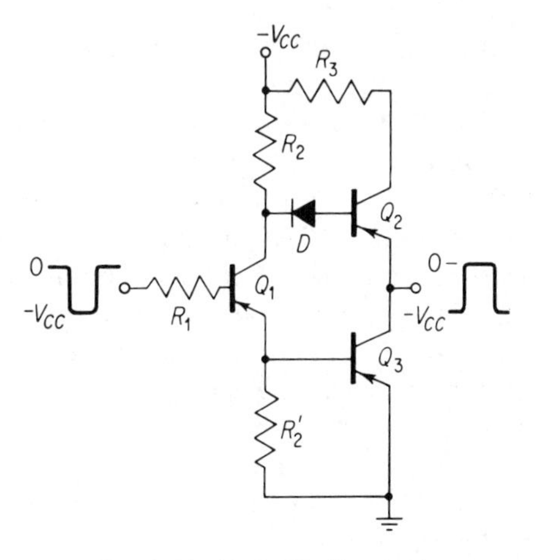

Fig. 9-29. Typical buffer circuit.

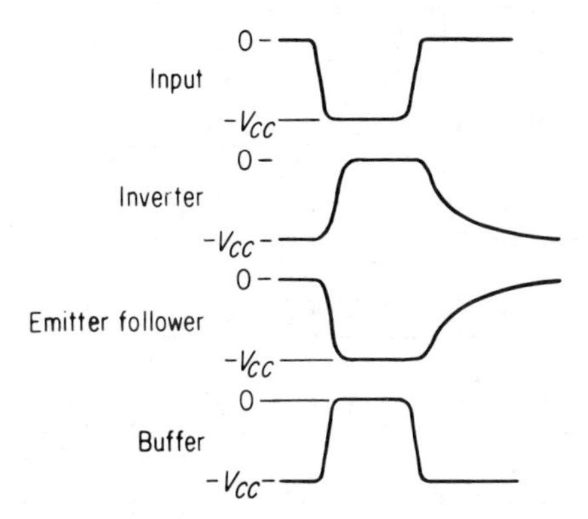

Fig. 9-30. Sketches of output waveforms for various circuit configurations.

buffer is seen to provide fast rise and fall times, as the fast turn-on characteristics of the transistors are utilized for both charging and discharging the load capacitance.

REGENERATIVE CIRCUITS

Another class of circuits useful in computer logic systems are regenerative circuits. These are circuits which will provide a level shift at the output (prescribed in time), for a given level shift at the input (triggered), or that will generate continuous level shifts at the output (free running). Such circuits are useful in transferring data (shifting), storing data, shaping pulses, delaying pulses or their edges, and clocking. A clock provides a pulse train to which all activities of a section of the computer are timed. Such computers are called *synchronous*, as all events are synchronized to the clock. An *asynchronous system* is one in which the data are allowed to propagate through the logic chains at their own pace.

The concept of clocking may be understood from Fig. 9-31 where two trains of "sloppy" pulses are to be OR gated. These pulses are the result of a loss in high-frequency response, which will show up as increases in the rise times, fall times, and delay times from various preceding stages. Notice that it is much easier to recognize a bit for the clocked than for the unclocked waveform. Notice also, if the clock rate is chosen to accommodate the worst system delays expected, that clocking provides at least one good pulse edge for each bit. If, in a given system, the majority of pulses arrive late, trailing edges will still be recognizable, and the successive stages may then be designed to switch on this good edge.

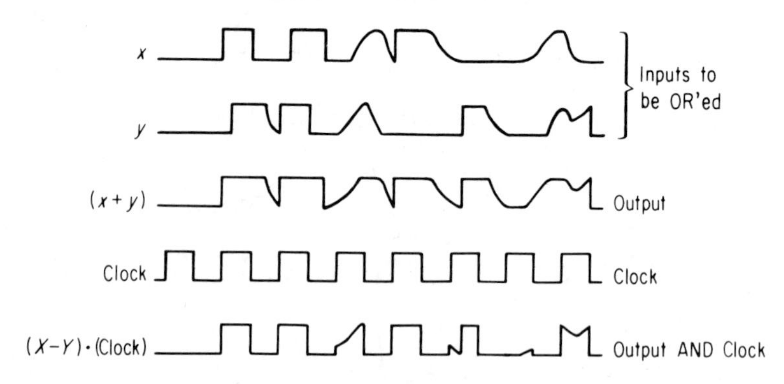

Fig. 9-31. Sketch of "sloppy" input pulses and clock pulse to illustrate clocking.

SR Flip-Flops

One of the simplest of the regenerative circuits is the set-reset flip-flop. This was discussed earlier when we looked at the forms and design of transistor logic. As mentioned, the flip-flop is formed by cross-coupling two gated inverters. Thus its design will not be considered. Since we have two kinds of gated inverters, NOR and NAND, we can form two types of flip-flops: one will flip on "1" inputs and the other on "0" inputs. These are illustrated in Fig. 9-32. Because of the back-to-back connection, if one of the inverter outputs is a "1," the other must by necessity be a "0"; the output of one of the inverters controls the output of the other. We have arbitrarily assigned a state to the flip-flops of Fig. 9-32: right side "1," left side "0." For this state, flip-flop *a*, which must normally have a "0" at *x* and *y*, must receive a "1" at *y* to flip; a "1" at *x* will permit it to retain its state. An example of this type is the RTL flip-flop. On the other hand, flip-flop *b*, which must normally have a "1" at *x* and *y*, must receive a "0" at *x* in order to change state. An

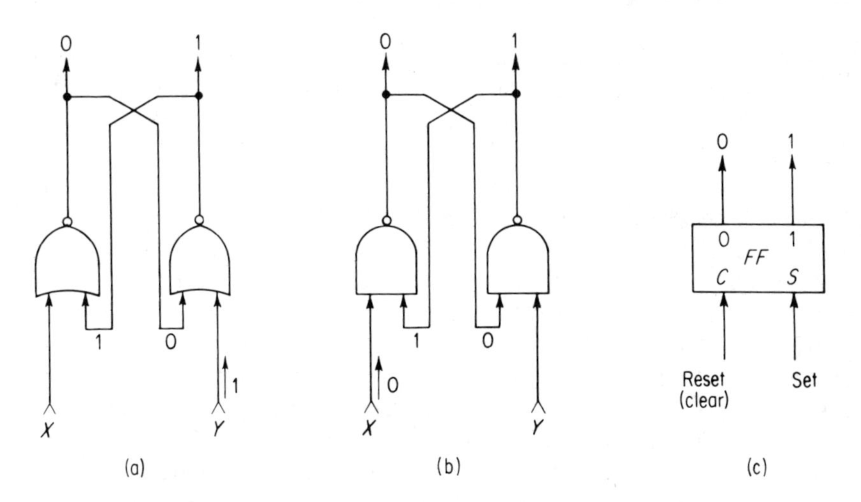

Fig. 9-32. Symbols of flip-flops. (a) "I" triggered; (b) "O" triggered; (c) symbol.

example of this type is the DTL flip-flop. Once flipped, the flip-flops must see a "1" at x and a "0" at y, for flip-flops a and b, respectively, in order to change back to their original state.

A flip-flop has two outputs, one being the complement of the other. For convenience, one of the outputs is usually referenced to the other. A flip-flop is said to be *set* when the side representing the "true" output contains a logical one; it is said to be reset (or cleared) when the side representing the "true" output contains a logical zero. Thus, the inputs are labeled *set* and *reset*. A possible RTL flip-flop is shown in Fig. 9-33, where the states are defined for "negative true" logic.

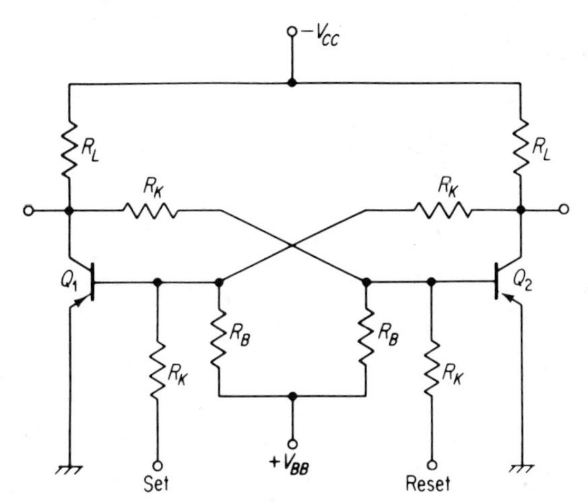

Fig. 9-33. SR flip-flop. For negative true logic ($-V_{CC}\equiv$logical 1), the set state is obtained for Q_1 ON and Q_2 OFF.

When power is initially applied to a flip-flop, it is most likely to assume the state where the transistor with the largest h_{FE} is conducting. A reset, or clear, is therefore required to insure that it is in its proper reference state. Referring to Fig. 9-33, assume that Q_2 has the largest h_{FE}. When power is applied, both Q_1 and Q_2 will begin to conduct, but Q_2 will conduct harder. As the collector of Q_2 decreases toward zero volts, the base drive to Q_1 will be reduced. Finally, the state will be reached where Q_2 is saturated and Q_1 is cut off by V_{BB} and R_B. Now suppose we wish to make sure that the flip-flop is reset. Applying a negative pulse (negative true) to the reset terminal succeeds in doing nothing more than saturating Q_2 harder for the duration of the pulse. A negative pulse at the set terminal will immediately sink base current from Q_1 and cancel the hold-off current from V_{BB} thus causing Q_1 to turn ON. The decrease in V_{CE} of Q_1 will then permit Q_2 to turn OFF. Notice that a hybrid type of logic is possible. Instead of turning an OFF side ON with a negative pulse, one could just as easily turn an ON side OFF with a positive pulse (if they are available in the system) and achieve the same end, although this would take more trigger voltage, since the gain of the transistor is not realized.

Another feature of the flip-flop is its complementing output. Recall the Exclusive-OR, Fig. 9-23, where one NOR input was x, y and the other NOR input was $\bar{x}$, $\bar{y}$. To avoid using the two inverters, we can use flip-flops as shown in Fig. 9-34, with the added ability to store a given x, y combination. This scheme may also be used to sample the states of two registers in that an output is realized only if the two flip-flops are in different states.

A flip-flop may readily be designed using the following method: this method, however, assumes very small loading and is not as good a design method as would result from a complete inverter design, where fan-outs account for loading.

OFF:

$$\frac{\underline{\underline{V_{BB}}}}{R_B} \geq \frac{\overline{V_{CE\,\mathrm{SAT}}}}{\underline{R_K}} + \overline{I_{CBO}} + \frac{\overline{I_{CS}}}{\beta_{CO}} \tag{9-21}$$

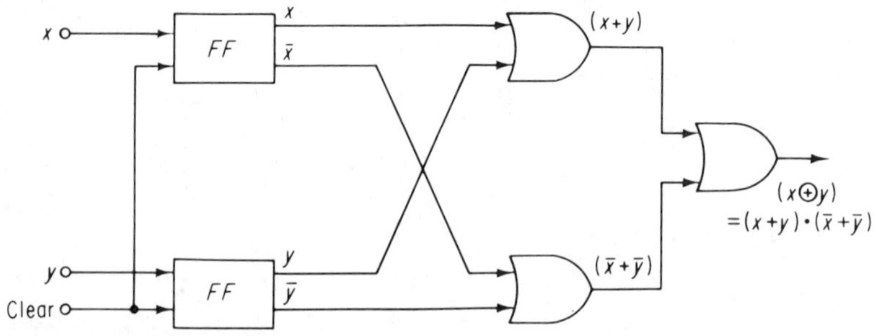

Fig. 9-34. Exclusive-OR sampling the state of two register elements.

ON:

$$\frac{\underline{\underline{V_{CC}}} - \overline{R_L}\,\overline{I_{CBO}} - \overline{V_{BE\,\text{ON}}}}{\overline{R_L} + \overline{R_K}} \geq \frac{\overline{V_{BB}} + \overline{V_{BE\,\text{ON}}}}{\underline{R_B}} + \frac{\overline{I_{CS}}}{\beta_C} \tag{9-22}$$

where

$$\beta_C \leq \beta_{CO} < h_{FE} \quad \text{and} \quad R_L = \frac{V_{CC}}{I_{CS}}$$

Usually the ratio taken for silicon transistor circuits is

$$4\beta_C \leq \frac{\beta_{CO}}{2} \leq h_{FE} \tag{9-23}$$

Since the flip-flop is a logic element capable of storage or memory, "words" consisting of binary numbers can be stored in a string of flip-flops called a *static register* and left there indefinitely until needed. Figure 9-35 shows a 4-bit register of flip-flops storing the word 1101, which was entered through the set or "data" inputs. The reset lines have all been made common to form the register "clear" line.

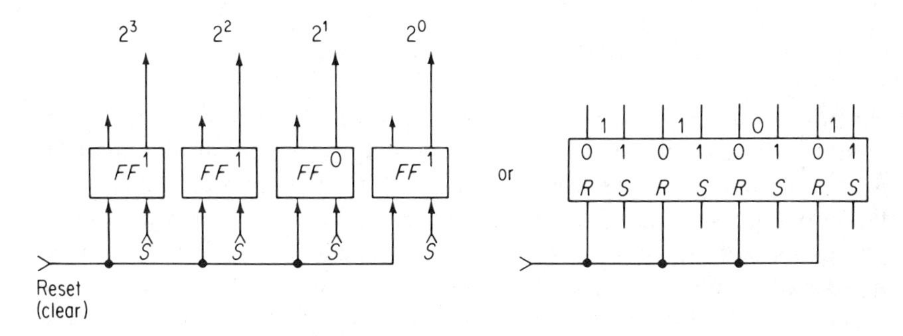

Fig. 9-35. Two symbolic representations of a register storing the word 1101.

Shift Registers and SRT Flip-Flops

The most common register form is not the aforementioned static register but the so-called shift register, one stage of which is often referred to as a *set-reset-triggerable flip-flop*. A *shift register*, unlike the static register, is a register in which the word or number stored can be shifted one bit to the left or right upon command. The simple shift register element then consists of a storage element (flip-flop) and logic gating which sense the bit of the previous stage and make the shift occur upon application of the command or clock signal. As such, the logic diagram for a shift register stage is as shown in Fig. 9-36.

This logic diagram states that, upon application of a logical "0" at the clock or shift line, the word stored in the previous register element will be

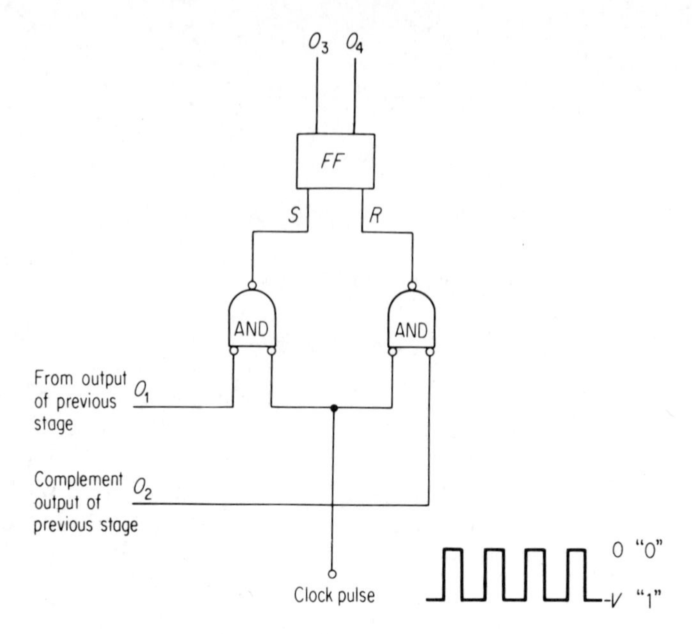

Fig. 9-36. Shift register stage.

shifted into this stage. This occurs because a logical "0" on one of the sense inputs and the clock input will set the flip-flop to its logical "1" state. That is, the "0" output of the AND gate turns "off" the appropriate side of the flip-flop if it was "on." Since in this logical scheme "0" inputs produce "1" outputs, a data line crossover is required when connecting stages. The shift register stage then reads into itself the state or number of the previous stage upon the application of each "0" going clock pulse. The last stage in a chain of such shift register stages presents a serial output of the data stored in the shift register.

A simple and widely used shift register circuit is shown in Fig. 9-37. Note the familiar flip-flop configuration and the AND gate formed by R_s and C_t. If both the data input (A) and the clock go to a logical zero at the same time, that side of the flip-flop (Q_3) will be turned off while the complementary data input (B) will be a logical one, thus blocking the clock pulse from reaching Q_4. Note that, since the A and B data inputs are always complements, the clock pulse can never be applied to Q_3 and Q_4 simultaneously, thereby eliminating the possibility of a fault condition. There is, however, a fault condition known as the "race" problem which can occur in a chain of shift register stages.

Consider two adjacent stages of a shift register as shown in Fig. 9-38. Assuming that a clock pulse on the clock line arrives at all stages simultaneously, note that stages one and two will want to change state simul-

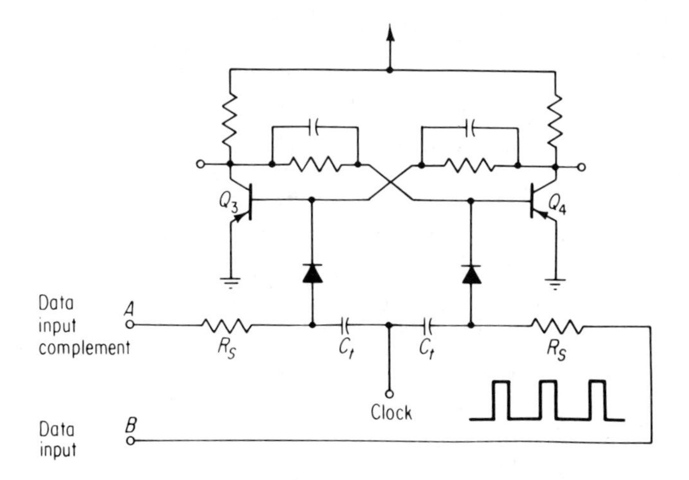

Fig. 9-37. Simple shift register circuit.

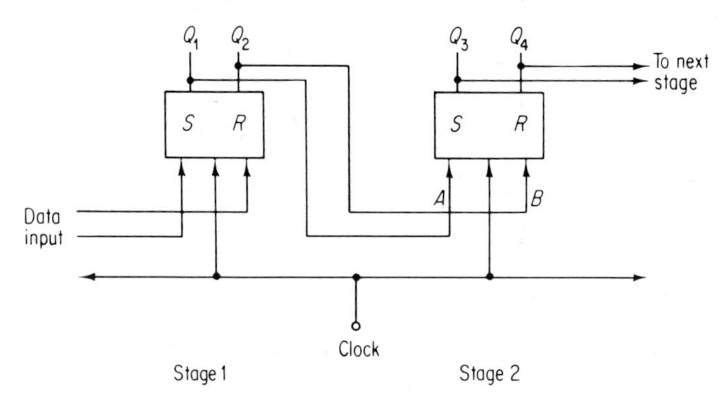

Fig. 9-38. Two adjacent stages of a shift register.

taneously. This means that data lines A and B may be changing at exactly the same time we are trying to read their information into the second stage. In other words, a race exists in trying to write the information of stage 1 into stage 2 before the state of stage 1 has changed. The severity of this race problem is dependent upon the relative transient responses of stages 1 and 2. Suppose the circuit of stage 1 has faster transient response than does stage 2. Then upon application of a clock pulse, the original information stored in stage 1 could be changed very quickly and then this new state would be shifted into the slower second stage. Such an occurrence would be a logical error. An obvious solution to this problem is to insert delay lines in the clock line. If the first solution is used (Fig. 9-39), a change in state of the first stage is delayed long enough so that the second stage receives the original information.

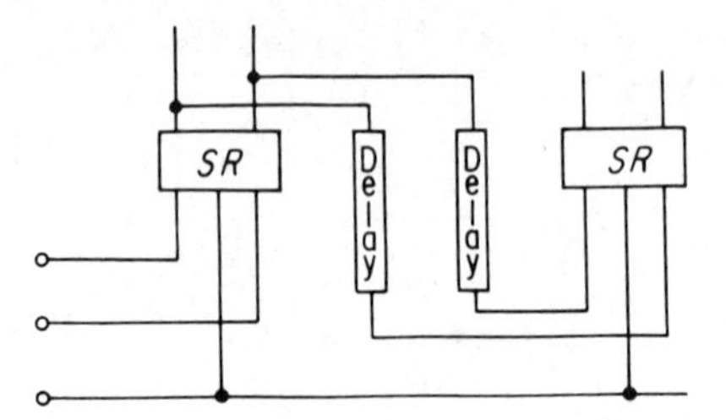

Fig. 9-39. One solution to the "race" problem in a shift register.

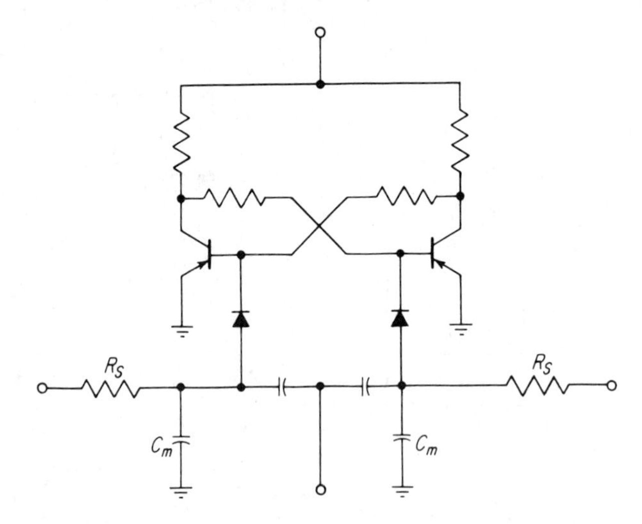

Fig. 9-40. The use of memory capacitors to eliminate the race problem.

A very simple method of implementing the scheme of Fig. 9-39 is using "memory capacitors" as shown in Fig. 9-40. Here the capacitors are used to store voltages representing the state of the previous stage. The length of time this state is stored is dependent upon the time constant $R_S C_M$.

If the second solution is used (Fig. 9-41), it merely means that the clock pulse arrives at the second stage and writes in the new information before arriving at the first stage. This is called having the clock flow (to the left) go against the data flow (to the right) and eliminates the possibility of a race problem.

Although using delay lines does solve the race problem, it does present new problems of cost, size, and slower system speed. Any delay introduced into the system limits the rate at which the register can shift, thus there will always be present a trade-off between system speed and system unreliability due to the race problem.

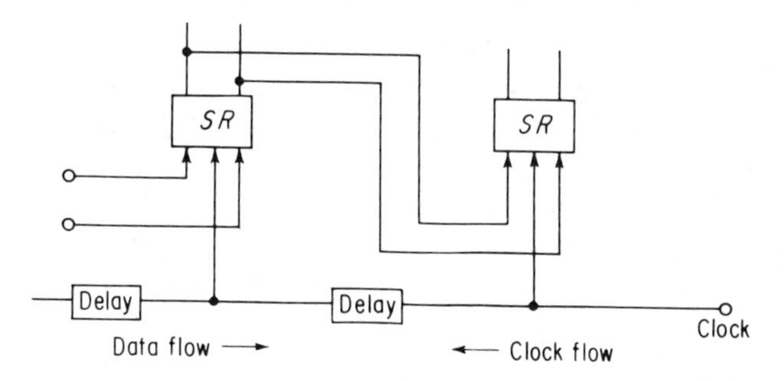

Fig. 9-41. An alternative scheme for elimination of the race problem.

A more radical way to eliminate the race problem is by employing a two-phase clock. In such a system, delay times are eliminated but a second rank of register stages is required. In essence this scheme, shown in Fig. 9-42, attains speed at the expense of hardware and is used only in high-speed data systems.

Usually this scheme is used with circuit forms, such as DCTL, DTL, and current switching, which have no capacitors and are inherently fast. The operation of the two-phase clock system is as follows: a bit appearing at the data input is written into the first stage of the data register upon application of a pulse from phase 1, ϕ_1, of the clock. Upon application of a pulse from phase 2, ϕ_2, of the clock, this bit is also written into the first stage of the slave register. Now when the second pulse from phase 1 is applied, a new bit is written into the first stage of the data register and the bit stored in the first stage of the slave register is written into the second stage of the data register. Notice that a race problem cannot exist since the second stage of the data

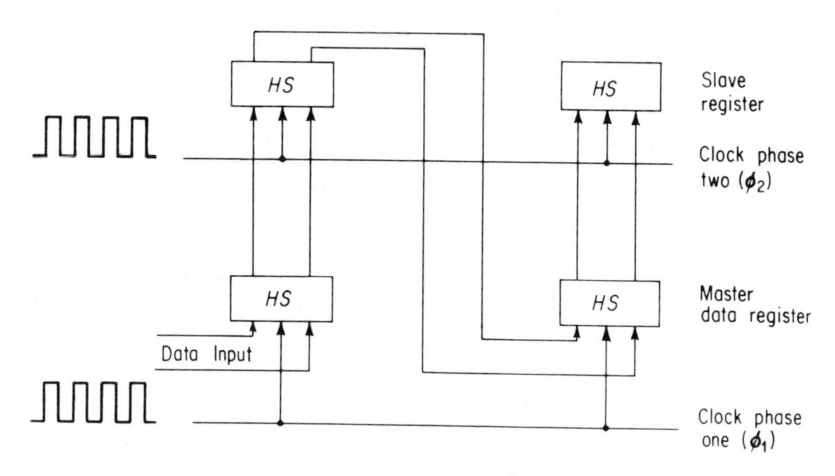

Fig. 9-42. Master-slave shift register.

register does not obtain its input data from the first stage of the data register, but rather from the slave register. This being the case, there is no possibility of erasing the number in a register stage before it is transferred to the next stage. In essence, two sets of data are always available at the end of a phase 1 clock pulse,—the new register word and the previous register word.

In the foregoing scheme a slave and a data register stage are considered one bit or one shift register element and each half is referred to as a *half-shift stage*.

Counters

A shift register stage may be connected to operate as a binary counter by connecting its data input lines to its own outputs. When this is done, the stage will change state upon application of each input clock pulse. That is, the stage will count, 0, 1 . . . 0, 1 . . . 0, 1 being triggered by each clock pulse. For this reason, such a connection of the shift register element is sometimes referred to as a *binary counter*. Figure 9-43 shows the counter connection. The AND gates act as input steering circuits in that they route the clock pulse to the correct side of the flip-flop so as to change the state of the counter. Again, the delay elements are required to overcome the race problem, which can be quite serious in this connection. Notice that the function of the AND gates is to cause the clock pulse to change the very state of the flip-flop that is controlling them in the first place. The delay elements must keep the pulses representing the original state of the flip-flop applied to the AND gates until the clock has been routed to the proper side of the flip-flop. A saving factor in all of this is that the flip-flop itself has delay due to the transistors. Thus, the output will not start its transition until after the clock pulse has been properly applied.

It is possible to trigger a counter stage on sine waves, square waves, pulse trains, or on random pulses. Furthermore, triggering can be on the

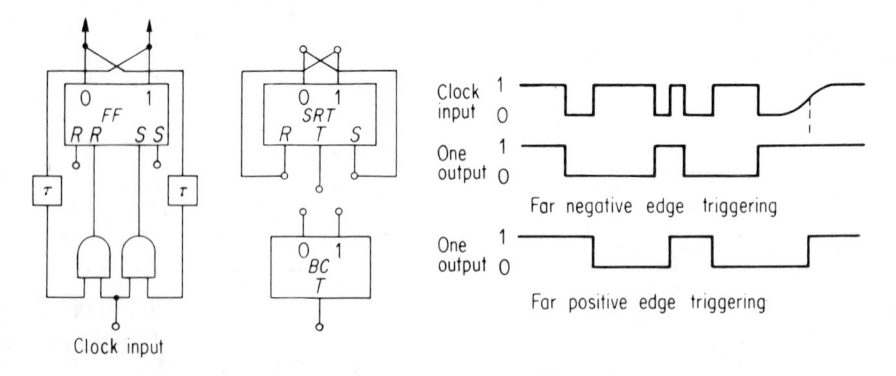

Fig. 9-43. Binary counter and possible "counting" waveforms.

leading or trailing edges of positive or negative pulses (or on positive or negative levels). One possible counter stage, made with discrete elements, is shown in Fig. 9-44. Because of its particular steering diode configuration, this binary will trigger on positive-going slopes. Thus, it will trigger on the leading edge of positive pulses or on the trailing edge of negative pulses.

One disadvantage of this counter circuit is that the maximum counting rate is limited by the time constant $R_T C_T$. This problem can be overcome to a large extent by placing a diode in parallel with R_T. Another way to overcome this problem is to use a triggering scheme which does not involve an RC time constant. Such a configuration is shown in Fig. 9-45 and is known

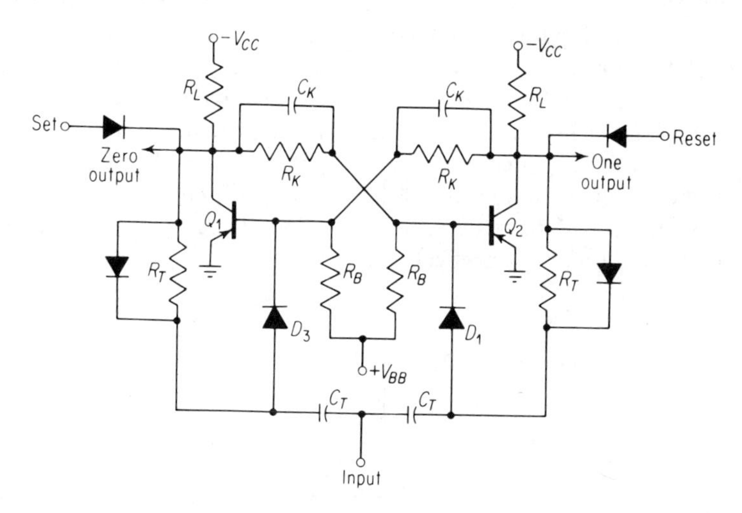

Fig. 9-44. A binary counter which triggers on positive-going pulses.

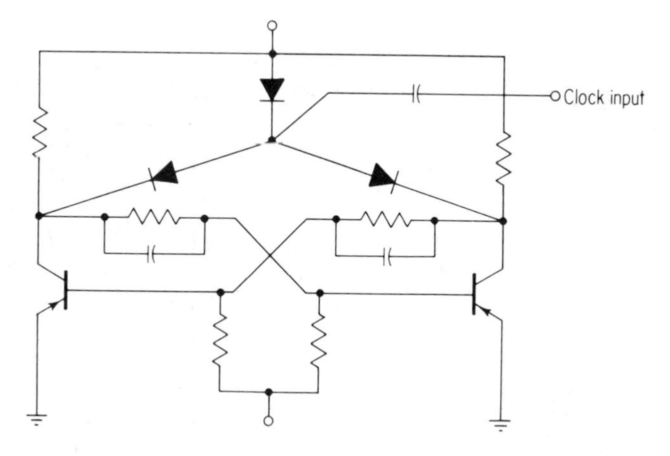

Fig. 9-45. Realization of a binary counter from a flip-flop by the use of collector triggering.

as *collector triggering*. Here the three diodes steer the positive input pulse to the proper OFF collector and cause the flip-flop to change state.

The binary counter counts by two; the output will have a repetition rate or frequency equal to one-half that of the input. It is sometimes referred to as a *frequency scaler*. A chain of cascaded binaries will provide an output frequency of

$$f_o = \frac{f_i}{(2)^n} \tag{9-24}$$

where n is the number of cascaded binaries and f_i is the input frequency. With straight binary counters, however, it is possible to achieve only cycles of 2, 4, 8, 16, . . . , 2^n input pulses. Suppose one wanted to count down by three instead of two or four. He would then need n (integer) binaries to count to $x = 3$, where $n > \log_2 x$. Thus, to count to three requires two binaries that must be gated in such a way that they complete a count cycle in three pulses instead of four. Such a scheme is shown in Fig. 9-46. Here, the diodes perform the inhibit functions as will be shown later. The process starts with both counters in their reset state, $A = B =$ ON. Here $\bar{A}$ and $\bar{B}$ or both are $-V_{CC}$ and the diode has no effect. The first pulse changes A to $-V_{CC}$ and

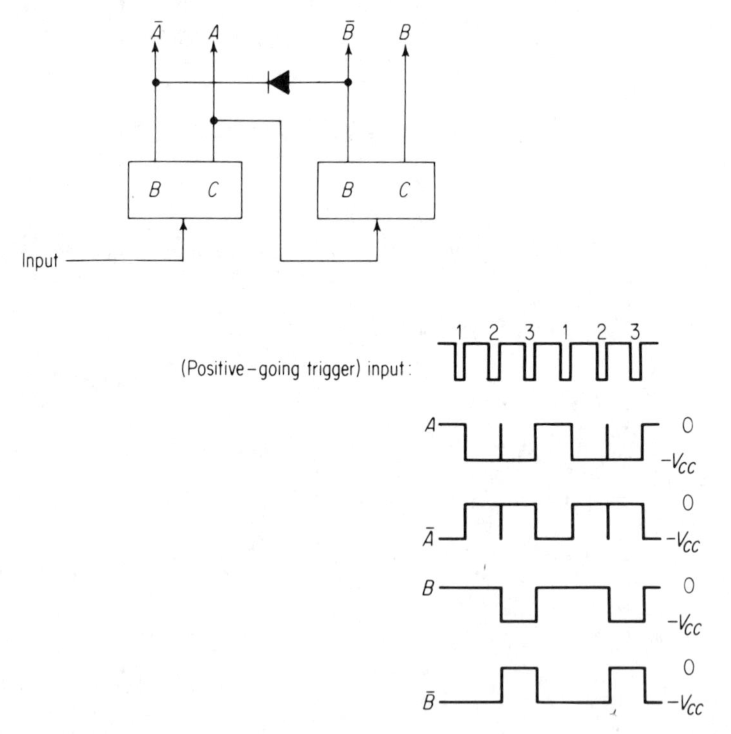

Fig. 9-46. Scale of three-binary counter.

$\bar{A}$ to $-V_{SAT}$. If the binaries trigger on positive going edges, B will not change state. $\bar{B}$ is at $-V_{CC}$ and $\bar{A}$ is at $-V_{SAT}$ so the diode is reverse-biased. The second pulse drives A to $-V_{SAT}$ which flips B. However, $\bar{B}$ is now at $-V_{SAT}$ and it pulls $\bar{A}$ back to $-V_{SAT}$ by the diode-clamping action. The third pulse flips A and B back to their original or reset state.

A more general case would be to AND gate the outputs of the chain of binaries such that one can select the count pulses desired and tell the counter when to recycle. This is shown in Fig. 9-47 for a scale-of-three counter.

Triggering the binary may be accomplished in many ways, two of which are illustrated in Fig. 9-48. The speed-up capacitors serve the two purposes of speeding-up the transient and providing memory. Memory is a basic requirement in the binary counter because the memory component aids in the triggering (inductors in series with the collector load would also serve as memory elements). Note that in general whenever a binary (or flip-flop) changes state, it goes through a point in time where each side is conducting

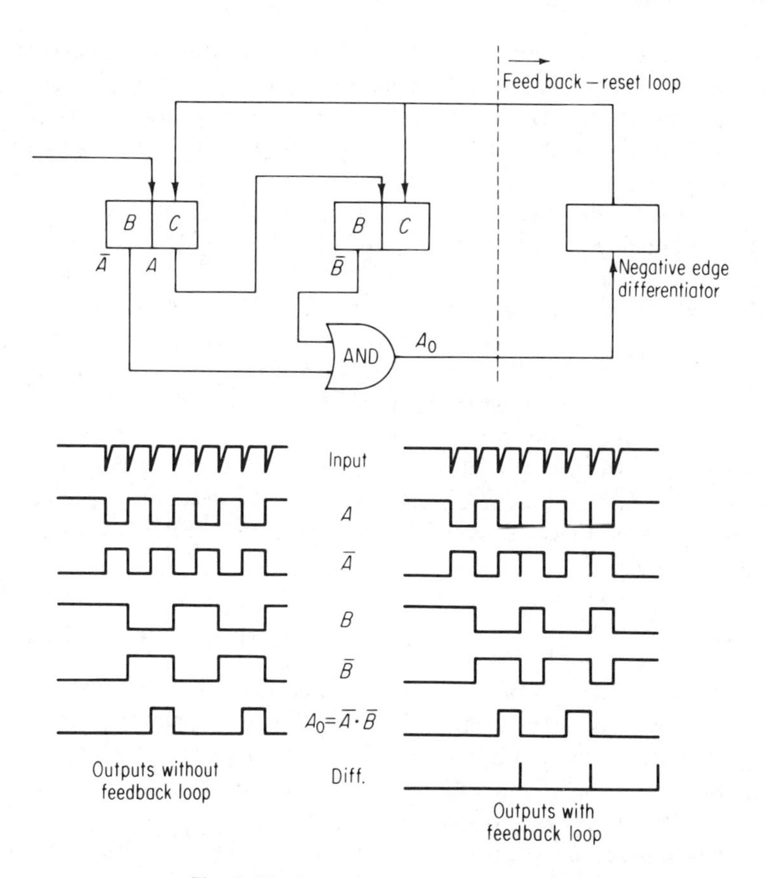

Fig. 9-47. General scale-of-three counter

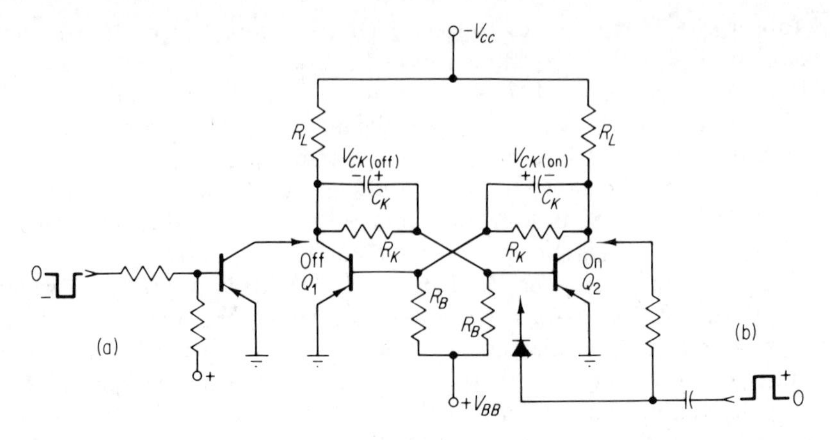

Fig. 9-48. Two possible trigger methods. (a) Grounding OFF collector; (b) turning ON bases OFF.

equally; unless some component "remembers" what state the binary was in, and thus the state to which it should go, there would be nothing to guarantee the correct change of state (in a flip-flop, this may be guaranteed by using a set pulse longer than the set time). The speed-up capacitors provide inertia which continues to aid triggering after the trigger pulse is removed. In the case of collector triggering, this is accomplished by the positive voltage ($V_{CC} - V_{BE}$) being placed on the base of the ON stage (Q_2) to turn it OFF. There is obviously an optimum value of C_K that will permit it to charge up to a suitable voltage ($\tau = RC$) before the next pulse comes along. If C is too large, the voltage will not build up fast enough; if C is too small, it will not store enough charge. In the case of base triggering, C is less critical; the collector of Q_2 goes negative which in itself helps to drive Q_1 ON. These problems become quite involved in binaries that are to count at varying rates. In fact, the most stringent test of a counter is to drive it with a slow train of fast double pulses; the double pulse resolution of a counter specifies how close in time two otherwise isolated pulses can come before they are counted as only one pulse.

JK Flip-Flops

In the preceding section we have discussed various shift register circuits all of which were basically using SRT types of flip-flops. Now we shall discuss a triggerable memory circuit, the JK flip-flop (6). This circuit offers a distinct advantage over the aforementioned SRT circuits in certain applications where the data inputs are not always complementary.

This advantage is most readily illustrated with the aid of Fig. 9-49 where the SRT and JK flip-flops and their truth tables are compared. The logical

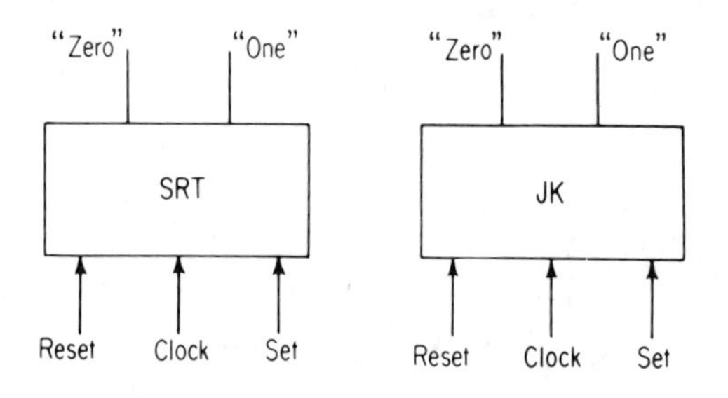

Conditions at time $t = n$			"One" output after clock $t = n + 1$	
Reset	Set	"One" Output	SRT	JK
0	1	1	1	1
0	1	0	1	1
1	0	1	0	0
1	0	0	0	0
1	1	1	1	1
1	1	0	0	0
0	0	1	?	0
0	0	0	?	1

Fig. 9-49. Truth table for SRT and JK flip-flops.

one outputs are given for time $t = n$. That is, the original state before application of a clock pulse and time $t = n + 1$ is given (just after the application of a clock pulse). Notice that the SRT and JK flip-flops have identical truth tables for the case where the data inputs are complementary, as well as for the case where logical ones are present at both data inputs simultaneously. For the case where logical zeros are present at both data inputs, the SRT flip-flop produces no predefinable output.

As an example, refer to Fig. 9-37 and note that with a logical "0" at both data inputs, the clock pulse is steered to both bases and both stages try to turn off. Thus, the output is indeterminate. For a logical "1" at both data inputs the clock is blocked from both bases so the flip-flop remains in its original state.

On the other hand, the JK flip-flop will produce a predefinable output if both inputs are a logical "0"; namely, the output will be complemented (as in a binary counter).

A relatively simple extension of the SRT flip-flop of Fig. 9-37 leads to the JK flip-flop shown in Fig. 9-50. Resistors R_T are chosen such that there is sufficiently negative voltage on one of the capacitors, C_T, to block the clock pulse from going to both bases when both inputs S and R are zero.

The logic diagram for the JK flip-flop is shown in Fig. 9-51 where NOR logic is assumed. For negative true NOR logic (RTL is one possible gate

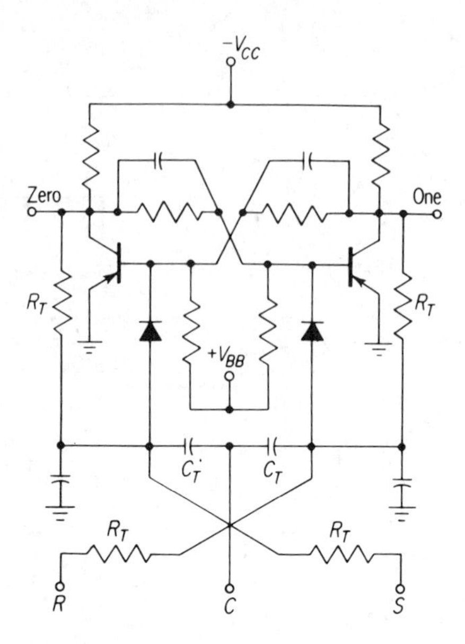

Fig. 9-50. RC-coupled JK flip-flop.

form), the timing diagrams included in Fig. 9-51 are applicable, although circuit delays are ignored. Notice that the clock is normally at 1 and that data are entered only on the 1 to 0 clock transitions. The timing diagrams include each of the four data input conditions and appear in the following order: $S = R = 1$; $S = 0$, $R = 1$; $S = 1$, $R = 0$; $S = R = 1$; $S = R = 0$; $S = R = 0$. It is very enlightening to verify the waveforms of Fig. 9-51 for it will become obvious that the "race" problem has been eliminated! Once the clock has gone to 0, the data are transferred, and any subsequent changes in the data are not gated through to the output. Since the clock must return to 1 to get "set up" for new data, the only requisite is that the data change be completed by some minimum time before the clock pulse arrives.

Since "memory" is afforded logically rather than electronically, the problems of limited pulse width, amplitude, and repetition rate usually encountered with the *RC* triggerable circuits of Fig. 9-50 are eliminated in the foregoing example. The price obviously is one of increased complexity and component count.

One-Shot Multivibrators

A family of regenerative circuits useful in pulse shaping and delaying are known as *delay multivibrators, monostable multivibrators,* or *one-shot* (*multivibrators*). The one-shot has a preferred or stable state in which it can rest for an indefinite time. Upon proper triggering—turning ON transistors OFF,

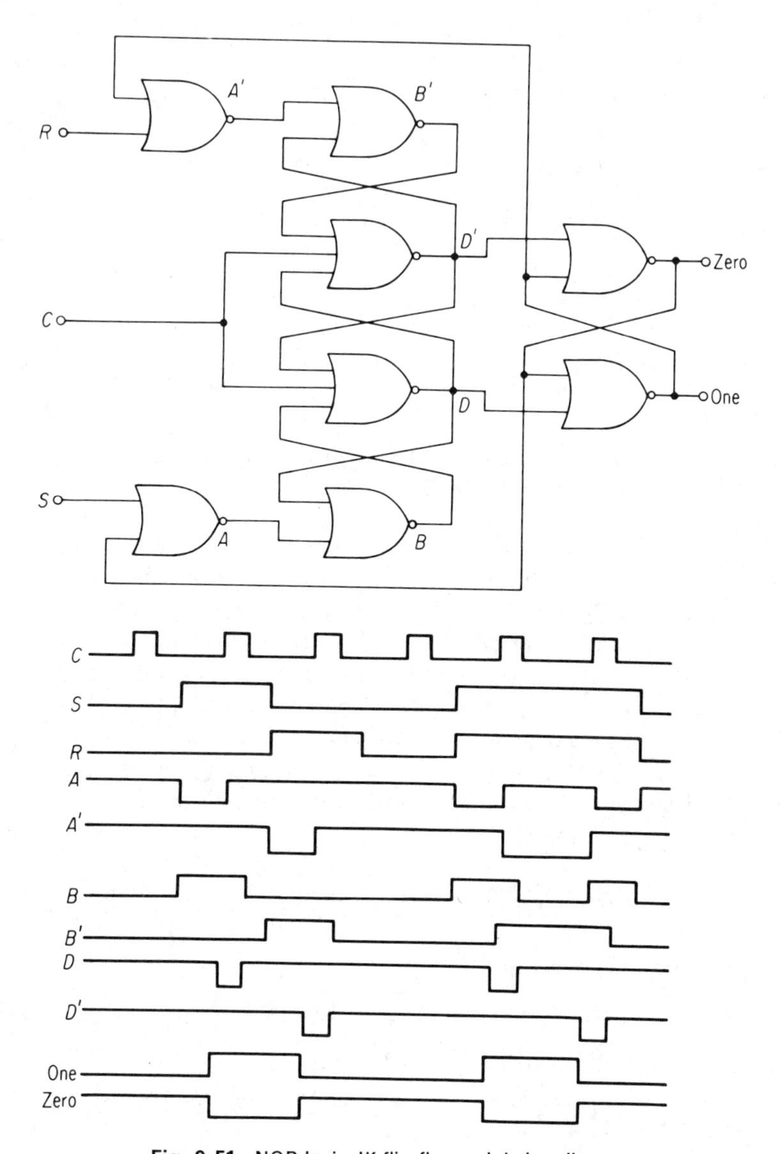

Fig. 9-51. NOR logic JK flip-flop and timing diagram.

OFF transistors ON, or bottoming OFF collectors—it can be switched into its unstable state for a predetermined time. Thus, short pulses may be stretched or narrowed or pulses of unequal width can be shaped to uniform width.

Ignoring triggering methods, consider the one-shot shown in Fig. 9-52. Resistors R_K and R_B along with V_{BB} are chosen such that Q_2 will saturate

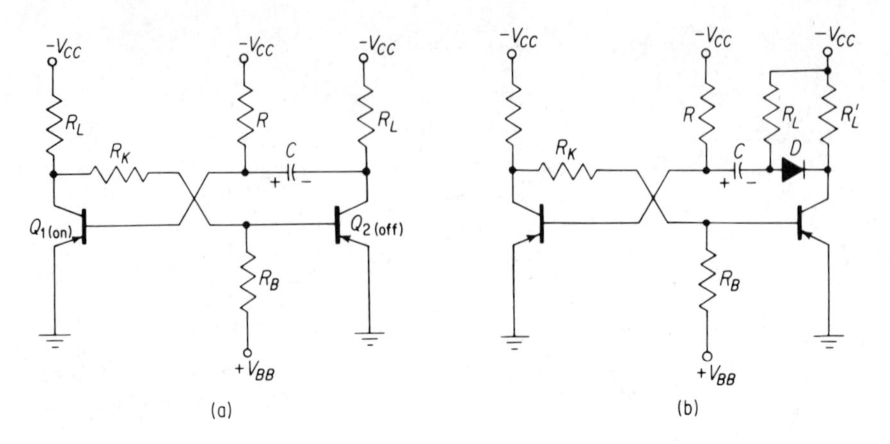

Fig. 9-52. One-shot or monostable multivibrator. (a) Common configuration long fall time at Q_2; (b) configuration to give fast fall time at Q_2.

R_L when Q_1 is OFF. Resistor R is chosen such that Q_1 will be ON during the "rest state" with Q_2 OFF. In this case, capacitor C is charged to $V_C = -(V_{BE\,\text{ON}_1} + I_{CBO_2}R_L) + V_{CC}$. If Q_1 is turned OFF in some acceptable manner, its collector will go negative enough to turn Q_2 ON. As the collector of Q_2 was near supply voltage and is going to ground, a potential of $V_C = V_{CC} - V_{BE\,\text{ON}_1} - I_{CBO_2}R_L - V_{CE\,\text{SAT}_2}$ will be coupled to the base of Q_1 to hold if OFF.* The voltage across R is thus $V_{CC} + V_C = 2V_{CC} - V_{BE\,\text{ON}_1} - I_{CBO_2}R_L - V_{CE\,\text{SAT}_2}$ and will decay to V_C at which time Q_1 will be biased ON again. Neglecting the transistor stored charge and related effects, the pulse width will be approximately:

$$\tau = RC \ln \left(\frac{2V_{CC} - V_{BE\,\text{ON}_1} - I_{CBO_2}R_L - V_{CE\,\text{SAT}_2}}{V_{CC} - V_{BE\,\text{ON}_1}} \right) \tag{9-25}$$

which, for large V_{CC}, becomes

$$\tau = RC \ln (2) = 0.69\ RC \tag{9-26}$$

If we include the effects of I_{CBO_1}, Eq. 9-25 becomes

$$\tau = RC \ln \left(\frac{2V_{CC} - V_{BE\,\text{ON}_1} - I_{CBO_1}R - I_{CBO_2}R_L - V_{CE\,\text{SAT}_2}}{V_{CC} - V_{BE\,\text{ON}_1}} \right) \tag{9-27}$$

Since the fall time of Q_2 (going OFF) will be limited by the R_LC time constant, it is often necessary to isolate the collector of Q_2 to realize a fast fall time. This is accomplished through the R'_L-diode network as shown in Fig. 9-52(b); R_L still serves to charge capacitor C but load current is now provided through R'_L.

*Assuming that the base-emitter diode does not break down.

The design equations may readily be written as

$$\frac{\underline{\overline{V_{BB}}}}{\underline{R_B}} \geq \frac{\overline{V_{CE\,\text{SAT}_1}}}{\underline{R_K}} + \overline{I_{CBO_2}} + \frac{\overline{I_{CS}}}{\underline{\beta_{CO}}} \quad (Q_2 \text{ OFF}) \tag{9-28}$$

$$\overline{I_{CS}} = \frac{\overline{V_{CO}}}{\underline{R_L}} \tag{9-29}$$

$$\frac{\overline{V_{CC}}}{\underline{R_L h_{FE_2}}} + \frac{\overline{V_{BB}} + \overline{V_{BE\,\text{ON}_2}}}{\underline{R_B}} \leq \frac{\underline{V_{CC}} - \overline{V_{BE\,\text{ON}_2}} - \overline{I_{CBO_1}}\,\overline{R_L}}{\overline{R_K + R_L}} \quad (Q_2\text{ON}) \tag{9-30}$$

From Eq. 9-26, if $R = V_{CC}/I_{CBO_1}$ the circuit will not operate. Thus,

$$\bar{R} < \frac{\underline{V_{CC}} - \overline{V_{BE\,\text{ON}_1}} - \overline{V_{CE\,\text{SAT}_2}} - \overline{I_{CBO_2}}\,\overline{R_L}}{\overline{I_{CBO_1}}} \quad (Q_1 \text{ OFF}) \tag{9-31}$$

$$\frac{\overline{V_{CC}}}{\underline{R_L h_{FE_1}}} \leq \frac{\underline{V_{CC}} - \overline{V_{BE\,\text{ON}_1}}}{\bar{R}} \quad (Q_1 \text{ ON}) \tag{9-32}$$

It was mentioned that the hold-OFF of Q_1 is accomplished by putting some momentary reverse bias on the base-emitter diode by means of capacitor C. The degree of hold-OFF voltage was calculated assuming the diode did not break down. Since hold-OFF voltages may be rather large (near the collector supply voltage), it is often necessary to protect the base-emitter diode from breakdown. The circuit of Fig. 9-53 shows one common protection method for the case where a capacitor is discharged into a transistor base.

Another variation of the multivibrator is the astable multivibrator shown in Fig. 9-54. This circuit, which is useful as a pulse source or clock, needs no triggering but will free-run at some frequency and duty cycle prescribed by the time constants R_1C_1 and R_2C_2 (a duty cycle not equal to 50 per cent indicates unsymmetrical ON and OFF times).

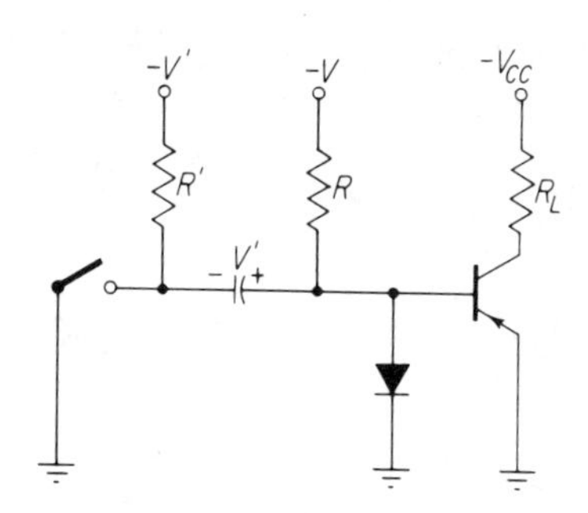

Fig. 9-53. Illustration of the use of a diode to protect the base-emitter diode from reverse breakdown.

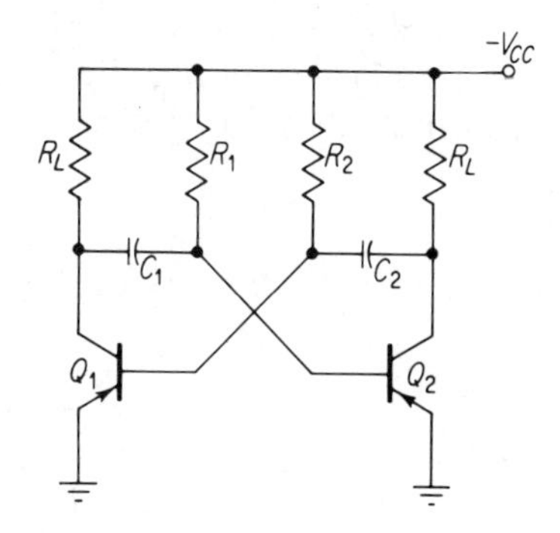

Fig. 9-54. Astable or free-running multivibrator.

Assume Q_1 conducts first when the power is applied; then, its collector will head toward zero. This positive-going pulse is coupled through C_1 to the base of Q_2 which will help to turn Q_2 OFF. At the same time, the collector of Q_2 will be going negative, and this is coupled to the base of Q_1, thereby increasing its turn-ON drive. Initially, capacitor C_1 was at zero voltage and it charges to $(V_{CC} - V_{CE\,SAT})$ through R_1 while Q_1 is ON. Soon this gets negative enough to turn Q_2 ON. As Q_2 goes ON, the resistor will immediately have $2V_{CC} - V_{BE\,ON}$ which will decay to $V_{CC} - V_{BE\,ON}$. Thus for large V_{CC}, the period is

$$\tau = T_1 + T_2 \cong (R_1C_1 + R_2C_2) \ln 2 \tag{9-33}$$

or

$$\tau = 2(R_1C_1) \ln (2) \tag{9-34}$$

for $R_1C_1 = R_2C_2$.

The pulse rise time may be calculated as

$$t_r = R_LC_1 \ln (10) = 2.3R_LC_1 \text{ (or } C_2) \tag{9-35}$$

The major dc design requirement is that $\overline{R_1}, \overline{R_2} < \underline{h_{FE}}\,\underline{R_L}$ to insure saturation. This gives the rise-time-to-period ratio (for $R_1C_1 = R_2C_2$) of

$$\frac{t_r}{\tau} = \frac{R_L \ln (10)}{2R_1 \ln (2)} = 1.66\frac{R_L}{R_1} = \frac{1.66}{\beta_C} \tag{9-36}$$

Equation 9-36 indicates that there is a minimum obtainable rise time in such a circuit since β_C must be less than, or equal to, the minimum h_{FE}. For example, with $\underline{h_{FE}} = 25$, $t_r/\tau > 1.66/25 = 6.64$ percent, and at 10 kHZ., this would mean $t_r > 6.64$ μsec.

The Blocking Oscillator

Another regenerative pulse-generating circuit is the blocking oscillator. It may be triggered or free-running and is usually employed where high-power pulses are required. The common-emitter blocking oscillator with emitter timing is shown in Fig. 9-55(a) and its simplified equivalent saturation circuit in Fig. 9-55(b). The emitter current is

$$I_e \approx \frac{V_{CC}}{(n+1)R_e} \tag{9-37}$$

and the inductance current is

$$I_L \approx \frac{nV_{CC}t}{(n+1)L} \tag{9-38}$$

Thus,

$$I_b = I_e - I_c = I_e - \left(\frac{I_b}{n} + I_L\right)$$

$$= \frac{n(I_e - I_L)}{n+1} = \frac{nV_{CC}}{(n+1)^2}\left\{\frac{1}{R_e} - \frac{nt}{L}\right\} \tag{9-39}$$

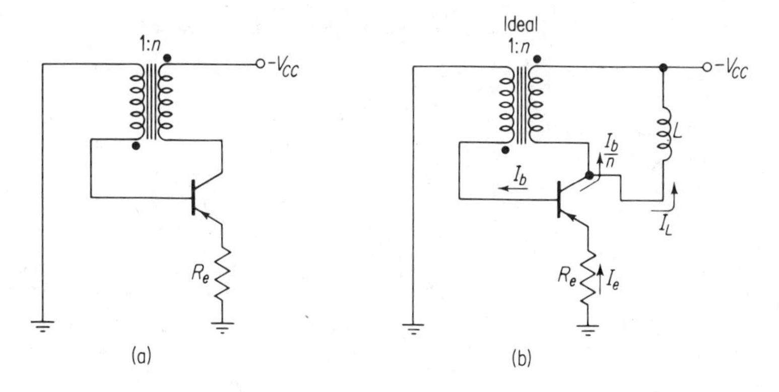

Fig. 9-55. Common-emitter blocking oscillator with emitter timing.

and

$$I_c = I_e - I_b = \frac{nV_{CC}}{(n+1)^2}\left\{\frac{1}{nR_e} + \frac{nt}{L}\right\} \tag{9-40}$$

Since the pulse will end when the transistor has reached the edge of saturation where $I_c = h_{FE}I_b$, the pulse width T may be calculated as

$$\left\{\frac{1}{nR_e} + \frac{nT}{L}\right\} = h_{FE}\left\{\frac{1}{R_e} - \frac{nT}{L}\right\} \tag{9-41}$$

or

$$T = \frac{L(nh_{FE} - 1)}{n^2R_e(h_{FE} + 1)} \approx \frac{L}{nR_e} \tag{9-42}$$

Thus, the unloaded pulse width is essentially independent of the transistor current gain. Generally, blocking oscillators are transformer coupled to their loads or are coupled through emitter followers. One such coupling method is shown in Fig. 9-56.

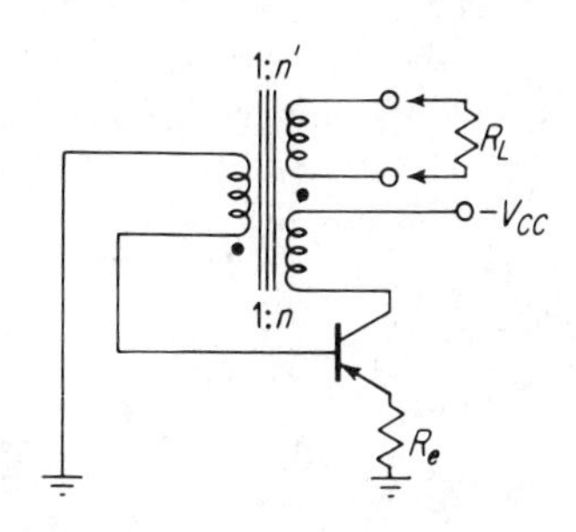

Fig. 9-56. One method of coupling pulses to the load.

The Schmitt Trigger

The Schmitt trigger is a wave-shaping or level-sensing threshold circuit. It finds common usage in pulse-forming amplifier-drivers and may also be used in level comparators.

A typical Schmitt trigger circuit is shown in Fig. 9-57. Its operation may be explained as follows: If we assume zero input voltage, any current flowing through R_E will develop a negative voltage at the emitters back-biasing Q_1. This negative voltage, produced by the current drawn by $Q_2(I_{E2})$, is enough to maintain the back bias on Q_1 and keep it OFF. A voltage

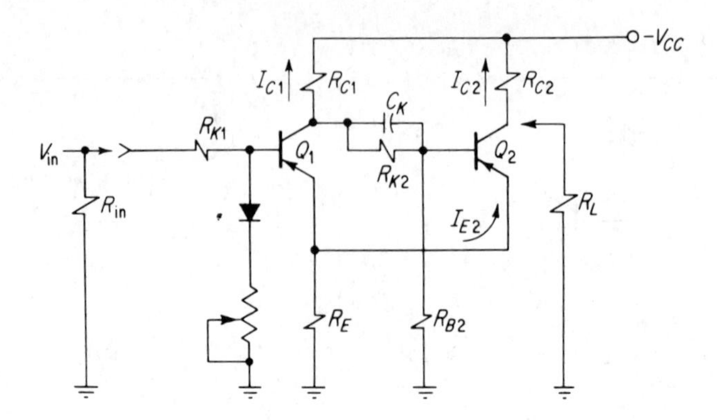

Fig. 9-57. Typical Schmitt trigger circuit.

divider, $R_{C1} - R_{K2} - R_{B2}$, forward-biases Q_2 and keeps it ON. Thus the output is low and approximately,

$$V_{\text{out low}} = (V_{CE\,\text{SAT}} + V_E) = V_{CE\,\text{SAT}} + I_{E2}R_E \approx V_{CE\,\text{SAT}} + I_{C2}R_L \tag{9-43}$$

Now, when a signal arrives and drives the input to Q_1 more negative than V_E, Q_1 will begin to conduct. This causes its collector voltage to drop and begins to cut OFF Q_2. Simultaneously, since $R_{C2} < R_{C1}$, we have $I_{C2} > I_{C1}$ and the emitter voltage falls. This aids the regenerative action to turn Q_2 OFF very quickly. The output voltage is then approximately

$$V_{\text{out high}} \approx \frac{V_{CC}R_L}{R_{C2} + R_L} \tag{9-44}$$

If the input voltage begins to drop enough to cut OFF Q_1 (note that V_E has dropped to V'_E), regenerative action will snap Q_2 back ON.

Let us calculate the turn ON and turn OFF voltages required at the input.

$Q_{1\,\text{ON}}$

from OFF: $V_{\text{ON}} \geq V_{BE\,\text{ON}_1} + V_E = V_{BE\,\text{ON}_1} + I_{E2}R_E = V_{BE\,\text{ON}_1} + I_{C2}R_E$

$$\approx V_{BE\,\text{ON}_1} + \frac{V_{CC}R_E}{R_{C2} + R_E} \tag{9-45}$$

$Q_{1\,\text{OFF}}$

from ON: $V_{\text{OFF}} \leq V_{BE\,\text{ON}_1} + V'_E = V_{BE\,\text{ON}_1} + I_{E1}R_E$

$$\approx V_{BE\,\text{ON}_1} + \frac{V_{CC}R_E(R_{K2} + R_{B2})}{R_{C1}(R_{K2} + R_{B2}) + R_E(R_{C1} + R_{K2} + R_{B2})} \tag{9-46}$$

That V_{OFF} must be smaller than V_{ON} is illustrated in Fig. (9-58a) which shows the negative resistance characteristic necessary for regenerative action. If

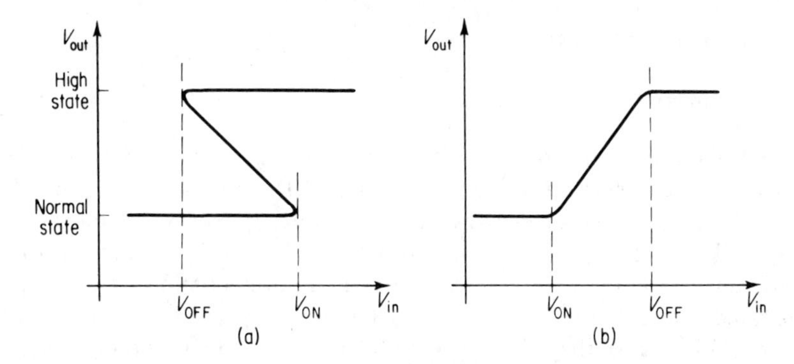

Fig. 9-58. Voltage transfer characteristics. The transfer characteristic (a) is for a regenerative amplifier; that of (b) a nonregenerative amplifier.

$|V_{OFF}| > |V_{ON}|$ the device would act as a simple linear amplifier, as shown in Fig. 9-58(b).

Notice the requirement that $|V_{OFF}| < |V_{ON}|$ results in a "dead" region in the input. A "squaring" of a triangular wave by a Schmitt trigger is illustrated in Fig. 9-59, where the effects of the dead region are noted. Because the output swing ΔV_{out} is proportional to the loading (as will be the rise time and frequency response),

$$\Delta V_{out} = \frac{V_{CC}R_L}{R_{C2} + R_L} - \left(V_{CE\,SAT} + \frac{V_{CC}R_E}{R_{C2} + R_E}\right) \tag{9-47}$$

it is often desirable to put an emitter follower or buffer between the output and the load.

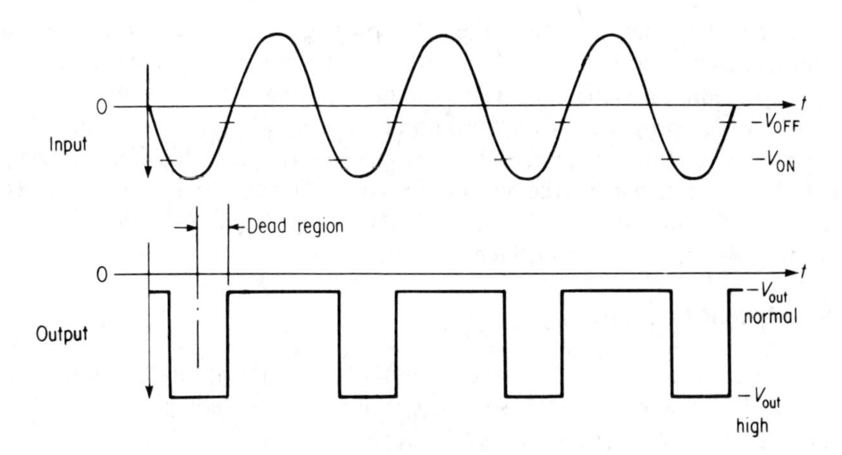

Fig. 9-59. Pulse squaring with a Schmitt trigger.

COUNTER SYSTEMS

Serial Counters

The chain of binary counters discussed previously performed *serial counting*, that is, each stage in the counter counted the output of the preceding stage. Since a propagation delay is introduced through each stage, the over-all delay will be accumulative. If the outputs of the various stages are to be gated, this accumulative delay can present a severe timing limitation, as illustrated in Fig. 9-60 (where a stage delay of one-quarter clock period was assumed for the negative-edge triggered binaries). The figure shows the actual and ideal (broken) timing waveforms for the serial counter. Notice that by stage 4 the output has completely lost its identity with the input, and that by stage 5 coincidence gating with stage 1 would be impossible.

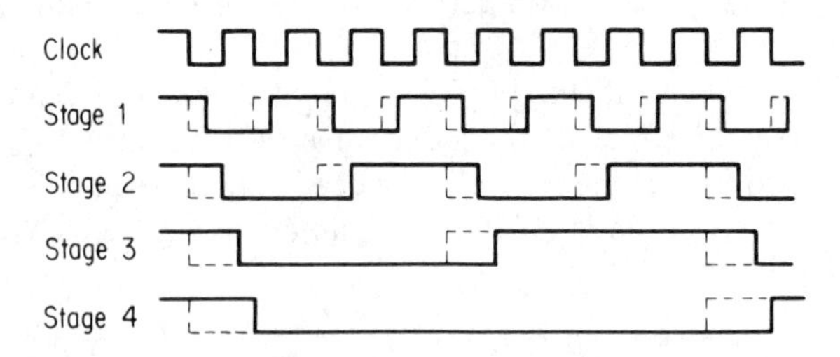

Fig. 9-60. Actual (solid) and ideal (broken) timing waveforms for the serial counter.

Parallel Counters

One system that overcomes the delay propagations associated with serial counters is the parallel counter, in which each binary is controlled by all those preceding it. This system has the disadvantage of complexity, a complexity which progresses with the number of stages. Figure 9-61 illustrates the parallel counter, composed of negative-edge triggered binaries and zero true AND gates, wherein the output of each AND gate forms the count pulse for the following stage, and its associated timing diagram. Notice that the outputs all retain a constant identity with the clock.

Serial/Parallel Converters

It is often necessary to convert a serial code into a parallel code or a parallel into a serial code. These code conversions are accomplished with the shift register. For serial-to-parallel conversion, serial data enter the first stage of the register. After a number of clock pulses equal to the number of bits per serial word, the word appears in parallel form at the register outputs.

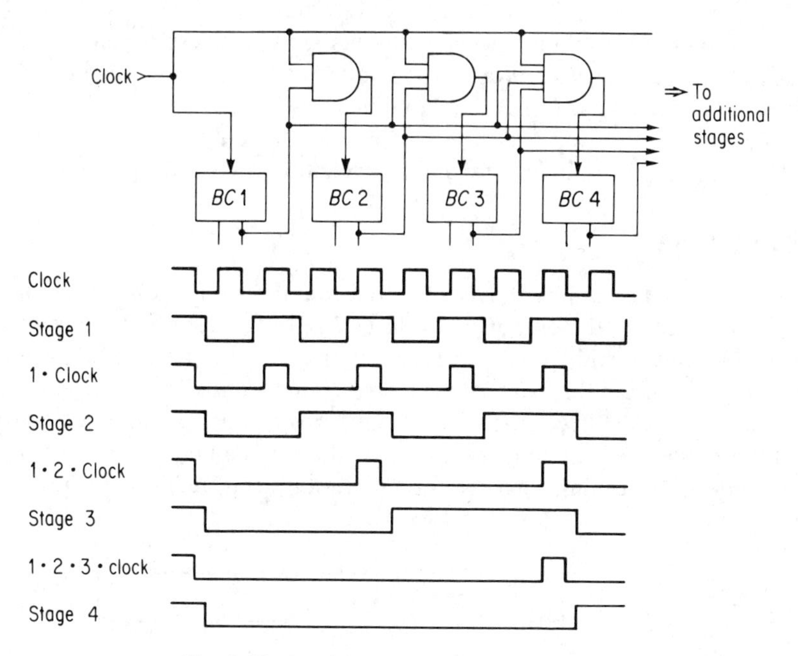

Fig. 9-61. Parallel counter and timing diagram.

For parallel-to-serial conversion, the word contained in the register exits the last stage of the register in serial form and is timed by the shift clock. A possible three-bit serial/parallel converter is illustrated in Fig. 9-62 where the zero false NAND gates are required to perform parallel read-in and read-out.

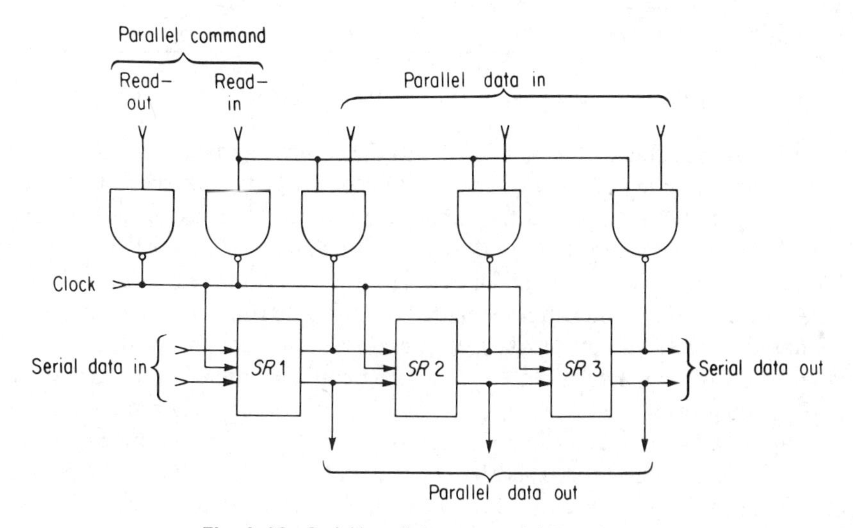

Fig. 9-62. Serial/parallel converter (shift register).

For zero command inputs, five gate outputs are true, thereby permitting the clock to shift the register. For a true command input on the read-out gate, the gate output is false (zero) thus stopping the clock.

Similarly, a true input on the read-in gate not only stops the clock but permits the parallel data to enter and set each flip-flop to its appropriate state.

Ring Counters

A shift register can be converted into a counter (or a clock phase generator) by connecting the output of the last stage to the data input lines of the first stage, thus forming a closed loop or "ring." If a logical "one" is preset into only one stage, this "one" will be propagated from stage to stage around the ring at the clock frequency f. Each stage has a "one" duration time of $1/f$ and the period of each stage is n/f. Thus, the ring counter counts the input (clock) by n. The timing diagrams for a three-stage ring counter are illustrated in Fig. 9-63.

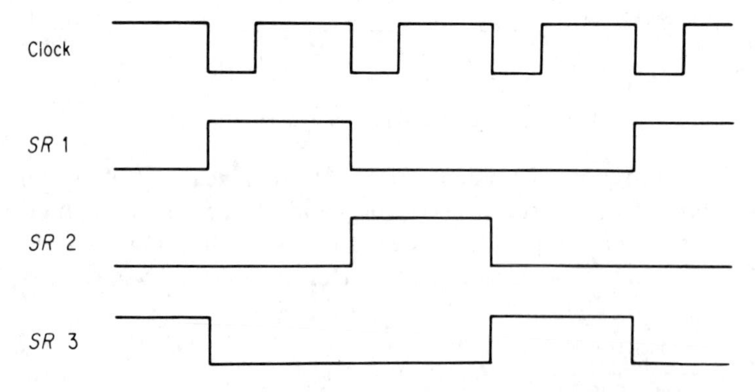

Fig. 9-63. Three-stage ring counter waveforms containing a single one.

It is also possible to preset a word into the counter and circulate the entire word. The example of Fig. 9-63 may be looked upon as circulating the word "100." Figure 9-64 illustrates the timing diagram for the case of a five-stage ring counter containing the word 10110. Circulation of a word other than 100 . . . is a method often used for generating the various phases of a clock.

A special case of the ring counter is the Möbius ring, so named because the data connections between last stage and first stage are inverted. No word is set into the ring. Instead, a "one" is progressively entered into each stage until all contain a "one." Then, the "ones" are progressively removed until all contain a "zero." The timing diagrams for a three-stage Möbius ring are shown in Fig. 9-65. Notice that the "one" duration is now n/f and the period of each stage is $2n/f$. Thus the Möbius ring counter counts the input (clock)

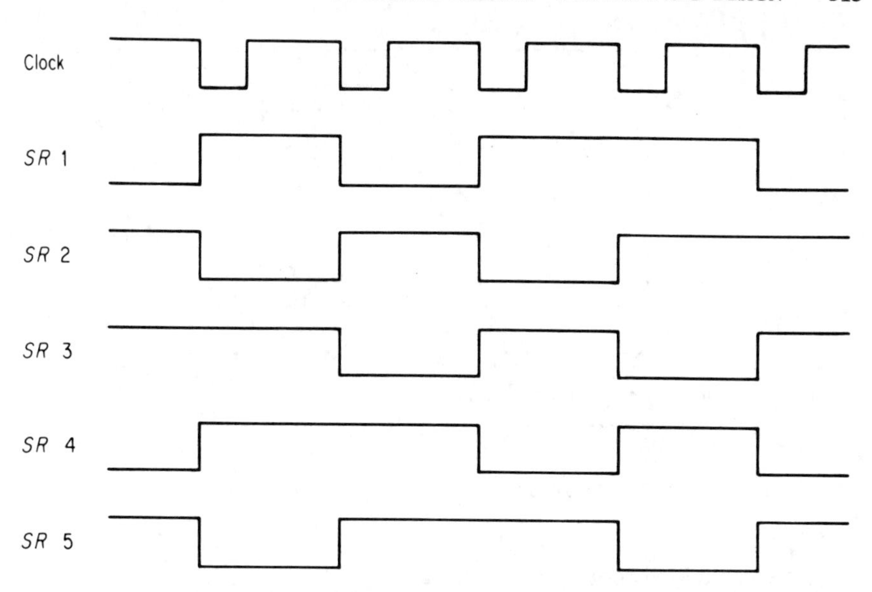

Fig. 9-64. Five-stage ring counter waveforms containing the word 10110.

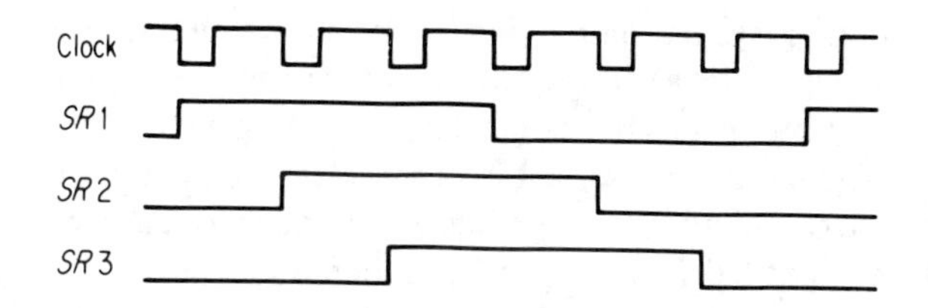

Fig. 9-65. Waveforms for a 3-stage Möbius ring.

by $2n$. This is a disadvantage in that one can count only by even numbers, but it is also an advantage in that only five stages are needed to count by ten.

The counting rate of each stage is only f/n, or $f/2n$ for the Möbius ring, a very desirable characteristic of the ring counter. If one were counting a 200 MHz pulse train, the first binary in a serial counter would be required to count at 200 MHz, whereas the individual stages of a five-stage Möbius ring would be required to count at only 20 MHz.

JK Flip-Flop Counter

This ring counter incorporates all the advantages of the Möbius ring counter and can count by odd numbers as well. The three-stage counter illustrated in Fig. 9-66 counts by five (note that in the dotted connection it is a Möbius counter and counts by six), utilizing the ability of the JK to complement when both data lines are at logical 0.

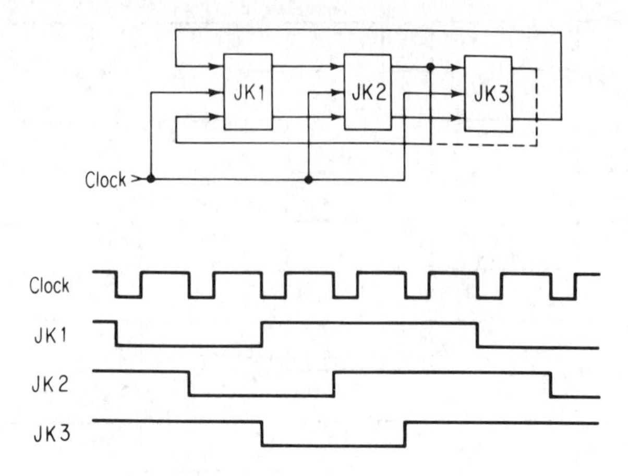

Fig. 9-66. JK flip-flop ring counter.

A SYSTEM EXAMPLE—THE PARALLEL ADDER

As an example of a system employing many of the circuits previously discussed, consider a parallel adder. A parallel adder, as the name implies, adds all bits of the two numbers simultaneously, whereas a serial adder adds them one at a time. The parallel adder is, therefore, faster than the serial adder, but instead of only one full adder, it requires a full adder for each bit.

A completely clocked four-bit parallel adder system is shown in Fig. 9-67. A reset clock phase, ϕ_R, is provided to clear both input registers before entering the numbers to be added. The numbers are then entered into the x and y registers, after which they are transferred into the full adders via the AND gates with the transfer-and-add phase, ϕ_{TA}. This phase, ϕ_{TA}, must last long enough to allow the carrier to propagate throughout the five bits (the extra register stage is needed to accommodate the possible carry "overflow") and to permit the sum bits to be set into the output register. The read-out phase, ϕ_{RO}, cannot occur until sufficient time for all carry bits to propagate has elapsed in order to prevent an erroneous setting of the output register which is cleared by ϕ_{OR}.

Four clock phases have been mentioned. These sequence and control the progress of the bits through the adder by allowing sufficient time for each event. An arbitrary set of a four-phase clock-timing sequence which satisfies the requirements is illustrated in Fig. 9-68. Each phase is derived from one "master clock," as shown in Fig. 9-69, where a clock frequency of $4f$ is counted down by two binary counters. AND gating of the appropriate outputs, designated nf and $\overline{nf}$ for "true" and "complement," then derives the desired clock phases.

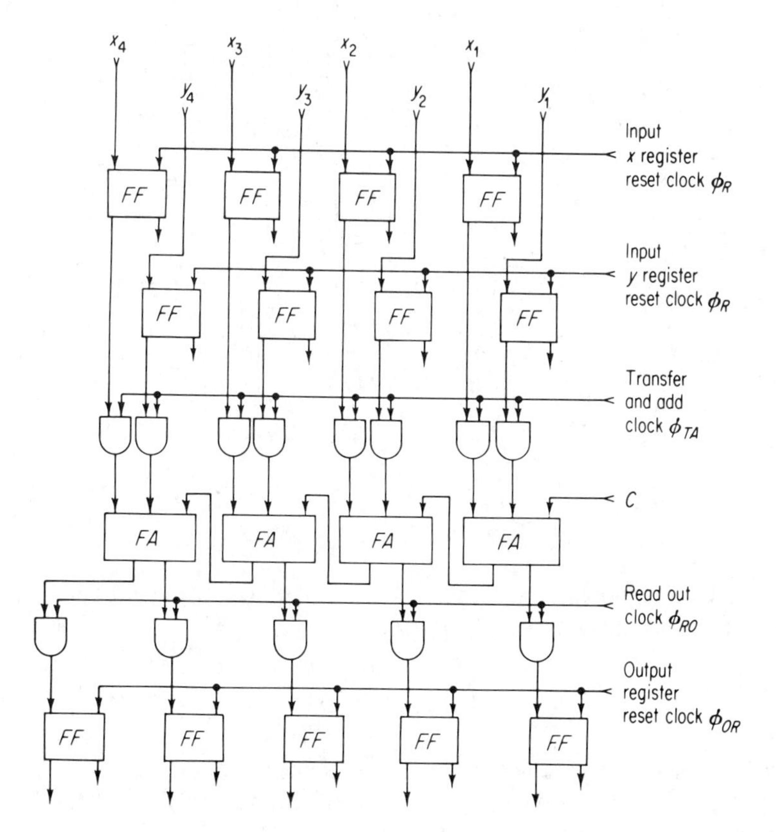

Fig. 9-67. Four-bit parallel adder system.

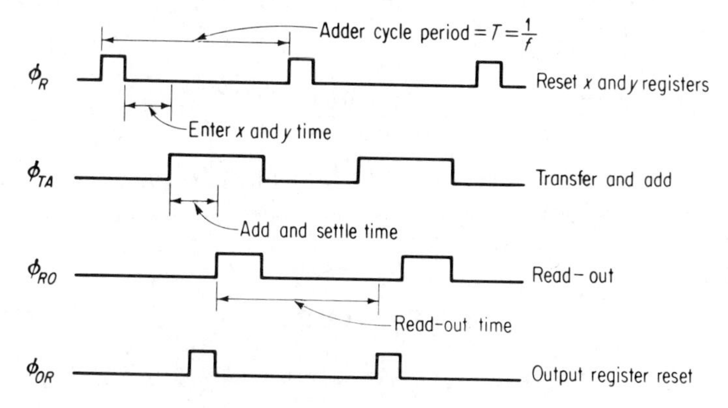

Fig. 9-68. Four-phase clock for full adder.

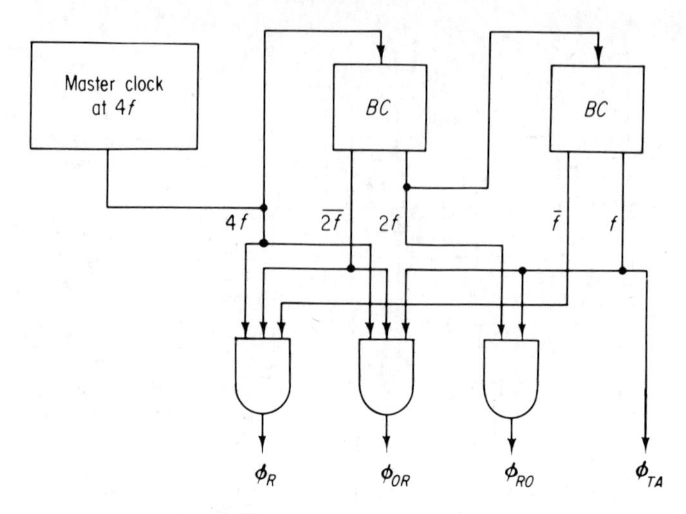

Fig. 9-69. Four-phase clock generator.

One final comment is in order. Such a system will also perform subtraction. This is illustrated by the following example:

$$\begin{array}{r} 10 = +\,001010 \\ -6 = -\,000110 \\ \hline 4 = +\,000100 \end{array}$$

Recall our discussion of the flip-flop: a complement output is always available (or a flip-flop may always be complemented). If we complement the binary number of 6 and "add" instead of subtract we get:

$$\begin{array}{r} +\,001010 \\ +\,111001 \\ \hline +1000011 \end{array}$$

which is 3 and an "overflow" bit. If we add the overflow bit (end-around carry) we get

$$\begin{array}{r} +\,000011 \\ +\,000001 \\ \hline +\,000100 \end{array}$$

which is the correct answer. The input to the first bit of the full adder of Fig. 9-67 (labeled C) is for such end-around carry bits required for subtraction.

EXERCISES

1. Convert the decimal number $(385)_{10}$ to its binary equivalent.

2. Using only NOR gates, design a half adder.

3. A form of nonsaturating circuit is the emitter-coupled logic circuit shown below. Write the logic equation expressing the output C in terms of the inputs A and B. Assume the highest input or output voltage level to be logically true.

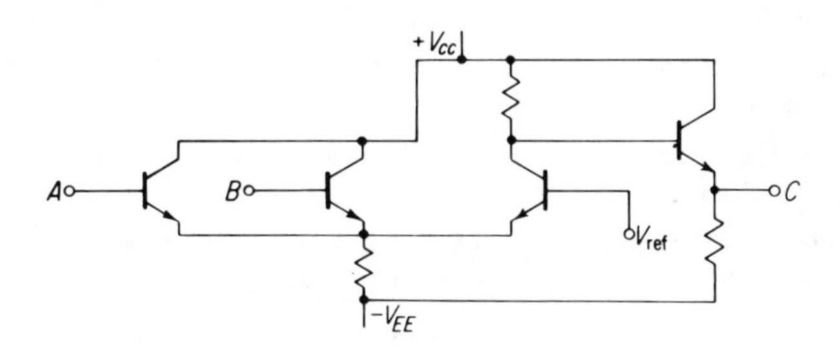

4. The ability of a logic circuit to reject spurious signals or noise is an important property which must be considered in its design. False signals, triggered by noise, cause errors in digital systems. Calculate the negative power supply noise margin of the circuit shown in Exercise 3. The negative noise margin is defined as the amount by which the supply voltage V_{CC} of this circuit may be reduced and still insure proper circuit operation. How would you compute the positive power supply noise immunity?

5. The triggering scheme shown in the shift register circuit of Fig. 9-35 has several disadvantages. One is the fact that noise introduced at the clock input can falsely trigger the flip-flop. Another difficulty with this circuit for some applications is the need for a faulty fast clock transition to accomplish triggering. In other words, this flip-flop is not sensitive to level changes. Make an *estimate* of the longest clock transition time that will trigger the flip-flop.

REFERENCES

1. M. SCHWARTZ, *Information Transmission, Modulation, and Noise*, New York: McGraw-Hill Book Company, 1959.

2. P. THOMAS and J. HILTEBEITEL, "Transistor Guide for Switching Circuit Designers," *Electrical Design News* (Nov. 1961).

3. P. THOMAS and J. HILTEBEITEL, "A Worst-Case Procedure for RTL NOR and NAND Gates"; *Solid State Design* (July, 1963).

4. A. K. RAPP and J. L. ROBINSON, "Rapid Transfer Principles for Transistor Switching Circuits," International Solid State Circuits Conference (Digest of Papers) (Feb. 1961).

5. T. C. BARTEE, *Digital Computer Fundamentals*, New York: McGraw-Hill Book Company, 1958.

6. M. PHISTER, Jr., *Logical Design of Digital Computers*, New York: John Wiley & Sons, Inc., 1958.

7. J. MILLMAN, and H. TAUB, *Pulse Digital and Switching Waveforms*, New York: McGraw-Hill Book Company, 1965.

8. E. J. MCCLUSKEY, *Introduction to the Theory of Switching Circuits*, New York: McGraw-Hill Book Company, 1965.

9. J. L. MOLL, "Large Signal Transient Response in Junction Transistors," *Proc. IRE* **42** (Dec. 1954) 1773–1784.

10. M. P. RISTENBATT, *Semiconductor Circuits, Linear and Digital.* Englewood Cliffs, N.J.: Prentice-Hall, Inc., 1975.

11. *Integrated Electronic Systems.* Westinghouse Defense and Space Center, Englewood Cliffs, N.J.: Prentice-Hall, Inc., 1970.

12. C. BELOVE, H. SCHACHTER and D. L. SCHILLING, *Digital and Analog Systems, Circuits and Devices: An Introduction.* New York: McGraw-Hill Book Company, 1973.

13. L. G. COWLES, *Transistor Circuits and Applications.* Englewood Cliffs, N.J.: Prentice-Hall, Inc., 1968.

14. E. J. ANGELO, JR., *Electronics: BJT's, FET's and Microcircuits.* New York: McGraw-Hill Book Company, 1969.

15. H. C. LIN, *Integrated Electronics.* Boston, Mass: Holden-Day Book Company, 1967.

10

Oscillators, Mixers, Converters, and Detectors

INTRODUCTION

Oscillators, mixers, converters, and detectors fall into a special category of circuit applications which usually utilizes some inherent nonlinearity in the transistor. Unlike other applications where nonlinearity is a disadvantage, it becomes essential here, particularly in the case of mixers, converters, and detectors.

Before proceeding to analyze this subject, let us review a few areas of application. Perhaps the most common application of these special circuits is in a superheterodyne receiver. Figure 10-1 shows a block diagram of such a receiver. The incoming modulated RF signal passes through an RF amplifier

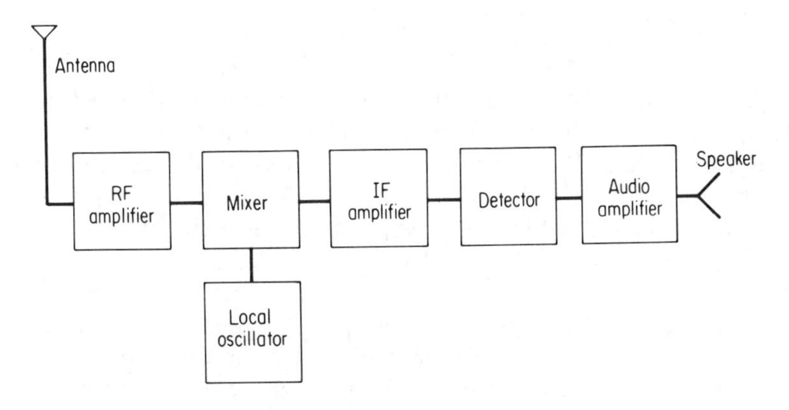

Fig. 10-1. Block diagram of typical superheterodyne receiver showing relative locations of the mixer oscillator and detector.

into the mixer. The mixer combines the RF signal with another signal from the oscillator and produces a beat frequency output whose frequency is the difference between the RF and oscillator frequencies. After further amplification in the IF amplifier, the signal passes through the detector which extracts the modulation information from the IF signal.

Oscillators are also used in many applications for the generation of high-frequency power. A common application is in transmitters. Let us now look in detail at each of these special circuits.

BASIC CONSIDERATIONS

The basic oscillator system is presented in the diagram of Fig. 10-2. This diagram shows that the oscillator is basically an amplifier with a feedback path such that some fraction of the output signal is fed back to the input. When the sum of the phase shifts around the feedback loop is 360 n degrees (n = integer), and the gain around the feedback loop is one, the system will oscillate. The *frequency of oscillation* is that frequency which makes the sum of the phase shifts zero; the *output power* is the difference between the output power of the amplifier and the feedback power required to obtain that output.

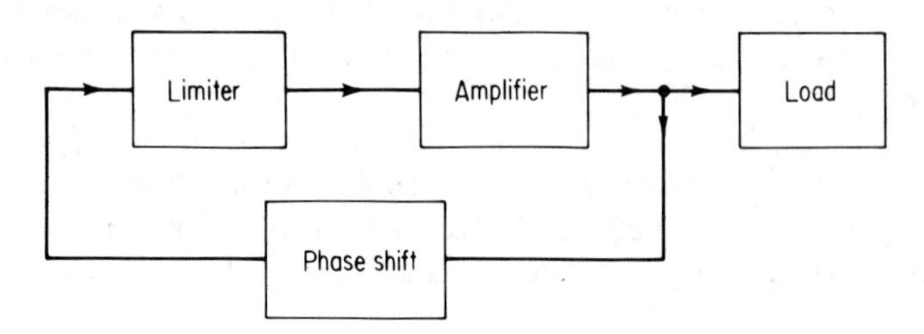

Fig. 10-2. Basic requirements of an oscillator system.

An oscillator must have associated with it some mechanism which limits its output. For oscillations to be self-sustaining, the loop gain must be equal to one; that is, the amplifier must supply all the losses of the circuit, including the load. For the oscillator to be self-starting, the loop gain must be greater than one initially and fall to one as the oscillations build up and the limiter reduces the gain in the feedback loop.

Where the transistor is operating as a large-signal amplifier, limiting action takes place in the transistor itself, since the transistor is operating in a region where the gain is decreasing with increasing drive. In some audio applications where the distortion of the sine wave output must be reduced to a very small value and the frequency over a wide range must be adjusted, it is necessary to operate the transistor as a small-signal amplifier. In this case limiting must take place in some external circuit element, generally a nonlinear resistance, such as a thermistor, a low wattage incandescent lamp, or an additional feedback loop, in which some of the output signal is rectified, and the resultant dc output is used to control the gain. Our discussion deals with the large-signal or self-limiting case only.

If the circuit is adjusted so that the proper gain and phase shift occur at only one frequency, then the circuit will oscillate at only that frequency and the output will be a sine wave. Where these conditions are satisfied over a band of frequencies, the circuit oscillates over the entire band and is called a *relaxation oscillator*. Examples of relaxation oscillators are multivibrators and blocking oscillators. Since these were covered in Chap. 9, only sine wave oscillators will be considered in this chapter.

The designer has available many different techniques to produce the required gain and phase-shift conditions. In general, low-frequency oscillators (up to about 100 KHz) use the phase-shift characteristics of RL or RC networks. At higher frequencies LC networks are used. Examples of each type of oscillator are now considered. The mathematical analysis of oscillator circuits is often based on small-signal conditions and the circuit analyzed by writing the loop equations. Then, the real parts of the loop equations are set equal to zero to determine the conditions for oscillation, and the imaginary parts are set equal to zero to determine the frequency of oscillation (1).* This text uses this approach.

LOW-FREQUENCY OSCILLATORS

Over the frequency range up to about 100 KHz, transistor sine wave oscillators are generally designed using some combination of resistance and capacitance in the feedback loop. Proper phase shift for regeneration may be provided by these resistors and capacitors, in which case the oscillator is termed a *phase-shift* oscillator.

Figure 10-3 is an example of a typical RC phase-shift oscillator. The design equations may be derived by referring to Fig. 10-3, where a low-frequency hybrid-pi equivalent circuit for the transistor has been combined with the ac circuit of the phase-shift oscillator. For simplicity, we have assumed that the input resistance is combined with the external feedback resistance.

The loop equations for Fig. 10-3(b) are

$$i_b(R_{\text{in}} + R + X_C + R - R_{\text{in}}) - i_1 R = 0 \quad (10\text{-}1)$$

$$i_1(2R + X_C) - i_b R - i_2 R = 0 \quad (10\text{-}2)$$

$$i_2(R + R_L + X_C) - i_1 R + \beta i_b R_L = 0 \quad (10\text{-}3)$$

*Numbers in parentheses refer to material listed at the end of the chapter.

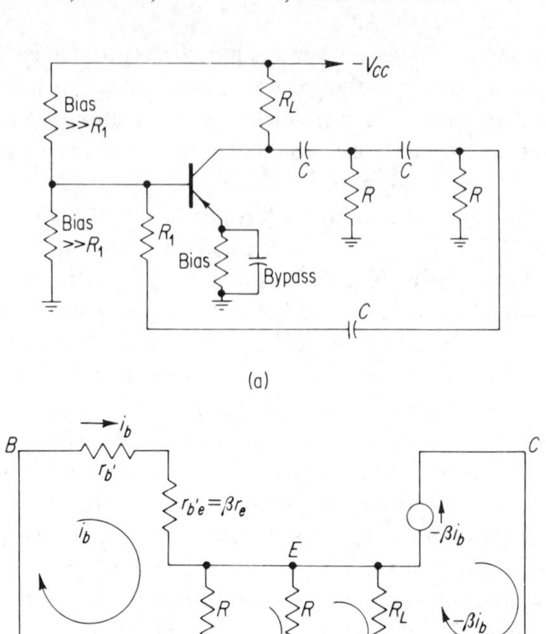

Fig. 10-3. Typical RC phase-shift oscillator. (a) With bias circuit; (b) ac circuit only with low-frequency hybrid-pi transistor representation.

where

$$R_{\text{in}} = r_{b'} + r_{b'e} \quad \text{and} \quad X_C = -j/\omega C.$$

Solving Eqs. 10-1–10-3 to express all currents in terms of i_b gives

$$3R^2(R + R_L) + 2R^2X_C + 4RX_C(R + R_L) + X_C^2(R + R_L) + 4RX_C^2 + X_C^3 - 2R^3 + \beta R^2R_L = 0 \quad (10\text{-}4)$$

To find the frequency of oscillation, set the imaginary terms of Eq. 10-4 equal to zero (odd powers of X_C) and obtain

$$2R^2X_C + 4RX_C(R + R_L) + X_C^3 = 0 \quad (10\text{-}5)$$

or, in a more useful form,

$$\omega^2C^2 = \frac{1}{6R^2 + 4RR_L} \quad (10\text{-}6)$$

and, therefore,

$$f = \frac{1}{2\pi C}\sqrt{\frac{1}{6R^2 + 4RR_L}} \quad (10\text{-}7)$$

To find the conditions required for oscillation, set the real parts of Eq. 10-4 equal to zero (even powers of X_C) and obtain

$$3R^2(R + R_L) + X_C^2(R + R_L) + 4RX_C^2 - 2R^3 + \beta R^2 R_L = 0 \tag{10-8}$$

Let us substitute for X_C^2 the terms from Eq. 10-6. Then we obtain

$$R^2 R_L(\beta + 3) + R^3 - (5R + R_L)(6R^2 + 4RR_L) = 0 \tag{10-9}$$

or

$$R^2 R_L(\beta - 23) - 29R^3 - 4RR_L^2 = 0 \tag{10-10}$$

Thus, to oscillate we must have

$$\beta \geq 23 + \frac{29R}{R_L} + \frac{4R_L}{R} \tag{10-11}$$

Note also that

$$R_1 = R - R_{\text{in}} = R - r_{b'} - \beta r_e \tag{10-12}$$

Thus R_1 should be calculated at the maximum expected value of β to insure that the circuit has sufficient feedback.

The preference for *RC* oscillator control in the audio and low-frequency RF range is based primarily on economic considerations. Although the *LC* oscillators to be discussed in the succeeding section operate equally well in this range, the values of inductance and capacitance required are often prohibitively large.

HIGH-FREQUENCY *LC* OSCILLATORS

In the frequency range from 100 KHz to 500 MHz, it becomes practical to use lumped inductance and capacitance as the frequency-controlling elements of the oscillator. Figure 10-4(a) shows the common-base transistor equivalent of the familiar Colpitts, Hartley, and modified Colpitts or Clapp oscillator circuits. Figure 10-4(b, c) shows the same three circuits in the common-emitter and common-collector configuration. (Bias networks have been omitted for simplicity.) A comparison of the respective circuits for common emitter, common collector, and common base will show that the choice of the ground point in the oscillator circuits is completely arbitrary and the three configurations are identical from the electrical standpoint. The same can be said for the Hartley and Colpitts circuits since it is immaterial whether feedback is obtained from an inductive or capacitive divider network.

These various configurations may all be used for high-frequency oscillators. One circuit, however, is used more often than the others for most high-frequency transistor oscillators. This circuit is shown in Fig. 10-5(a). The ac circuit is shown in Fig. 10-5(b). Figure 10-5(c) shows the same circuit in conjunction with the high-frequency hybrid-pi equivalent circuit. Notice that this is essentially a Colpitts type of oscillator with the base-emitter capacitance, $C_{b'e}$, being used as one of the tapped capacitors, and

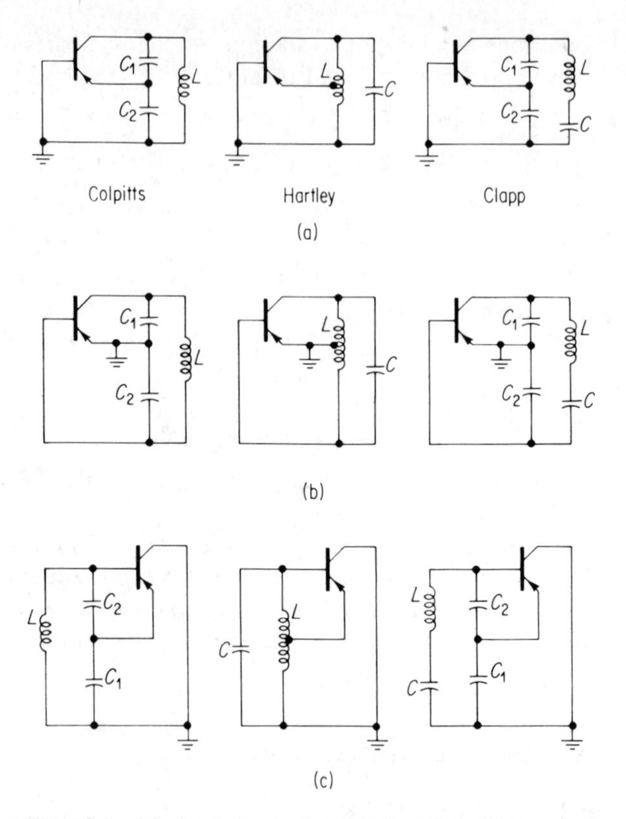

Fig. 10-4. AC circuits for basic transistor LC oscillators. (a) Common-base; (b) common-emitter; (c) common-collector.

$r_{b'}$ representing a series loss resistance. This circuit may be analyzed by referring to Fig. 10-5(c).

Let the impedance of the tank circuit be Z_T where

$$Z_T = \frac{j\omega L}{1 - \omega^2 LC} \tag{10-13}$$

Then the loop equations are

$$i_b(r_{b'} + X_{C_{b'e}} + X_{C_1} + Z_T) - i_1 X_{C_{b'e}} + g_m v_{b'e} X_{C_1} = 0$$

$$i_b(r_{b'} + Z_T) + i_1 X_{C_{TC}} = 0 \tag{10-14}$$

Solving these equations and remembering that

$$v_{b'e} = X_{C_{b'e}}(i_b - i_1) \tag{10-15}$$

gives

$$r_{b'} + X_{C_{b'e}} + X_{C_1} + Z_T + \frac{X_{C_{b'e}}(r_{b'} + Z_T)}{X_{C_{TC}}} + g_m X_{C_1} X_{C_{b'e}}\left[1 + \frac{r_{b'} + Z_T}{X_{C_{TC}}}\right] = 0 \tag{10-16}$$

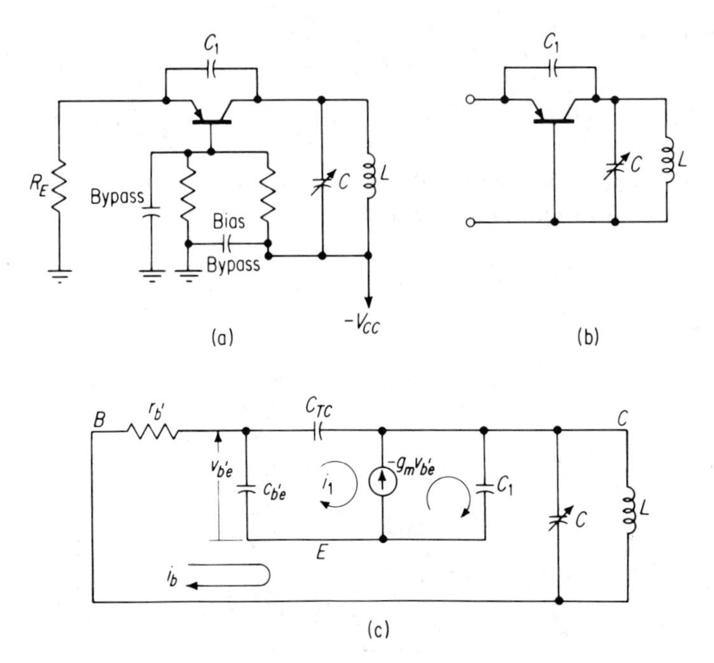

Fig. 10-5. (a) Typical common-base oscillator circuit; (b) ac circuit only; (c) ac circuit combined with the high-frequency hybrid-pi circuit.

To find the frequency of oscillation, set the imaginary terms equal to zero.

$$X_{C_{b'e}} + X_{C_1} + Z_T + Z_T\frac{X_{C_{b'e}}}{X_{C_{TC}}} + g_m X_{C_1}\frac{X_{C_{b'e}} r_{b'}}{X_{C_{TC}}} = 0 \tag{10-17}$$

Substituting for Z_T and the reactances gives, finally:

$$\frac{-1}{\omega C_{b'e}} - \frac{1}{\omega C_1} + \frac{\omega L}{1-\omega^2 LC}\left(1 + \frac{C_{TC}}{C_{b'e}}\right) - \frac{g_m r_{b'} C_{TC}}{\omega C_1 C_{b'e}} = 0 \tag{10-18}$$

Multiplying by $\omega C_{b'e}$, we obtain

$$\frac{\omega^2 L(C_{b'e} + C_{TC})}{1-\omega^2 LC} = \frac{g_m r_{b'} C_{TC}}{C_1} + \frac{C_{b'e}}{C_1} + 1 \tag{10-19}$$

or

$$\omega^2 LC = \frac{1 + \dfrac{1}{C_1}(C_{b'e} + g_m r_{b'} C_{TC})}{1 + \dfrac{1}{C_1}(C_{b'e} + g_m r_{b'} C_{TC}) + \dfrac{C_{b'e} + C_{TC}}{C}} \tag{10-20}$$

and thus,

$$f = \frac{1}{2\pi\sqrt{LC}}\sqrt{\frac{1 + \dfrac{1}{C_1}(C_{b'e} + g_m r_{b'} C_{TC})}{1 + \dfrac{1}{C_1}(C_{b'e} + g_m r_{b'} C_{TC}) + \dfrac{C_{b'e} + C_{TC}}{C}}} \tag{10-21}$$

Notice that if

$$C \gg C_{b'e} + C_{TC}$$

$$f = \frac{1}{2\pi\sqrt{LC}}$$

To find the conditions necessary for oscillation to occur, set the real terms of Eq. 10-16 equal to zero and obtain

$$r_{b'}\left(1 + \frac{X_{C_{b'e}}}{X_{C_{TC}}}\right) + g_m X_{C_1} X_{C_{b'e}}\left(1 + \frac{Z_T}{X_{C_{TC}}}\right) = 0 \tag{10-22}$$

which reduces to

$$g_m = \frac{r_{b'}(C_{b'e} + C_{TC})\omega^2 C_1}{1 + \dfrac{C_{TC}\omega^2 L}{1 - \omega^2 LC}} \tag{10-23}$$

However,

$$f_T = \frac{1}{2\pi C_{b'e} r_c} = \frac{g_m}{2\pi C_{b'e}}$$

Then, for the circuit to oscillate

$$f_T \geq \frac{2\pi f^2 C_1 r_{b'}\left(1 + \dfrac{C_{TC}}{C_{b'e}}\right)}{1 + \dfrac{C_{TC}\omega^2 L}{1 - \omega^2 LC}} \tag{10-24}$$

which reduces to

$$f_T \geq \frac{2\pi f^2 C_1 r_{b'}\left(1 + \dfrac{C_{TC}}{C_{b'e}}\right)}{\dfrac{1 - \omega^2 LC\left(1 + \dfrac{C_{TC}}{C}\right)}{1 - \omega^2 LC}} \tag{10-25}$$

However, if $C_{TC} \ll C$, which is often the case in many high-frequency oscillators, then the condition for oscillation reduces to

$$f_T \geq 2\pi f^2 C_1 r_{b'}\left(1 + \frac{C_{TC}}{C_{b'e}}\right) \tag{10-26}$$

If we substitute the value for f_{max}, the maximum frequency of oscillation (unity power gain) for a transistor, given by Eq. 5-7, into Eq. 10-26, we obtain

$$f_{max} \geq \frac{f}{2}\sqrt{\frac{C_1(C_{b'e} + C_{TC})}{C_{b'e}C_{TC}}} \tag{10-27}$$

Notice that as C_{TC} increases, a higher f_T is required for oscillation. This may be explained by noting that the feedback through C_{TC} is degenerative. Therefore, gain is required to overcome this degeneration.

The circuit just analyzed is commonly referred to as a *common-base oscillator* and is useful through a very wide frequency range. At very high frequencies, most transistors will oscillate in this configuration without external feedback due to internal and case capacitances. At lower frequencies, usually in the region under 50–100 MHz, difficulty may be encountered in obtaining reliable operation. This is owing to improper phase shift and may be corrected by tapping the feedback end of C_1 down from the top of the tank circuit (L). That is, not resonating the complete LC of the tuned circuit for feedback.

The various circuits illustrated earlier may all be analyzed in a similar manner. In the case of the Hartley oscillator, care must be taken to include the mutual inductance term between the two halves of the inductance.

PRACTICAL DESIGN CONSIDERATIONS

Although there may sometimes be nothing to recommend one circuit configuration over another electrically, certain features do make a circuit more or less desirable from a construction standpoint. When selecting a circuit, some difficulties will be avoided if the following points are observed.

1. The element used to adjust the frequency of the tuned circuit should not be involved in establishing the feedback ratio; that is, a Hartley or Clapp circuit should be used for capacitive tuning and a Colpitts circuit with inductive tuning.
2. A circuit should be used which will allow the shaft of the tunning capacitor to be grounded.
3. Since a tapped coil adds to cost, the Colpitts circuit is preferred over the Hartley circuit.

To a good approximation the frequency of operation of the Colpitts and Hartley circuits is given by the familiar $f_0 = 1/2\pi\sqrt{LC}$ where C is the parallel combination of tuning capacitance, distributed coil capacitance, stray circuit capacitance, and short-circuit input and output capacitances of the transistor at the frequency in question. The frequency of oscillation of the Clapp circuit is given by $f_0 = 1/2\pi\sqrt{LC}$, where C in this case is the series tuning capacitance only. The transistor capacitances are swamped by the capacitor divider and need not be considered if

$$\frac{(C_1)(C_2)}{C_1 + C_2} \gg C$$

The power output of a transistor oscillator is best determined experimentally, since a mathematical analysis involves nonlinear network analysis which is difficult. A good check on oscillator operation may be made by plotting power output minus power input versus power output for the transistor operated as an amplifier, that is with the feedback loop open, while maintaining constant bias conditions. The value of power output less power input for which the curve becomes maximum is the maximum power available from an oscillator. If the phase and magnitude of feedback are correct and circuit loss is negligible, this power output should be obtained as an oscillator.

ULTRAHIGH-FREQUENCY OSCILLATORS

At frequencies above 500 MHz, lumped capacitors and inductors become impractically small and lossy and the designer must resort to distributed networks for tuning and coupling.

The impedance (Z_S) at the input of a lossless transmission line is given in many textbooks (2) by the expression:

$$Z_S = Z_0 \frac{(Z_R + jZ_0 \tan \beta l)}{(Z_0 + jZ_R \tan \beta l)} \tag{10-28}$$

where Z_0 = characteristic impedance of the line
Z_R = terminating impedance of the line
$\beta = 2\pi/\lambda$, the phase constant
λ = wavelength
l = length of line

When the line is terminated in a short circuit Eq. 10-28 becomes

$$Z_S = jZ_0 \tan \beta l \tag{10-29}$$

Several important facts can be deduced from this: the impedance seen looking into a lossless short-circuited transmission line is purely reactive. The sign of the reactance, plus or minus, depends upon λ and/or l since the tangent function is positive from 0° to 90° and negative between 90° and 180°. Therefore, the reactance will be inductive from 0 to $\frac{1}{4}$ wavelength and capacitive from $\frac{1}{4}$ to $\frac{1}{2}$ wavelength. At $\frac{1}{4}$ wavelength the impedance is infinite, corresponding to an antiresonant (parallel-tuned) circuit, and at $\frac{1}{2}$ wavelength the impedance is zero, corresponding to a series-tuned circuit. A similar analysis can be made for a transmission line terminated in an open circuit. The short-circuited line, however, is generally preferred for mechanical rigidity and the relative ease of obtaining an adjustable short circuit.

Where the short-circuited transmission line is to form an antiresonant circuit with an external capacitive reactance, the line is made shorter than $\lambda/4$ for parallel resonance. In contrast to the lumped-constant resonant circuit the transmission line is known as a *distributed-constant* type.

Figure 10-6 is a diagram of an oscillator using transmission-line rather than lumped-constant techniques (5). Coaxial lines are used throughout. The system will oscillate at a frequency determined by the length of the center conductor of the coaxial cavity, the output capacitance of the transistor, and the capacitive loading of the coupling probe.

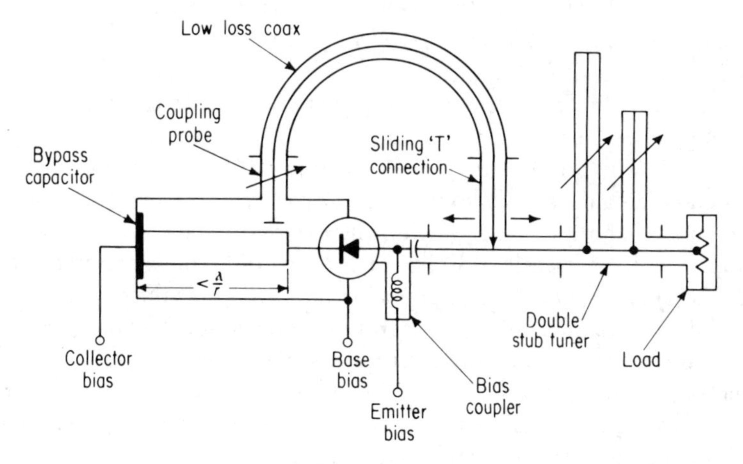

Fig. 10-6. UHF transistor oscillator using coaxial lines as part of the tuned circuits. Frequency of operation 0.5 to 2 GHz.

The capacitive coupling probe couples energy out of the cavity and the double stub tuner matches the load to the input and allows proper division of output and feedback energy.

FREQUENCY STABILITY OF VARIABLE-FREQUENCY OSCILLATORS

In the preceding discussion of high-frequency oscillators, the frequency of oscillation was dependent upon some combination of L and C (lumped or distributed) in the feedback loop to provide the required phase shift. These circuits are useful because the frequency of oscillation can be changed easily by changing the LC combination. Once set to a particular frequency, however, the frequency stability of the oscillator depends upon the stability of phase shift. Mechanical vibrations and shock, or variations in temperature and transistor biasing, will affect the frequency of oscillation to the degree that the reactive components in the feedback loop are affected.

Frequency stability can be defined as the ratio of relative change in frequency to the relative change in the factor causing this frequency change (3).

Thus voltage stability is

$$D_{f_v} = \frac{\Delta f/f}{\Delta v/v} \tag{10-30}$$

The measures taken to improve oscillator stability are in effect measures which insulate the circuit from changes in its environment.

Since instability in oscillator frequency stems from variations either in the transistor itself or in the external circuit parameters, it is convenient to consider these two effects separately.

Changes in the complex transistor parameters occur with changes in operating point due to changes in power supply voltage or changes in temperature. The problem may be handled in either (or both) of two ways: (1) by stabilizing the supply voltage and providing bias temperature stability, (2) by insulating these changes from the external circuit.

Supply voltage is easily stabilized by means of a voltage reference diode. Methods for stabilizing the bias point for variations in temperature have been discussed in detail in Chap. 3. Temperature stabilization may often be accomplished by means of a temperature-sensitive resistor. The temperature characteristics for the resistor can be determined by inserting a variable resistor into the circuit and varying this resistance, as the temperature of the circuit is varied over the required range, in order to keep the frequency constant. The resistance curve arrived at will not necessarily maintain a constant bias but will change the bias in such a way that a constant frequency is maintained. Take care to determine that all transistors of the particular type used follow the same compensation curve.

It can be shown that the frequency of oscillation can be made relatively independent of changes in transistor parameters by several methods. One way of reducing the effects of changes in the reactive component of the input and output impedance of a transistor is by swamping these reactances with much smaller reactances in the external circuit. An application of the method is shown in the Clapp oscillator circuits of Fig. 10-4.

When the value of C_1 is much larger than C_{out} and C_2 is much larger than C_{in}, changes in C_{in} and C_{out} have relatively little effect on tuned-circuit resonance. Another method of reducing the effect of transistor input and output reactance on frequency is by tapping into a lower impedance point on the tuned circuit by means of a tap on the inductor or by using a capacitor divider. A significant improvement in frequency stability can be realized by using the minimum possible amount of feedback necessary to produce the desired output power. In the Colpitts circuits this is accomplished by using the highest possible C_2 to C_1 ratio, and in the Hartley circuit by tapping the inductance as close to the base end as possible and/or using the lowest value of coupling capacitor possible. This increase in frequency stability is the result of two factors. First, since the variation of input capacitance is much more severe with changes in operating point than is the output capacitance, stability is improved by reflecting less of this variation across the tuned circuit. Second, the tuned-circuit loaded Q is improved by reducing the loading due to input resistance.

The measures taken to stabilize the frequency-determining elements in the circuitry external to the transistor are identical to those employed in vacuum tube circuits. These are

1. Mechanical rigidity of all leads and components in the circuit so that the position of leads and components remains unchanged with respect to the ground plane and to each other under mechanical vibration, shock, or temperature change.
2. The use of temperature-stable coils and capacitors in the tuned circuits.
3. The use of temperature-sensitive capacitors to correct small changes in inductance with temperature.

AMPLITUDE STABILITY

The amplitude of the oscillator output power is determined by the characteristic of the nonlinearity providing the limiting action. In general, any factor which changes the amplifier gain of the transistor will change its output as an oscillator. Changes in the output may occur as a result of changes in supply voltage or of temperature effects on the bias point. The output may be stabilized in these cases by the methods discussed previously.

In the case of variable frequency oscillators, where the frequency is to be variable over a range greater than a few per cent of an octave and where the amplitude must be held relatively constant over that range, the problem becomes more complex. In an earlier chapter the power gain of an amplifier transistor was found to fall off at the rate of 6 db per octave at frequencies above approximately $f_{max}/100$. At frequencies within this range, the oscillator power output also decreases with increasing frequency, as would be expected. The actual manner in which the output power varies with frequency is not a simple 6 db/octave relationship, but is governed by the manner in which the large-signal characteristics vary with frequency.

One method for reducing the variation in power output over a range of frequencies is to optimize the feedback at the high-frequency end of the tuning range (or above the high-frequency end of the range if necessary). This method usually takes the form of experimentally determining the value of feedback capacitor and/or coil tap which will give the desired results.

CRYSTAL OSCILLATORS

In the preceding sections, the frequency of oscillation was controlled by some combination of lumped inductance, L, and capacitance, C, and the frequency stability of the oscillation was found to depend to a large degree on the stability of these L's and C's. Many applications require very stable single-frequency operation. In such cases, a quartz crystal is used as the frequency-determining element.

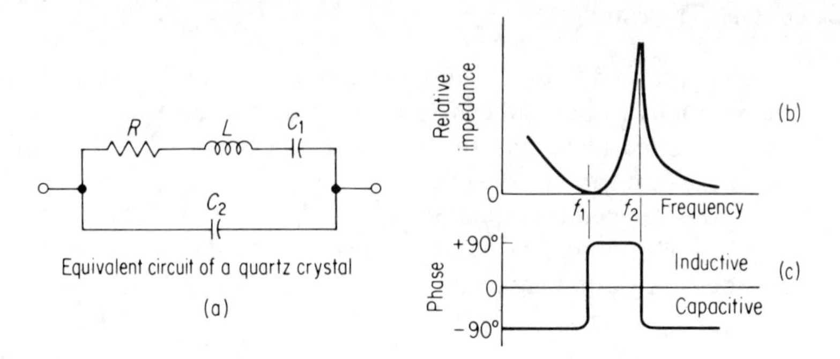

Fig. 10-7. Characteristics of quartz crystals. (a) Equivalent circuit; (b) amplitude response; (c) phase response.

Figure 10-7 shows an equivalent circuit of the quartz crystal along with its impedance and phase characteristics (7). The values of R, L, and C_1 are functions of the mechanical properties of the quartz plate, whereas C_2 is the capacitance between the crystal electrodes and between wire leads. In the relative impedance diagram of Fig. 10-7, f_1 is the series-resonant mode of the crystal and occurs at a value $\omega_1 = 2\pi f_1$ which makes

$$\omega_1 L = \frac{1}{\omega_1 C_1} \tag{10-31}$$

so that the impedance is low, real, and has a value R (R is very small). Here, f_2 is the parallel-resonant mode and occurs at a frequency where the series RLC branch is slightly inductive and resonates with C_2. The frequency difference between f_1 and f_2 is usually less than 1 per cent.

In order to show the relative values of L, C_1, C_2, and Q involved, these values are given in Table 10-1 below for a typical low-frequency crystal.

TABLE 10-1

Series resonance	427.4 KHz
Parallel resonance	430.1 KHz
$L = 3.3$ henrys	
$C_1 = 0.042$ pf	
$C_2 = 5.8$ pf	
$Q = 23{,}000$ (approximately)	

Note the extremely large value of Q (23,000) given in this example. This value is typical of most crystals and is in large part responsible for the excellent frequency stability of the crystal oscillator circuit.

From the equivalent circuit of the quartz crystal, it is apparent that a crystal may be substituted at any point in the transistor oscillator circuit where a series- or parallel-tuned circuit would ordinarily be employed. In addition, the crystal may also be used in its series-resonant mode as an extremely selective feedback path or bypass element.

Figure 10-8 shows four common-base oscillator circuits where a crystal is used in its series- and parallel-resonant modes as a tuned circuit, and in its series-resonant mode as a feedback coupling element and bypass element. (Biasing networks are omitted for convenience.)

In addition to the fundamental series and parallel resonances determined by the mechanical dimensions of the crystal blank and holder capacitance, some crystals also exhibit piezoelectric activity at frequencies above this

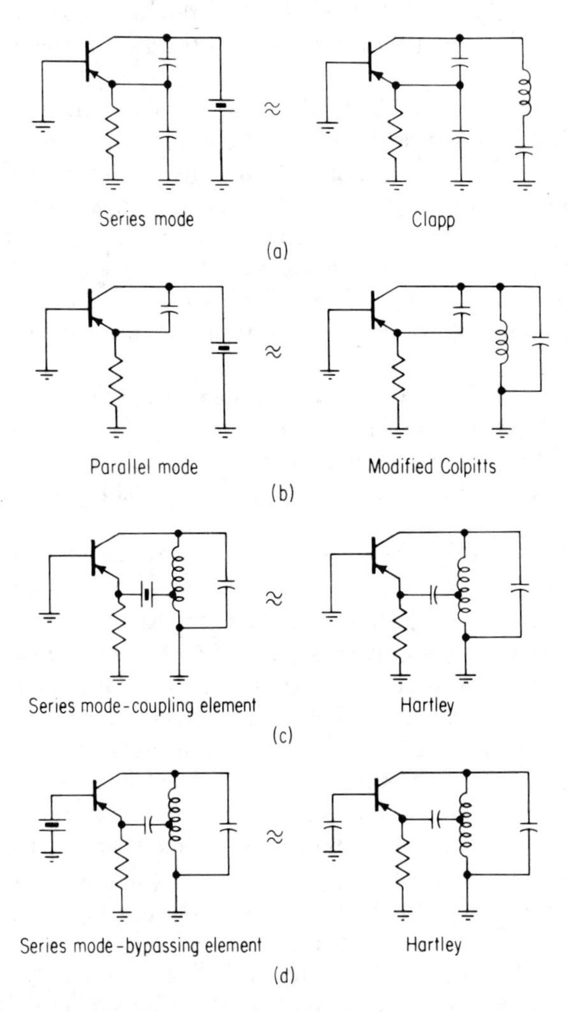

Fig. 10-8. Typical ac circuits of crystal oscillators and their LC equivalents.

fundamental. These frequencies are approximately at odd harmonic multiples of the fundamental. The term *overtone* is used because of the inexact harmonic relationship involved since nonlinear terms are responsible rather than a simple Fourier expansion. Thus the third overtone is approximately three times the fundamental, the fifth overtone approximately five times the fundamental, etc. The overtone oscillator has the advantage of generating a sinusoidal waveform directly at the frequency where it will be used, thus improving system efficiency and reducing complexity by eliminating the need for frequency multiplier stages.

Since it is impractical to grind the quartz crystal to frequencies higher than approximately 20 MHz, overtone operation is indicated when it is necessary to obtain crystal control of the oscillator at frequencies beyond this frequency.

Crystal activity or efficiency decreases as the order of the overtone increases, so that for circuits in Fig. 10-8(a, b), the crystal will oscillate at its point of maximum activity which is its fundamental frequency. By careful selection of bypass capacitors and coupling capacitors, it is sometimes possible to obtain third overtone operation with these circuits. The capacitors are chosen to provide sufficient feedback energy for oscillation on the third overtone and insufficient feedback for oscillation on the fundamental. Overtone operation above the third overtone is not recommended for these circuits.

In the general case, overtone oscillators employ a tuned circuit in addition to the crystal somewhere in the feedback loop in order to provide the selectivity and to prevent oscillation at the fundamental or a lower-order overtone. The circuits of Fig. 10-8(c, d) provide excellent operation up to the seventh overtone.

In order to obtain stable crystal control at the seventh and higher overtones it is usually necessary to neutralize the shunt capacitance of the crystal holder shown as C_2 in the equivalent circuit. At the higher overtone frequencies the value of capacitive reactance of C_2 becomes sufficiently small so that it is of the same order of magnitude as the impedance of the series-resonant arm. When this condition occurs, the crystal loses control and the tuned circuit becomes the frequency-controlling element.

Neutralization of the shunt-holder capacitance is easily accomplished by means of a small inductor shunting the crystal as shown in Fig. 10-9 and tuned to resonance with C_2 at the overtone frequency of interest. Crystals intended for overtone operation are ground specifically for that service in order to maximize activity and guarantee the resonant frequency. Where extreme frequency accuracy is necessary, the mode of operation, series or parallel, should be specified along with the frequency tolerance. The oscillator circuit in which the crystal is used should also be similar to the circuit in which the crystal is tested during manufacture.

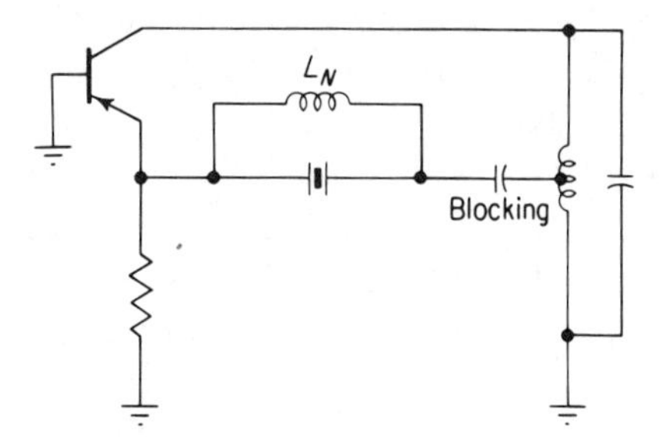

Fig. 10-9. Typical ac circuit of overtone crystal oscillator showing neutralization of crystal case capacity with inductor L_N.

MIXERS AND CONVERTERS

Mixers and/or converters are used in all superheterodyne receivers and in many other types of electronic equipment. They may be described as any device which, when two signals of different frequencies are applied to the input, produces one or more different frequencies at the output. In general, the output frequency will be either the sum or the difference of the two input frequencies.

The terms *mixer* and *converter* are often used interchangeably. It has, however, become general practice to use the term *mixer* when referring to a device where both the input signals originate from other sources (that is, a separate oscillator is used). The term *converter* refers to those devices where one of the input signals is generated within the mixing device itself (that is, a self-oscillating mixer). This notation will be used in this text.

Mixer Theory

Let us consider several devices as shown in Fig. 10-10. In device X, the transfer characteristic is

$$i_{\text{out}} = k_1 e_{\text{in}} \tag{10-32}$$

Let us now apply two input signals of different frequency to the input of this device:

$$e_{\text{in}} = A \sin \omega_s t + B \sin \omega_l t \tag{10-33}$$

where ω_s is the frequency of the input signal
ω_l is the frequency of the local oscillator.

Then,

$$i_{\text{out}} = k_1(A \sin \omega_s t + B \sin \omega_l t) \tag{10-34}$$

Note that, with a linear transfer characteristic, the only frequencies that

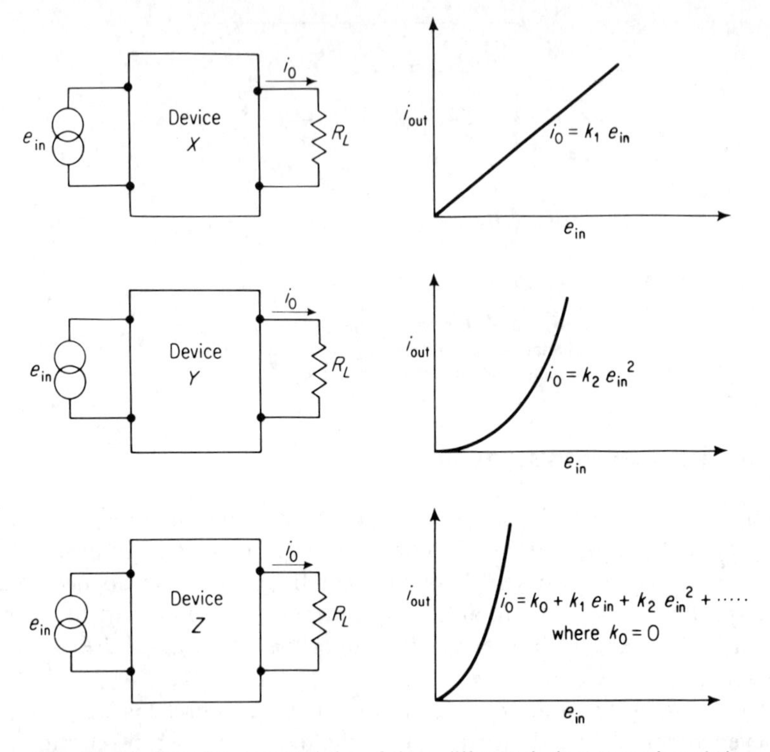

Fig. 10-10. Transfer characteristics of three different devices to analyze their utility as mixers.

exist in the output are those which existed at the input. Thus, a device with a linear transfer characteristic cannot be used as a mixer.

Let us now look at device Y in Fig. 10-10. Here the transfer characteristic is given by

$$i_{out} = k_2 e_{in}^2 \tag{10-35a}$$

If we now apply the input signal given by Eq. 10-33, we find that

$$i_{out} = k_2(A \sin \omega_s t + B \sin \omega_l t)^2 \tag{10-35b}$$

Then

$$i_{out} = k_2(A^2 \sin^2 \omega_s t + 2AB \sin \omega_s t \sin \omega_l t + B^2 \sin^2 \omega_l t) \tag{10-36}$$

From a simple trigonometric identity, we find that

$$\sin^2 \omega t = \tfrac{1}{2}(1 - \cos 2\omega t)$$

$$\sin \omega_1 t \sin \omega_2 t = \tfrac{1}{2}[\cos (\omega_1 - \omega_2)t - \cos (\omega_1 + \omega_2)t]$$

Thus Eq. 10-36 may be written as

$$i_{out} = \frac{k_2 A^2}{2}(1 - \cos 2\omega_s t) + ABk_2[\cos (\omega_l - \omega_s)t - \cos (\omega_s + \omega_l)t] + \frac{k_2 B^2}{2}(1 - \cos 2\omega_l t) \tag{10-37}$$

From Eq. 10-37 it may be seen that when two signals of different frequency are applied to the input of a device with a square-law transfer characteristic, the output contains the sum and difference frequencies, as well as the second harmonics of the input signals. The sum and difference signals are the desired results of the mixing action, illustrating that a square-law device may be used as a mixer.

In practice, tuned circuits would filter out all but the desired intermediate frequency (IF). The IF is usually the difference frequency although the sum is sometimes used.

Up to this point we have assumed that the amplitude coefficients, A and B in Eq. 10-33, were not time varying. Let us now assume that the input signal at radian frequency ω_s is amplitude-modulated as follows:

$$A = C(1 + m \cos \omega_m t) \tag{10-38}$$

where m = the modulation index and is between 0 and 1

ω_m = radian modulation frequency and is much less than ω_s and ω_l.

The output current now becomes

$$i_{\text{out}} = k_2[C(1 + m \cos \omega_m t) \sin \omega_s t + B \sin \omega_l t] \tag{10-39}$$

which may be expanded in a similar manner. If all terms which are not near the difference frequency $(\omega_l - \omega_s) \pm \omega_m$ are assumed to be eliminated by tuned circuits in the output, Eq. 10-39 reduces to

$$i_{\text{out}} = k_2 BC[\cos (\omega_l - \omega_s)t + \tfrac{1}{2}m \cos (\omega_l - \omega_s - \omega_m)t + \tfrac{1}{2}m \cos (\omega_l - \omega_s + \omega_m)t] \tag{10-40}$$

However, from trigonometry

$$\cos (\omega_1 - \omega_2)t + \cos (\omega_1 + \omega_2)t = 2 \cos \omega_1 t \cos \omega_2 t$$

and thus Eq. 10-40 reduces to

$$i_{\text{out}} = k_2 BC[(1 + m \cos \omega_m t) \cos (\omega_l - \omega_s)t] \tag{10-41}$$

Therefore, the difference frequency (the IF frequency) will have impressed upon it the same modulation signal which was present on the incoming signal. Herein lies the fundamental concept of a superheterodyne receiver: The modulation present on the incoming RF signal can be directly transferred to a lower IF signal by a mixer or converter. A similar result would have been obtained if we had chosen the sum frequencies.

Let us now look at device Z in Fig. 10-10. The transfer characteristic of this device is nonlinear and may be expressed in terms of a series expansion as expressed by

$$i_{\text{out}} = k_0 + k_1 e_{\text{in}} + k_2 e_{\text{in}}^2 + k_3 e_{\text{in}}^3 + \dots \tag{10-42}$$

If we again apply an input voltage as shown in Eq. 10-33, we can see that the first two terms of Eq. 10-42 will not contribute to mixing action. The third term will, however, cause square-law mixing. If the higher-order terms are analyzed as the product of $(e_{in})^2(e_{in})^n$, we can intuitively see that they will also produce difference frequencies.

Thus any nonlinear device with at least square-law "curvature" will serve as a mixer. In general, the transfer characteristics of practical devices are not square-law and are better represented by Eq. 10-42. Thus the output from most mixers or converters will contain many different frequencies. This requires good selectivity in the output circuit to separate the desired IF frequency from the others.

Transistor Mixers

When a transistor is operated such that one or both of the input signals swing the output current into its nonlinear region, mixing will occur. For example, if one signal is of sufficient amplitude to swing the transistor from cutoff to a high output current, the transfer characteristic will be very nonlinear over this range and mixing will occur. This is illustrated in Fig. 10-11. Figure 10-12(a) is a circuit of a transistor mixer in which a relatively high-power local oscillator signal is fed to the emitter and a small RF signal is fed to the base. The difference intermediate frequency appears across the tuned circuit in the collector. This is called *emitter injection* of the local oscillator

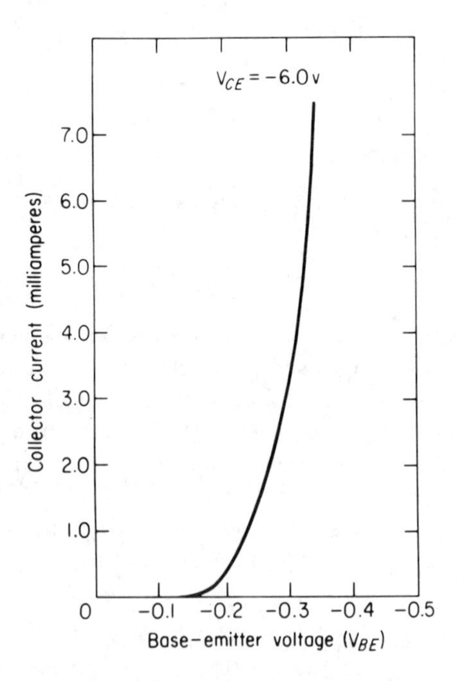

Fig. 10-11. DC transfer characteristics of 2N2399.

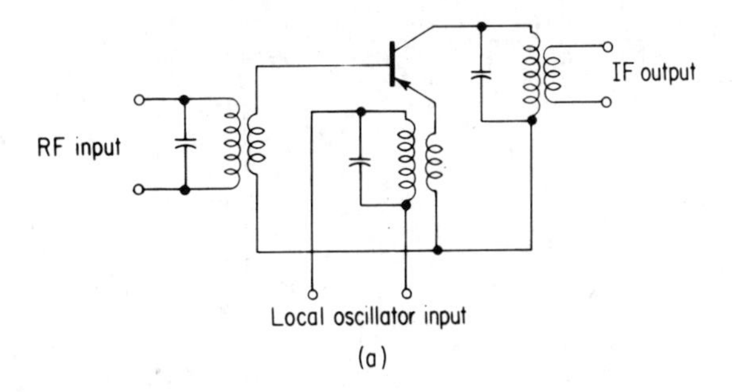

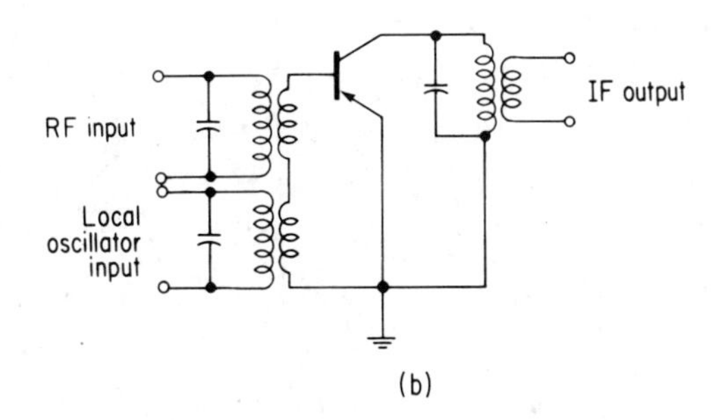

Fig. 10-12. Typical ac circuits of transistor mixers showing (a) emitter and (b) base injection of the local oscillator signal.

signal. In Fig. 10-12(b) both the local oscillator signal and the RF signal are fed to the base. This is called *base injection.* These circuits show the local oscillator as transformer coupled, although *RC* coupling may also be used.

CALCULATION OF CONVERSION GAIN

The calculation of conversion gain is not simple because the device must operate over the nonlinear region of the transfer characteristic. Thus a small-signal analysis is not valid. By making certain assumptions it is, however, possible to arrive at an approximate expression for conversion gain. One can analyze conversion gain by assuming that the collector current can be represented by a square wave switched at the local oscillator frequency and derive an expression for conversion gain in terms of hybrid-pi parameters. As mentioned earlier, any network which has a nonlinear transfer characteristic with a term to the second power can be used as a mixer. Such a network is shown in Fig. 10-13(a). A possible relation between i_2 and e_1 is shown in Fig. 10-13(b).

In a mixer, two separate signals comprise e_1. The two signals are shown superimposed in Fig. 10-14. One, designated as l, is a relatively large signal

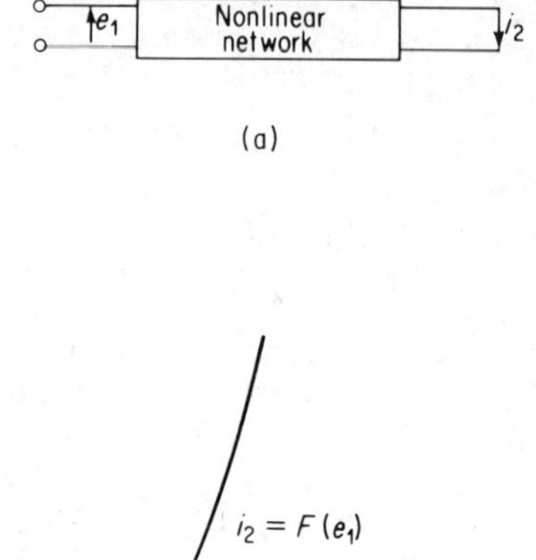

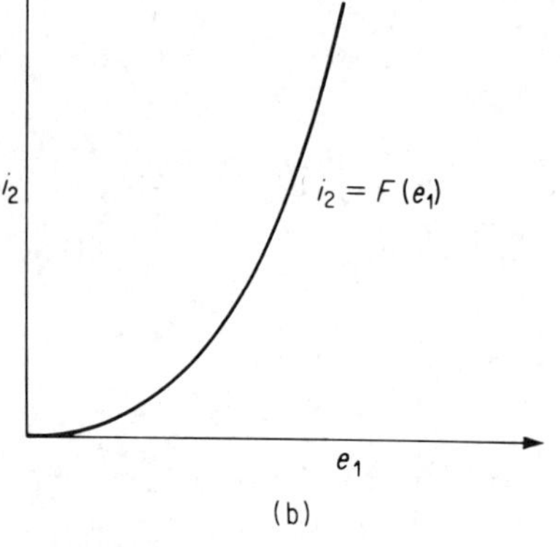

Fig. 10-13. (a) Black box representation of nonlinear network and (b) its transfer characteristic.

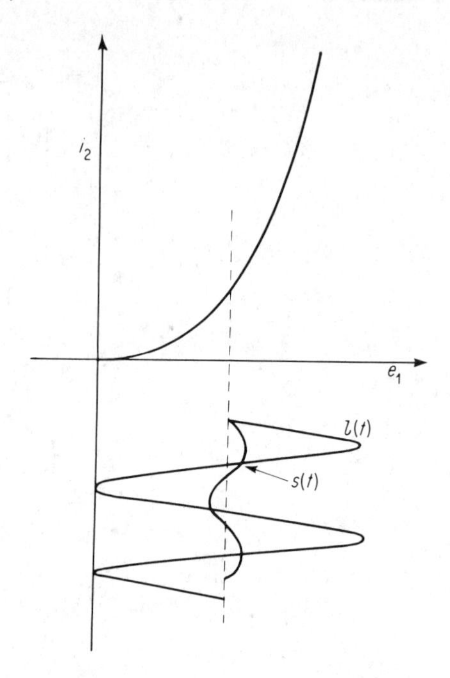

Fig. 10-14. Typical mixer transfer characteristic with the two input signals shown superimposed. The RF signal, $s(t)$, is much smaller than the oscillator signal (t).

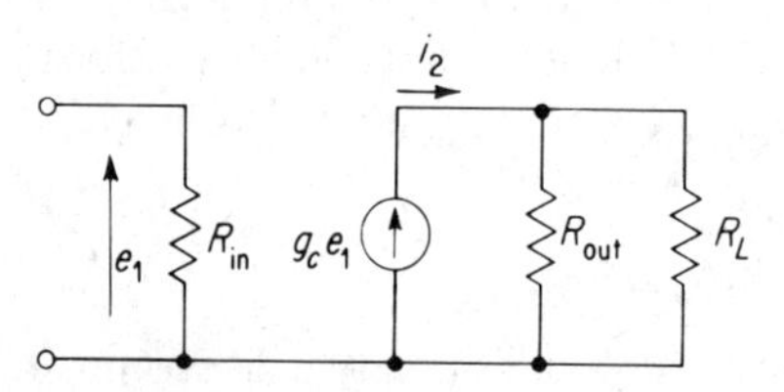

Fig. 10-15. Equivalent circuit of mixer used in analysis of conversion gain.

swinging over a large portion of the characteristic. The other signal, called s, is relatively small. Its amplitude is limited so that the slope of the characteristic is assumed not to change at the limits of its swing.

A circuit representation of the device is given in Fig. 10-15.

The power gain, or, more correctly, the conversion power gain, is defined as

$$C_G = \frac{\text{power out at IF}}{\text{power in at RF}} \tag{10-43}$$

From circuit theory it follows that the conversion gain of the device of Fig. 10-15 is

$$C_G = |g_c|^2 \left(\frac{R_{\text{out}}}{R_{\text{out}} + R_L}\right)^2 R_L R_{\text{in}} \tag{10-44}$$

where g_c = conversion transconductance
R_{out} = output resistance of the device at the IF range
R_L = load resistor on the device
R_{in} = input resistance of the device at the RF range.

Calculation of Transistor Conversion Transconductance

To apply to transistors the theory just presented, it is necessary to get an analytic expression for the extrinsic i_c/v_{be}, and to express the application of the local oscillator signal $l(t)$ so that the transistor can be evaluated as a linear network with time-varying components amplifying the RF signal.

A simplified hybrid-pi high-frequency equivalent circuit which provides a means for this analysis is shown in Fig. 10-16. This simplification assumes that the susceptance $\omega C_{b'e}$ is much greater than the conductance $1/r_{b'e}$, and that R_L is much smaller than R_{out}. C_{TC} can be neglected because it will be assumed that the input will be shorted to the IF signal voltages appearing at the output in the mixer circuit. This effectively neturalizes the transistor at the IF range. From Fig. 10-16

$$\frac{I_c}{V_{be}} = \frac{g_m}{1 + j\omega_s r_{b'} C_{b'e}} \tag{10-45}$$

where I_c and V_{be} = complex current and voltage transforms of i_c and v_{be}. This is the desired expression for the extrinsic transconductance. If I_E is assumed to be a square wave having values $I_E = 0$ and $I_E = I_{E\,max}$, the analysis is simple since $g(t)$ is alternately zero or $g(t)_{max}$. This is illustrated in Fig. 10-17. Thus

$$g_c = \frac{1}{\pi} \frac{g_m}{1 + jr_{b'} g_m \dfrac{\omega_s}{\omega_T}} \tag{10-46}$$

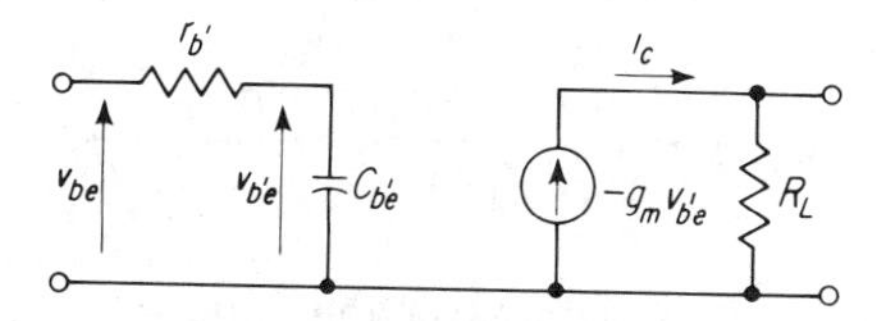

Fig. 10-16. High-frequency hybrid-pi circuit used for conversion gain analysis.

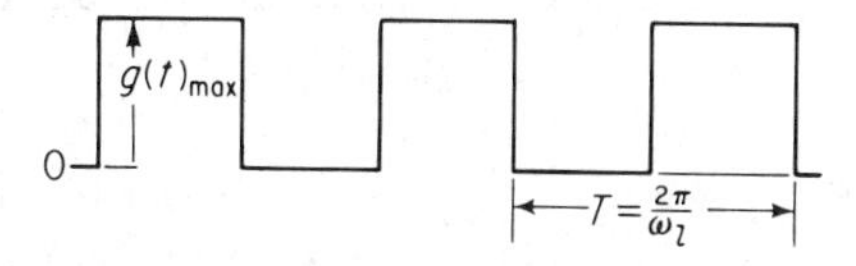

Fig. 10-17. Assumed local oscillator waveform as a function of time.

The conversion transconductance is g_c—it is the ratio of the short-circuit output current at IF to the input voltage at RF

Experimental Measurements

Experimental measurements made at a number of frequencies verify the preceding derivation. There exists, however, a consistent disagreement between the values calculated assuming square-wave analysis and the values measured. It can be assumed that this is owing to the deviation of I_E from a square wave. Actually, I_E can have many waveshapes, depending on bias, local oscillator drive, and frequency. Since the local oscillator signal is a large voltage, it is assumed that the variation of g_c due to the departure of the I_E waveshape from "squareness" can be accounted for by a minor modification of Eq. 10-46. This equation is of the form

$$g_c = k_c g$$

where

$$k_c = \frac{1}{\pi} \quad \text{(for square wave } I_E\text{)}$$

and

$$g = \frac{g_m}{1 + jr_{b'}g_m \dfrac{\omega_s}{\omega_T}}$$

One now attempts to find a k_c which will give a good "fit" with the data. A value of k_c can be calculated for each transistor for the measured power gain at each RF and IF frequency. From these calculations we find that the best k_c lies between 0.375 and 0.4. The conversion gain can be recalculated for all the devices using a k_c of 0.4. Thus conversion gain may be calculated at high frequency with reasonable accuracy from a modified form of Eq. 10-46 as

$$g_c = \frac{0.4g_m}{1 + jr_{b'}g_m \dfrac{\omega_s}{\omega_T}} = \frac{0.4g_m}{1 + jr_{b'}g_m \dfrac{f_s}{f_T}} \qquad (10\text{-}47)$$

Figure 10-18 shows how the conversion gain and noise figure vary with oscillator injection voltage and collector current. Note that the performance of a mixer is a function of the injection level, and the optimum injection level must be used if one expects to duplicate the calculated results from Eq. 10-47.

Tuned-Circuit Design

The design of tuned circuits for the mixer follows the procedure outlined in Chap. 5 for the small-signal tuned amplifier. The input transformer should provide a conjugate match of the RF signal source to the input

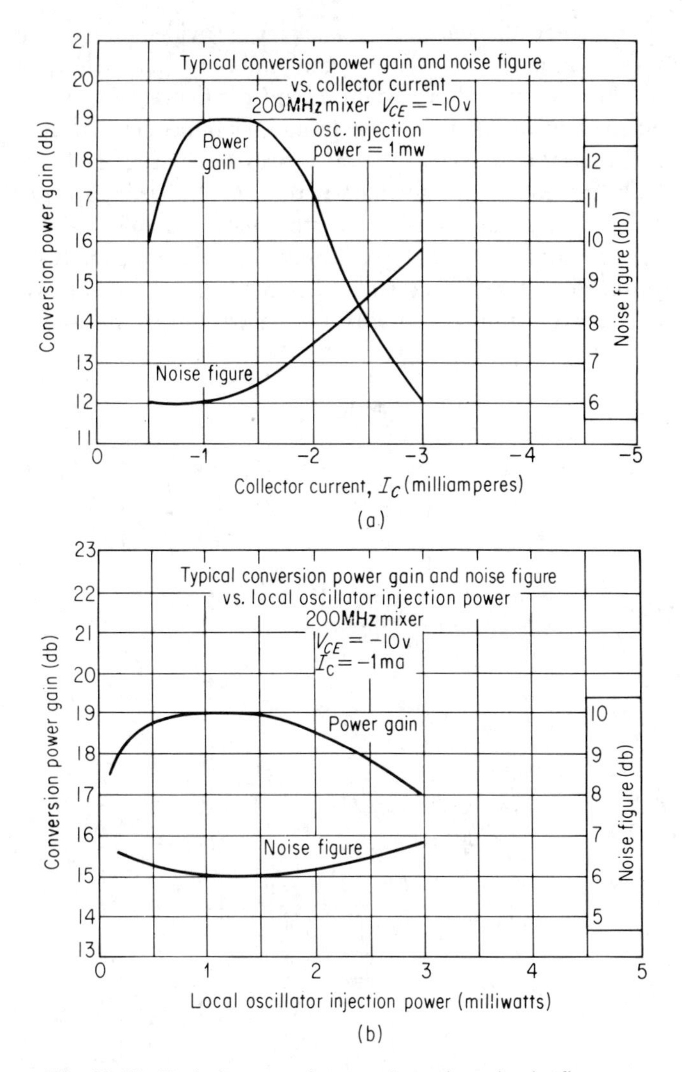

Fig. 10-18. Typical curves of conversion gain and noise figure as a function of collector current and local oscillator injection power for 2N2399.

impedance (as measured with a short-circuited output) of the transistor at the RF frequency in order to accomplish maximum power transfer. The output transformer usually provides a conjugate match approaching the output impedance (as measured with short-circuited input) of the transistor at the IF frequency (12). The output impedance at the IF frequency is often very high and it may not be possible to obtain a match in some applications. The local oscillator coupling network is best determined experimentally.

It is unnecessary to neutralize the mixer, provided that the output is effectively short-circuited for the RF signal and the input is effectively short-circuited for the IF signal. In certain circuit configurations, however, care must be taken to insure that the base-to-emitter impedance is low at the IF frequency. If this is not the case, the mixer will not be effectively neutralized and the problems normally associated with unneutralized stages will result. At high frequencies this can cause a substantial loss of conversion gain.

Figure 10-19 shows several typical mixer circuits. In Fig. 10-19(a, b), a series trap, tuned to the IF frequency, has been added to effectively short-circuit the base to the emitter at the IF frequency. This trap should be tuned

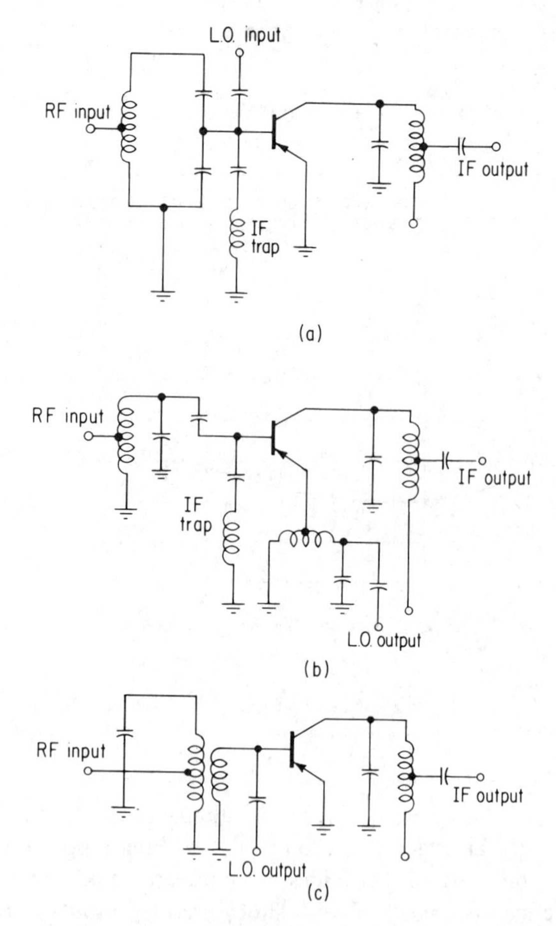

Fig. 10-19. Three typical mixer circuits (ac only) showing use of series trap for neutralization. No trap is usually needed in circuit (c) because the secondary of the input transformer is a low impedance at IF.

to resonance before placing it in the circuit, since, otherwise, attempts to adjust it in the circuit will usually result in the stage oscillating.

Converters

The local oscillator function can be included in the mixer circuit, thereby eliminating the oscillator transistor. This circuit is called a *converter*, *autodyne*, or *self-oscillating mixer*, to avoid confusion with a mixer in which the mixing and local oscillator functions are performed by separate transistors.

The advantage of this circuit is cost reduction through elimination of the oscillator transistor. The disadvantage is that there is less isolation between

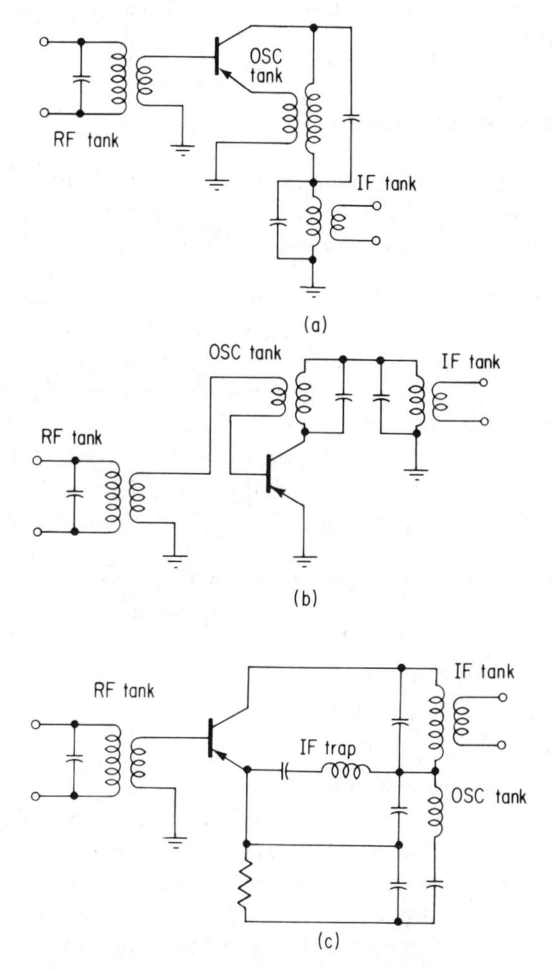

Fig. 10-20. Typical converter circuits. (a) Emitter injection; (b) base injection; (c) Clapp-type oscillator circuit.

the RF signal and the oscillator frequency, with the result of increased oscillator pulling for strong RF signal levels and also increased oscillator radiation.

Figures 10-20(a, b) are examples of converter circuits using emitter and base injection, respectively. In these circuits the oscillator is transformer coupled. Figure 10-20(c) is an example of a nontransformer-coupled emitter-injection circuit based on the Clapp oscillator.

The operation of these circuits can be easily understood if we assume that the RF and IF tank circuits are short circuits at the oscillator frequency, that the oscillator and IF tank circuits are short circuits at the RF frequency, and that the oscillator and RF tank circuits are short circuits at the IF frequency. The analysis is then similar to the mixer and oscillator theory presented earlier.

TRANSISTOR DETECTORS

Transistors operated essentially at cutoff can be made to perform as AM detectors. This detection method has several advantages over the semiconductor diode method in common use (3); namely,

1. The signal power level where detection changes from square law to linear law is slightly lower for transistors than for diodes.
2. The power gain of the transistor can be utilized in order to provide a considerable gain for the stage rather than a net loss as in the case of the diode detector.
3. Automatic gain control of transistor RF amplifier stages requires considerably more power than that normally required in vacuum tube circuits. This AGC power may be supplied by the transistor detector.

The gain of the transistor detector is defined by (3):

$$P_G = 10 \log \frac{2P_{\text{out}}}{m^2 P_{\text{in}}} \tag{10-48}$$

where P_{out} = audio power output
P_{in} = unmodulated power input
m = modulation index.

For good rectification efficiency, the quiescent operating point must be on the most nonlinear portion of the dynamic diode characteristic as shown in Fig. 10-21. A look at the base characteristics shows that approximately 0.25 volt and 0.6 volt initiation voltage is required for germanium and silicon, respectively, before appreciable collector current change occurs. An experimental analysis of transistor detectors shows that the collector current waveform is as shown in the figure. That is, a series of half sine wave pulses

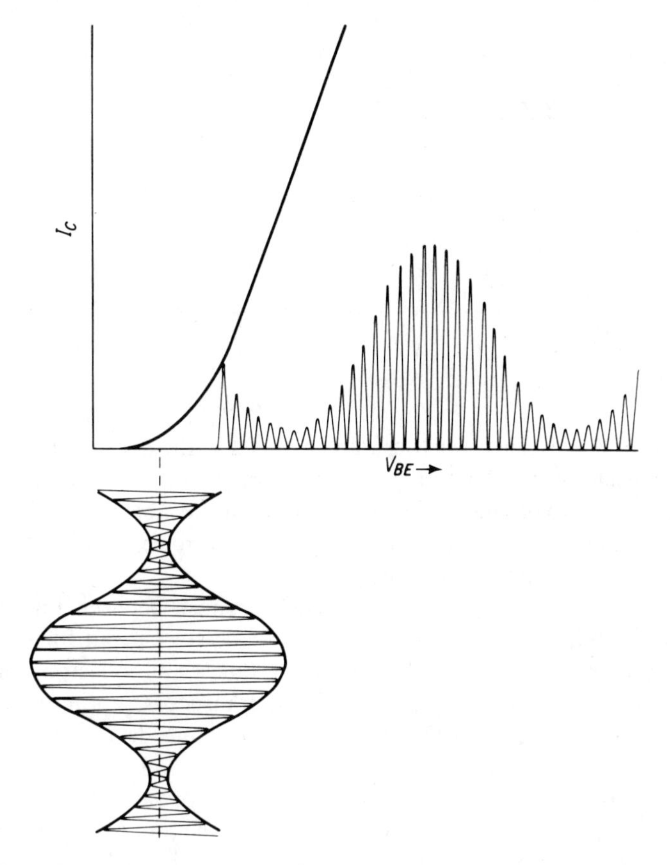

Fig. 10-21. Transfer and rectification characteristics of a transistor detector.

at the incoming carrier frequency. Thus, the transistor detector is not, as sometimes stated, a diode detector (base-emitter diode) and an audio amplifier, but rather a detector and amplifier at the carrier frequency with audio restoration occurring because of the filter network at the output. Therefore, the detector transistor must be a good high-frequency amplifier at the frequency of the incoming signal.

Figure 10-22 is the basic detector circuit. C_1 is a capacitive bypass at the carrier frequency. C_2 has a high reactance at the highest audio frequency, and in conjunction with R_L, forms the filter network for recovering the audio modulation. R_1 and R_2 establish bias for efficient detection. The audio frequency output is developed across R_L.

The transistor detector stage is usually optimized for gain at low signal levels since detection at high levels does not present a problem. The bias point for optimum detection efficiency shifts with temperature so that a

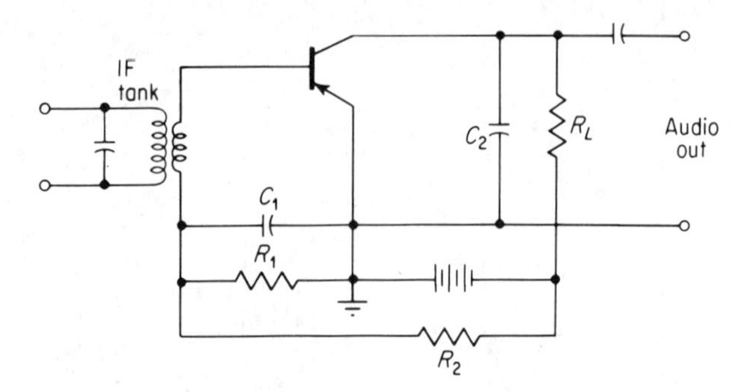

Fig. 10-22. Typical transistor detector circuit.

temperature-sensitive resistor is usually included in the detector bias network to maintain efficiency. The temperature characteristics for this resistor are best determined experimentally by means of a variable resistor, which is varied to maintain maximum efficiency as the circuit is subjected to the required ambient temperature range.

Proper biasing of the detector is a compromise between low collector current for maximum nonlinearity and high collector current for good gain. For germanium transistors, the optimum quiescent collector current is usually between 25 and 50 μa. For silicon transistors, optimum collector current is usually between 100 and 150 μa. For planar devices, the optimum current is somewhat lower.

Figure 10-23 shows a method for obtaining Forward AGC (automatic gain control) control voltage from a transistor detector. An NPN transistor is used to provide a negative-going control signal. The resistors, R_5 and R_6,

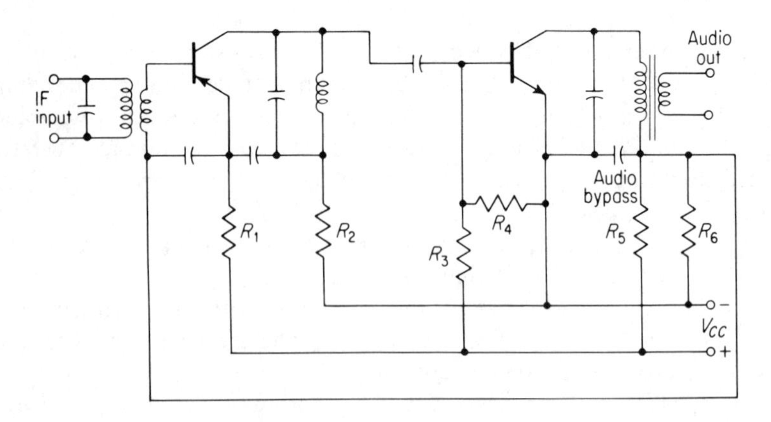

Fig. 10-23. Circuit showing the use of a transistor detector for generating Forward AGC control voltage.

establish the quiescent bias point for the RF amplifier with the detector transistor effectively in shunt with R_6 for control purposes. The series combination of R_1 and R_2 makes up the Forward AGC dropping resistance.

Superregenerative Detectors

Superregenerative detectors are commonly used at very high and ultrahigh frequencies because of their simplicity and high gains.

Regenerative amplification is limited to the point where oscillations begin. Beyond this point, further amplification ceases. A superregenerative detector overcomes this by introducing an alternating voltage of frequency between 25 and 250 KHz. This process is termed *quenching*, and results in a blocking action similar to the action taking place in blocking oscillators. This is done so as to vary the operating point of the transistor detector. The detector can oscillate only during the time when the varying operating point is in the region where oscillations take place. The regeneration can be greatly increased because the oscillations are constantly interrupted. The result is that the amplified signal is built up to a very large value. A sensitivity of 1 or 2 microvolts is not uncommon.

Amplification is greater when the difference between the quench frequency and the signal frequency is increased because the signal has more time to build up during the half cycle when no quench signal is present.

There is a limit as to how far the quenching can be carried out because, during the time there is no quenching, the input selectivity is that which corresponds to that of the Q of the tuned circuit alone. Quench frequencies considered optimal are 75 KHz for 30 MHz, 150 KHz for 60 MHz, and about 250 KHz for 150 MHz.

The selectivity of a superregenerative detector is poor as compared to a regenerative type, but, it is less affected by ignition noise, for example, when used in an automotive radio, and has some AVC (automatic volume control) action.

Most superregenerative detectors are self-quenched and the quenching frequency is determined by the amount of feedback and the time constant in the bias circuit.

Figure 10-24 shows a transistorized superregenerative detector with excellent dc stability. The properly phased feedback voltage is supplied by the feedback winding L_2. Capacitor C_2 provides some external control of the feedback. Capacitor C_1 and resistors R_1 and R_2 determine the time constant of the bias circuit. The tank circuit elements, L_3 and C_3, determine the received frequency. Antenna coupling coil L_1 is adjusted for the best sensitivity. The superregenerative detector action is illustrated in Fig. 10-25.

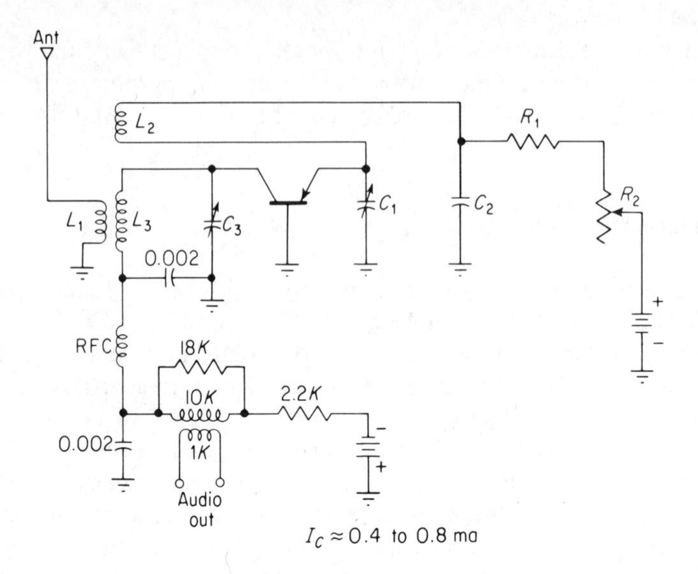

Fig. 10-24. Typical superregenerative detector circuit.

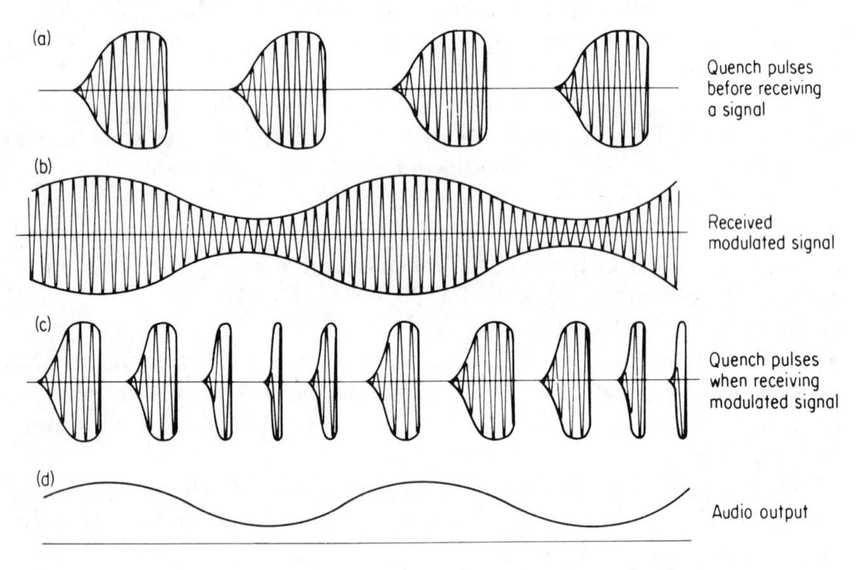

Fig. 10-25. Typical waveforms present in a superregenerative detector.

EXERCISES

1. An audio oscillator design is required to produce an output at 10 KHz. Design a simple *RC* oscillator which will meet this requirement.
2. For the oscillator design of Exercise 1 what value of load resistance will minimize the common-emitter current gain requirements of the transistor? For this value of load resistance what β is required?

3. Analyze the common-collector Clapp oscillator to determine the frequency of oscillation and starting conditions. How would you design the oscillator to minimize frequency drift with respect to temperature?

REFERENCES

1. J. MILLMAN, *Vacuum Tube and Semiconductor Electronics.* New York: McGraw-Hill Book Company, 1958.
2. W. C. JOHNSON, *Transmission Lines and Networks.* New York: McGraw-Hill Book Company, 1950.
3. R. F. SHEA, *Transistor Circuit Engineering.* New York: John Wiley & Sons, Inc., 1957.
4. G. K. MCAULIFFE, *Electronic Equip. Eng.*, October, 1961, p. 39.
5. L. A. WELDON, Application Lab Report 728, Philco Corp., Lansdale Division.
6. T. L. MARTIN, *Ultrahigh Frequency Engineering.* Englewood Cliffs, N.J.: Prentice-Hall, Inc., 1950.
7. F. E. TERMAN, *Radio Engineer's Handbook.* New York: McGraw-Hill Book Company, 1943.
8. A. L. GARTNER, *Transistors: Principles, Design and Applications.* Princeton, N.J.: D. Van Nostrand, Inc., 1960.
9. W. A. EDSON, *Vacuum Tube Oscillators.* New York: John Wiley & Sons, Inc., 1953.
10. K. A. PULLEN, *Handbook of Transistor Circuit Design.* Englewood Cliffs, N.J.: Prentice-Hall, Inc., 1961.
11. S. SEELY, *Electron Tube Circuits.* New York: McGraw-Hill Book Company, 1950.
12. M. JAVID and E. BRENNER, *Analysis of Electric Circuits.* New York: McGraw-Hill Book Company, 1959.
13. P. E. GRAY and C. L. SEARLE, *Electronic Principles—Physics, Models and Circuits.* New York: John Wiley & Sons Inc., 1969.
14. D. T. COMER, *Large-Signal Transistor Circuits.* Englewood Cliffs, N.J.: Prentice-Hall, Inc., 1967.
15. N. HOLONYAK, JR., *Integrated Electronic Systems.* Englewood Cliffs, N.J.: Prentice-Hall, Inc., 1970.
16. M. S. GHAUSI, *Principles and Design of Linear Active Circuits.* New York: McGraw-Hill Book Company, 1965.
17. E. J. ANGELO, JR., *Electronics: BJT's, FET's and Microcircuits.* New York: McGraw-Hill Book Company, 1969.
18. S. SEELY, *Electronic Circuits.* New York: Holt, Rinehart & Winston, Inc., 1968.
19. C. BELOVE, H. SCHACHTER and D. L. SCHILLING, *Digital and Analog Systems, Circuits and Devices: An Introduction.* New York: McGraw-Hill Book Company, 1973.

11

Class C Amplifiers, Frequency Multipliers, and High-Frequency Design Techniques

INTRODUCTION

In this chapter we depart from our usual format and present an essentially empirical approach to several special high-frequency design problems. Both Class C amplifiers and frequency multipliers involve nonlinear transistor operation similar to those described in the previous chapters. Analysis in this domain is at best approximate, and without a basic understanding of high-frequency construction techniques, it is impossible to fabricate operating circuits. Hence, the first part of this chapter contains an empirical discussion of Class C amplifiers and frequency multipliers.

It has often been said that high-frequency techniques are more art than science. The latter part of this chapter presents many practical hints on the art of high-frequency design. These techniques are useful for all high-frequency circuits and need not be confined to those circuits described in this chapter.

CLASS C OPERATION

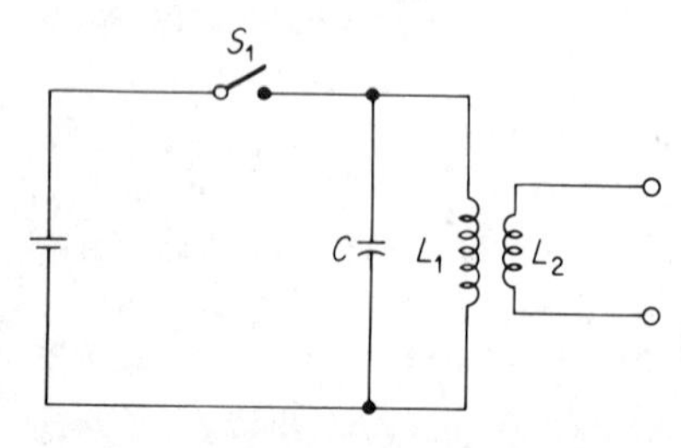

Fig. 11-1. Simple LC resonant circuit.

The operation of transistorized Class C power amplifiers and frequency multipliers can best be introduced by first analyzing a simple *LC* resonant circuit as shown in Fig. 11-1. The battery is used as the source of dc power with switch S_1 controlling the rate and amount of power being applied to the *LC* circuit. This *LC* network is generally called a *tank circuit* because of its ability to store electrical energy.

When S_1 is closed and opened abruptly, a burst of electrical energy is applied to the tank circuit from the dc source. The result is an ac signal appearing across the secondary winding, L_2. This conversion of dc to ac power is caused by the transfer of energy from the inductance to the capacitor and back at their resonant frequency. Thus by shocking a tank circuit with a dc pulse of a short duration, oscillations or ringing will be produced whose frequency will be mainly determined by the value of the LC combination. The oscillations would continue indefinitely were it not for circuit losses. The losses in this case are the leakage resistance of the capacitor and the loss resistance of the coil. The higher the value of circuit loss, the sooner the oscillations are damped out. Hence, high Q tank circuits are a necessity in Class C service, especially in frequency multipliers.

Figure 11-2 illustrates a tank circuit whose output is kept at the same level by injecting excitation pulses with S_1 in phase with the output signal. Thus, any losses are compensated for by additional energy insertion. This would be the case in straight-through amplifiers (that is, where the input and output frequency are the same).

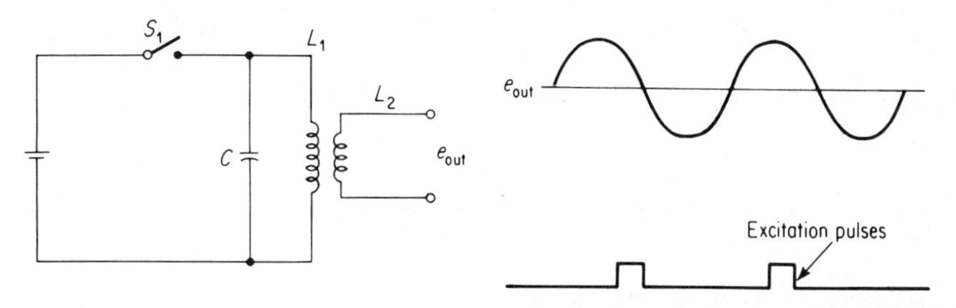

Fig. 11-2. LC tank circuit output voltage when the excitation pulses occur at the fundamental frequency.

Figure 11-3(a) illustrates the output of a low Q tank circuit receiving excitation pulses every third cycle, and Fig. 11-3(b) illustrates a high Q tank circuit. The circuit of Fig. 11-3(b) is more efficient than the one shown in Fig. 11-3(a) because the amplitude of each succeeding ac cycle diminishes at a slower rate. Another effect which may decrease the efficiency of the tank circuit is the width of the excitation pulses. This form of tank circuit loading becomes more troublesome at higher frequencies where the excitation pulse width approaches the duration of each individual ac cycle. This can interfere with the ability of the tank circuit to produce pure sine waves and, therefore, as the frequency of the oscillation desired becomes higher, the driving pulses should become progressively narrower in order to avoid this difficulty.

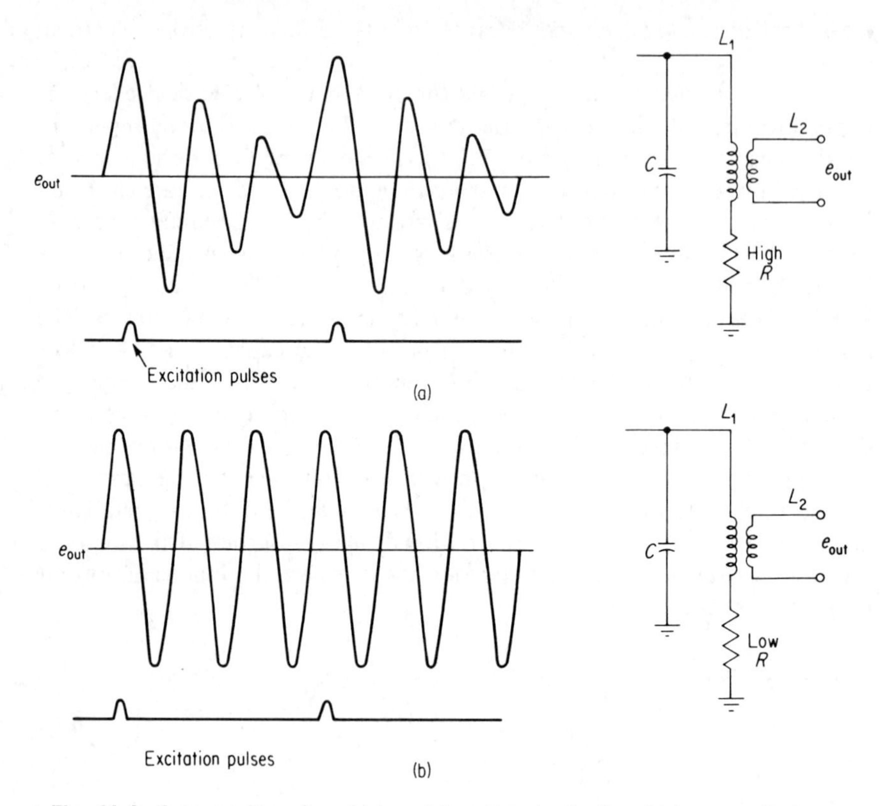

Fig. 11-3. Output voltage from high and low Q tank circuits which are excited every third cycle.

Tuned Class C RF Power Amplifiers

The Class C amplifier that we are concerned with uses a transistor to accomplish the switching action described earlier. By biasing the transistor well beyond collector current cutoff, and then driving the input with sufficient excitation voltage, so that the input current flows for only a short period of time, there will be developed in the output circuit, pulses of output current which will cause the tuned circuit to ring (oscillate). As the bias is increased beyond the cut-off value, the conduction angle decreases; this results in an increase of efficiency. A conduction angle of 90° to 120° is desirable for straight-through operation. In the preceding, the *conduction angle* is defined as the portion of the RF input cycle during which collector current flows. Figure 11-4 illustrates the operation of a Class C RF amplifier by waveform analysis. Notice the spreading effect of the collector current pulse due to storage time. By using transistors with low storage times, this effect can be minimized.

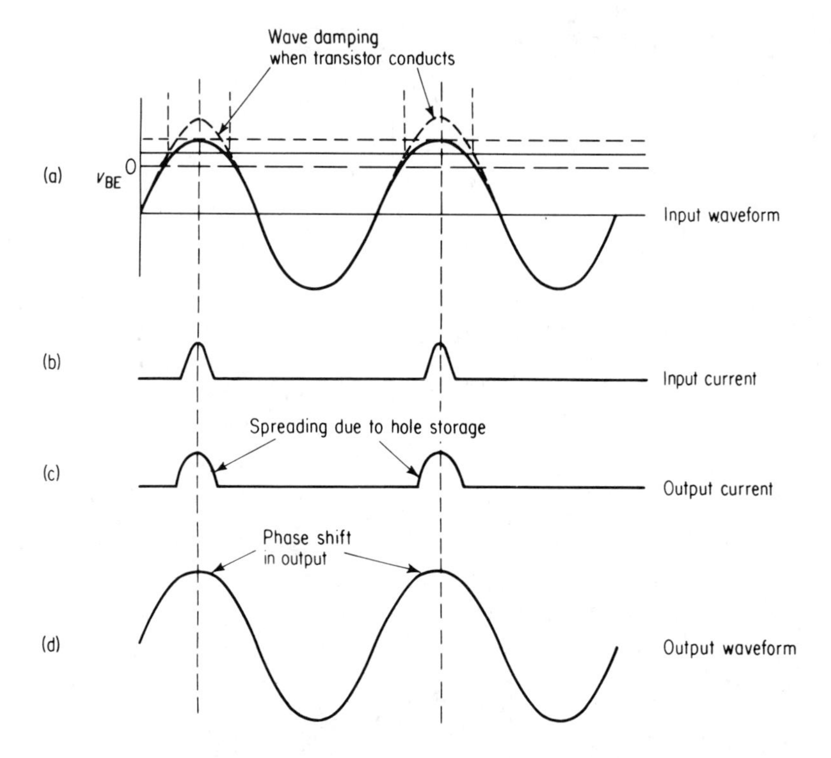

Fig. 11-4. Current and voltage waveforms for an NPN Class C transistor amplifier.

Class C amplifiers are employed chiefly where high output efficiencies are desired. The output tuned circuit produces the sine wave output voltage as shown in Fig. 11-4(d) where storage effects have shifted the phase of the output voltage slightly with respect to the excitation voltage.

Although only the common-base connection is shown in the accompanying figures, the general discussion also includes the common-emitter configuration.

A single-stage Class C tuned amplifier is illustrated in Fig. 11-5(a). The output tank circuit is tuned to the excitation frequency. Figure 11-5(b) shows the same stage with neutralization. Transistors can be paralleled in order to increase the power output; however, this requires that the driving power also be increased.

A push-pull circuit is shown in Fig. 11-6(a). This type of connection requires that the transistors be matched in order to distribute the dissipation evenly. Where neutralization is required, the circuit in Fig. 11-6(b) may be used. At very high frequencies, where it is desirable to reduce the capacitance across the tank circuit, inductive neutralization is often used. An inductively neutralized circuit is shown in Fig. 11-7. In this circuit, the inductance forms

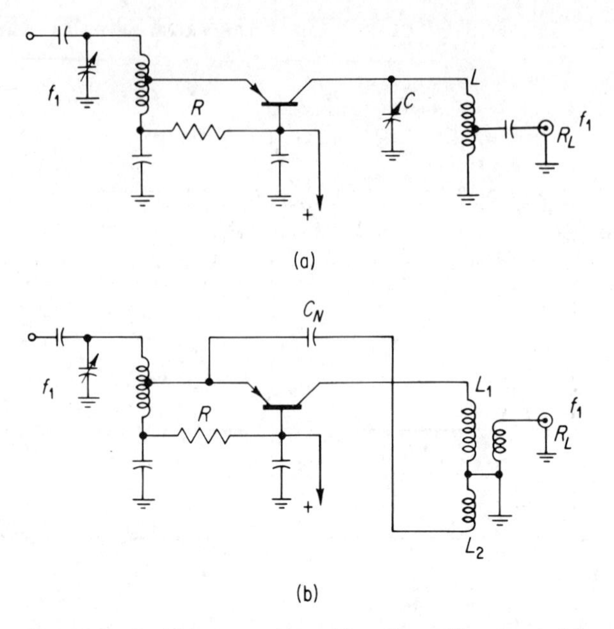

Fig. 11-5. Typical common-base Class C amplifier stages. (a) Neutralized; (b) unneutralized.

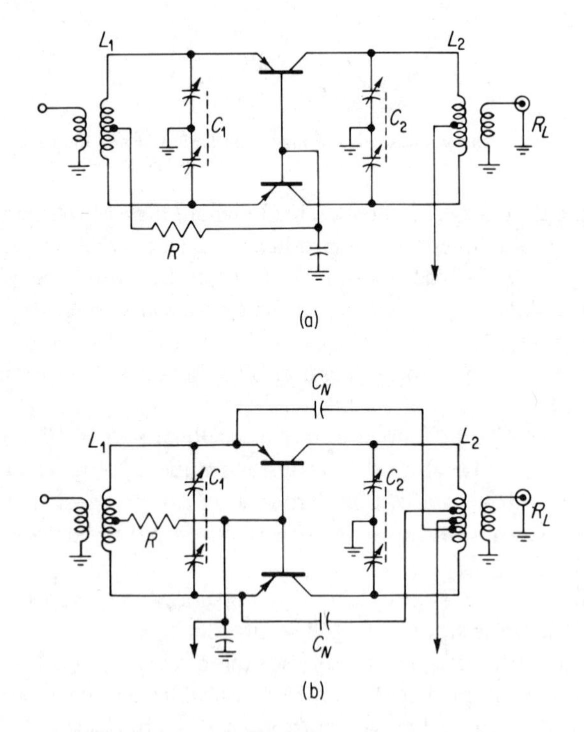

Fig. 11-6. (a) A push-pull Class C circuit; (b) with neutralization.

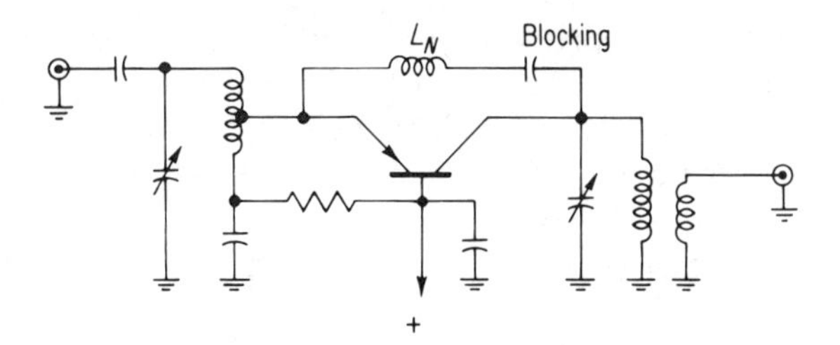

Fig. 11-7. Inductively-neutralized common-base amplifier.

a parallel resonant circuit with the feedback capacitance, and thereby effectively minimizes the feedback.

CLASS C RF POWER AMPLIFIER DESIGN

The essential requirements for the design of the tuned circuits in a Class C stage are a high load impedance, high Q coils and capacitors (and hence low loss), and a good circuit layout. A fair approximation of the load impedance R_L at the collector is given by

$$R_L = \frac{V_{CC}^2}{2P_{ac}} \tag{11-1}$$

This assumes that the peak ac voltage is equal to the supply voltage. Notice that this equation is the same as the one derived in Chap. 5 for Class B stages. In fact, the equation describes the maximum ac power output for a given transformer coupled circuit and supply voltage. Notice, however, that this value of R_L does not necessarily match the transistor for maximum gain. In fact it usually does not. If we look at Eq. 11-1 from another viewpoint, it may be seen that for a specified output power, there is one and only one supply voltage which will provide this output power and maximum gain from the stage (that is, at this supply voltage R_L will be equal to R_{out}). In an actual circuit it may not be possible to use this voltage because of the voltage breakdown rating of the transistor. This does, however, point up one area of compromise in the design of high-frequency Class C amplifiers. In low-frequency circuits the small loss in gain due to mismatch at the output is usually not serious because of the large gain available. In high-frequency Class C amplifiers, however, where the gain may be only 4–10 db per stage, the loss of 1 or more db due to mismatch is a serious compromise.

Once the load impedance is determined, the output network can be developed. We shall assume that the dynamic output resistance of the transistor is much greater than the required value of R_L (the usual case in all but very high power applications). In the event it does approach the value of R_L then, the output network must be modified to include this additional loading of the tank circuit.

The tank circuit design that follows is intended for power levels of about 25 milliwatts. Low-level tank circuits may be designed as described in Chap. 5. For small signals, it is often possible to design for matched conditions and then increase the driving power for the desired power output.

Class C Tank Circuit Design

The efficiency of the tank circuit is determined by the ratio of the loaded to the unloaded Q, and its maximum value is related to these Q's by the expression given in Eq. 5-32 for the optimal interstage gain:

$$\text{efficiency (tank circuit)} = \left(1 - \frac{Q_l}{Q_u}\right)^2 \times 100\% \tag{11-2}$$

Therefore, a 10:1 unloaded to loaded Q ratio (a reasonable value) results in an efficiency of 81 per cent, or in approximately a 20 per cent loss of power.

The basis for impedance matching lies in the fact that, at a given frequency, a parallel circuit has an equivalent series circuit and vice versa. The following simplified relationships will be useful in designing Class C tank circuit matching networks. Referring to the two RX circuits of Fig. 11-8, the expressions of the Q of each circuit are given by

$$Q_S = \frac{X_S}{R_S} \tag{11-3}$$

$$Q_P = \frac{R_P}{X_P} \tag{11-4}$$

If the two circuits have the same $Q(Q_S = Q_P)$, then we can find the relationship between the various impedances which will present the same terminal impedance (Z_T) at a specified frequency. This may be done by equating the two impedances as

$$Z_T = R_S + jX_S = \frac{jR_P X_P}{R_P + jX_P} \tag{11-5}$$

By equating the real and imaginary terms of Eq. 11-5 and substituting Eqs.

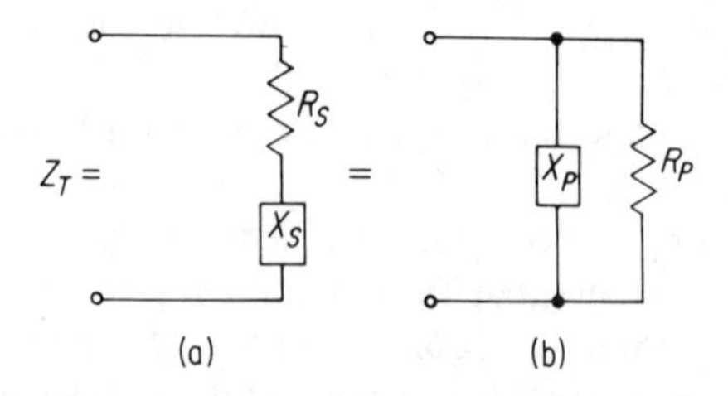

Fig. 11-8. Simple series and parallel RX networks which have the same impedance at one frequency.

11-3 and 11-4, the following expressions, Eqs. 11-6–11-9, may be obtained for conversion from one circuit to the other.

$$R_P = R_S(Q^2 + 1) \tag{11-6}$$

$$X_P = \frac{R_P}{Q} = \frac{R_S(Q^2 + 1)}{Q} \tag{11-7}$$

$$R_S = \frac{R_P}{Q^2 + 1} \tag{11-8}$$

$$X_S = QR_S = \frac{QR_P}{Q^2 + 1} \tag{11-9}$$

When the values of resistance and reactance fulfill these equations, the two circuits will have the same impedance at the specified frequency.

Although the impedance-matching techniques described in Chap. 5 may be used for Class C amplifiers, it is often desirable to resort to other matching techniques such as L or pi networks. A number of factors must be considered in choosing the matching network. Among these are the total allowable tank capacitance, the impedance to be transformed, the action of the matching network in filtering unwanted harmonics out of the output, and the Q of the tank circuit.

Expressions may be developed for the impedance transformation for the general case in each of these networks, but it is often easier to understand the transformation by considering these as special cases of series-to-parallel, or parallel-to-series circuit transformations. Several examples of this approach will now be discussed. In all cases the capacitors are considered lossless. If they have losses they may be lumped with R_S or R_P. High-quality capacitors are usually used in these applications to minimize losses.

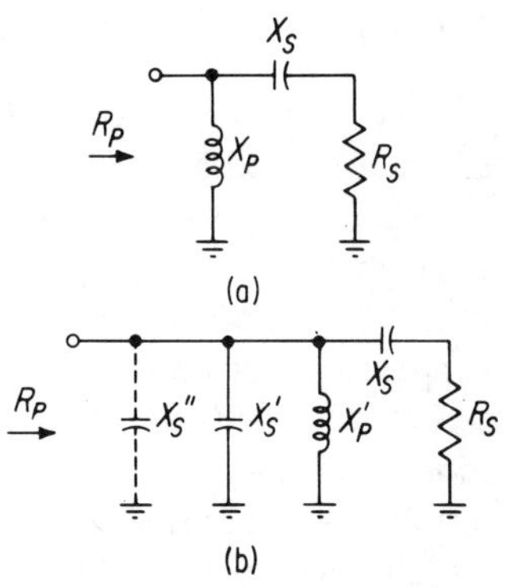

Fig. 11-9. L-type matching networks illustrating schematics used for calculation when: (a) $Q_L = Q_m$; (b) $Q_L > Q_m$.

The L network shown in Fig. 11-9(a) provides a simple means of impedance transformation. If the required collector load impedance and terminal resistance are known, the following procedure is used in determining the tank circuit values:

From Eqs. 11-6, 11-7, and 11-9, we obtain

$$Q_m = Q_l = \sqrt{\frac{R_P}{R_S} - 1} \tag{11-10}$$

$$X_S = Q_m R_S = Q_l R_S \tag{11-11}$$

$$X_P = \frac{R_P}{Q_m} = \frac{R_P}{Q_l} \tag{11-12}$$

The loaded circuit Q (Q_l) will be the same as the value of Q used in effecting a match to the load (Q_m).

If the loaded Q is to be a value that is higher than the Q used in effecting a match, then the following procedure is used. [See Fig. 11-9(b).]

The design equations for an L circuit designed for a $Q_l > Q_m$ are

$$Q_m = \sqrt{\frac{R_P}{R_S} - 1} \tag{11-13}$$

$$X_S = Q_m R_S \tag{11-14}$$

$$X'_P = \frac{R_P}{Q_l} \tag{11-15}$$

The new value of inductive reactance necessary to provide the desired Q_l for a given load impedance is X'_P. Then,

$$X'_P = X_S + X'_S + X''_S \tag{11-16}$$

where X'_S = reactance of the capacitor required to keep the tank circuit in resonance as a result of X_P being reduced in value to X'_P

X''_S = reactance of the output capacitance of the transistor.

Pi Networks

The pi network (Fig. 11-10) can be used to couple a transmitter to an antenna as well as for an interstage transformer.

When the collector circuit of the amplifier is coupled directly to the load through the pi network, the network transforms the load resistor into an impedance much higher in value. The pi network has excellent harmonic suppression as compared to the conventional coupling networks because the series inductor (L_1) forms a low-pass filter. This greatly attenuates all harmonics. But for the very reason that it will not suppress the fundamental, it is not desirable for frequency multipliers.

The pi network calculations can best be handled as two separate L networks. Figure 11-11 shows how a pi may be broken down into two equivalent L networks.

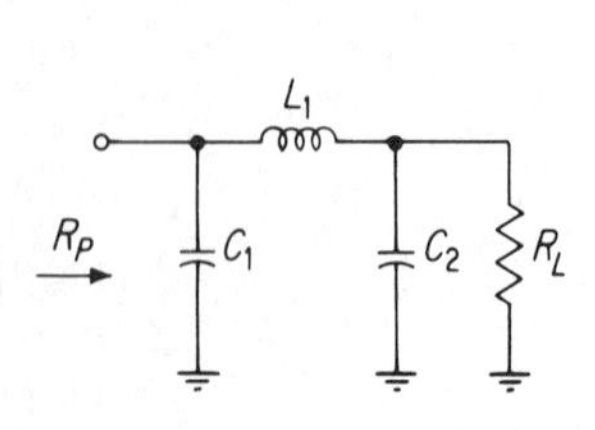

Fig. 11-10. Pi matching network.

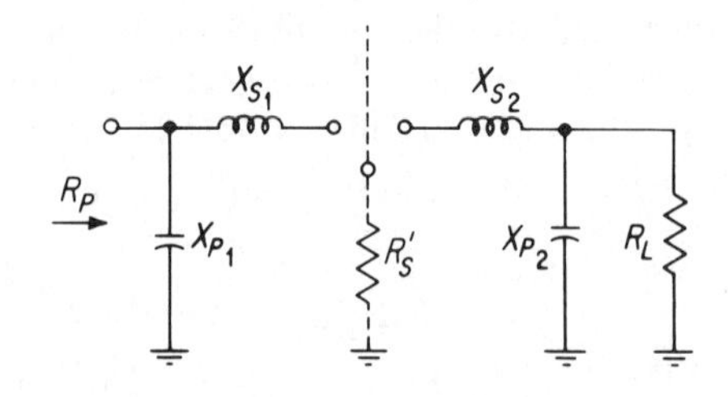

Fig. 11-11. Pi network separated into two L sections for ease of calculation. R'_S is an imaginary resistor used in the calculations.

Solve Network 1 as follows:

$$R'_S = \frac{R_P}{Q^2 + 1} \tag{11-17}$$

$$X_{S_1} = Q_l R'_S \tag{11-18}$$

$$X_{P_1} = \frac{R_P}{Q_l} \tag{11-19}$$

Q_l is the desired circuit Q, and R'_S is an imaginary series load resistor. This resistance is used only in the calculations and is not real. Its value will always be lower than either R_P or R_L. Now solve Network 2 as follows (note that Network 2 is drawn backwards from Network 1):

$$Q_m = \sqrt{\frac{R_L}{R'_S} - 1} \tag{11-20}$$

where Q_m is the Q necessary to effect a match from R'_S to R_{P_2}, and R_{P_2} is the load resistor. Then,

$$X_{S_2} = Q_m R'_S \tag{11-21}$$

$$X_{P_2} = \frac{R_L}{Q_m} \tag{11-22}$$

After the values for X_{P_1}, X_{S_1}, X_{S_2}, and X_{P_2} have been found, X_{S_1} and X_{S_2} are added together to form a single inductor.

Transformer Coupling

Where fairly high RF power is to be developed, it may be advantageous to resort to transformer coupling where the collector can be tapped down (connected as in an autotransformer) on the primary winding in order to maintain a high Q_l. For high-power designs, where the load impedance becomes low, the L and pi networks may yield values that may be impractical to fabricate. It is generally desirable to tap the collector as near to the top end of the tank circuit as possible, but the resultant size of the coil will dictate how far down the collector tap (or connection) is made.

Figure 11-12 shows a typical transformer-coupled RF stage for high-power applications. Note that the required reflected load resistance R_P is less than the actual load resistance R_L. If we assume a high unloaded Q and an output resistance of the transistor much higher than R_P, we may design the transformer in the following manner: Combining Eqs. 11-1 and 11-4 gives

$$X_L = \frac{V_{CC}^2\left(\frac{n_1}{n_2}\right)^2}{2P_{\text{out}}Q_l} \tag{11-23}$$

Then,

$$R_T = R_P\left(\frac{n_1}{n_2}\right)^2 \tag{11-24}$$

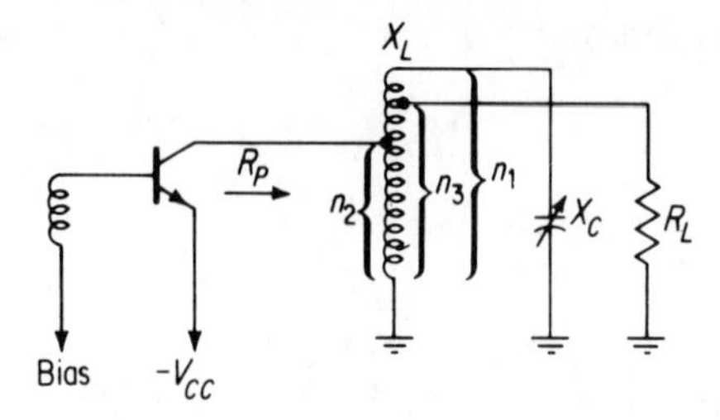

Fig. 11-12. Typical transformer-coupled high-frequency power amplifier.

(assuming the coefficient of coupling $k = 1$), where R_T is the impedance across the coil due to the transformation. The load tap may now be found from

$$R_L = R_T\left(\frac{n_3}{n_1}\right)^2 = R_P\left(\frac{n_3}{n_2}\right)^2 \tag{11-25}$$

OPERATION OF FREQUENCY MULTIPLIERS

The difficulty of obtaining VHF quartz crystals presents a serious problem to the VHF and UHF circuit designer. VHF overtone crystals become quite difficult to adjust as the frequency of oscillation is increased and they are very susceptible to damage.

One solution to this problem lies in the use of frequency multipliers, where a lower-frequency crystal controlled oscillator is employed and the higher-order harmonics are generated by the process of frequency multiplication. Output frequencies up to five times the input frequency have been successfully generated in a single multiplier stage.

A transistorized frequency multiplier is essentially a Class C amplifier with the output tank circuit tuned to a harmonic of the input signal. The transistors act as an amplifier for the input signal. The output tank circuit is not excited every cycle, as was the case with the standard Class C amplifier, but rather every second, third, fourth, etc., cycle depending upon the harmonic. This is illustrated in Fig. 11-13.

The efficiency of a frequency multiplier depends to a large extent on the conduction angle. As the conduction angle is decreased, the efficiency increases. One must, however, trade efficiency and output power to a certain extent. On the one hand, the conduction angle must be kept small to avoid loading the tank circuit because the transistor is held on for too large a portion of the output cycle. On the other hand, it is desirable to supply as much energy as possible to the tank circuit during conduction to improve the power output. Usually this compromise is best arrived at experimentally.

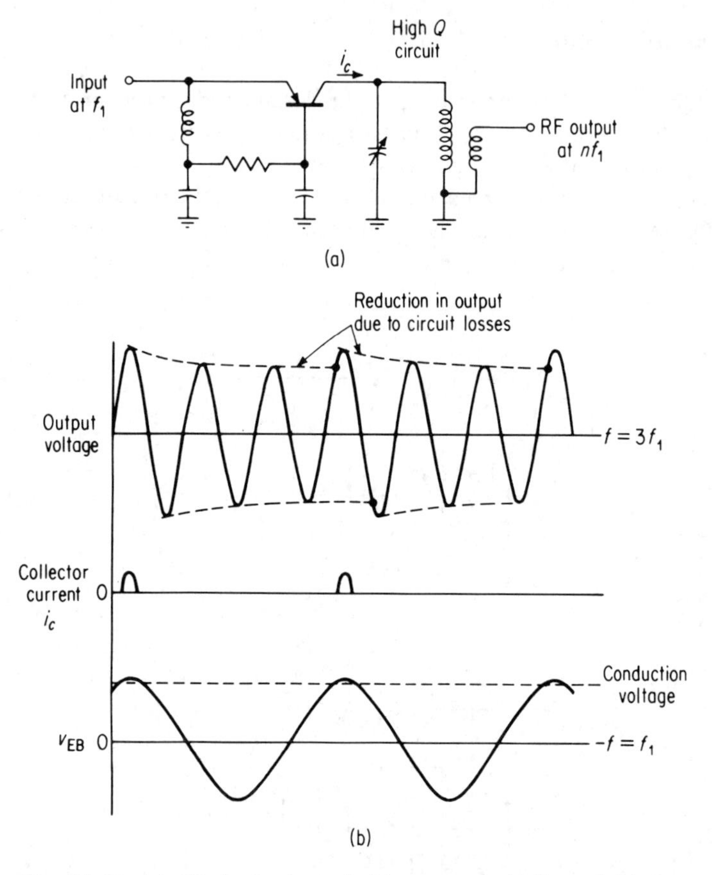

Fig. 11-13. (a) AC circuit of a typical frequency multiplier and (b) the current and voltage waveforms of a tripler. Conduction voltage is about 0.2 volt for germanium and 0.7 volt for silicon transistors.

In all cases the conduction angle should decrease as the order of multiplication increases. It should also be apparent that the output tank circuit must have a high Q, particularly with high-order multipliers, in order to avoid damping of the output wave between excitation pulses. (The effect of a low Q tank circuit was illustrated in Fig. 11-3.)

If the Class C power gain of a transistor amplifier is known, its power gain as a frequency multiplier can be roughly approximated from the equation

$$PG_n = \frac{PG_1}{n^2} \tag{11-26}$$

where n = harmonic to which the output is tuned
PG_n = multiplier power gain
PG_1 = Class C power gain of the amplifier at the fundamental frequency.

Basic Multiplier Circuits

Frequency multiplier circuits may be designed in a number of configurations. Although only the common-base connections are shown in the following illustrations, other configurations may be used.

When only even-order harmonics are desired, the circuit shown in Fig. 11-14(a) may be used. This is often called a *push-push circuit.* It is quite efficient because more excitation pulses are supplied to the tank circuit.

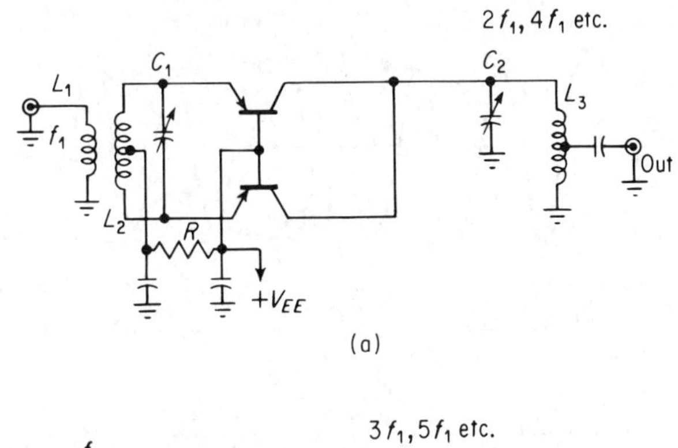

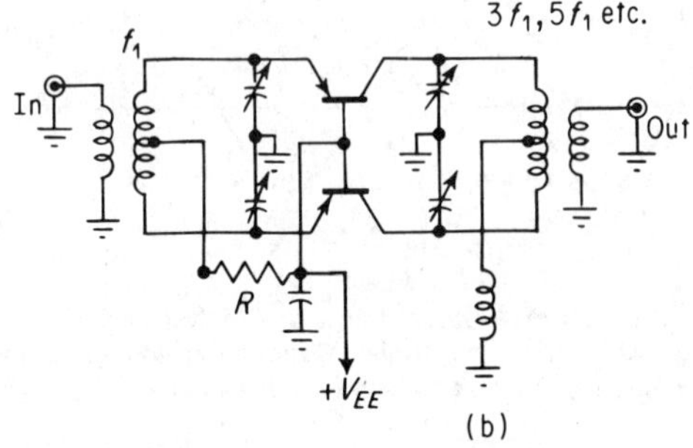

Fig. 11-14. Typical frequency multiplier circuits. (a) Push-push circuit; (b) push-pull circuit.

For suppression of even-order harmonics, the push-pull circuit shown in Fig. 11-14(b) should be used. Here, the even harmonics are canceled owing to the push-pull action of the stage and only odd-order frequency multiplication is possible.

By altering the shape of the exciting voltage from its usual sine waveform at the exciting frequency, it is possible to decrease the conduction angle and thus increase the efficiency without resorting to increases in the excitation voltage and bias.

The conduction angle may be decreased by adding some properly phased third harmonic voltage to the input. This circuit is illustrated in Fig. 11-15(a).

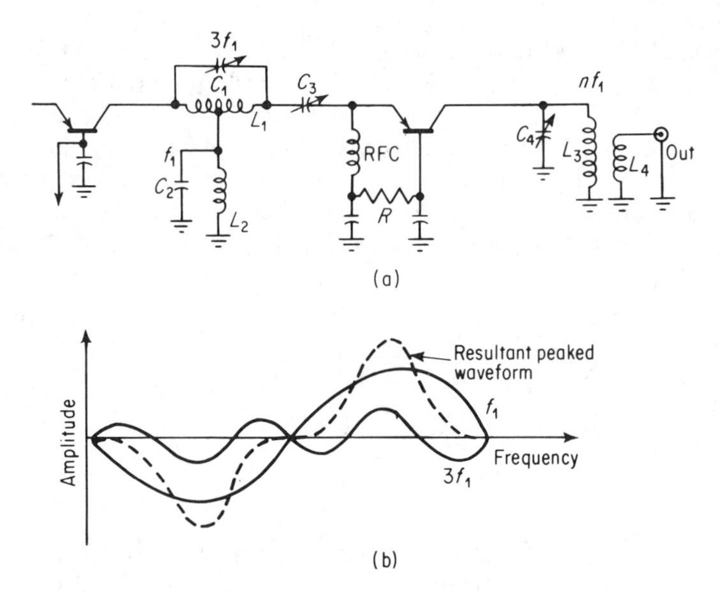

Fig. 11-15. (a) Frequency multiplier circuit which adds third harmonic to the fundamental frequency; (b) input signal showing the resulting waveform.

The result of adding the third harmonic voltage to the fundamental is shown graphically in Fig. 11-15(b). An excitation voltage with a peaked waveform will result, which reduces the conduction angle for a given bias level.

VHF CIRCUIT AND WIRING TECHNIQUES

Circuit Layout

As has been noted earlier, the design and construction of high-frequency circuits, particularly those using tuned circuits, is as much art as science. We include in this section a number of practical suggestions on how to minimize problems due to poor circuit layout. Observing all these hints is not guaranteed to solve every high-frequency circuit problem, but it will reduce the number of problems and give some insight into the nature of the effects caused by improper layout or choice of components.

Instability is frequently encountered in high-frequency work. This is the result of inadequate isolation between input and output, causing an excessive amount of feedback. Generally, there are two design approaches to the problem of inadequate isolation and reduction of feedback. One approach requires a compact unit where the designer must resort to complicated shielding, layout, and filtering. The other employs a larger unit and uses proper placement of the components to achieve the same end.

Occasionally a circuit develops instability by going into oscillation, not at the frequency of interest but at a much higher frequency. The heavy-lined portions in Fig. 11-16(a) illustrate the possible path for such a parasitic oscillation. L_2 merely acts as an RF choke to the high frequency, and the lead from the collector terminal to the tank circuit is series resonant with capacitor C_3. C_2 and part of L_1 form a series-resonant circuit to about the same frequency as the output. By changing the value of C_2 or the tap on L_1, this effect can be minimized.

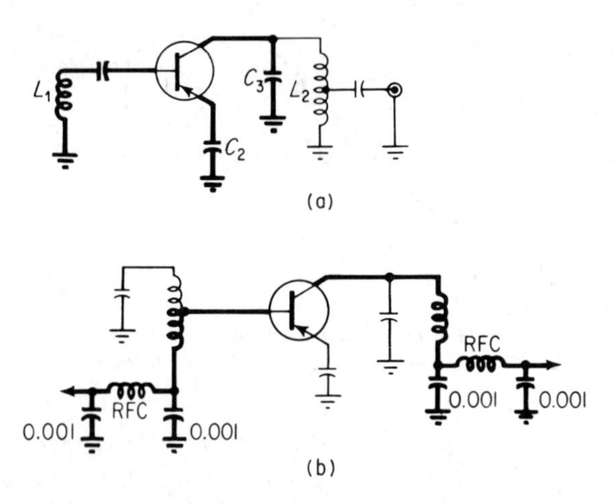

Fig. 11-16. Possible causes of parasitic oscillations in tuned amplifier circuits. (a) High-frequency parasitic circuit; (b) low-frequency parasitic circuit.

Very low-frequency parasitic oscillation can be developed by free use of high Q RF chokes as shown in Fig. 11-16(b). The heavy lines trace out the possible path to set up the oscillation. By removing one of the RF chokes or loading them to very low Q's with paralleled resistors, this form of parasitic oscillation can be eliminated.

Figure 11-17(a) illustrates a UHF circuit where the base terminal is mounted on a common partition made of reasonably thick copper. Capacitors are fabricated from part of the tank inductors L_1 and L_2. The teflon pieces are used as the dielectric material. At first glance, this technique may look undesirable. This would be true if a frequency of about 30–100 MHz is considered, as shown in Fig. 11-17(c). Here, the partition will become a common coupling between the input and output circuits.

At frequencies above 200 MHz this may not be true. The skin effect begins to become important, and as a result, the input and output base return circuits will be isolated as shown in Fig. 11-17(b).

Instability due to feedback in the RF portion can be minimized by proper placement of the RF coils, and by phasing of the windings for minimum

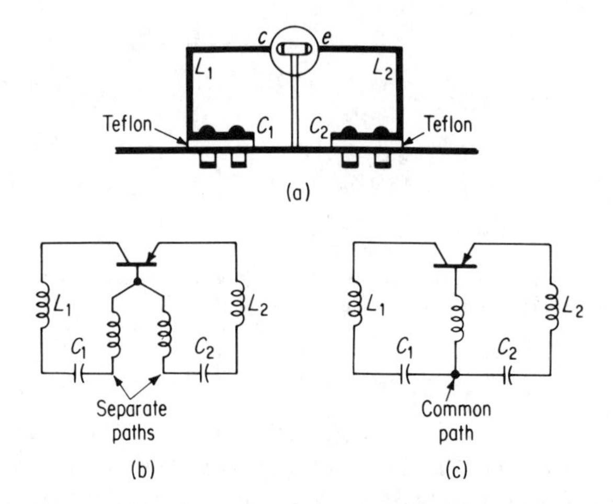

Fig. 11-17. UHF circuit illustrating how skin effect can isolate circuits using a common return path at high frequencies. (a) Actual circuit; (b) equivalent circuit for $f > 200$ MHz; (c) equivalent circuit for $f < 200$ MHz.

undesired coupling between the components. Figure 11-18(a) illustrates the wrong way to place the components; Fig. 11-18(b) illustrates the right way. Complete shielding from external fields is usually necessary. This may be achieved by building the high-frequency circuit on a metal plate and then fitting it into a chassis or metal box.

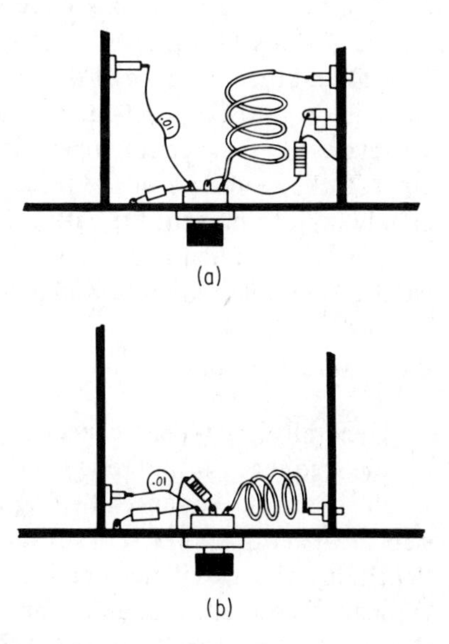

Fig. 11-18. Sketch showing (a) poor layout of components and (b) improved circuit.

At ultrahigh frequencies, the tank inductors are an appreciable part of $\frac{1}{4}$ wavelength, thereby making it a good radiator. By complete shielding, the radiated power is returned to the tank circuit; hence more power will be delivered to the output connector. The power leads that leave the chassis or box should go through a low-pass filter, or at least through a good grade feed-through capacitor, in order to contain the signal in the box. The only way that the signal should leave the box is through the output connector. Figure 11-19(a) shows a low-pass filter using an *LC* circuit; Fig. 11-19(b) shows an *RC* low-pass filter.

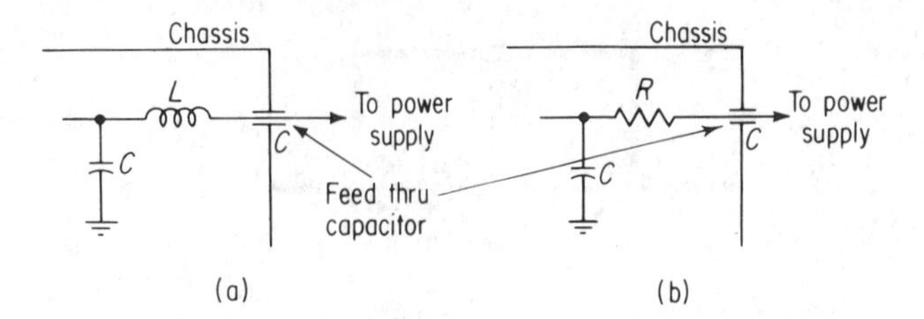

Fig. 11-19. (a) Typical LC and (b) RC low-pass filters for supply leads to high-frequency circuits.

Lead Dress

Circuit designers tend to overlook lead impedance when wiring components onto transistor sockets. The transistor is considered a low-impedance device but that does not necessarily mean that higher-frequency wiring techniques should be overlooked. In fact, this could be just the opposite. A piece of hook-up wire with its inherent distributed inductance can cause serious RF losses if not kept as short as possible.

Consider a 10 MHz tank circuit connected to the collector by a piece of wire having a self-inductance of about 0.1 microhenry. The load impedance is 2000 Ω and the reactance of the piece of wire is 6.3 Ω. This reactance is quite small in proportion to 2000 Ω so that in most cases it can be neglected. When, however, the tank circuit operates at 200 MHz with a load impedance of 2000 Ω, the reactance of the piece of wire is 125 Ω. This reactance is sufficiently high to prevent the entire signal at the collector from being applied to the tank circuit; hence, a loss in gain results. Also, this inductive component can cause a phase lag which may have undesirable effects.

Capacitors and Coils

Generally, the type of capacitor that can be used in a given circuit depends on the frequency of interest. Capacitors have some series inductance which results in series resonance at a frequency above which they cannot be used. A paper capacitor has a series-resonant frequency ranging from 1–10 MHz, depending on capacitance and lead length. A mica capacitor will have series-resonant frequencies ranging from 10–100 MHz. A good capacitor to use at the higher frequencies is the ceramic type; its series-resonant frequencies run as high as 500 MHz. Special ceramic capacitors are now available which are useful up to about 1200 MHz.

When paralleling capacitors to increase the total bypass capacitance, be careful that the resultant combination is not parallel resonant at the frequency of operation. Figure 11-20 shows how to check the parallel-resonant frequency. When wiring capacitors into a circuit, use the shortest lead possible

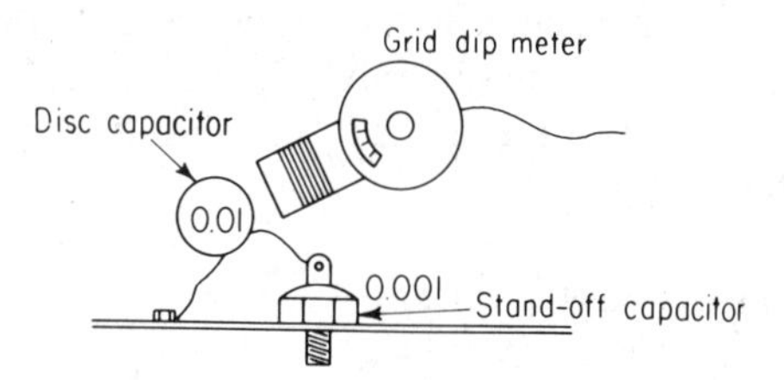

Fig. 11-20. Use of a grid-dip meter to check the parallel-resonant frequency of capacitors.

in order to reduce the lead inductance. Because the coils form an important part of any higher-frequency circuit, special consideration must be given to their fabrication.

Where coils are to be wound on forms, the material should be of high-frequency grade, such as ceramic. High unloaded Q's are usually desirable. In certain broad-band applications, it is necessary to minimize the shunt capacitance in order to keep the gain high. By winding the turns with sufficient spacing, the distributed capacitance can be minimized. The core should mesh with the low potential end and should be insulated from the chassis ground. In cases where neutralization is employed, a third winding should be added.

Avoid the low-frequency technique of using the secondary winding for neutralizing as well as for feeding the input of the succeeding stage. In order to provide a phase reversal, the ground end of the secondary winding is reversed, thereby causing the ground end to mesh with the high potential end of the primary and increasing the total distributed capacitance to ground. Figure 11-21 illustrates the three-winding method as compared to the two-winding method.

Another point worth mentioning is that once the value of C_N is determined, it is desirable not to disturb the position of the core with respect to

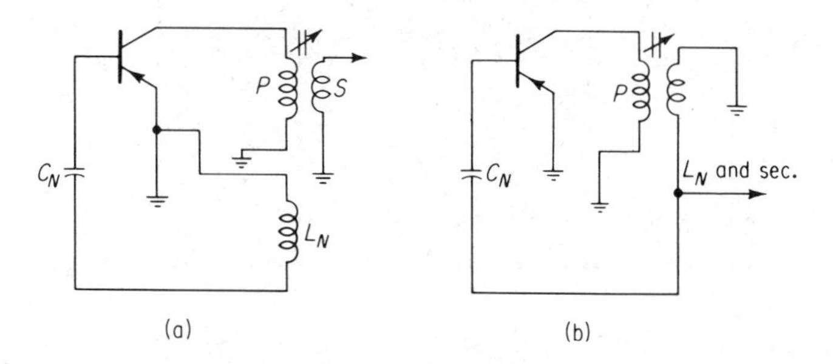

Fig. 11-21. Neutralization using two (b) and three (a) transformer windings. Three windings are preferred at high frequencies to reduce distributed capacitance and tuning interaction.

the neutralizing winding, since any change in the value of L_N will upset the neutralization if C_N remains the same. It would also be desirable to resonate the primary winding with the insulated core, just barely meshing with the collector end of the primary.

Choice of Circuit Elements (Fig. 11-22)

30–75 MHz range: In this portion of the radio frequency spectrum, most amplifiers are similar to those of the lower frequencies. Coils should be silver-plated and the turning capacitors should be of the high-Q type. Small butterfly and glass piston types are recommended.

75–450 MHz range: The transition from lumped property components to the distributed types begins in this portion of the radio frequency spectrum. Where lower electrical efficiency and instability are not especially important,

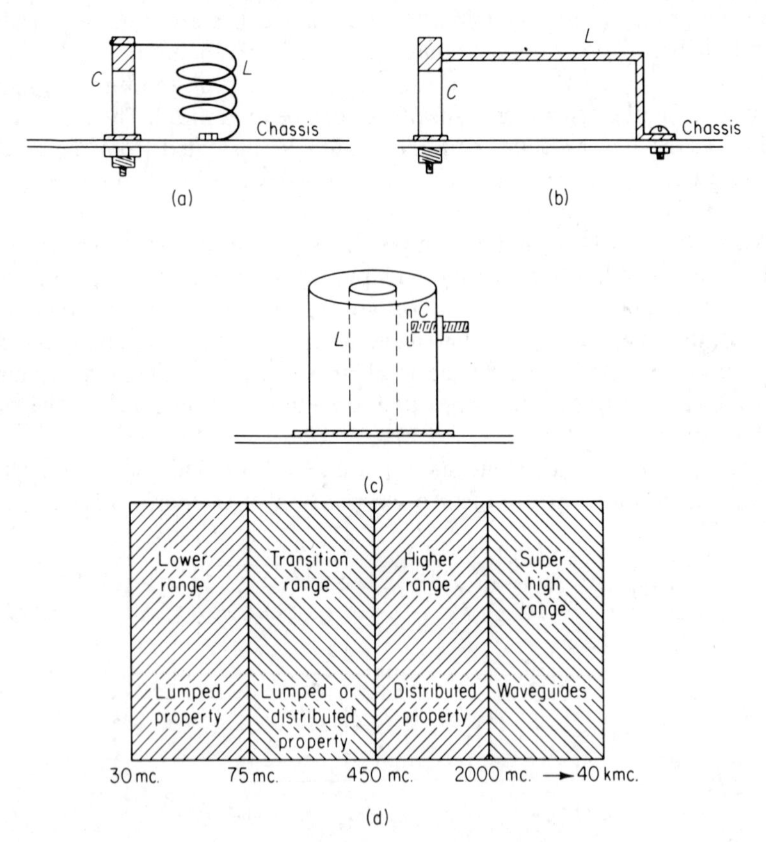

Fig. 11-22. Examples of LC elements. (a) Lumped; (b) distributed—stripline; (c) coaxial; (d) an approximate frequency chart illustrating the frequency range in which each is used.

lumped components can still be used. When efficiency and stability are important, distributed tank circuits are required. Use silver-plated and highly polished tank elements where possible. Any discontinuities in the tank components tend to lower the value of the unloaded Q.

450–2000 MHz range: At frequencies above 450 MHz, distributed tank circuits are used almost exclusively. The strip-line, Fig. 11-22(b), and coaxial types, Fig. 11-22(c), are those most commonly used, since their size at these frequencies is not objectionable. Most UHF receiving and transmitting circuits employ strip-line tank circuit.

MICROWAVE CIRCUIT BIASING

Often, the least considered factor in microwave transistor circuit design is the bias network. Considerable effort is spent in measuring s-parameters, noise figure, while the same resistor topology is used to bias the transistor. Since the cost per db of microwave gain or noise figure is so high, the circuit designer cannot afford to sacrifice rf performance by inattention to dc bias considerations. At low frequencies, emitter resistor stabilization with negative current feedback is used for dc stability. At microwave frequencies, the bypass capacitor becomes a problem since a good rf bypass at the design frequency often introduces low frequency instability and gives rise to bias oscillations.

In low-noise amplifier applications, even if a capacitor could be chosen to provide effective rf and low frequency emitter bypass, any small series emitter impedance at the operating frequency would reflect in a large noise figure degradation. Most microwave circuits, designed for best gain or lowest noise figure, will require that the emitter lead be dc grounded as close to the package as possible keeping the emitter series feedback at an absolute minimum.

In microwave transistors a small component of reverse current flow is a conventional I_{CBO} term but the major contributor is a surface current that flows across the top of the silicon crystal lattice. This surface current is a more linear function of temperature than the I_{CBO} current. The total reverse current, made up of I_{CBO} and the surface current, increases at a rate much less than that which would be expected from an I_{CBO} current of the same magnitude. A typical reverse current versus temperature relationship for a microwave transistor is shown in Fig. 11-23. The data applies to a collector-base voltage of 10 v.

A look at the four scattering-parameters and noise figure of microwave transistors reveals that s_{21}^2 and noise figure stand out as the most sensitive parameters to small changes in bias. Also, both of these parameters are stronger functions of collector current (I_C) than of collector to emitter voltage (V_{CE}). This means that if we know something about how s_{21}^2 and noise figure

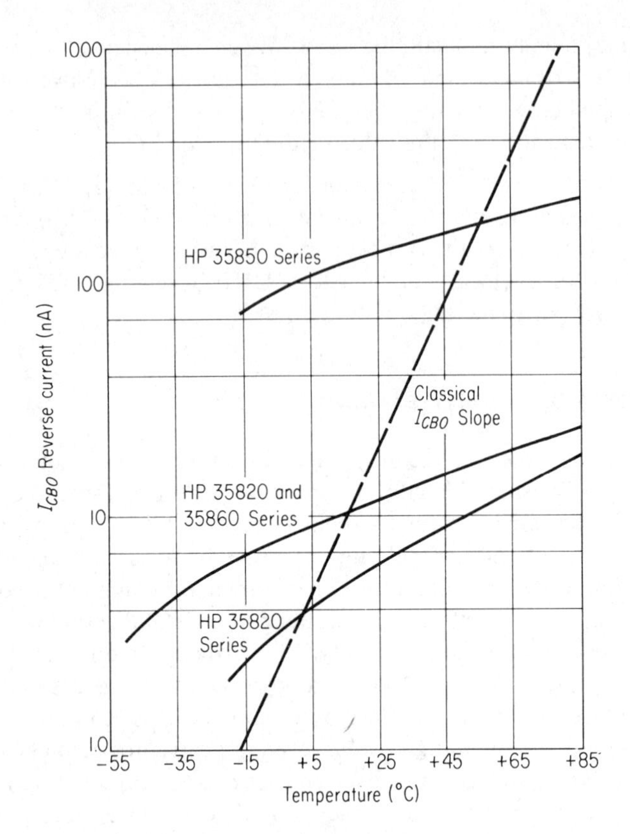

Fig. 11-23. The effect of surface current in microwave transistors is shown in a typical reverse current vs temperature for a microwave transistor. The total reverse current, I_{CBO} is the sum of surface and junction currents.

change with collector current and with temperature, some constraints can be placed on the bias network to minimize changes in rf performance over a specified temperature range.

The typical data shown in Fig. 11-24(a) is normalized to per cent change in both gain and collector current. The noise figure change is plotted in db. Although the absolute gain and noise figure are frequency dependent, their sensitivity with respect to collector current (I_C) can be considered frequency independent. Notice that a 20 per cent increase in collector current has a very small effect on either a transistor biased for minimum noise figure or a transistor biased for maximum gain.

Next, a look at some typical changes in noise figure and gain as a function of temperature [Fig. 11-24(b)] shows that both NF and gain degrade with increasing temperature. We see, for example, that a bias network that can hold the quiescent point such that the current does not increase more than 20

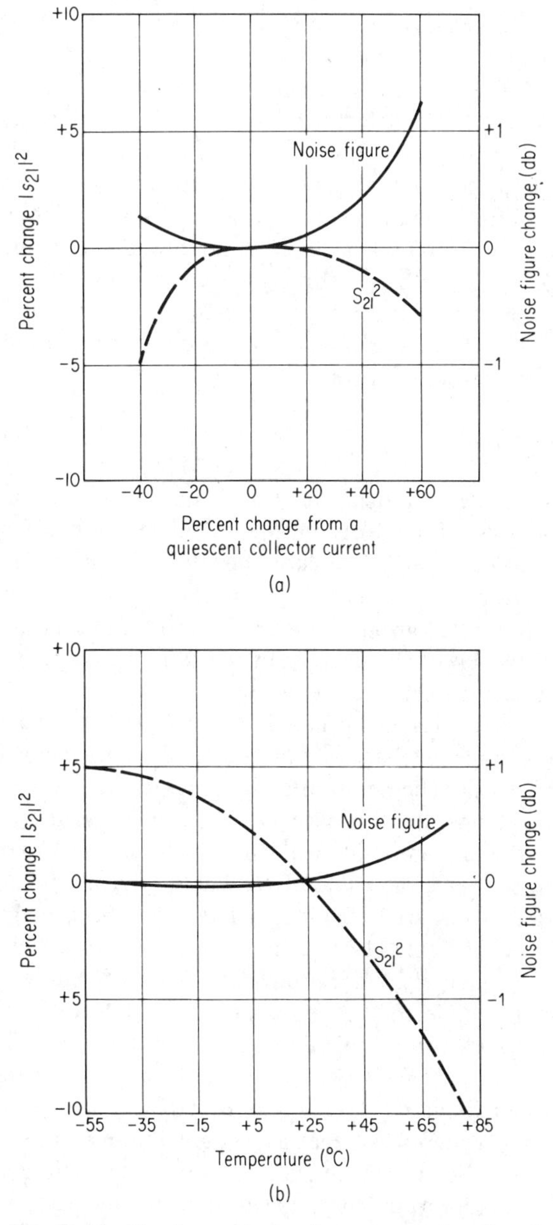

Fig. 11-24. The changes in rf performance as a function of collector current variation (a) and the temperature (b) show how circuit stability is affected by the bias network.

per cent to 60°C, will have a 5 per cent degradation in gain or a 0.3 db increase in noisc figure at 60°C due to transistor changes alone. Some temperature compensation could be designed into the bias circuitry by using lower values of collector current at 25°C and allowing the temperature sensitivity of the bias network to offset the temperature sensitivity of the transistor.

It should be pointed out that each amplifier function has a different bias requirement. For example, transistors used in gain stages in which the noise figure or the saturated output power are not critical have a much more relaxed bias stability requirement than a low noise front-end transistor. This can be seen in Fig. 11-24a since s_{21}^2 has a broad maximum compared to noise figure. A transistor biased for high linear output power must hold its quiescent point such that the 1 db compression point is not degraded with temperature and so that the maximum power dissipation in the device is not exceeded with increasing temperature.

Suggested Bias Circuits

Three bias circuit topologies are shown in Table 1 along with general expressions for collector currents and the calculated dc bias stability factors for V_{EB}, I_{CBO} and h_{FE}. Emitter resistor stabilized circuits are not considered since most microwave circuit designs, for reasons of noise figure, gain and rf stability, will require a dc ground emitter. The grounded emitter nonstabilized bias circuit (Table 11-1A) finds very little use in microwave circuit design since it exhibits the least dc bias stability.

The circuits in Tables 11-1B and 11-1C find wide-spread usage as bias networks. The voltage feedback circuit uses fewer components and is almost as temperature stable. The addition of R_{B1} and R_{B2} to the voltage feedback circuit does two things. First, it makes all the element values lower in resistance and therefore is compatible with thin/thick film resistor values. In the voltage feedback circuit, the value of R_B would typically be in the range of 30 to 100 k, values that are difficult to achieve in hybrid integrated circuits. Second, the circuit of Table 11-11c can be considered to have a constant base current source through R_B. On a production basis, this allows for trimming to initially set the collector current to some desired value. The collector current cannot be measured directly since the current in R_C is made up of base current, base bias network current, and collector current. However, since I_c is proportional to V_{CE}, monitoring V_{CE} while adjusting R_B can accommodate any value of h_{FE} that is encountered in transistors.

Differences in collector current stability for each topology are compared in Fig. 11-25. It is important to point out that, for the sake of comparison, each circuit was used to bias the transistor at a common quiescent point. This data is typical for frequently encountered microwave bias circuits and is valuable for relative comparisons. Notice that for each of the circuits, the

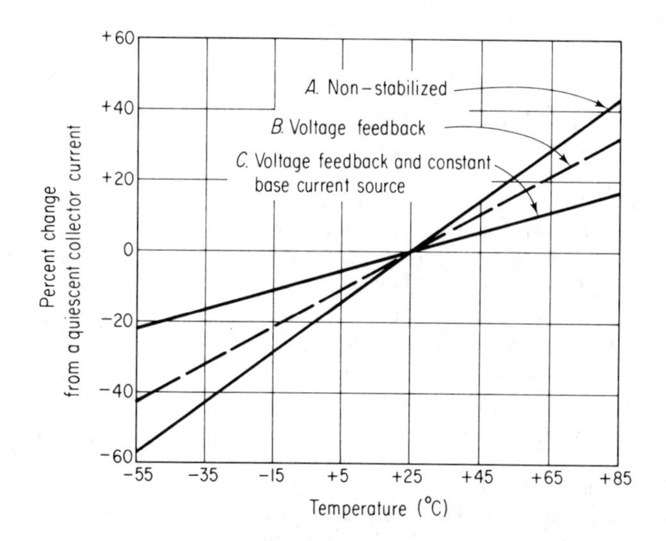

Fig. 11-25. These three bias circuits are commonly used in microwave applications. The graph shows the percent change from a nominal quiescent collector current as a function of temperature (normalized to 25°C).

collector current is a positive linear function of temperature. And from Fig. 11-24(a), we see that noise figure and gain are negative functions of both collector current and junction temperature.

EXAMPLE

Here is an example of how to calculate the resistor values for a voltage feedback, constant base current source bias circuit. The resistor values are calculated and the nearest 1% tolerance resistor values are shown.

1. Determine supply voltage available ($V_{CC} = 20$ v) and transistor bias operating point (10 v, 10 mA). Assume $I_{CBO} = 0$, $V_{BE} = 0.7$ v.
2. Select V_{BB} to be 2 v to ensure constant base current source.
3. Knowing the measured value of h_{FE}, (or assume 50), calculate base current $I_B\text{ (mA)} = I_C\text{ (mA)}/h_{FE} = 10/50 = 0.2$ mA.
4. Calculate R_B knowing $V_{BE} = 0.7$ v and $V_{BB} = 2$ v

$$R_B = \frac{V_{BB} - V_{BE}}{I_B} = \frac{2 - 0.7}{0.2 \times 10^{-3}} = 6.5\ \text{K}\Omega\ (\text{use } R_B = 6.81\ \text{k}\Omega)$$

5. Calculate R_{B2} assuming $I_{BB} = 1$ mA

$$R_{B2} = \frac{V_{BB}}{I_{BB}} = \frac{2}{10^{-3}} = 2\ \text{k}\Omega\ (\text{use } R_{B2} = 1.96\ \text{k}\Omega)$$

6. Now calculate R_{B1} knowing I_B, I_{BB}, V_{BB} and V_{CC}

$$R_{B1} = \frac{V_{CE} - V_{BB}}{I_{BB} + I_B} = \frac{10 - 2}{1.2 \times 10^{-3}} = 6.66\ \text{k}\Omega\ (\text{use } R_{B1} = 6.19\ \text{k}\Omega)$$

TABLE 11-1 Using the Stability Factors

First calculate the stability factors for V'_{BE}, I_{CBO}, and h_{FE} as shown below. Then, to find the change in collector current at any temperature, multiply the change from 25°C of each temperature dependent variable with its corresponding stability factor and sum:

$$\Delta I_C = \frac{\partial I_C}{\partial I_{CBO}} \cdot \Delta I_{CBO} + \frac{\partial I_C}{\partial V'_{BE}} \cdot \Delta V'_{BE} + \frac{\partial I_C}{\partial h_{FE}} \cdot \Delta h_{FE}$$

$$= SI_{CBO} \cdot \Delta I_{CBO} + SV'_{BE} \cdot \Delta V'_{BE} + Sh_{FE} \cdot \Delta h_{FE}$$

It would appear to be an easy task to further analyze the individual stability factors for minimums in terms of the external circuit resistor values. This is not too easily done since all the factors are interrelated. The stability factors must be considered simultaneously since an optimum set of resistor values to minimize one parameter could grossly increase another.

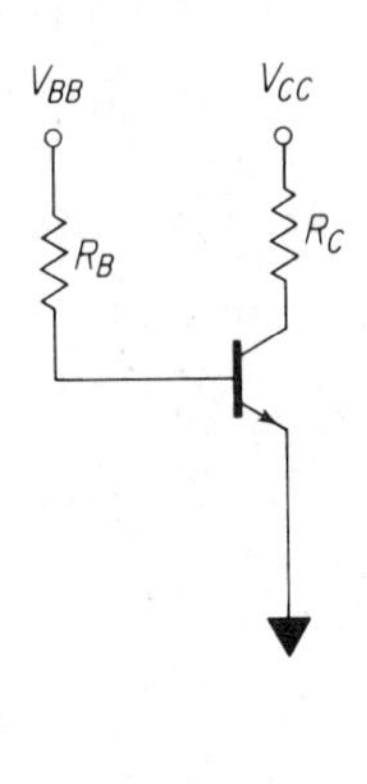

(a)

A. Non-Stabilized

	A. Non-Stabilized
Collector current at any temperature (I_C):	$\frac{h_{FE}(V_{BB} - V'_{BE})}{(h_{ie} + R_B)} + I_{CBO}(1 + h_{FE})$
I_{CBO} Stability factor: $S_{ICBO} = \left.\frac{\partial I_C}{\partial I_{CBO}}\right\vert_{h_{FE}, V'_{BE}=\text{Constant}}$	$1 + h_{FE}$
V'_{BE} Stability factor: $S_{V'_{BE}} = \left.\frac{\partial I_C}{\partial V'_{BE}}\right\vert_{I_{CBO}, h_{FE}=\text{Constant}}$	$\frac{-h_{FE}}{h_{ie} + R_B}$
h_{FE} Stability factor: $S_{hFE} = \left.\frac{\partial I_C}{\partial h_{FE}}\right\vert_{I_{CBO}, V'_{BE}=\text{Constant}}$	$\frac{V_{BB} - V'_{BE}}{h_{ie} + R_B} + I_{CBO}$

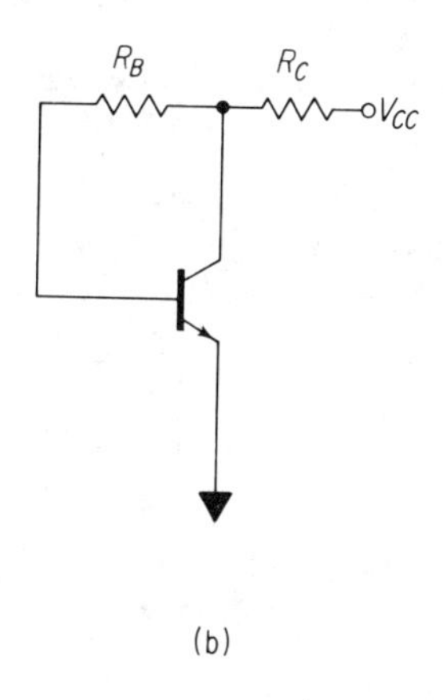

(b)

R_{B1} R_C V_{CC} R_B R_{B2}

(c)

B. Voltage Feedback	C. Voltage Feedback and Constant Base Current Source
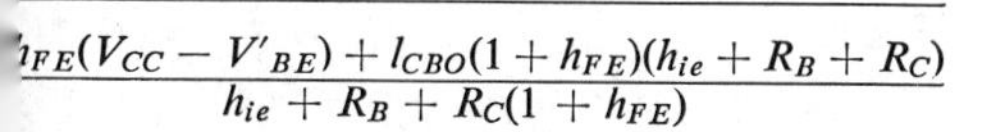 $\frac{h_{FE}(V_{CC}-V'_{BE})+I_{CBO}(1+h_{FE})(h_{ie}+R_B+R_C)}{h_{ie}+R_B+R_C(1+h_{FE})}$	$h_{FE}\left\{\frac{-V_{BE}A-R_{B2}[R_C I_{CBO}(1+h_{FE})-V_{CC}]}{(R_B+h_{ie})A+R_{B2}(h_{FE}R_C+R_C+R_{B1})}\right\}$ $+ I_{CBO}(1+h_{FE})$
$\frac{1+h_{FE})(h_{ie}+R_B+R_C)}{h_{ie}+R_B+R_C(1+h_{FE})}$	$(1+h_{FE})-\left(\frac{R_{B2}H_{FE}R_C(1+h_{FE})}{A(R_B+h_{ie})+R_{B2}(h_{FE}R_C+R_C+R_{B1})}\right)$
$\frac{-h_{FE}}{_{ie}+R_B+R_C(1+h_{FE})}$	$\frac{-h_{FE}A}{(R_B+h_{ie})A+R_{B2}(h_{FE}R_C+R_C+R_{B1})}$
$\frac{h_{FE}R_C+R_B+h_{ie}+R_C)(V_{CC}-V'_{BE}+KI_{CBO})}{D^2}$ $-R_C\left[\frac{(h_{FE}(V_{CC}-V'_{BE}+KI_{CBO})+KI_{CBO}}{D^2}\right]$ here $K = h_{ie}+R_B+R_C$ and $D = h_{FE}R_C+R_B+h_{ie}+R_C$	$h_{FE}\left\{\frac{R_{B2}R_C(V_{CC}-B)-I_{CBO}(C+R_{B2}R_{B1})}{D^2}\right\}$ $+\left\{\frac{B-R_{B2}[R_C I_{CBO}(1+h_{FE})-V_{CC}]}{D}\right\}$ where $A = R_{B1}+R_{B2}+R_C$ $B = -V'_{BE}(R_{B1}+R_{B2}+R_C)$ $C = (R_B+h_{ie})(R_{B1}+R_{B2}+R_C)$ $D = C+R_{B2}(h_{FE}R_C+R_C+R_{B1})$

7. Calculate R_C knowing I_C, $I_{BB} + I_B$, V_{CC} and V_{CE}

$$R_C = \frac{V_{CC} - V_{CE}}{I_C + I_{BB} + I_B} = \frac{20 - 10}{11.2 \times 10^{-3}} = 0.893 \text{ k}\Omega \text{ (use } R_C = 909 \text{ k}\Omega\text{)}$$

8. The general equation for collector current can now be used to check the design using actual resistor values.
9. After the circuit is designed, R_B may be adjusted to obtain an exact value of I_C.

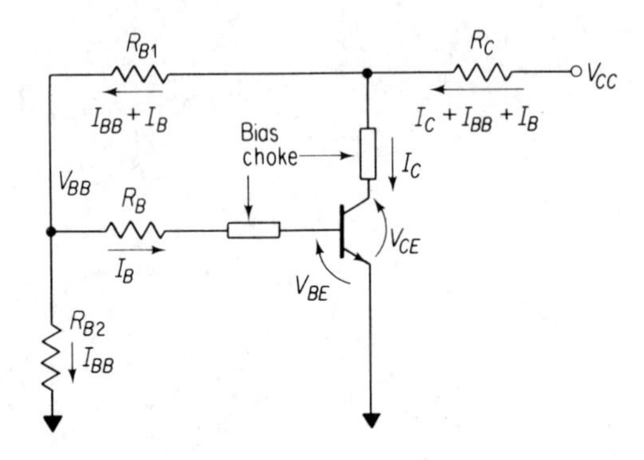

EXERCISES

1. Calculate the current which flows in a tank circuit when the tank circuit is excited by a step input voltage.

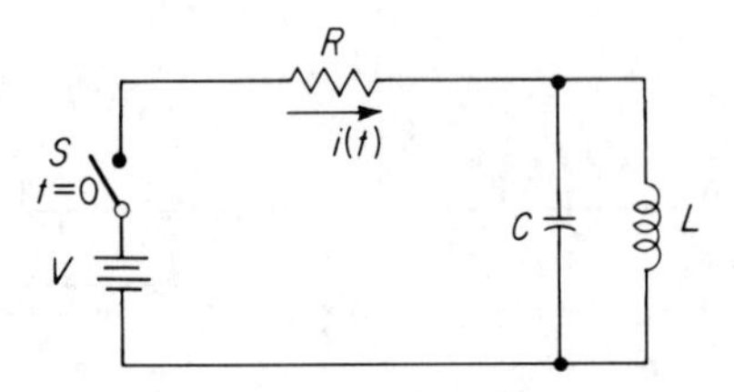

2. Given the following common-base Class C amplifier circuit:

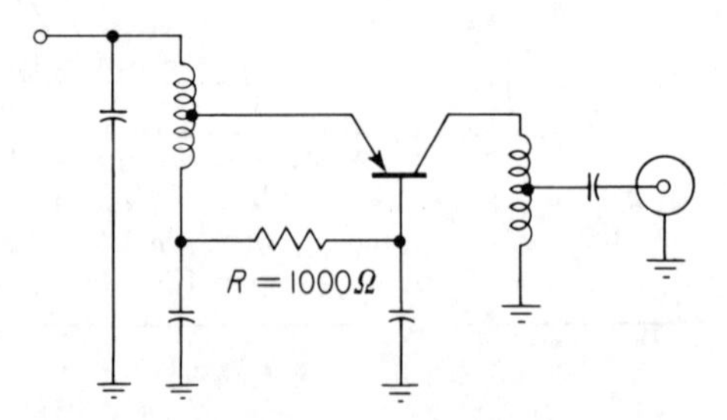

The collector current is 1.85 ma, V_{pp} is 4.4 volts, and a germanium transistor is being used. Find the conduction angle.

3. The output stage shown in the figure is to be designed for Class C amplifier operation.

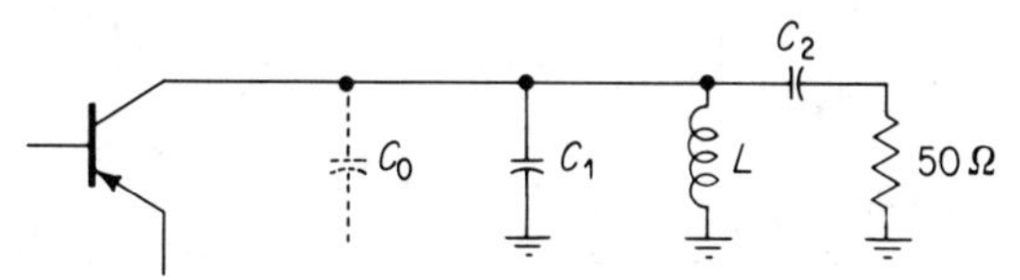

Given a supply voltage of 20 volts, output power of 50 mw, frequency of 30 MHz, load Q of 10, and C_{out} of 1 pf, calculate R_L, C_1, C_2, and L.

REFERENCES

1. J. F. PIERCE, *Transistor Circuit Theory and Design.* Columbus, Ohio.: Charles E. Merrill Books, Inc., 1963.
2. T. L. MARTIN, JR., *Ultrahigh Frequency Engineering.* Englewood Cliffs, N.J.: Prentice-Hall, Inc., 1950.
3. L. B. ARGIMBEAU, *Vacuum Tube Circuits and Transistors.* New York: John Wiley & Sons, Inc., 1956.
4. R. F. SHEA *et al., Transistor Circuit Engineering.* New York: John Wiley & Sons, Inc., 1957.
5. J. D. RYDER, *Electronic Fundamentals and Application.* Englewood Cliffs, N.J.: Prentice-Hall, Inc., 1976.
6. A. W. LO *et al., Transistor Electronics.* Englewood Cliffs, N.J.: Prentice-Hall, Inc., 1961.
7. J. R. MILLER *et al., Communications Handbook.* Dallas, Tex.: Texas Instruments, Inc., 1965.

12

Transistor Noise Characteristics

INTRODUCTION

The term *noise*, as applied to electronic circuitry, refers to any spurious signals which ultimately appear at the output terminals of a system. The noise may be produced by sources external to the electronic circuitry, and/or by internal noise sources in the equipment itself.

External noise may arise from such sources as atmospheric lightning discharges, automobile ignition systems, X-ray machines, and motors. Noise due to external sources may be minimized by proper shielding of either the noise source, the equipment, or of both.

Internal noise may arise from such sources as undesired oscillations, power supply ripple, and microphonics. Even if all the foregoing sources of noise are eliminated, there remain present in any electronic equipment two fundamental sources of noise which ultimately limit the useful amplification to be obtained in the equipment.

Perhaps the most fundamental source of spontaneous fluctuations, or noise, in electrical circuits is "thermal" noise. Consider the carbon resistor shown in Fig. 12-1. At any temperature above absolute zero, the conduction electrons in the resistor are in a state of random motion, due to thermal

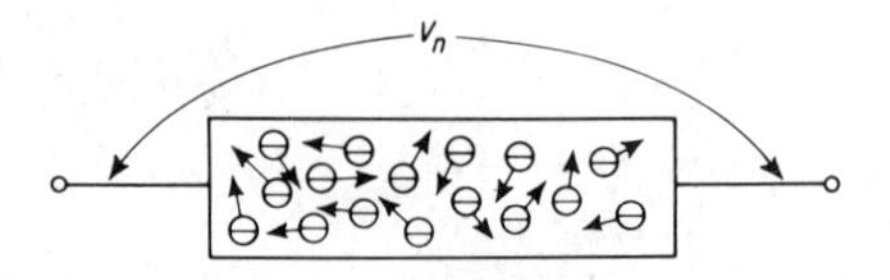

Fig. 12-1. Illustration of random motion of conduction electrons in a carbon resistor.

agitation. Even with no external current flowing in the device, an instantaneous voltage appears at the terminals of the resistor, owing to the random motion of the electrons. In other words, at a particular instant of time, if there are more electrons with a component of velocity toward the left than toward the right side of the resistor, the right-hand terminal is instantaneously more positive than the left-hand terminal. Of course, the average value of the voltage appearing between the resistor terminals is zero, in keeping with the principle of conservation of energy; hence, it is convenient to use the rms value of the thermal noise voltage. The square of the average thermal noise voltage appearing between the terminals of a resistance (R) can be shown to be (1):*

$$\overline{v_n^2} = \int_0^\infty 4kTR\frac{hf}{kT}\left[\frac{1}{e^{hf/kT}-1}\right]df \qquad (12\text{-}1)$$

where h is Planck's constant ($h = 6.62 \times 10^{-34}$ joule-sec), f is frequency in hertz and all other symbols have their previously defined meanings. For all practical frequencies, Eq. 12-1 reduces to

$$\overline{v_n^2} = 4kTR\,\Delta f \qquad (12\text{-}2)$$

where Δf is the equivalent noise bandwidth in hertz. In going from Eq. 12-1 to Eq. 12-2, the frequency integral from zero to infinity was replaced by the equivalent noise bandwidth, Δf. In order to justify this substitution, the equivalent noise bandwidth must be strictly defined as

$$\Delta f = \frac{\int_0^\infty G(f)\,df}{G_0} \qquad (12\text{-}3)$$

where $G(f)$ is the power frequency response of the system, and G_0 is the peak value of $G(f)$. The equivalent noise bandwidth may be represented graphically as an ideal square bandpass having the same peak amplitude and the same area as the actual power frequency response, as shown in Fig. 12-2.

The rms noise voltage output of an impedance may also be obtained from Eq. 12-2 by replacing R by the real part of the impedance, Re[Z], since that is the resistive component.

The other fundamental noise source present in electronic circuitry is called *shot* noise (2), and is caused by the random entrance and exit of charge carriers passing through a region which is otherwise depleted of free charge carriers. Perhaps the best-known example of shot noise is the temperature-limited vacuum tube diode. In the tube diode, electrons are "boiled" off the cathode by a high-temperature filament. Although the time-averaged emission rate is constant, instantaneous fluctuations occur. This random emission of electrons from the tube cathode produces an rms fluctuation in the plate

*Numbers in parentheses refer to material listed at the end of the chapter.

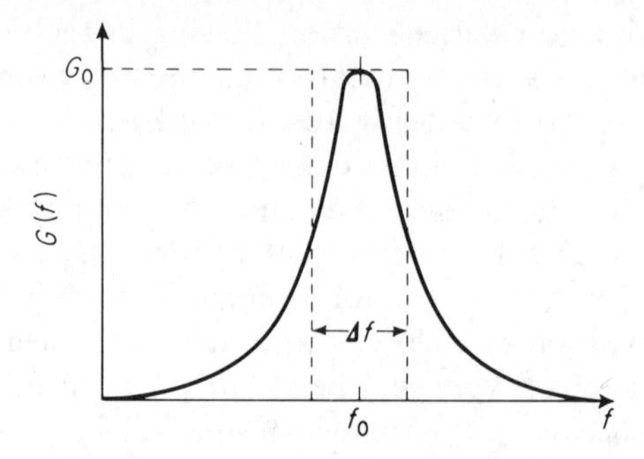

Fig. 12-2. Graphical representation of the average noise bandwidth as a square wave whose amplitude is G_0 and bandwidth is Δf.

current. The square of the equivalent rms noise current has been shown (2) to be

$$\overline{i^2} = 2qI_p\,\Delta f \tag{12-4}$$

where I_p is the average plate current in amperes.

Shot noise is not limited to vacuum tubes. The random passage of charge carriers through the depletion region of a semiconductor diode also produces an equivalent rms shot noise current. The square of this shot noise current is also given by Eq. 12-4, where I_p is the average, or dc current through the diode.

Although Eqs. 12-3 and 12-4 were developed only for the thermal noise of a resistor and the shot noise of a diode, respectively, their use is not limited to only these elements. For instance, a generator whose output contains a noise voltage may be said to have an equivalent noise resistance equal to

$$R_{eq} = \frac{\overline{v_o^2}}{4kT\Delta f} \tag{12-5}$$

where $\overline{v_o^2}$ is the mean square output noise voltage of the generator. One can also consider the case of an electronic receiver which is driven from a 75 Ω antenna, where, in addition to the desired signal received from the antenna, atmospheric background radiation is also presented to the receiver input terminals. Therefore, the 75 Ω antenna has a noise output voltage greater than the thermal noise voltage of a 75 Ω resistance. In such cases, it is often convenient to consider the antenna temperature to be the variable, and the antenna is said to have an equivalent noise temperature equal to

$$T_{eq} = \frac{\overline{v_o^2}}{4k\Delta f(75)} \tag{12-6}$$

Thus, the equivalent noise temperature of the antenna will be higher than the actual ambient temperature, if the antenna noise-output voltage is greater than the thermal noise voltage of a 75 Ω resistance. In addition to these sources of noise, semiconductors have an additional type of noise. This is thought to be due to the diffusion of carriers into and out of traps or recombination centers, and is generally called $1/f$ noise since its energy varies inversely with the frequency. This is troublesome only at low frequencies and can now, in most cases, be reduced to such a low level by suitable fabrication techniques that it is not generally included in design specifically, but is lumped in with other sources of noise.

In addition to the effective noise bandwidth, two other quantities are of interest in representing the noise performance of electronic equipment. These quantities are noise factor and noise figure. According to the IEEE Standards on Noise (3), these two terms may be used interchangeably, but this text prefers *noise figure*. Noise figure, commonly expressed in decibels (db), is ten times the decimal logarithm of the noise ratio. That is, the *noise figure* of a transducer, at a specified input frequency, is defined by the IEEE as "ten times the logarithmic ratio of

1. the total noise power per unit bandwidth at a corresponding output frequency available at the output terminals to
2. that portion of (1) engendered at the input frequency by the input termination."

In effect, this definition of noise figure amounts to a comparison of the actual noise power output of a transducer to the noise power output of an identical noisefree equivalent transducer. An equivalent definition of noise figure is "the degradation of the signal-to-noise ratio produced by the transducer, as expressed in db."

NOISE IN TRANSISTOR CIRCUITS

In several of the preceding chapters, we described the usefulness of small-signal equivalent circuits in analyzing transistor circuits. Since noise voltages and currents are very small signals, equivalent-circuit techniques lend themselves to the analysis of noise in transistor circuits. As in the preceding chapters, two approaches to equivalent noise circuits are possible:

1. The "black box" equivalent circuit, containing two ideal noise sources, as shown in Fig. 12-3(a).
2. A "physical" equivalent noise circuit, in which the sources of noise are related to the physical properties of the transistor which gives rise to noise in the device.

As a simple example of this latter approach, consider the approximate equivalent circuit of Fig. 12-3(b). In this circuit, the thermal noise produced

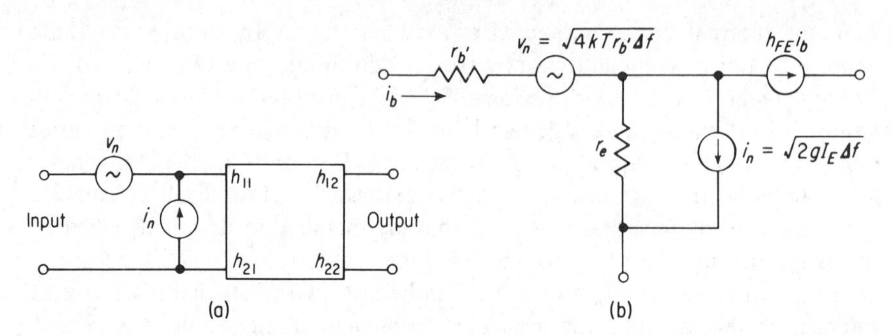

Fig. 12-3. Two methods of analyzing noise in transistors; (a) the black-box approach and (b) the equivalent circuit.

by the base resistance, $r_{b'}$, and the shot noise produced by the emitter diode current, I_E, have been included in the equivalent circuit. These noise sources are thus related to physical parameters of the transistor. The equivalent circuit of Fig. 12-3(b) is by no means a complete representation of a "noisy" transistor, but is merely an example of the approach to be used in relating internal noise sources to the physical properties of the device. These noise sources are, therefore, functions of the operating conditions imposed on the transistor.

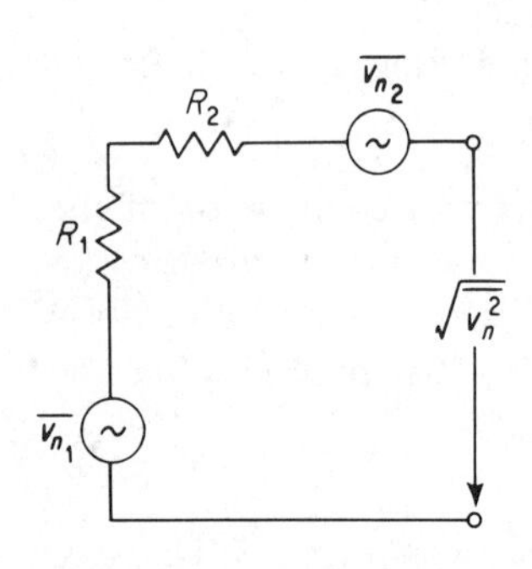

Fig. 12-4. Illustration of the noise generators of two resistors for demonstrating the effect of correlation.

For all practical purposes, pure thermal and shot noise generators are independent of frequency. In both vacuum tubes and transistors, however, there are other secondary effects which result in variations in the total noise output of the device with frequency. One such effect is the "correlation" between two or more equivalent noise generators in a device. Consider the case of two resistors in series, as shown in Fig. 12-4. The square of the sum of the thermal noise output voltages is

$$\overline{v_n^2} = \overline{(v_{n_1} + v_{n_2})^2} = \overline{v_{n_1}^2} + 2\overline{v_{n_1}v_{n_2}} + \overline{v_{n_2}^2} \tag{12-7}$$

But $\overline{v_{n_1}}$ and $\overline{v_{n_2}}$ are average values of the noise voltages, and since $\overline{v_{n_1}}$ and $\overline{v_{n_2}}$ are completely independent of each other, or are not correlated, the average value of the product $\overline{v_{n_1}}$ and $\overline{v_{n_2}}$ is identically zero. The total rms noise output voltage is, therefore, equal to

$$\overline{v_n^2} = \overline{v_{n_1}^2} + \overline{v_{n_2}^2} = 4kT(R_1 + R_2)\,\Delta f$$

which is the same noise output voltage as would be obtained if R_1 and R_2

were replaced by a single resistor whose value was $(R_1 + R_2)$. If, however, $\overline{v_{n_1}}$ and $\overline{v_{n_2}}$ were somewhat correlated, their cross product $(\overline{v_{n_1}v_{n_2}})$ would not in general be zero. Thus, if the degree of correlation between two noise sources is a function of frequency, the total noise output of the circuit containing the two noise generators would also be frequency-dependent.

Another phenomenon which results in a noise variation with frequency is "conductance modulation." For example, in a vacuum tube cathode, electrons are ideally emitted in a completely random manner. From a microscopic point of view, however, the region surrounding the area which has just lost an electron has a very slight local positive charge, so that the probability of emission from this region is slightly less than from another region. In effect, the conductance of the cathode material is modulated very slightly by this effect, producing a low-frequency component of noise in addition to the normal shot and thermal noise. This excess low-frequency component of noise has a $1/f$ frequency spectrum, and is referred to as $1/f$ or *excess* noise. Excess or $1/f$ noise is also present in transistors as discussed earlier, but since a satisfactory analytical explanation of the $1/f$ noise observed in transistors is not available, it is generally treated on an empirical basis.

THE USE OF EQUIVALENT NOISE CIRCUITS (5, 6, 7)

Perhaps the simplest and most easily understood equivalent noise circuit for a common-emitter transistor is that shown in Fig. 12-5. In this circuit, the voltage generator e_b represents the thermal noise of the base spreading resistance $r_{b'}$, the current generators i_{ne} and i_{nc} represent the shot noise of the emitter and collector diodes respectively. The values of these noise

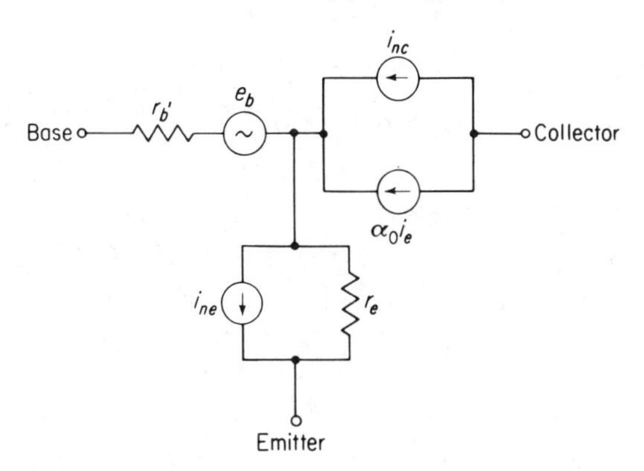

Fig. 12-5. Common-emitter equivalent circuit showing location of noise generators e_b, i_{nc}, and i_{ne}.

generators are given by

$$\overline{e_b^2} = 4kTr_{b'}\,\Delta f \tag{12-8}$$

$$\overline{i_{ne}^2} = 2qI_E\Delta f = \frac{2kT\,\Delta f}{r_e} \tag{12-9}$$

$$\overline{i_{nc}^2} = 2qI_C\Delta f = 2q\,|\alpha|\,I_E\,\Delta f \tag{12-10}$$

where we have used earlier equivalences, such as the definition of r_e and α, and neglected I_{CBO}. Note that both the collector and emitter shot noise generators are dependent on the emitter current, I_E, and are, therefore, at least partially correlated. Rather than carry partially correlated noise quantities through our calculations, and constantly have to account for cross products of the two generators, it is convenient at this point to replace the collector noise generator with one whose value represents only the uncorrelated portion of the original collector noise generator. The emitter diode will then exhibit full shot noise, as given in Eq. 12-9, and the collector diode will exhibit shot noise due to that portion of the collector current which arises from charge carriers which do not cross the emitter depletion layer. The value of this new collector noise generator (4) is given by

$$i_{nc}^2 = 2q(1-\alpha_0)I_C\,\Delta f\left(1+\frac{f_1}{f}\right)\left[1+\left(\frac{f}{f_2}\right)^2\right] \tag{12-11}$$

The last two terms of Eq. 12-11 are introduced to account for the degree of correlation between the emitter and collector noise generators being a function of frequency. Figure 12-6 shows the measured noise figure of a typical transistor as a function of frequency and defines the frequencies f_1 and f_2. This frequency-dependence of noise figure is due to the equivalent

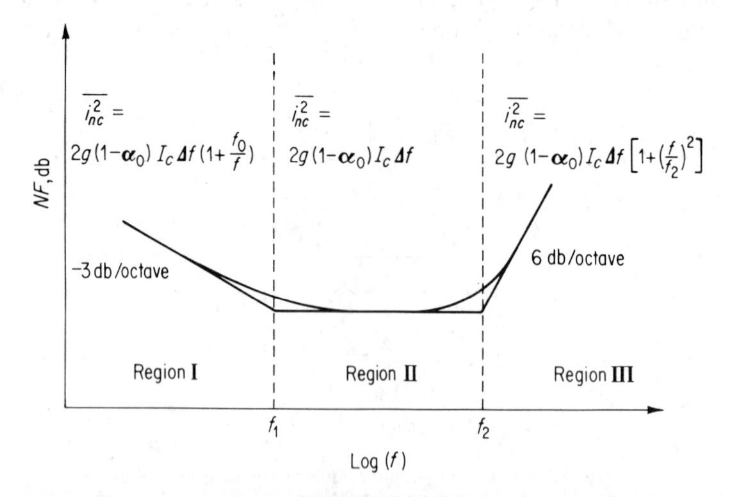

Fig. 12-6. Typical curve of transistor noise figure as a function of frequency. Note the terms contributing to i_{nc} in each region.

collector noise source, which has the values indicated in Fig. 12-6 for the various regions of interest.

Returning to the equivalent circuit of Fig. 12-5, and substituting the values for the noise generators, the transistor noise figure may be obtained. This is done as follows: Since according to the IEEE definition, the noise figure is the ratio of two output powers, it is independent of any load resistance. Thus, the noise figure may be calculated as the ratio of

1. the total short-circuit noise output current squared, to
2. the short-circuit output current squared, due to the input thermal noise of the source resistance.

For low noise, a transistor should have a high value of $h_{FE} = \alpha_0/(1 - \alpha_0)$, and a low base-spreading resistance. Transistor noise performance at very low or very high frequencies depends on the frequency parameters f_1 and f_2. Nielsen (4) has shown that f_2 is approximately equal to $\sqrt{1 - \alpha_0} f_\alpha$, where f_α is the alpha cut-off frequency. To date, there is no theory which allows prediction of f_1, although it is generally between 100 Hz and 10 KHz, and is roughly proportional to f_α. Both f_1 and f_2 are also functions of the transistor operating point.

An alternative approach to the calculation of noise figure involves the use of a black-box equivalent circuit, as shown in Fig. 12-7. This approach consists of replacing the amplifier by an equivalent noiseless amplifier, and representing the total amplifier noise by two generators at the input terminals, e_n and i_n. Since the noiseless amplifier produces the same gain and bandwidth as the actual amplifier, the ratio of the noise power outputs of the actual and ideal amplifiers is equal to the ratio of the noise powers at the amplifier input terminals. Since the input noise powers both appear as voltages across the amplifier input impedance, the input noise power ratio may also be expressed as the square of the input noise voltage ratio. By inspection of Fig. 12-7 the total input noise voltage is

$$v_T = \frac{e_n + e_g}{R_g + Z_{\text{in}}} Z_{\text{in}} + \frac{i_n R_g Z_{\text{in}}}{R_g + Z_{\text{in}}} \tag{12-12}$$

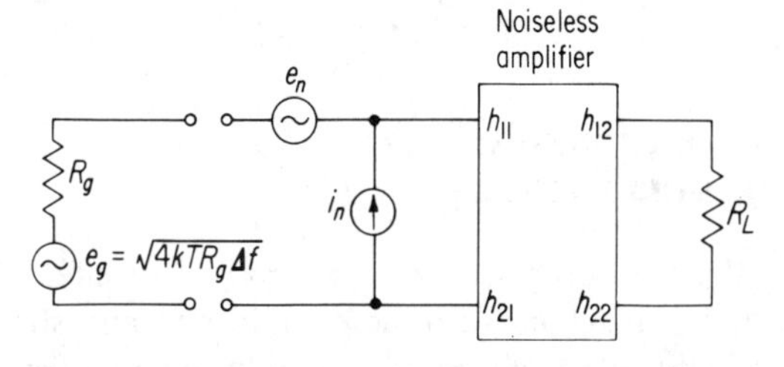

Fig. 12-7. Circuit used for black-box analysis of transistor noise figure.

and the input noise voltage due to the source resistance is

$$v_R = \frac{e_g Z_{\text{in}}}{R_g + Z_{\text{in}}} \tag{12-13}$$

where Z_{in} is the amplifier input impedance. The noise figure is the square of the ratio of Eq. 12-12 to Eq. 12-13, or

$$F = \frac{\overline{v_T^2}}{v_R^2} = 1 + \frac{\overline{e_n^2}}{4kTR_g\,\Delta f} + \frac{\overline{i_n^2}R_g}{4kT\,\Delta f} + \frac{2\gamma \overline{e_n i_n}}{4kT\,\Delta f} \tag{12-14}$$

where γ is the correlation coefficient between e_n and i_n.

In addition to the foregoing relations, differentiation of Eq. 12-14 with respect to R_g yields the interesting result that the optimum source resistance is given by

$$R_{gm} = \frac{e_n}{i_n} \Longrightarrow \overline{e_n^2} = R_{gm}^2 \overline{i_n^2} \tag{12-15}$$

so that the correlation coefficient becomes $\gamma = (r_{b'} + r_e)/R_{gm}$. At first glance, these expressions appear to be just another way of relating noise to the usual transistor parameters. Returning to the black-box equivalent circuit of Fig. 12-7, we see that $\overline{i_n^2}$ represents the equivalent noise at the input of the amplifier with an open-circuited input (R_g infinite), whereas $\overline{e_n^2}$ represents the equivalent noise at the input of the amplifier with a short-circuited input ($R_g = 0$). These two quantities are thus relatively easy to measure as a function of operating conditions and frequency. Given curves of $\overline{i_n^2}$ and $\overline{e_n^2}$ versus emitter current, collector-to-emitter voltage, and frequency, the noise performance of a transistor is completely specified. The noise figure is, therefore,

$$F = 1 + \frac{1}{4kT\,\Delta f}\left[\frac{\overline{e_n^2}}{R_g} + \overline{i_n^2}(R_g + 2r_{b'} + 2r_e)\right] \tag{12-16}$$

and the optimum source resistance (squared) and resultant minimum noise figure are, respectively,

$$R_{gm}^2 = \frac{\overline{e_n^2}}{\overline{i_n^2}} \tag{12-17}$$

$$F_m = 1 + \frac{(1+\gamma)R_{gm}\overline{i_n^2}}{2kT\,\Delta f} \tag{12-18}$$

METHODS OF MEASURING TRANSISTOR NOISE FIGURE

As described in the preceding section, one method of effectively measuring transistor noise figure is to measure the open-circuit and short-circuit noise power outputs as a function of operating conditions and frequency. Several alternative methods of measuring transistor noise factor directly are also in general use. Such methods almost always compare transistor output under two different conditions.

One of the simplest ways to measure noise factor directly is by the "hot resistor" technique. In the circuit of Fig. 12-8 the total noise power output is equal to the power gain of the amplifier times the thermal noise power of the source resistance plus the equivalent noise power input of the amplifier. At room temperature (T_0), then, the total noise output power is

$$P_{\text{out}} = \frac{A_v^2}{R_L}[4kT_0R_g\,\Delta f + \overline{e_{ni}^2}] \tag{12-19}$$

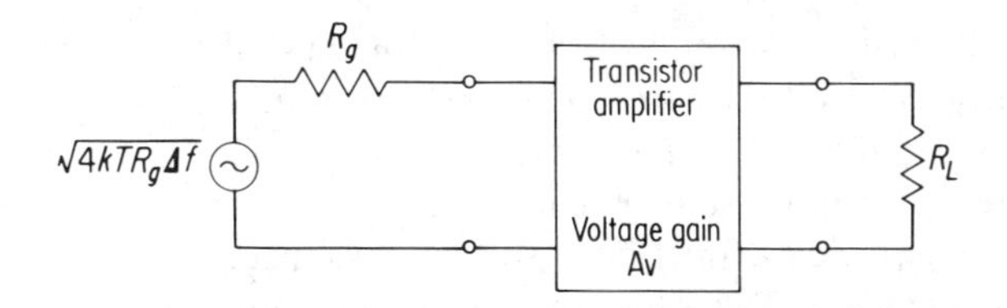

Fig. 12-8. Measurement of noise figure using the hot resistor method.

where $\overline{e_{ni}^2}$ is the equivalent noise input voltage squared. Now suppose the temperature of the source resistance *only* is increased until the noise power output doubles its room temperature value. At this elevated temperature (T_1), the total noise output power is

$$P_1 = \frac{A_v^2}{R_L}[4kT_1R_g\,\Delta f + \overline{e_{ni}^2}] \equiv 2P_{\text{out}} \tag{12-20}$$

These relationships may be solved for the equivalent noise power input of the amplifier and give $\overline{e_{ni}^2} = 4kR_g\,\Delta f(T_1 - 2T_0)$. By definition, the amplifier noise figure is the ratio of the total noise power output to the noise power output due to the source resistance (at room temperature) alone. Thus,

$$F = \frac{4kT_0R_g\,\Delta f + \overline{e_{ni}^2}}{4kT_0R_g\,\Delta f} = \frac{T_1 - T_0}{T_0} = \frac{\Delta T}{T_0} \tag{12-21}$$

The noise figure may be obtained directly from Eq. 12-21, where $T_0 = 300°\text{K}$, and ΔT is the increase in the temperature of the source resistance required to double the total room temperature noise power output of the amplifier. Note that this method requires no knowledge of either the amplifier power gain or the system bandwidth, provided that these quantities remain constant during the measurements. In addition, since the additional noise produced by heating the source resistor has the same waveform and frequency spectrum as the source resistor noise at room temperature, any convenient instrument, such as a peak-reading VTVM, may be used to make the comparison.

If the effective noise bandwidth (Δf) of the system extends into either region I or region III of Fig. 12-6, the equivalent amplifier noise will be a function of frequency. The value of $\overline{e_{ni}^2}$ obtained by the foregoing method will thus be an average value over the bandwidth used. In such cases, the noise figure obtained is called the *wide-band noise figure*. If the system bandwidth is less than 20 per cent of the center frequency the noise figure obtained is essentially that at a single frequency, and is called the *spot noise figure*. In accord with the definition of effective noise bandwidth, the average noise figure is defined as

$$\bar{F} = \frac{\int_0^\infty G(f)F(f)\,df}{G_0\,\Delta f} \tag{12-22}$$

The "hot resistor" method of noise factor, described earlier, consists of introducing a controlled source of noise at the amplifier input terminals. Thus, instead of heating the source resistance so as to increase its noise output, external "noise" sources may be used to obtain noise figure. Since noise figure is explicitly a function of source resistance, this value of resistance must remain constant. Two other methods of measuring noise figure directly are shown in Fig. 12-9. In Fig. 12-9(a), a vacuum tube diode having a high plate resistance is shunted across the source resistor. By varying the filament voltage, the plate current of the tube may be controlled, thereby producing a controlled source of noise (in this case, shot noise) suitable for a comparison measurement of noise factor. In Fig. 12-9(b), a single-frequency sine wave generator is connected in series with the source resistor.

The noise power output is first measured with the signal generator input shorted. The input sine wave signal is then increased until the output power is doubled. The voltage (e_g) at the input required to double the output power is the equivalent input noise voltage. The noise figure may then be calculated from the approach used to obtain Eq. 12-21. The result is

$$F = \frac{e_g^2}{4kTR_g\,\Delta f} \tag{12-23}$$

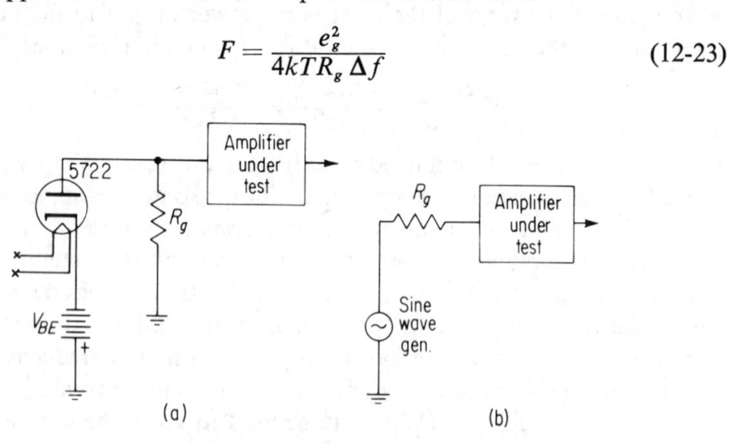

Fig. 12-9. Two methods of measuring noise figure. (a) Noise diode method; (b) signal generator.

Both the noise diode method and the sine wave generator method are limited, however. As previously described, the vacuum tube diode output noise current has a very low-frequency component which is not predictable. This method is, therefore, limited to relatively high-frequency noise factor measurements. The sine wave comparison method is widely used, but since a single-frequency sine wave is being compared with random noise signals spread over the effective noise bandwidth, an accurate measurement of the system bandwidth is required. A true rms or power meter must be used with this method in order to provide an accurate comparison of relative powers.

The high-level measurement of the amplifier power gain is not widely used because of the possibility of nonlinear amplifier gain at the relatively higher signal levels.

NOISE IN LOW-FREQUENCY AMPLIFIERS

Thus far, the terms *transistor noise figure* and *amplifier noise figure* have been used more or less interchangeably. Actually, a distinction should be made between the two terms. Amplifier noise is always greater than transistor noise, because the former includes the thermal noise introduced by bias resistors and other amplifier components, whereas the latter term does not. Equations 12-8–12-18 are applicable only to transistor noise figure, although the equivalent noise circuit of Fig. 12-5 could easily be modified to take other circuit elements into account. [The results of such calculations have been described in the literature (8, 9) and will not be discussed here].

Transistor noise figure, as used here, generally refers to common-emitter transistor noise figure. To a very close approximation, however, the noise figures of the three possible configurations are identical, at frequencies up to approximately f_α. Thus a knowledge of common-emitter noise figure behavior is useful in transistor circuit design of other connections.

Since all the low-frequency amplifier parameters are relatively independent of collector-to-emitter voltage, the low-frequency noise figure should be nearly independent of voltage. Figure 12-10 shows typical variations of noise figure with operating point.

As mentioned previously, amplifier noise figure is always greater than transistor noise figure, since the presence of base bias and emitter resistors can only increase the noise figure. If any net bias resistance from base to ground is kept much greater than the *optimum* source resistance, and any emitter resistance is kept much smaller than the *optimum* source resistance, the effects of these elements will be negligible.

In connection with the design of low-noise, wide-band transistor amplifiers, the choice of transistor must also be considered. It may be argued that this choice is primarily dictated by gain-bandwidth requirements, rather than

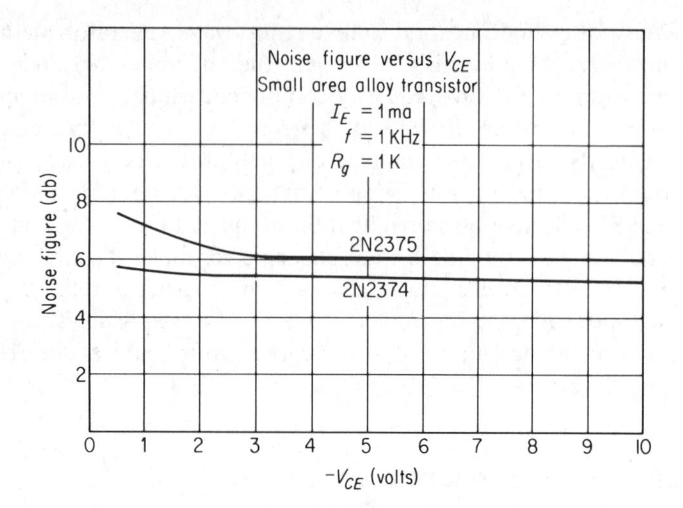

Fig. 12-10. Noise figure vs V_{ce}.

by noise behavior. It turns out, however, that the higher-frequency transistors generally have a higher f_1, resulting in higher low-frequency noise, and thereby necessitating a compromise. In choosing a suitable transistor for wide-band low-noise amplifiers, however, the choice should be made with the desired operating point in mind. For instance, at very low emitter current and low collector-to-emitter voltage, a transistor such as the Micro-Alloy Transistor actually has a greater gain-bandwidth product than many diffused-base transistors, even though the latter type is generally considered to be a "higher-frequency" transistor. Thus one must choose a transistor for wide-band applications with the operating point considerations in mind.

In summary, the design considerations for low-noise operation of low-frequency amplifiers are as follows:

1. The operating point should be chosen such that the emitter current is on the order of 200 μa or less, with reasonably low collector-to-emitter voltage. The choice of a transistor for wide-band applications should be made with the operating point considerations in mind.
2. The effect of base bias and emitter resistors on the common-emitter noise factor may be generalized so as to apply to any of the three transistor amplifier configurations. Any shunt resistance across the input terminal should be much greater than the optimum source resistance. Any unbypassed resistance from the common terminal to ground should be much smaller than the optimum source resistance. If possible, the amplifier should be operated from the optimum source resistance.

3. Attention to the foregoing considerations, applicable to either narrow-band or wide-band transistor amplifiers, will result in optimum noise performance of the amplifier.

Low-Noise Tuned Amplifiers

The design of low-noise tuned amplifiers is most often done experimentally. The source resistance and operating point should be adjusted to obtain optimum noise figure at the frequency of interest. This sometimes results in a loss of gain. The following suggestions are useful in determining the optimal circuit:

1. The optimal source resistance is usually less than the input impedance.
2. At very high frequencies, the transistor operates best when driven from an inductive source. Thus, the input coupling networks of Fig. 12-11(a) will usually result in a slightly improved noise figure over the circuits of Fig. 12-11(b). The reason for this is not completely understood but the effect has been noted in a number of different transistor types.

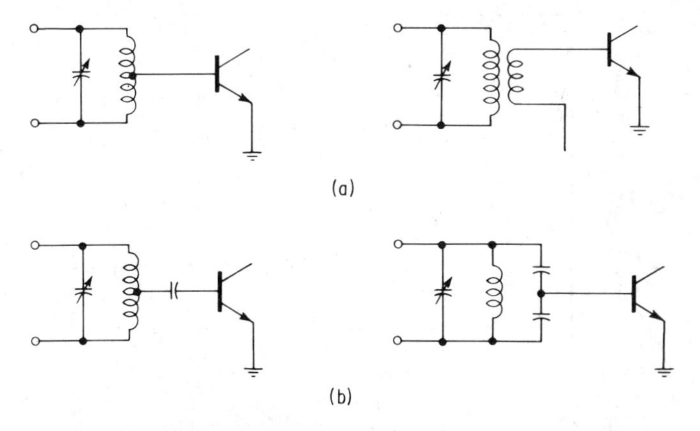

Fig. 12-11. Transistor ac input circuits showing: (a) preferred low-noise circuits; (b) capacitive coupling.

3. In amplifiers where the gain is low, the second-stage noise must be considered. In fact, sometimes a better system noise figure may be obtained by trading gain for noise in the first stage. This trade is easily calculable and is given by:

$$F_0 = F_1 + \frac{F_2 - 1}{PG_1} \tag{12-24}$$

where F_0 = over-all system noise figure
F_1 = noise figure of the first stage
F_2 = noise figure of the second stage
PG_1 = power gain of the first stage.

Note that all the terms in Eq. 12-24 should be expressed as power ratios, not in decibels.

4. Earlier in this chapter we noted that the effective noise bandwidth must be known when using several of the methods for measuring noise figure. The system bandwidth must also be known when relating the sensitivity of a receiver to its noise figure. Figure 12-12 shows a plot of the ratio of effective noise bandwidth to 3 db bandwidth per stage of a number of identical cascaded single-tuned stages. If these ratios are used for double-tuned circuits, they will give a pessimistic answer.

   Figure 12-12 shows a plot of system noise figure versus sensitivity for various noise bandwidths. Notice that significant improvements in sensitivity can be made by making the noise bandwidth (usually the IF bandwidth in a receiver) much smaller, even though the noise figure of the RF transistor has not changed.

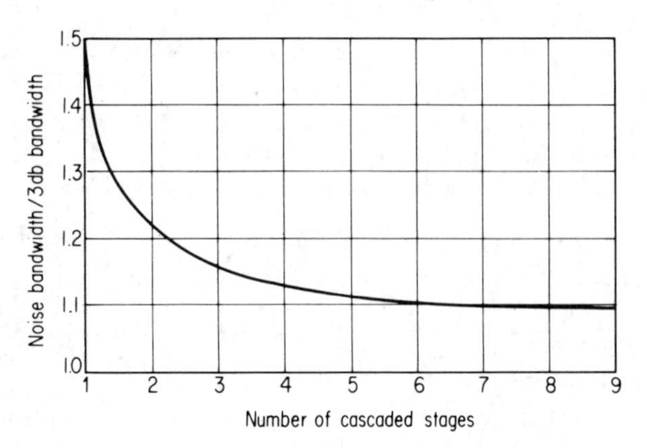

Fig. 12-12. Effective noise bandwidth normalized to 3 db bandwidth vs number of cascaded synchronously-tuned stages.

5. The optimal current at a given frequency is best determined experimentally, or by reference to the manufacturer's characteristic curves. Usually the optimal current varies from device to device; hence an average value should be picked. Normally the variation of noise figure with current is much less than is the variation of gain. Thus, some compromise between the two can usually be made.
6. In general, the noise figure of high-frequency transistors is fairly independent of collector voltage. On most diffused types, 6 volts is sufficient bias voltage.

7. The noise figure of both the common-emitter and common-base stage are the same at high frequency.

Mid-Frequency Amplifier Noise Figure

In the mid-frequency region between f_1 and f_2, the noise figure of most transistors is about 3 db under matched conditions. In fact, if we assume that the input resistance of a transistor contributes noise in the same way as the source resistance, and that we are in the range where the terms containing f_1 and f_2 are negligible, the matched transistor noise figure should be exactly 3 db.

In designing transistor amplifiers in this region, the designer will usually find that the best techniques to use are those somewhat between the methods used on high- and low-frequency amplifiers. At the low-frequency end, operation at lower currents is required with an increase in current as the frequency is increased. (Since the computational techniques are somewhat more complicated, however, and are similar in nature, we do not describe them here.)

EXERCISES

1. Show that the maximum noise power available from a resistor R is $P_{\max} = kT \Delta f$ and is independent of R.

2. A low-frequency PNP germanium alloy junction transistor is being used in a low-level audio preamplifier. This transistor has the following characteristics at an operating point of $I_E = 1.0$ ma and $V_{CE} = -5$ volt:

$$\beta_0 = 100$$
$$r'_b = 200\ \Omega$$
$$f_1 = 500\text{ Hz}$$
$$f_2 = 1\text{ MHz}$$

Calculate the source resistance that the preamplifier should utilize to minimize the noise figure. What noise figure is obtained for this value of source resistance? What happens to the noise figure if a source resistance ten times above or below this optimum source resistance is used for the amplifier?

3. For the amplifier referred to in Exercise 2, construct the "black-box" equivalent noise circuit by computing $\overline{e_n^2}$ and $\overline{i_n^2}$. Normalize the results to the noise bandwidth Δf. What is the value of the correlation coefficient?

4. A wide-band amplifier is being designed to amplify low-level signals in the frequency range of 0.5–2.0 MHz. This frequency range lies within the band (between f_1 and f_2) for the diffused-base transistor selected. The amplifier circuit selected is shown herewith.

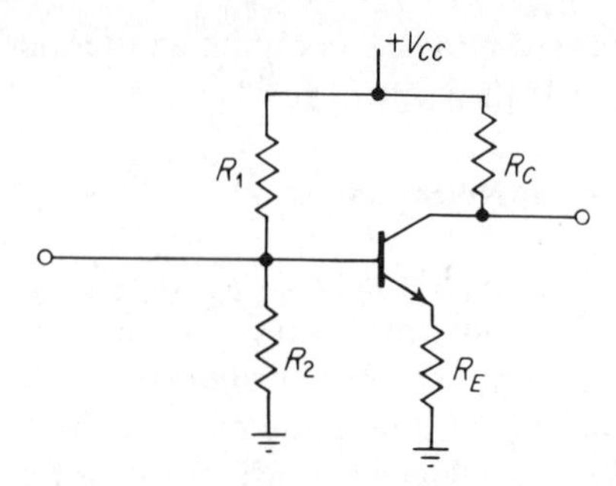

Calculate a general expression for the mid-band noise figure for this circuit. Under what conditions do R_1, R_2, and R_E *not* influence the over-all transistor noise figure?

5. Prove that the amplifier noise figure measured by use of the noise diode method may be calculated by use of the expression

$$NF = 10 \log_{10}\left(\frac{qI_D R_g}{2kT}\right)$$

where I_D is the diode current.

REFERENCES

1. J. B. JOHNSON, *Phys Rev.*, **32** (1928), 97.
2. W. SCHOTTKY, *Ann. Physik*, **57** (1918), 541.
3. "IRE Standards on Methods of Measuring Noise in Linear Two-Ports, 1959," *Proc. IRE*, **48**, No. 1 (January, 1960) 60.
4. E. G. NIELSEN, "Behavior of Noise Figure in Junction Transistors," *Proc. IRE*, **45**, No. 7 (July, 1957) 458.
5. J. W. HALLIGAN, "Noise Figure Behavior of a Practical Transistor Amplifier," Application Lab, Report 640, Philco Corp., Lansdale, Pa.
6. H. F. COOKE, "Transistor Upper Noise Corner Frequency," *Proc. IRE*, **49**, No. 3 (March, 1961), 648.
7. J. A. EKISS, and J. W. HALLIGAN, "A Theoretical Comparison of Average and Spot-Noise Figure in Transistor Amplifiers," *Proc. IRE*, **49**, No. 7 (July, 1961), 1216.
8. R. D. MIDDLEBROOK, "Optimum Noise Performance of Transistor Input Circuits," *Semiconductor Products*, **1**, No. 4 (July/August, 1958), 14.
9. J. W. HALLIGAN, "Effect of External Base and Emitter Resistors on Noise Figure," *Proc. IRE*, **48**, No. 5 (May, 1960), 936.
10. A. VAN DER ZIEL, *Noise*. Englewood Cliffs, N.J.: Prentice-Hall, Inc., 1956.

13

Tunnel Diode and Other High-Frequency Diode Characteristics and Applications

INTRODUCTION

Tunnel diodes are PN junction devices which possess ultrahigh frequency capabilities. This chapter describes some of the unique characteristics and some of the potential applications of the tunnel diode. (The avalanche diode and Gunn oscillator are also discussed briefly.) Owing to its ac negative resistance, the tunnel diode can serve as the active element in amplifiers, oscillators, and switches.

Backward diodes provide very high forward conductance and majority carrier operation, two attributes which are very desirable for rectifiers.

THEORY OF THE TUNNEL EFFECT

In Chap. 1 we considered the conduction mechanism in a PN junction diode and learned that:

1. There is a region within the diode which is devoid of free charge known as the *depletion layer*.
2. With zero applied voltage across the diode there is a potential barrier at the junction so that the net current is zero.
3. If a reverse bias is applied, that is, an external battery which aids the potential barrier, there will be a small saturation current until a breakdown voltage is reached.
4. When the diode is forward-biased, the potential barrier at the junction is overcome and it becomes possible for large numbers of positive carriers from the p side to cross the junction to the side of n-type material

and vice versa. Ideally, the injection current which flows under these conditions is given by $I = I_S(e^{qV/kT} - 1)$.

In heavily doped PN junctions there is, in addition to the normal injection current, a tunneling current which flows by virtue of the very narrow depletion region and the principle of quantum mechanical tunneling (1).* Mitchell derived simple first-order, approximate expressions for tunnel current as a function of applied voltage in terms of the peak current and voltage (2). A good approximation is achieved by use of

$$I = \frac{I_P}{V_P} V \exp\left(1 - \frac{V}{V_P}\right) \tag{13-1}$$

The quantities I_P and V_P, the peak current and voltage, are defined as shown in Fig. 13-1.

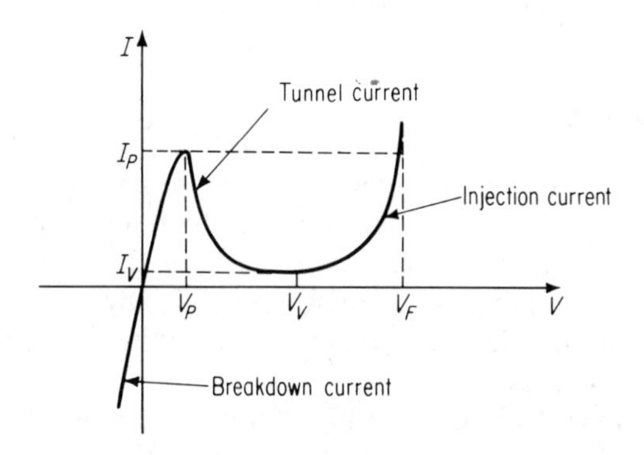

Fig. 13-1. Tunnel diode I-V characteristics showing breakdown, tunnel, and injection currents.

The breakdown voltage of a PN junction diode is a direct function of the resistivity of the material. For highly doped material with low resistivity, breakdown occurs at very low reverse-bias voltages.

Thus, in summary, the composite tunnel diode *I-V* characteristic (Fig. 13-1) consists of:

1. A breakdown current at a very low reverse bias which proves extremely useful in backward diodes, as we shall see later.
2. A tunnel current which reaches a peak value at a small forward bias.
3. An injection current which is dependent on qV/kT as it is in conventional diodes and transistors, rather than on the doping density.

*Numbers in parentheses refer to material listed at the end of the chapter.

We have not as yet placed any numerical figures on the I-V characteristic of Fig. 13-1. To do so we must know three things: (1) the material of which the diode is made, (2) its junction area, (3) the impurity density. The material determines the characteristic tunnel diode voltages to a first-order approximation. They are presented in Table 13-1. The doping density determines the peak current density, J_P. The peak current itself is then simply AJ_P, where A is the junction area. The peak current can be as low as a few microamperes, as it is in backward diodes, or as high as several hundred amperes for energy-conversion applications. Most tunnel diodes in present-day use have peak currents in the 1–20 ma range.

TABLE 13-1

Material	V_P (mv)	V_V (mv)	V_F (mv)
Germanium	55	320	475
Silicon	65	420	720
Gallium arsenide	150	500	970
Gallium antimonide	40	225	500
Indium arsenide	25	65	180

$T = 25°C$

Thus far we have considered only the static characteristics of the tunnel diode but mentioned in passing that the depletion region is very narrow. This fact implies extremely high-frequency capabilities since the carriers do not have too large a region of high field to traverse. Oscillations in excess of 100 Gc have been reported (4) but all tunnel diodes cannot oscillate at such frequencies, or switch in fractions of a nanosecond, because their small signal ac characteristic equivalent circuit possesses three parasitic elements. The junction, having a very thin depletion layer between two sides of appreciable area, gives rise to a shunt capacitance; the external leads must have some inductance; although the diode material resistivity is low, there is a series resistance, R_S. These parasitic elements are connected to the diode resistance in the ac equivalent circuit of Fig. 13-2. If the applied voltage is $V_P < V_{\text{applied}} < V_V$, the diode resistance, r, is negative. That is, a slight increase in applied voltage will cause a reduction in the diode current. This implies negative resistance which enables amplification and oscillation as we show later.

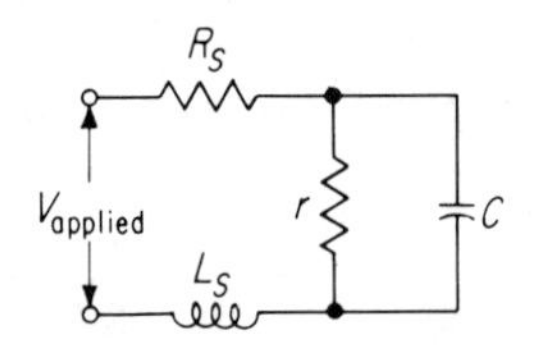

Fig. 13-2. Tunnel diode ac equivalent circuit.

PRACTICAL TUNNEL EFFECT DEVICES

Backward Diodes

Upon examination of the complete *I-V* characteristic of a tunnel diode, one notes that the reverse conductance is much higher than that due only to an injection current. When a conventional germanium diode conducts a forward current of several milliamperes, its voltage drop is about 250 millivolts (mv). A germanium tunnel diode conducting a reverse current in the same range would have a voltage drop of less than 100 mv. This is a very desirable characteristic in rectifiers. If the tunnel current is kept low by proper area etching and doping, we will indeed have a useful rectifier. Figure 13-3 compares the *I-V* characteristics of a backward diode and a conventional diode. The parameters which must be specified for a backward diode are

1. Peak point current, I_P, which should be held to a specified maximum as it is a leakage current.
2. Forward voltage, V_F, which is specified at a forward current, I_0. V_F should have a maximum limit to guarantee that the diode does indeed offer high forward conductance.
3. Reverse voltage, V_R, which is specified at a reverse current, $-I_0$. This parameter should have a minimum limit to insure that a useful range of input voltage will be rectified.
4. Capacitance at the valley voltage.
5. Series inductance which is primarily determined by the geometry of the package and lead length.

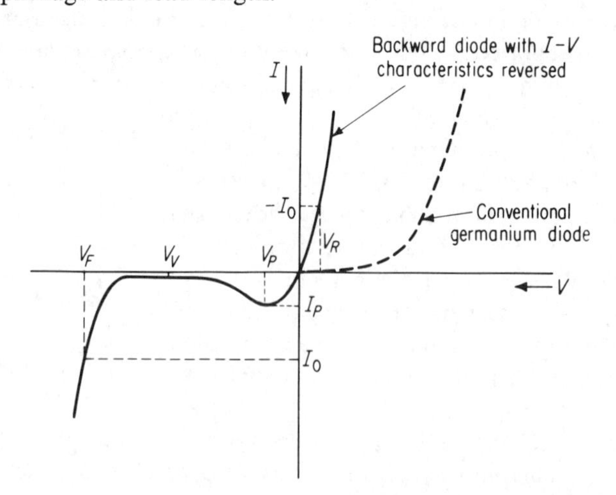

Fig. 13-3. Static characteristics of germanium backward and conventional diodes.

The switching speed of a backward diode is essentially controlled by the shunt capacitance because there is no minority carrier storage which introduces the reverse recovery time in conventional rectifier diodes. Its properties make it extremely useful in tunnel diode circuits.

Types of Tunnel Diodes

Tunnel diodes have been fabricated from many types of semiconductor material. The choice of a material can be dictated by the following six considerations:

1. Peak-to-valley current ratio (I_P/I_V). In most applications this should be as high as possible. Examples include low noise amplification and efficient oscillator operation.
2. Voltage swing. This may be expressed as ($V_F - V_P$) in millivolts, or as the ratios (V_V/V_P) and (V_F/V_P), depending upon the application.
3. Merit factor I_P/C. This should be as high as possible to obtain high-speed performance in *any* type of circuit. It is dependent upon the material and its doping levels, because both I_P and C are directly proportional to the junction area.
4. Temperature range for storage and operating.
5. Reliability. Long-term stability of parameters, indestructibility in severe environments, predictability of parameter variations with temperature are covered by the broad term *reliability*.
6. Cost. Popular materials like germanium are least expensive, whereas a rather unusual type, such as gallium antimonide, is more costly.

Table 13-2 compares four types of tunnel diodes of different materials.

TABLE 13-2

Material	I_P/I_V	$V_F - V_P$	V_V/V_P	V_F/V_P	I_P/C	Temperature Range (storage; °C)
Germanium	8	420 mv	6	9	0.110 ma/pf	−65–+100
Silicon	4	655	6.5	11	0.002	−85–+200
Gallium arsenide	10	880	4.5	10	0.13	−65–+175
Gallium antimonide	15	470	5	10	0.60	−55–+100

Typical room temperature values are given in each case and are based on diodes having peak currents of 3–5 ma.

PARAMETER VARIATIONS WITH TEMPERATURE

In general, the tunneling region of the tunnel diode is relatively independent of temperature. Hence, the peak current can display a zero temperature coefficient for a limited, but appreciable, range of temperatures around room temperature. Empirical data for diodes of various impurity concentrations are given in the form of plots of percentage change of I_P versus temperature (5).

On the other hand, the valley current shows a definite increase with temperature. However, it is roughly a 10 per cent increase for a 20°C change in temperature, which is considerably less than the thermal rise in I_{CBO} of a transistor.

TABLE 13-3

Characteristic		Rate of Decrease (mv/°C)
Transistor	V_{BE}	2
Tunnel diode	V_P	0.08
	V_V	0.6
	V_F	1

All the tunnel diode characteristic voltages decrease as the temperature increases. Although V_F is a forward injection voltage similar to the V_{BE} of a transistor, its decrease occurs at half the rate of V_{BE}. Typical rates of decrease for the characteristic voltages in germanium devices are given in Table 13-3.

Just as peak current is relatively stable with temperature changes, whereas valley current increases with temperature, other external agents, such as nuclear radiation, have little effect on I_P but cause a drastic increase in I_V. It has long been known, for example, that thin mechanical base width transistors are best for radiation resistance. It should not be surprising to find a heavily doped, narrow depletion layer device to be similarly inclined to resist high radiation fluxes. Reports from Battelle Memorial Institute (6) indicate that irradiation effects become noticeable at an integrated fast neutron flux of 10^{14} n/cm^2 in silicon and 10^{16} n/cm^2 in germanium and gallium arsenide. The increase of I_V continues as the flux levels are increased by two orders of magnitude until the I_P/I_V is lowered to unity. Thus, these devices could be used in nuclear instrumentation with reasonable tolerances.

RATINGS AND CHARACTERISTICS

Although tunnel diodes are two-terminal devices, their many uses and unique characteristics require careful consideration of absolute maximum ratings, specifications, and figures of merit.

DC Characteristics

The current-voltage characteristic of a tunnel diode over its entire operating range is sketched in Fig. 13-4.

The maximum allowable current in the forward or reverse direction is limited to be several times greater than the peak current, I_P. Safe operation at or below this maximum current is proved by life tests. Since the tunnel diode is a two-terminal element, and the voltage corresponding to $I_{\text{forward(max)}}$ is single-valued, as shown in Fig. 13-4, the power dissipation is dependent upon the current. Therefore, an absolute maximum rating must be given for current alone, a parameter over which the circuit designer has control.

Three points on the characteristic curve are of prime importance, especially in switching applications. They are (V_P, I_P), (V_V, I_V), and (V_F, I_P). A rapid and reasonably accurate measurement of these coordinates may be made with a transistor curve tracer.

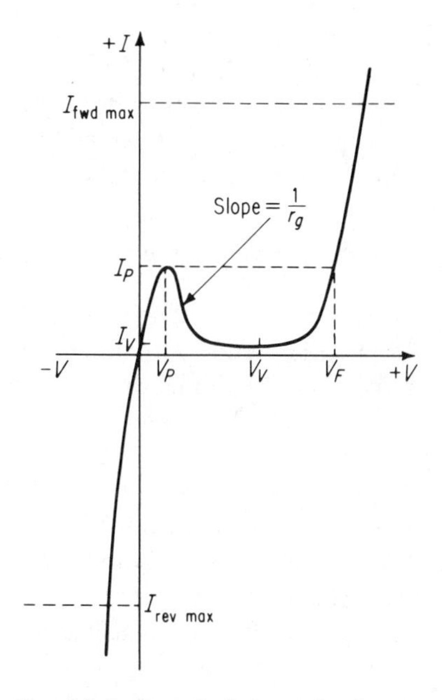

Fig. 13-4. Tunnel diode static characteristic defining important current and voltage points.

AC Equivalent Circuit

The small-signal equivalent circuit for a tunnel diode, biased in the negative resistance region, was shown in Fig. 13-2.

The shunt capacitance, C, is the sum of the diode's barrier or junction capacitance, C_d, and the case header capacitance, C_{case}.

An impedance bridge may be used for measurement of the total capacitance. An external variable bias supply is required, and a stable null for $R = \infty$ may be obtained with the bias voltage equal to V_V for the diode. The variation of C_d with applied voltage is of the same form as that for an abrupt PN junction:

$$C \sim \frac{1}{V^{1/2}} \tag{13-2}$$

The series inductance, L_S, is external to the PN junction. As it is due to the package and lead length alone, measurements of L_S are generally made on an inactive sample.

The series or bulk resistance, R_S, is obtained from measurement of the diode voltage for a large reverse current. Since this current is far in excess of $I_{\text{reverse (max)}}$, the current must be applied for a very brief interval of time and the measurement is thus done on a pulsed basis.

The value of r_d is obtained by measuring the steepest negative slope of the tunnel diode *I-V* characteristic and calculating its reciprocal.

Factors Affecting Circuit Reliability

The key parameter in nearly every tunnel diode switching circuit is I_P. Worst-case analyses of several logic schemes have shown that the peak currents of the tunnel diodes employed must be matched to within ± 3 per cent in order to permit operation with practical supply and component tolerances.

For worst-case design of many switching circuits, it is necessary to have minimum and maximum limits on the voltages V_P, V_V, and V_F. The ratio I_P/I_V is a figure of merit useful for predicting the extent of the dynamic range over which the tunnel diode may be used for linear service. The manufacturer generally specifies a minimum peak-to-valley current ratio.

Factors Affecting Fast Switching Performance

The designer of a switching circuit is almost invariably concerned with the speed limitations of his devices. The tunnel effect proceeds at relativistic velocities, but the shunt capacitance of practical tunnel diodes requires a finite charging time in order to enable switching across the negative resistance region. Assume a tunnel diode biased with a current source equal to I_P so that a short current pulse can initiate switching from (V_P, I_P) to (V_F, I_P). An approximate *characteristic* curve and loadline for this operation is sketched as Fig. 13-5. Under these conditions the tunnel diode shunt capacitance, C, is being charged by a current $(I_P - I_V)$. The switching time is determined by the time it takes this current to charge the capacitance to V_F. The resulting switching time is, therefore, very closely predicted by the expression:

$$t = \left(\frac{V_F - V_P}{I_P - I_V}\right)C \tag{13-3}$$

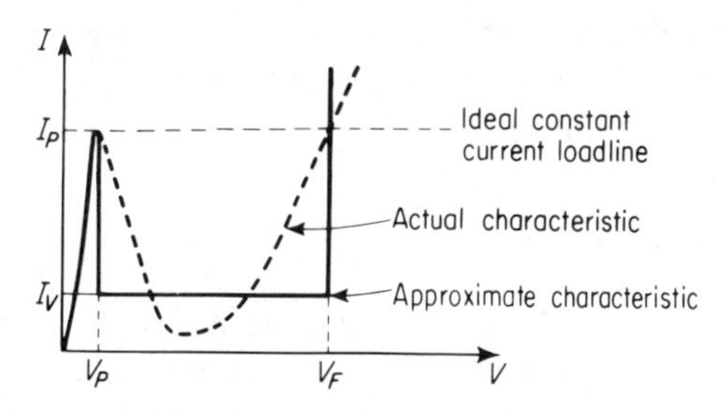

Fig. 13-5. Approximate I-V characteristic used in calculating switching time.

Now both I_P and C_d are directly proportional tothe PN junction area. Since we wish I_P to be large and C_d to be small to minimize t, the ratio I_P/C_d will be a good measure of the merit of the material used for fabricating a tunnel diode. It has the dimensions of ma/pf or volt/nsec. This figure of merit is also applicable to small-signal evaluations and it has been noted that for operation above 1 GHz, an I_P/C_d ratio of at least 0.1 ma/pf is required.

Packaging and external inductances should be kept to a minimum since increasing L_S will result in undesirable overshoots and delays, which increase the over-all switching time.

Factors of Importance in Small-Signal Applications

The maximum frequency of oscillation of a tunnel diode will be limited by either $f_{\max}$, the resistive cut-off frequency, or f_{SR}, the self-resonant, "package-limited" frequency. The smaller of the two frequencies determines the maximum useful frequency of the diode.

The resistive cut-off frequency from simple network analysis is

$$f_{\max} = \frac{1}{2\pi r_d C}\sqrt{\frac{r_d}{R_S} + 1} \tag{13-4}$$

The real part of the admittance is zero at this frequency and positive above it, precluding amplification. If $f_{\max}$ is higher than f_{SR}, it will be extremely difficult to measure the resistive cut-off frequency. The inductance of the external bias circuit will tend to produce relaxation oscillations, so that special jigs must be provided to maximize f_{SR}. Superregenerative detectors and spectrum analyzers have been used to detect the weak output at $f_{\max}$.

The self-resonant frequency can be determined as

$$f_{SR} = \frac{1}{2\pi}\sqrt{\frac{1}{L_S C} - \frac{1}{(r_d C)^2}} \tag{13-5}$$

At this frequency the imaginary part of the tunnel diode equivalent circuit admittance becomes zero. For $f < f_{SR}$, the input impedance is capacitive.

If low-loss tuning elements are available, sinusoidal oscillations are possible at frequencies above f_{SR} with series tuning.

The efficiency of a tunnel diode Class A oscillator can be analyzed to give a general expression:

$$\text{percentage of efficiency} = 25\left(1 - \frac{R_S}{r_d}\right)\left\{1 - \left(\frac{f}{f_{\max}}\right)^2\right\} \tag{13-6}$$

Note that for an idealized tunnel diode, the maximum efficiency is 25 per cent. As f approaches $f_{\max}$, the efficiency and output reduce to zero.

In Fig. 13-6 we have shown a general schematic for a negative resistance oscillator circuit. The tunnel diode parameters R_S, L_S, and C_d are lumped into R, L, and C, respectively. This circuit was analyzed by Cunningham (8), and it can be shown that the frequency of oscillation is

$$f = \frac{1}{2\pi}\sqrt{\frac{1}{LC}\left(1 - \frac{R}{r_d}\right)} \tag{13-7}$$

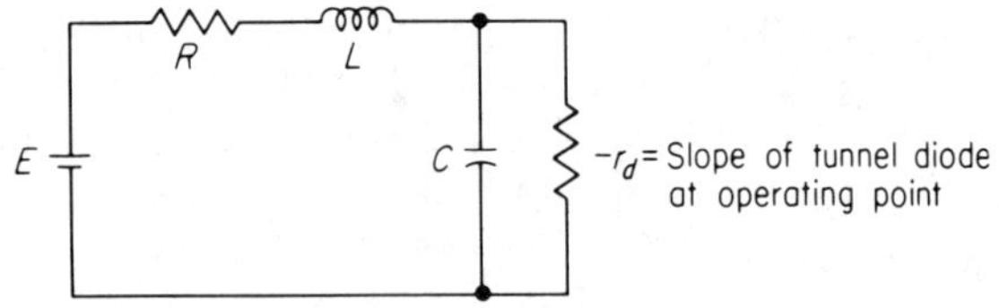

Fig. 13-6. A generalized tunnel diode oscillator.

Oscillators are designed for infinite gain and zero bandwidth. It is certainly possible to use the tunnel diode's negative resistance to provide voltage or current amplification over bandwidths as high as several hundred megacycles. The figure of merit which may be used to measure the frequency response of the tunnel diode is the gain-bandwidth product:

$$\sqrt{A_T} \cdot BW = \frac{1}{2\pi r_d C} \tag{13-8}$$

This is a constant which depends on only two of the tunnel diode's parameters. In order to construct an amplifier which operates with this gain-bandwidth product, the generator and load resistances must each be matched and equal to $\frac{1}{2} r_d$.

A major reason for choosing the tunnel diode for amplification at gigacycle frequencies is its low noise figure. An expression for the *minimum* noise figure can be derived as

$$F = 10 \log\left(1 + \frac{qIr_d}{2kT}\right) \tag{13-9}$$

where I is in milliamperes, and r_d is the corresponding dynamic resistance in ohms. A high I_P/I_V ratio is most desirable in minimizing F to about 3 db.

The conditions under which this best noise figure is realized are $R_{\text{source}} = r_d$, and $R_L \longrightarrow \infty$. These, of course, reduce the gain-bandwidth of the amplifier.

Under conditions of maximum gain-bandwidth, the noise figure is increased by 3 db.

THE MODES OF TUNNEL DIODE OPERATION

Circuits which use a negative resistance device as the active element are classed as astable, monostable, and bistable. This refers to the number of positive resistance regions which the circuit loadline intersects in its quiescent state. Typical loadlines for the three modes of tunnel diode operation are shown in Fig. 13-7.

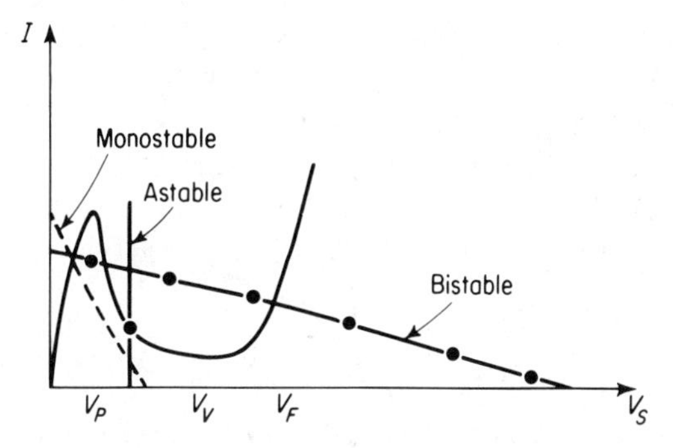

Fig. 13-7. Required dc loadlines for astable, monostable, and bistable operation.

The *astable* circuits require a supply voltage, V_S, in the range $V_P < V_S < V_V$, and loadline whose slope is steeper than that of the diode negative resistance at the intersection. This insures astability, that is, *no* crossing of either positive resistance region. Amplifiers and oscillators are applications in this class.

The dashed line labeled *monostable* represents the quiescent bias of a "one-shot" multivibrator as it awaits the arrival of a trigger pulse. The restrictions on V_S and on load resistance are simply that their combination places the loadline in a position which intersects one, and only one, of the positive resistance portions of the diode curve. In several monostable applications, V_S is greater than V_V or V_F, and the high-voltage positive resistance region is crossed under quiescent conditions.

If V_S is greater than V_V, and the load resistance is sufficiently high to have the loadline intersecting both positive resistance regions, we have *bistable* operation. The presence of parasitic capacitance and inductance prevents the diode from assuming a *stable* operating point where the loadline intersects the negative resistance region. In many circuits within this class, the bias supply is treated as a current source. Applications include binary counters, memory cells, and logic building blocks.

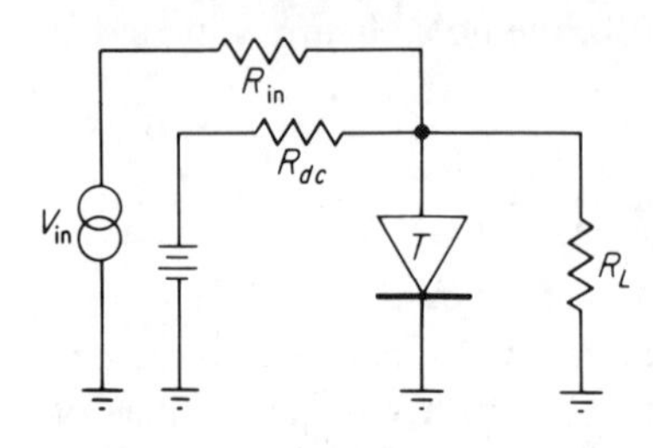

Fig. 13-8. A generalized tunnel diode bias circuit.

In discussing the loadlines and biasing arrangements, we have assumed a generally applicable circuit form as shown in Fig. 13-8. The parallel combination of R_{in}, R_{dc}, and R_L may be treated as an equivalent load, R_P. This shunts the tunnel diode ac equivalent circuit as pictured in Fig. 13-2. If we consider the sum of R_P and R_S and call it R_T, we find that the tunnel diode may be biased stably in the negative resistance region only if $L/r_dC_d < R_T < R$. This is so because the condition for oscillation is satisfied if $R_T \leq L/r_dC_d$, and bistable operation results if $R_T > r_d$. A graphic representation of this situation is shown in Fig. 13-9 (12). The inductance L is the sum of L_S and any additional series circuit connector inductances.

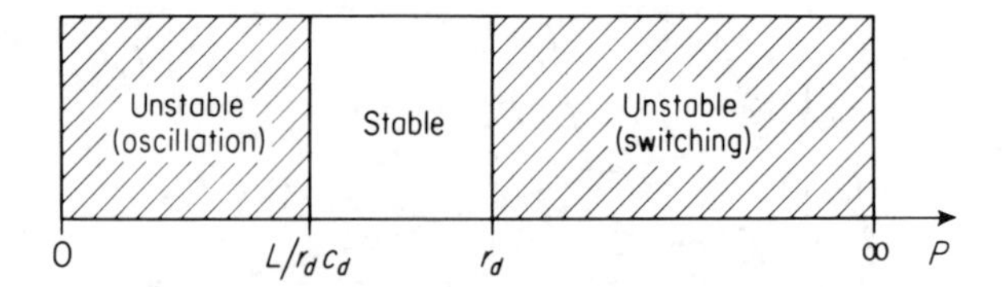

Fig. 13-9. Graphic illustration of allowable bias resistance.

For high-frequency amplifiers we can see a serious problem. The gain-bandwidth product is given by Eq. 13-8, and high-frequency requirements demand r_dC_d to be as small as possible. However, it is important to make (L/r_dC_d) considerably smaller than r_d so that there will be a reasonable range of R_T which can be used. A circuit designer would be in serious trouble if he found that his load resistor has to be $17.2\ \Omega \pm 0.1$ per cent. Therefore, to reduce L/r_dC_d without sacrificing high-frequency capabilities, the device designer must provide a package with the lowest possible inductance. Furthermore, the circuit designer must provide extremely low-inductance connections to the diode in the equipment.

SMALL-SIGNAL APPLICATIONS OF THE TUNNEL DIODE

Amplifiers

Having considered the characteristics of a tunnel diode relevant to its use in an amplifier, we should naturally want to know how the tunnel diode, a two-terminal device, facilitates amplification. First, we know that with positive conductances alone, the maximum transfer of power from a generator to a load occurs when the load conductance G_L equals the generator conductance, G_S. This maximum power is $i^2/4\, G_S$, where i is the current supplied by the generator. If we define the transducer gain or available power gain as the ratio of load power to the power available from the source ($i^2/4\, G_S$) we find that it is at most unity for perfect matching. By introducing a negative conductance, as shown in Fig. 13-10, we may obtain a gain

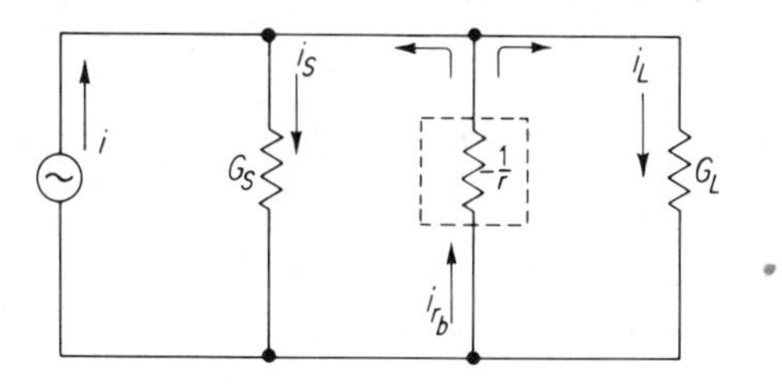

Fig. 13-10. The negative resistance provides gain in a tunnel diode amplifier.

greater than unity. The negative conductance serves to make the ac current i_L greater than $i/2$, thus achieving a power gain. The power in the load is

$$\frac{i_L^2}{G_L} = \frac{i^2 G_L}{\left(G_S + G_L - \frac{1}{r_d}\right)^2} \tag{13-10}$$

and the transducer gain is

$$A_T = \frac{\dfrac{i^2 G_L}{\left(G_S + G_L - \frac{1}{r_d}\right)^2}}{\dfrac{i^2}{4G_S}} = \frac{4G_S G_L}{\left(G_S + G_L - \frac{1}{r_d}\right)^2} \tag{13-11}$$

If $1/r_d = (G_S + G_L)$, A_T becomes infinite and instability results. Therefore, $G_S + G_L$ must be slightly larger than $1/r_d$ to achieve a reasonably high gain with stability.

The choice of the operating point on the tunnel diode characteristic determines $-1/r_d$, but it is often dictated by a desire to minimize noise. It is primarily the fact that tunnel diode amplifiers at high frequencies may offer lower noise figures than transistors, that makes tunnel diode amplifiers attractive. Equation 13-9 gives the best possible or minimum noise figure of a tunnel diode.

If we add a capacitor, C, and inductor, L, to the parallel amplifier of Fig. 13-10 we obtain the tuned-circuit transducer gain as

$$A_T = \frac{4G_S G_L}{\left(G_S + G_L - \frac{1}{r_d}\right)^2 + \omega^2 C^2 \left(1 - \frac{\omega_0^2}{\omega^2}\right)^2} \tag{13-12}$$

where ω is the signal frequency and ω_0 is $1/\sqrt{LC}$. At resonance $\omega = \omega_0$ so Eq. 13-12 reduces to Eq. 13-11.

A wide-band tunnel diode amplifier constructed within a coaxial housing, and with a short-circuited transmission tuning stub for shunt peaking, was described by King and Sharpe (11). The circuit and its transfer characteristic are illustrated in Fig. 13-11.

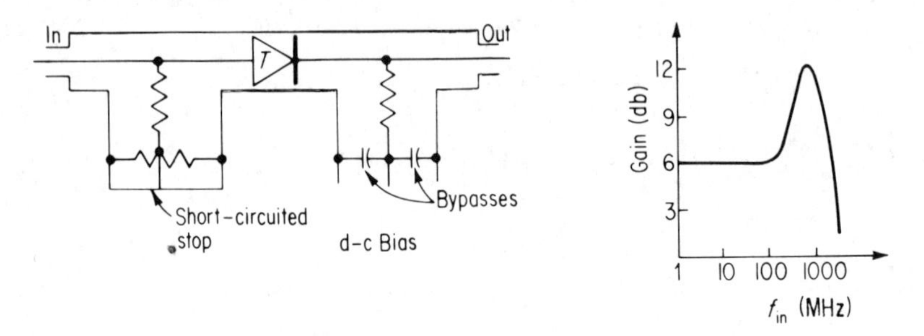

Fig. 13-11. A wide-band tunnel diode amplifier.

From time to time there appears in the literature a paper describing the operation of a tunnel diode amplifier at some frequency well in excess of 1 GHz. Invariably these amplifiers employ a tunnel diode selected for exceptionally low L_S/r_dC. The diode is an integral part of the reflection cavity, and circulators or directional couplers are used to separate input from output (12).

In designing any practical tunnel diode amplifier or oscillator, remember that the negative resistance characteristic is nonlinear, bias-dependent, and temperature-dependent. The nonlinear nature of the ac negative conductance may be readily seen by differentiating Eq. 13-1 to obtain

$$g_d = \frac{1}{r_d} = \frac{dI}{dV} = \frac{I_P}{V_P}\left(1 - \frac{V}{V_P}\right) \exp\left(1 - \frac{V}{V_P}\right) \tag{13-13}$$

If the applied signals are large enough to produce a significant change in g_d over one cycle there will be nonlinear amplification, or AGC action, which may or may not be desired. Similarly, if V is the quiescent bias, we see that instability in V causes a considerable shift in g_d which in turn alters the transducer gain, if G_S and G_L are kept constant.

The negative conductance of the tunnel diode will decrease slightly with temperature because of the thermal decrease in V_P, if V is greater than the inflection point voltage (roughly 150 mv for germanium).

Oscillators

By proper choice of L and C, it is possible to obtain a stable output from a sine wave oscillator whose frequency is given approximately by

$$\omega = \sqrt{\frac{1}{L_T(C + C_T)} - \frac{1}{r_d^2 C(C + C_T)}} \tag{13-14}$$

for $R_S \ll r_d$.

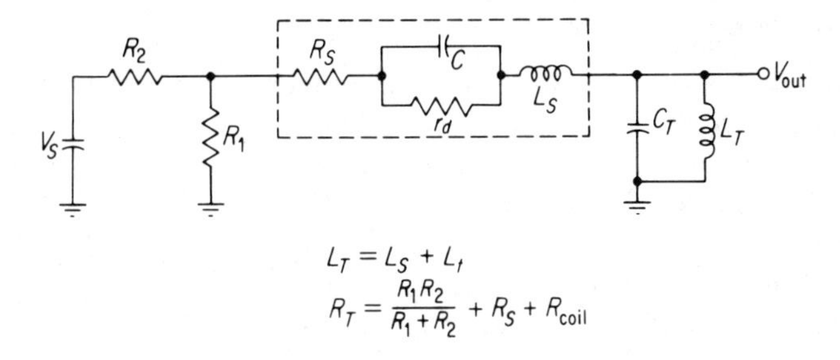

Fig. 13-12. A tunnel diode oscillator circuit.

An oscillator circuit with the tunnel diode shown in its network representation form is given in Fig. 13-12. Since the voltage range for which the tunnel diode dynamic resistance is negative is limited, there is a severe limitation on the power output. Furthermore, this output power and the oscillator efficiency both go down as the operating frequency approaches f_{max}. Dermit (13) has analyzed the power output of a tunnel diode oscillator, finding that

$$P_{out} = \frac{V^2}{8r_d^2}(r_d - R_S)\left\{1 - \left(\frac{f}{f_{max}}\right)^2\right\} \tag{13-15}$$

The maximum power output is given by

$$P_{out(max)} = \frac{V^2}{8L}\left(\frac{1}{\omega} - R_S C\right) \tag{13-16}$$

when operation is limited to the linear portion of the negative resistance characteristic. Note that there is another useful figure of merit given by the product of power and inductance. As inductance is lowered, power output increases until the impedances become so low that coupling becomes a major problem.

Frequency stability in a tunnel diode oscillator is achieved by control of the negative resistance and shunt capacitance. At oscillator frequencies considerably lower than f_{max}, the circuit parameters swamp out variations in the tunnel diode parameters. At higher frequencies, however, it is essential to control the bias voltage and temperature so that g_d is kept constant, as we showed when we considered amplifiers. The diode capacitance increases as the bias is shifted upward, but shows negligible change with temperature.

A crystal-controlled oscillator of the type shown in Fig. 13-13 removes the burden of frequency stability from the tunnel diode and simplifies the design.

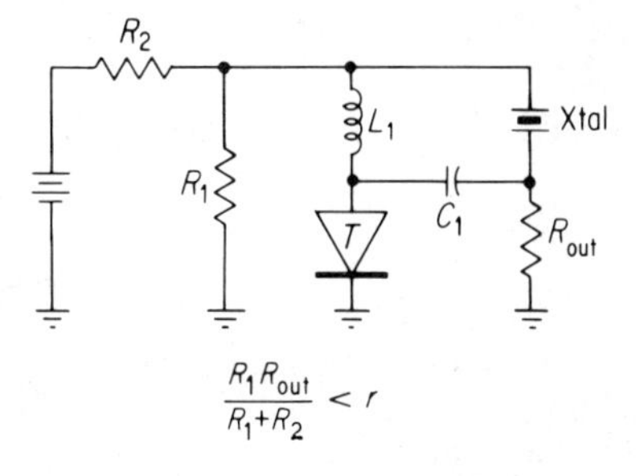

Fig. 13-13. A crystal-controlled tunnel diode oscillator.

Frequency Conversion

The tunnel diode is not limited to operation on a single frequency but can be effectively used for frequency conversion. One of the earliest tunnel diode applications was a down-converter with gain (14). The quiescent bias point is in the low-voltage, positive resistance region, with a low impedance load-line. A local oscillator drives the operating point over the peak and into the negative resistance region for amplification of the incoming signal. The advantage of this form of converter is that the noise figure is independent of the ratio of input frequency to output frequency. This is because the tunnel diode's nonlinear resistance, rather than the nonlinear reactance of a parametric converter, is utilized. The requirements of a tunnel diode for this service are similar to those for amplifier use.

An *autodyne converter* is one in which the tunnel diode is self-oscillating. Here we may use unstable tunnel diodes, that is, those having

$$\frac{L_S}{r_d^2 C} > 3 \tag{13-17}$$

The circuit described by Sterzer *et al.* (15), is shown schematically as Fig. 13-14. The signal frequency was 1.665 GHz and the output frequency was 30 MHz.

The feasibility of using a tunnel diode for frequency multiplication has been demonstrated (16). The portion of the *I-V* characteristic around the peak point resembles the parabola $i = KV^2$, where K is a proportionality constant. Therefore, if we bias at the peak point and apply a sinusoidal voltage, $V = \sin \omega t$, we obtain an output current,

$$i = KV^2 = K \sin^2 \omega t = \frac{K}{2}(1 - \cos 2\omega t) \tag{13-18}$$

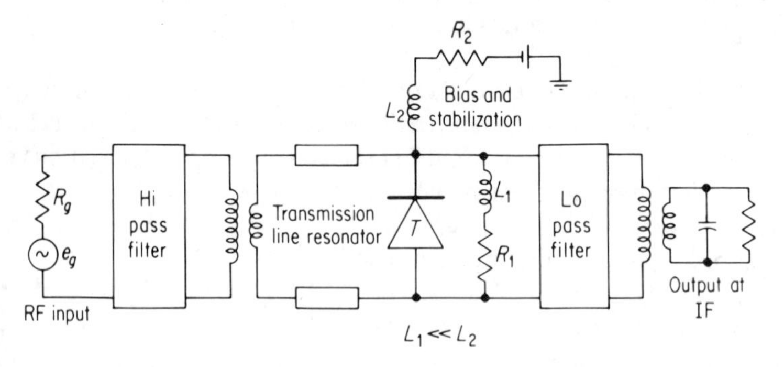

Fig. 13-14. An autodyne converter after Sterzer et al.

Superregenerative Circuits

Superregeneration can be used to produce large voltage gains in tunnel diode amplifiers. Very careful biasing at a point near the peak or near the valley (in a positive resistance region) is required. An external quench oscillator drives the operating point into the negative resistance region. If the conditions stipulated by Jordan and Elco (17) are fulfilled, oscillations will build up to an amplitude which is dependent upon the RF input. Bogusz and Schaffer (18) have described an externally quenched superregenerative amplifier and diode detector which handled signals as low as 40 μv (50 per cent modulation), and provided a voltage gain from input to detected audio output of 250.

MONOSTABLE OPERATIONS OF THE TUNNEL DIODE

Monostable operations encompass the largest variety of tunnel diode circuits and perhaps the greatest volume of tunnel diode usage. One of the more common applications of monostable operation is the threshold detector.

Threshold Sensing Applications

Undoubtedly the simplest possible tunnel diode circuit is that shown in Fig. 13-15. It has no quiescent bias supply but simply responds to the applied input.

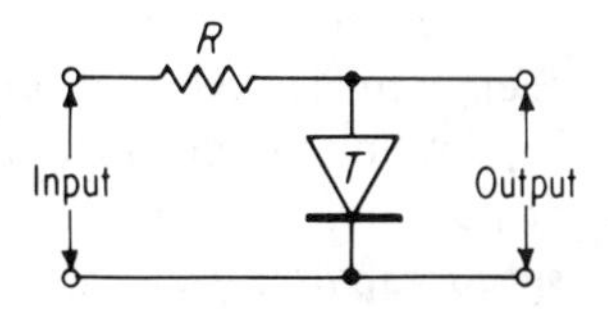

Fig. 13-15. The threshold detector.

In this monostable application, the loadline has a fixed slope corresponding to $-1/R$, but its position is entirely dependent upon the amplitude of the input which is the only source voltage. Figure 13-16 shows a family of loadlines of a resistor, R, superimposed upon the tunnel diode I-V curve. The dashed line which intersects the peak point is obtained when the input voltage is at V_T, the threshold level. As the input voltage rises from zero to V_T, it has values such as V_1, V_2, V_3, etc. The instantaneous loadlines for these voltages are shown. Although the input

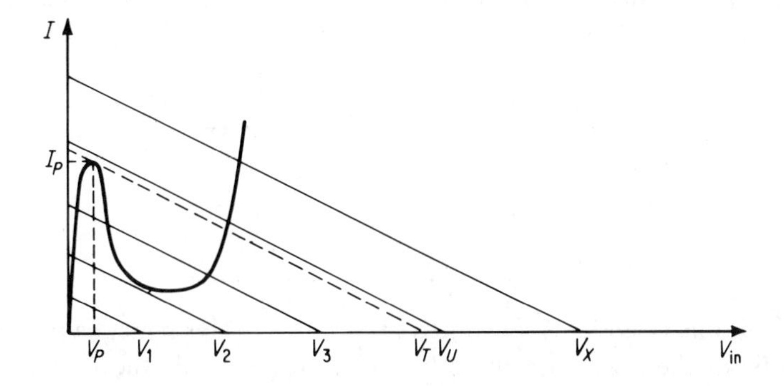

Fig. 13-16. Various dc loads for different input voltages. An output occurs when the input voltage exceeds V_T, the threshold voltage.

range from V_3 to V_T provides bistable conditions, there is no auxiliary input available to supply current to the tunnel diode in excess of I_P, so that "early" switching is precluded. Switching does occur when V_T is exceeded. For an input of V_U we have a tunnel diode voltage which is about sevenfold greater than V_P. V_T is equal to $V_P + I_P R$. This relationship makes it easy to choose a tunnel diode and resistor combination for a desired threshold voltage.

Usually V_T is much larger than V_P, so that the precision with which this circuit can sense the threshold level is based largely upon the tolerance and temperature dependence of the resistor, R, and the tunnel diode's peak current.

This simple circuit is useful in nuclear instrumentation where the energy of particles under study is related to the amplitude of the pulses they produce.

The threshold-sensing property of the simple resistor-tunnel diode series combination may be applied to a variety of interesting uses. The generation of an electrical analog of a mathematical function generally involves the use of conventional rectifier diodes with accurately controlled bias supplies. A tunnel diode function generator (19) was constructed by connecting sixteen resistor-tunnel diode combinations to a common input. A portion of each tunnel diode voltage is made part of a weighted sum to produce an output voltage proportional to the square root of the input amplitude. The advantages of this technique are the elimination of bias supplies and the high-speed capability of the tunnel diodes. With 58 tunnel diodes it is possible to achieve an accuracy of ± 1 per cent.

Combined with transistor circuits, threshold sensors can also be used to perform analog-to-digital conversion (20) and to protect the transistors from damage due to excessive transient currents.

TUNNEL DIODE FULL BINARY ADDER

The full adder performs the basic operation of binary addition, including the "carry" function generated by any previous stage. The basic addition table is as follows:

Inputs	Sum Bit
$0 + 0$	0
$0 + 1$	1
$1 + 0$	1
$1 + 1$	0 with a carry of 1

A four-bit binary adder is illustrated in Fig. 13-17. This adder cannot handle sign bits for the binary words to be added; it adds only the magnitudes of the numbers. This same basic configuration may be extended to any number

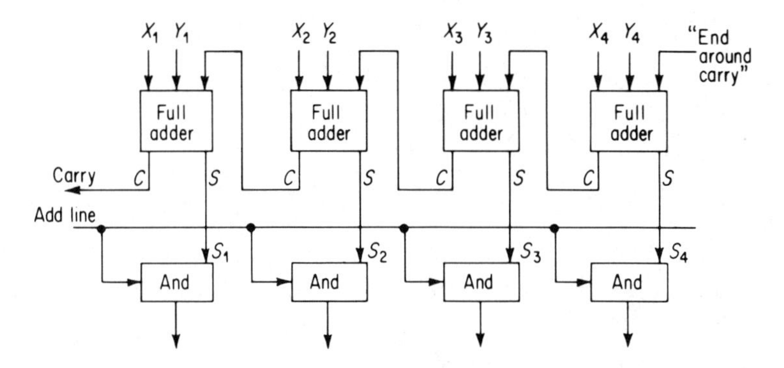

Fig. 13-17. A four-bit, parallel, binary full adder.

of bits. For example, a seven-bit adder would contain seven full adders. Note that the carry line could be used to enable the four-bit adder to have a five-bit output.

The addition of binary numbers is commonly done by making use of diode-transistor logic. A full binary adder usually consists of 20 diodes, 9 transistors, and 20 resistors. A simple tunnel diode circuit has been reported that can perform full binary addition using one tunnel diode and six resistors. (21). The circuit, which uses a bridge, is illustrated in Fig. 13-18.

This circuit operates with no quiescent bias. The inputs at x, y, and z are pulses, so that the voltage amplitudes V_x, V_y, and V_z should take the two discrete values zero and V'. In practice, however, a tolerance margin on these levels will necessarily have to be applied.

In analyzing this simple circuit, let us first look at the I-V characteristic of the tunnel diode in this application. This is shown in Fig. 13-19 with the three operating points indicated as A, B, and C. Points A, B, and C are

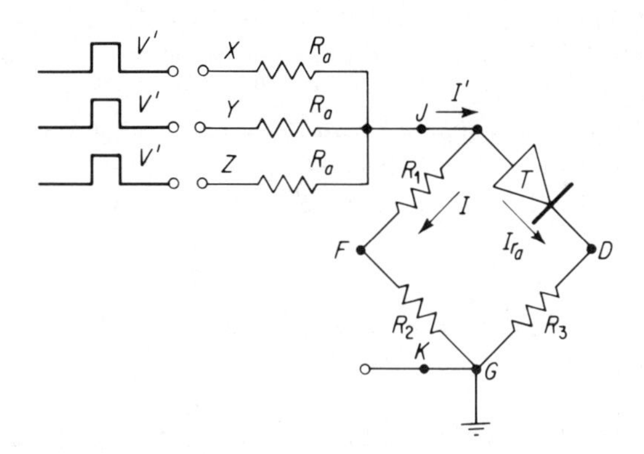

Fig. 13-18. A tunnel diode circuit which performs full binary addition.

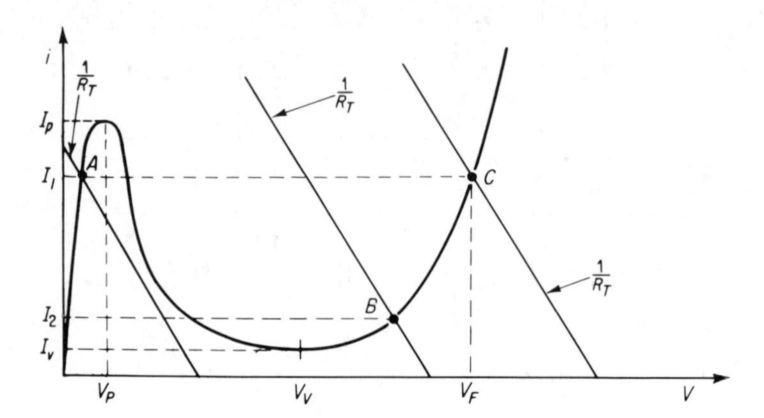

Fig. 13-19. Biasing of tunnel diode in the binary full adder.

achieved by one, two, and three inputs, respectively. R_T is the Thevenin equivalent resistance of the circuit connected to the diode.

Referring to Fig. 13-18, whether there are one, two, or three inputs, the voltage V_{DG} has the property of *Sum*, and the voltage V_{DF} has the property of *Carry*. This is based on having the proper values of all circuit components and input voltage. Two necessary requirements are (1) that $R_1 = V_1/I_1$; (2) that $R_2 = R_3$. With one input applied, the voltage V_{DG} across R_3 is I_1R_2 and the voltage V_{DF} is equal to zero. At operating point A, the resistance that the tunnel diode presents to the bridge is R_1. Therefore, the bridge is balanced and V_{DF} is equal to zero.

With two inputs, the voltage V_{DG} across R_2 is equal to I_2R_2. This value will be 5–20 times lower than I_1R_2 when using a tunnel diode with a high peak-to-valley ratio. The higher the peak-to-valley ratio, the greater the difference between V_{DG} for one input, and V_{DG} for two inputs. For three inputs, the voltage V_{DG} will again be equal to I_1R_2, since the tunnel diode is operating at point C. With two and three inputs, the resistance that the tunnel diode presents to the bridge is V_2/I_2 and V_3/I_1. Since both of these values are greater than R_1, the bridge will be unbalanced and there will be an output voltage between D and F. From this it can be seen that the voltage V_{DG} across R_2 has the property of *Sum* for all three inputs, and the voltage V_{DF} has the property of *Carry*.

DIGITAL APPLICATIONS OF THE TUNNEL DIODE

The bistability of a tunnel diode supplied by a current source enables the performance of ONE-ZERO type of binary operations which are required for the building blocks of digital computers.

The Tunnel Diode as a Memory Element

Use of the tunnel diode as a memory element is based upon the existence of bistable operating voltages when it is supplied with a current larger than its valley current, I_V, and smaller than its peak current, I_P. Upon application of a bias current I_1, where $I_V < I_1 < I_P$, the tunnel diode voltage will be in the 10–50 mv region of Fig. 13-20(a), the characteristic for a germanium tunnel diode with $I_P = 1$ ma. Additional current, I_A, supplied as a pulse will cause the tunnel diode voltage to increase above V_P if $I_1 + I_A > I_P$. When I_A is removed, the operating point is in the 350–450 mv region. Figure 13-20(b) shows the biasing for the loadline indicated on Fig. 13-20(a). The write-in and read-out pulse sequence and the resulting tunnel diode voltages are shown in Fig. 13-20(c). The reset pulse current I_B must have an amplitude greater than $(I_1 - I_V)$.

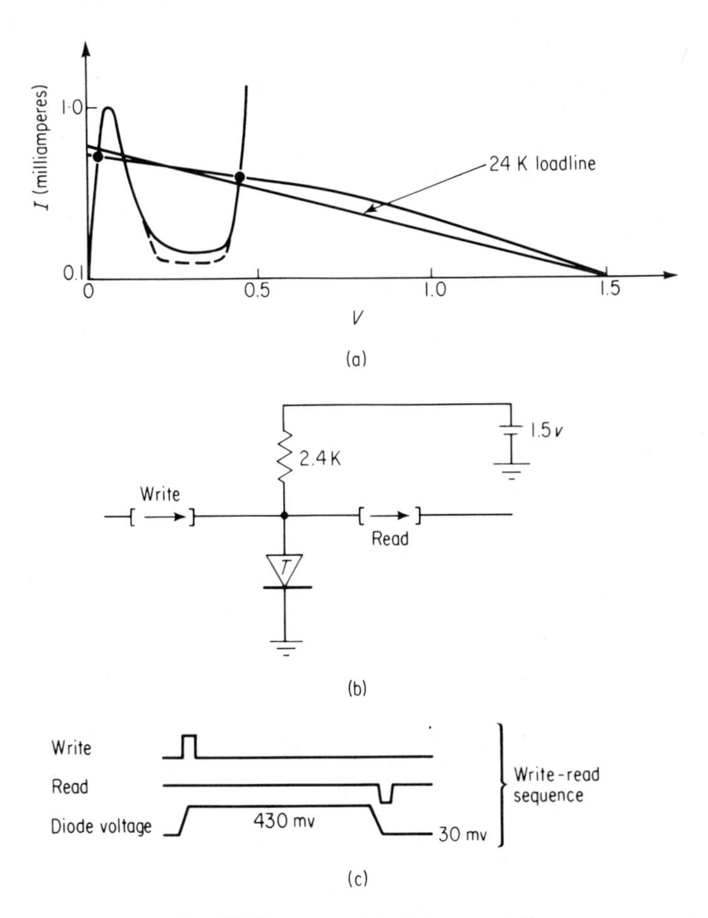

Fig. 13-20. A tunnel diode memory cell.

The simplest form of read-out from a tunnel diode memory is destructive. The capacitor and transformer provide a sinusoidal voltage read-out as shown in Fig. 13-21(a). Selection lines X_i and Y_j directly couple the write-in and read-out pulses to the tunnel diode. Each line carries half the required pulse current, thus making possible coincident current selection which is needed in a large memory matrix, as shown in Fig. 13-21(b). The bias, X, and Y lines are generally word-organized for economy in a large system. The minimum current for a half-select write pulse is determined by a worst-case equation:

$$I_{HS\,(\min)} = \tfrac{1}{2}\,(\text{overdrive factor})(I_{P\,\max} - I_{1\,\min}) \tag{13-19}$$

For fast switching, 10 per cent overdrive is generally sufficient. Thus, the overdrive factor is 1.10 $I_{P\,\max}$. Using a supply voltage and resistors with ± 5 per cent tolerances,

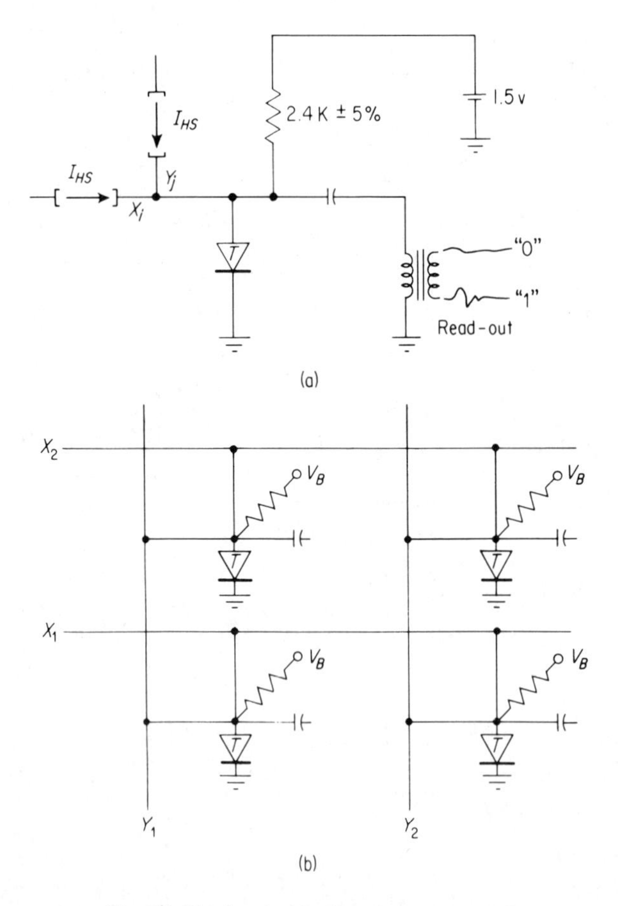

Fig. 13-21. A coincident-current memory cell.

$$I_{1\,\min} = \frac{V_{\text{bias}}}{\bar{R}} = \frac{1.425}{2520} = 0.566 \text{ ma}$$

The resulting minimum I_{HS} is 0.263 ma.

The maximum current for a half-select pulse is that which when added to $I_{1\,\max}$ just produces $I_{P\,\min}$. $I_{HS\,\max} = I_{P\,\min} - I_{1\,\max} = 0.284$ ma, for the diode with peak current tolerance of ± 2.5 per cent and ± 5 per cent resistor and supply tolerances.

THE TUNNEL DIODE AS A DECISION-MAKING ELEMENT

The tunnel diode used for the performance of a logical function in a digital computer is likely to have its output and input coupled to and from other tunnel diode circuits. This introduces the need for unilateralization, or provision for flow of information in one direction. The memory circuit with its sinusoidal read-out had no such requirement.

The circuit of Fig. 13-22 represents an OR gate in rudimentary form. Tunnel diode C is biased to switch to its high state if either A or B is high. If, for example, A is high and B is initially low, C will be switched to its high state. This gives rise to a current through R_B which is sufficient to cause diode TD_B to go into its high state. "Sneak paths," such as this, are in the direction opposite to the desired flow of information. A simple nonlinear element which would inhibit sneak paths is clearly desirable.

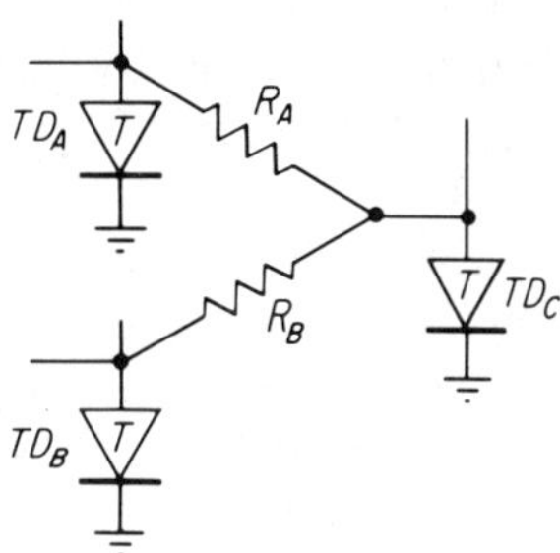

Fig. 13-22. A tunnel diode OR circuit.

THE TUNNEL DIODE IN A BINARY COUNTER

One of the most important digital applications of the tunnel diode is its use in flip-flops or binary counters. The circuit of Fig. 13-23 is an inductor-coupled flip-flop.

Under steady-state conditions, the 400 mv between point A and ground is sufficient to maintain one diode in its high state. The difference in diode currents, 0.6 ma, flows through the inductor, producing a negligible voltage drop.

The application of a negative pulse having an amplitude in the 0.3–0.4-volt range at the input terminal will initiate a change of state of both diodes. First, the increased voltage at A is sufficient to maintain both diodes in their

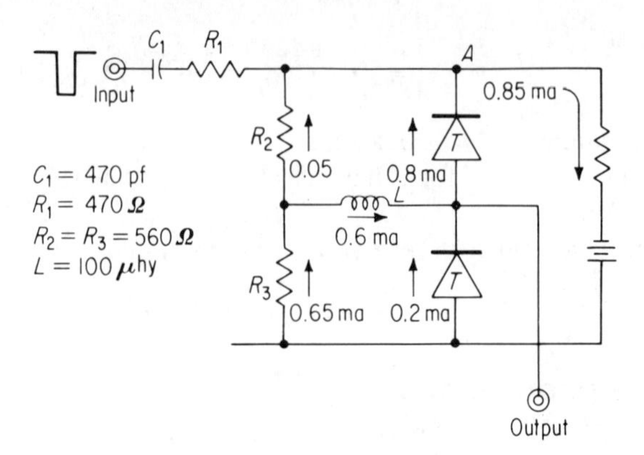

Fig. 13-23. A tunnel diode flip-flop.

high states. The induced voltage across the inductor, however, forces the diode which had originally been high to the low state.

The L/R_2 time constant should be minimized to promote high-speed operation. Since R_2 is determined by the diode dc characteristics, L must be minimized. Below 100 μhy, however, the reduced coupling retards the switching of both diodes.

This type of flip-flop is attractive because it is triggered by pulse heights of the order of tunnel diode voltage swings. This makes it possible to connect several flip-flops to form shift registers and ring counters as shown in Figs. 13-24 and 13-25. The flip-flop of Fig. 13-23 was operated with the high-speed performance of Table 13-4. This indicates a maximum input pulse rate of about 15 MHz.

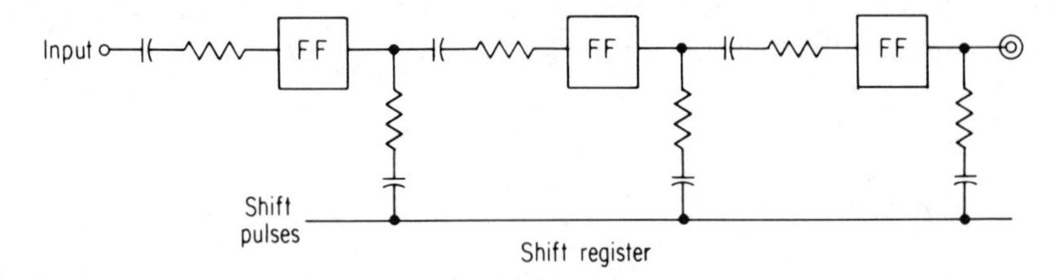

Fig. 13-24. Connection of the flip-flop as a shift register.

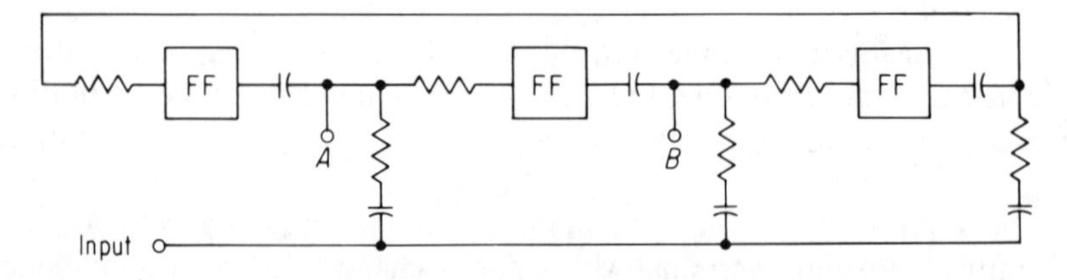

Fig. 13-25. A tunnel diode ring counter.

TABLE 13-4

Input Pulse Amplitude (mv)	Turn-on Time (ns)	Turn-off Time (ns)
200	45	18
300	37	23
330	37	26
360	37	30
400	37	33

Combinations of tunnel diodes with each other, and with transistors and other active devices, widen the scope of tunnel diode digital applications. N series-connected tunnel diodes produce a composite I-V characteristic which may have between $N + 1$ and $2N$ positive resistance regions, depending upon their relative peak and valley currents. Such combinations can be used for pulse-counting and staircase generation with transistor resets (22).

DIODE CIRCUITS

Diodes play important roles in a large variety of communication, computer, and power-rectification circuits. Applications involving high-frequency signals and little power include waveshaping circuits, such as clippers and clampers; logic circuits, such as AND and OR gates and choppers; and detectors for demodulation of amplitude-modulated signals. In such circuits, the diodes are usually physically small and, if semiconductor diodes are used, may be parts of complex integrated circuits. Power rectification equipment, on the other hand, may employ considerably larger diodes (often called *rectifiers*) which are not required to handle frequencies greater than a few dozen, or perhaps a few hundred, hertz. The principles of circuit operation and analysis presented here do not depend upon the type of diode used.

WAVESHAPING CIRCUITS

Circuits which modify the shape of an input voltage (or, occasionally, current) waveform, typically by disconnecting the input from the output for a portion of the wave, or by short-circuiting the output for a portion of the wave, are called *clippers*. In the simplest cases, the circuits include only diodes and resistors, and perhaps fixed sources. More complicated circuits include capacitances (either as fixed components or stray) and, rarely, inductors. Other diode circuits attempt to change the average value (dc level) of a waveform without serious change in the waveshape; these are called *clampers* or *dc restorers*.

Clippers

Figure 13-26 shows several simple diode-resistor clipping circuits. In each case, diodes are in series with the signal (between input and output), in shunt with the output, or both. The operation of such circuits is *most easily* determined by assuming that each diode is an ideal open circuit when it is reverse-biased, and an ideal short circuit when it is conducting. The operation is *more accurately* determined by allowing for each diode's forward and reverse resistances, R_f and R_r, and the cut-in voltage, V_γ.

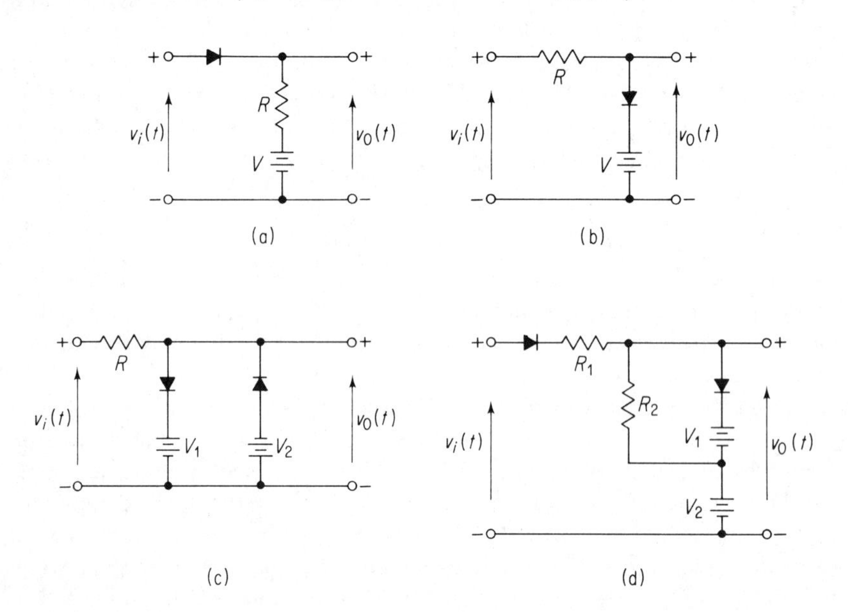

Fig. 13-26. Simple clipping circuits.

The analysis of the operation of the circuit of Fig. 13-26a is typical. Assume first that the diode is ideal. Whenever $v_i > V$, the diode conducts. Since it is ideal, it then acts as a short circuit, and $v_o = v_i$. Whenever $v_i < V$, the diode is reverse-biased and acts as an open circuit; then $v_o = V$, since there is no current flowing in the circuit and no voltage drop across R. The effect of this circuit upon an arbitrary input waveform is shown in Fig. 13-27. Note that V can be positive, zero or negative. Note also that if the diode were reversed, then v_o would equal v_i whenever the input was negative with respect to V, and that v_o would equal V at all other times.

Assume now that the diode is more accurately represented by a small voltage source V_γ in series with a small resistance R_f, when it is conducting;

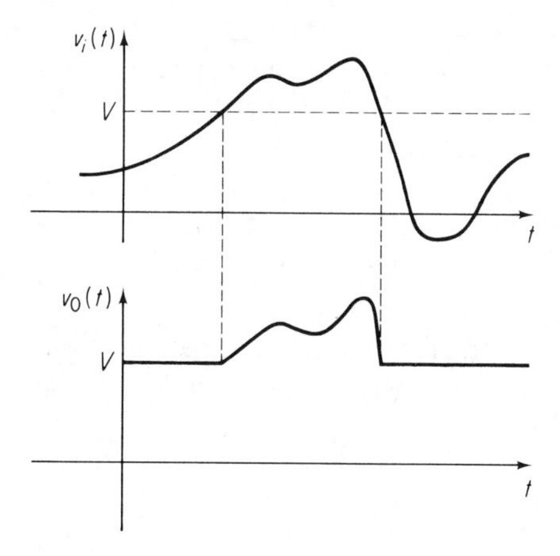

Fig. 13-27. An example of the operation of the circuit of Fig. 13-26(a), assuming an ideal diode. Here, $V > 0$.

and by a large resistance R_r otherwise. (The cut-in voltage V_γ is the *break* point between conducting and nonconducting states.) An analysis of the circuit of Fig. 13-26(a) leads to the conclusion that the clipping action is not quite perfect—that is, not all of v_i is transmitted to the output when it should be, and not all of v_i is absent from the output when it should be. It is not difficult to show that when the diode is conducting (*on*),

$$v_o = V + \left(\frac{R}{R + R_f}\right)(v_i - V - V_\gamma)$$

and that when the diode is *off*,

$$v_o = V + \left(\frac{R}{R + R_r}\right)(v_i - V)$$

The effect of including R_f, R_r and V_γ in the analysis is shown in Fig. 13-28.

The analysis of other clipping circuits, such as are shown in Figs. 13-26(b), (c) and (d), is similar. It is first necessary to determine *when* each diode is in

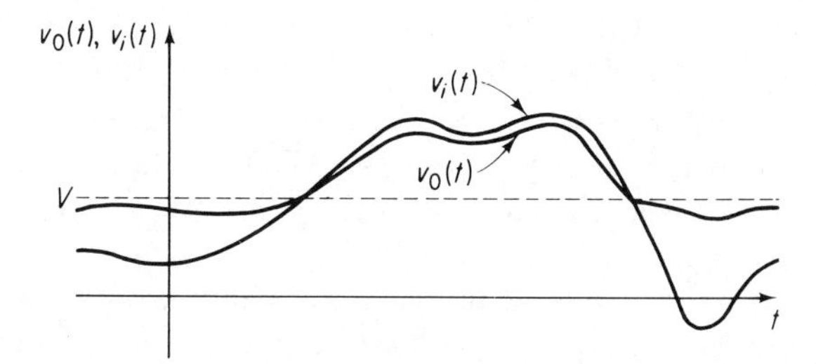

Fig. 13-28. An example of the operation of the circuit of Fig. 13-26(a), when the diode is not ideal. Here, $V > 0$.

its conducting and nonconducting states, and then to determine the effect of the circuit upon the input voltage waveform for each case. If the diode is assumed to be ideal, then for the circuit of Fig. 13-26(b), $v_o = v_i$ whenever $v_i < V$, and $v_o = V$ otherwise.

Figure 13-26(c) shows a double-clipper. If the diodes are assumed to be ideal, then neither diode will conduct whenever $-V_2 < v_i < V_1$. Then the output voltage will equal the input voltage. Otherwise, $v_o = V_1$ when $v_i > V_1$ and $v_o = -V_2$ when $-V_2 < v_i$.

It can be shown that for optimum circuit performance with non-ideal diodes, R in Figs. 13-26(a), (b) and (c) should be approximately equal to $\sqrt{R_r R_f}$.

Transfer Characteristics

In circuits such as those shown in Fig. 13-26, in which there are no energy-storing elements such as capacitors, each value of the input voltage will result in one and only one value of the output voltage. It is often helpful to plot v_o versus v_i; the result can be used to graphically determine v_o for any given v_i. The derivation of the transfer characteristic for the circuit of Fig. 13-26(a) will serve as an example.

The equations presented above, for a circuit with non-ideal diodes, represent two straight lines when v_o is plotted against v_i. These lines intersect at $v_i = V + V_\gamma(R + R_r)/(R_r - R_f)$. To the left of the intersection, the line corresponding to a nonconducting diode applies; to the right of the intersection, the other line applies. From the equations, it is seen that the slopes of the lines are, respectively, $R/(R + R_r)$ and $R/(R + R_f)$.

The resulting graph is shown in Fig. 13-29. Also shown there is the use of the transfer characteristic to determine the waveshape of v_o if the waveshape of the input voltage is given. The characteristic may be considered to be a bent mirror which *reflects* input voltages, and the output waveshape is constructed point-by-point for increasing values of time.

Clamping Circuits

It is sometimes necessary to add or subtract a dc voltage to a given waveform, without greatly changing the *shape* of the wave. Sometimes it is desirable to *shift* a waveform so that a particular part of it—say, the positive peaks—is maintained at a specified voltage. Circuits which do this are called *clamping circuits.* When the circuit is used to restore a dc level which has been lost from a waveform as it passed through a capacitor or a transformer, it is referred to as a *dc restorer*; in many television receivers, such a circuit attempts to hold the *black level* constant despite changes in the content of the picture, and to provide a black background when a scene has been faded-out.

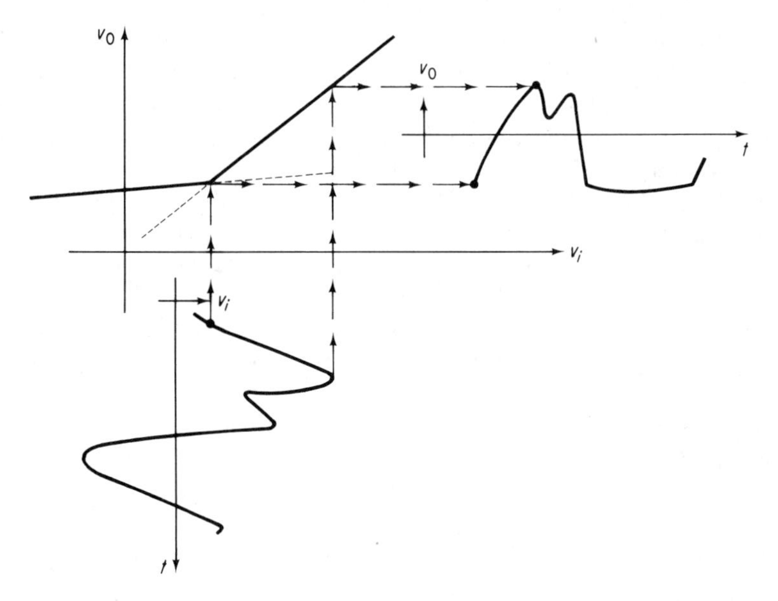

Fig. 13-29. The transfer characteristic for the circuit of Fig. 13-26(a), and an example of its use to find V_0 for a given V_i.

The principle of a clamping circuit is simple: remove the dc level, if any, contained in the waveform, and establish a new one that is suitable. The most common method of removing the dc level of the input wave is to pass the signal through a capacitor. Thereafter, the most positive or most negative portion of the waveform is *clamped* to a specified reference voltage. Two circuits to do this are shown in Fig. 13-30.

The operation of the circuit of Fig. 13-30(a) is easy to describe if the diode is assumed to be ideal. Whenever v_o attempts to go positive, the diode conducts. Then current flows from the signal source through the capacitor and the diode, and the capacitor charges to the peak value of the input voltage. Then, when the input voltage begins to fall, the output voltage will also fall, and the diode will be reverse-biased and will stop conducting. The charge which has been placed on the capacitor is trapped there, and the capacitor acts like a fixed voltage source. The diode will not conduct again, and the

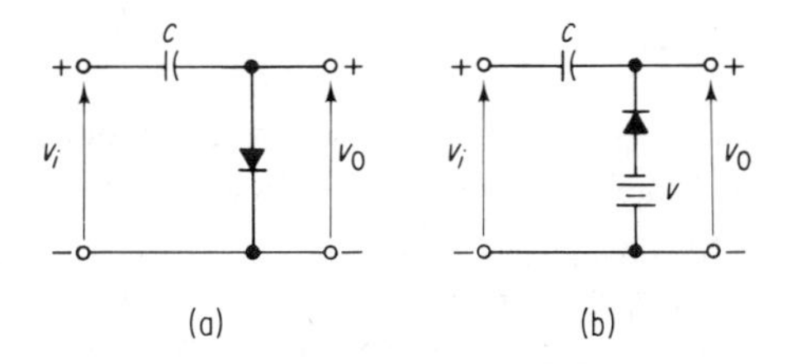

Fig. 13-30. Two clamping circuits.

charge on the capacitor will not change, until the input goes more positive than before; if the input is a periodic wave, then the charge on the capacitor will be determined during the first cycle and will not change thereafter.

If, instead, the diode is not ideal, but can be represented by R_f in series with V_γ (when it is conducting) or by R_r (when it is supposed to be *off*), then current will flow through the capacitor both when the output voltage is positive and when it is negative, and the charge on the capacitor will grow and shrink. When the diode was assumed to be ideal, v_o could never go positive, but if the diode is not ideal, v_o will exceed V_γ and will depend upon the waveshape of v_i. When the diode is *off*, the waveshape of v_o will not be exactly the same as the waveshape of v_i because of the (usually slight) charging of the capacitor due to current flowing through the diode's R_r. The operation of this circuit is illustrated in Fig. 13-31, where a sinusoidal input voltage is assumed, and the output voltage is shown for the case when the diode is ideal, and when it is non-ideal.

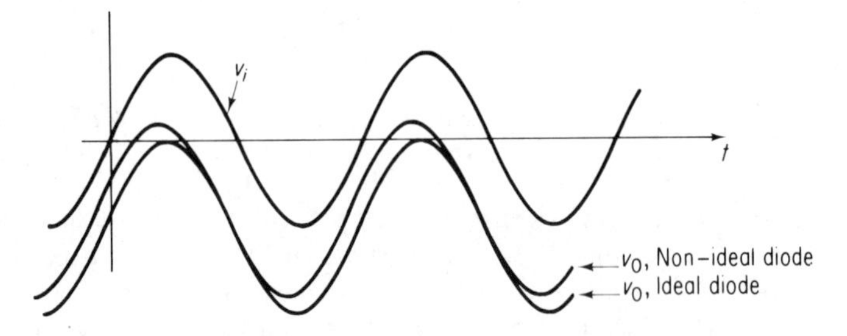

Fig. 13-31. Waveforms for the circuit of Fig. 13-30a.

If it is important that the *shape* of the wave be altered as little as possible, then it is necessary that the capacitor be large enough so that the circuit's two time-constants, R_rC and R_fC, both be large compared to the period of the input signal.

Figure 13-30(b) shows another clamping circuit. Here, the most negative part of the output is *clamped* to $-V$, and, if the capacitor is sufficiently large, the *shape* of v_o will be identical with the shape of v_i.

BACKWARD DIODE

A germanium backward diode has the characteristic depicted in Fig. 13-32. When introduced into each coupling line of the OR gate as in Fig. 13-33, flow of current from output to input is reduced to a few microamperes. The desired current flow, however, is unaffected. The backward diode can thus be used as a coupling element to enable unilateralization of tunnel diode

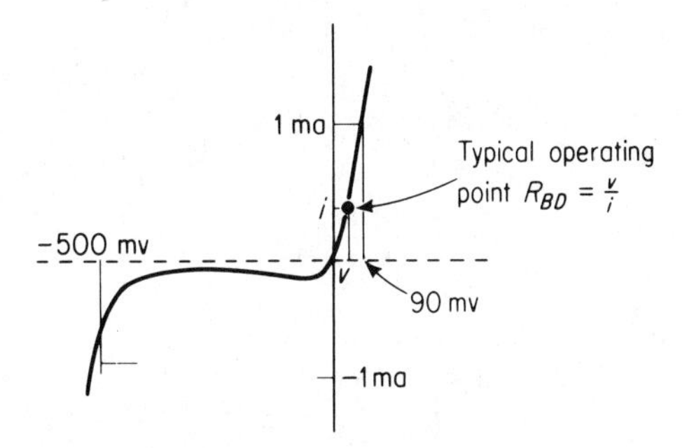

Fig. 13-32. Definition of the backward diode series resistance.

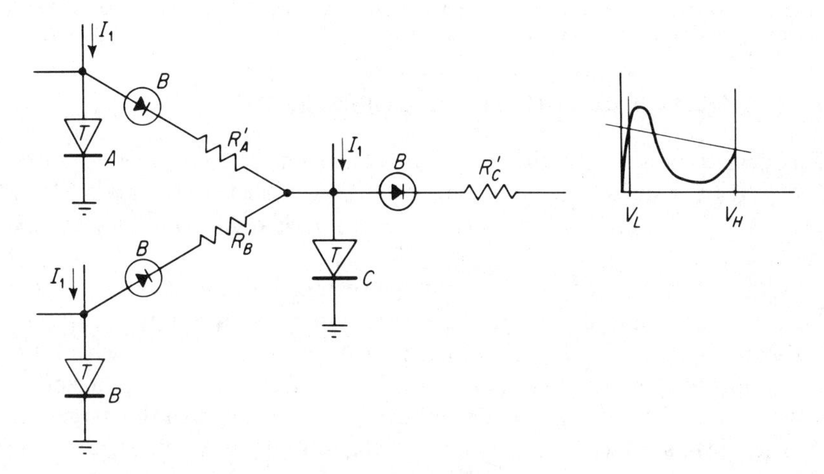

Fig. 13-33. The application of backward diodes as coupling elements.

logic. It can be used as the only coupling element (that is, no series resistor) in an OR gate with 5 ma peak current tunnel diodes.

For input signals of less than 500 mv peak-to-peak amplitude, a germanium backward diode will be considerably more efficient than conventional germanium rectifiers because of the low forward drop. Figure 13-34 shows what outputs may be expected from rectifiers and dc restorer circuits using the two types of diodes.

A new and promising application of the backward diode results from its low noise at audio frequencies. It is then very well suited for use as a microwave mixer (23). It offers as much as 30 db improvement in noise figure over point contact diodes in certain applications that require an intermediate frequency (IF) of less than 1 MHz. The backward diode is preferred over a tunnel diode as a low-noise down converter because the former is unconditionally stable for all source and load conductances.

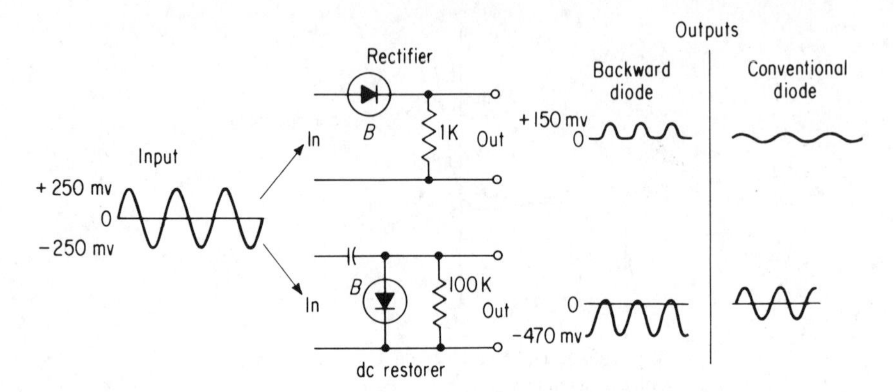

Fig. 13-34. The backward diode used as a dc restorer. The backward diode operates at much lower voltage levels than the conventional diode.

THE SILICON CONTROLLED RECTIFIER (SCR)

This device is essentially the solid state equivalent of a gas filled thyratron. The SCR behaves like a gated diode; that is, once turned on it approximates a forward biased diode. However, a gate current is required to trigger it into the ON state.

With reference to Fig. 13-35 the gate triggering action occurs when a small current is passed through the base region p_3. This current is carried by holes and is similar to injection into an *npn* transistor thereby permitting a large transistorlike current to flow across the effective transistor formed by $n_4p_3n_2$. Thus the resulting anode-cathode current is considerably larger than the gate current. Once initiated however, the resulting electron current across the n_4p_3 (emitter) and n_2p_3 (collector) junctions results in transistor action in the effective transistor $p_1n_2p_3$ with p_1n_2 acting as an effective emitter junction for holes. Thus, the regenerative action continues and the gate current ceases to have any effect. If no gate current exists, the device is held in the OFF state by two back biased junctions in series until the voltage is great enough to cause minority carrier flow to turn the device on. The action then is similar to that for a given gate current. The gate then allows control of the turn on voltage required to initiate conduction. Thus, a small current can be used to control a large current and the device is thus a controlled switch or rectifier.

The time required to switch an SCR from the forward blocking or OFF state to forward conducting or ON state is of the order of the transit time of minority carriers across the base region n_2p_3 and is thus on the order of 0.1–1 μsec. Once the SCR is on the only way to bring it to the OFF state is to reduce the anode current to a value below the critical value I_{ON}, the holding current. When this is done, the SCR takes about 1–10 μsec to return to the blocking state. Thus, the minimum switching time of an SCR is on the order of 1 μsec,

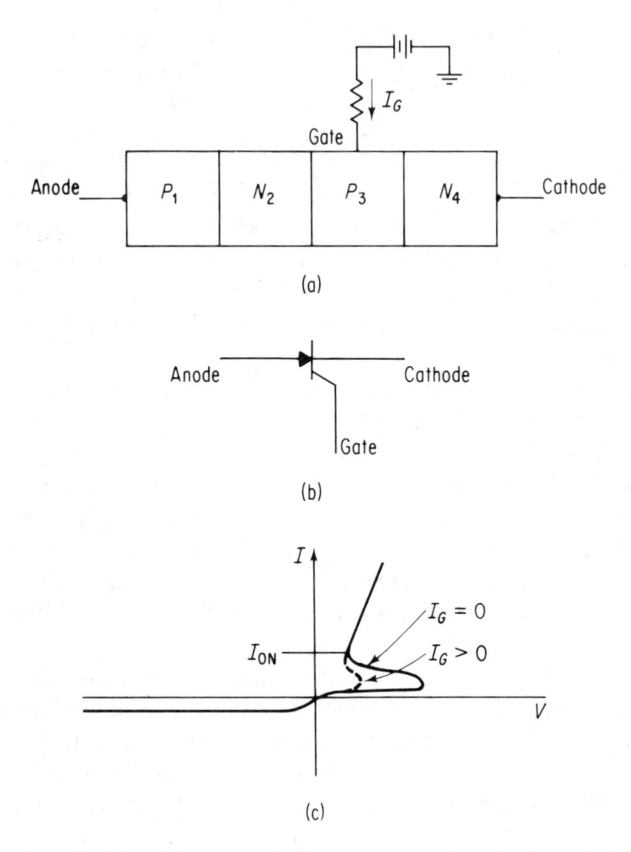

Fig. 13-35. SCR (a) physical arrangement; (b) circuit symbol (c) I-V characteristic.

which restricts it to low frequency applications. However, it is capable of controlling currents in excess of 1000 amperes, and can withstand voltages in excess of 1000 vòlts before turning on (in the absence of gate current), thus making it extremely useful in handling large power. Thus, it is often useful in motor controllers, inverters, rectifiers, light dimmers etc., and other applications where switching or control of hundreds or thousands of watts of ac are necessary. In certain applications these can be activated by optical signals impressed on the gate area of the device as well, thus providing an additional control mode.

AVALANCHE DIODES

It has recently been discovered that PN junction diodes operating in the reverse-biased direction are capable of microwave oscillation and amplification (25). The theory of these devices is rather complicated but stems from the negative resistance inherent in two-carrier flow under space charge conditions. That is, if a PN junction diode is biased in the reverse-biased direction, and well into the avalanche region so that the carriers, both holes and electrons, all drift essentially at a limiting velocity determined by the material used, an instability develops. Thus, we have a rather narrow depletion region of the order of 10^{-6} meter (1 μm) traversed by carriers at velocities of the order of 10^5 m/sec, resulting in a transit time of the order of 10^{-11} sec. This, when suitably exploited by placing the device in an appropriate resonant cavity, can cause oscillation in the microwave region, since the inverse of the transit time is essentially the frequency. That is, any small disturbance will grow and couple strongly to the carrier stream within the device, resulting in amplification.

The typical diode of gallium arsenide, when biased with 30–50 volts and correspondingly 10–6 ma, for a total dc power input of the order of 300 mw, will oscillate in the frequency range 10–15 GHz, at output powers of the order of 10–20 mw, when mounted in reduced height X-band waveguide. For silicon diodes, one generally uses a PIN structure as proposed by Read, but the principle of operation is similar. The intrinsic material permits longer and much better controlled transit times, however, so that operation at lower frequencies (1–3 GHz), and considerably larger output powers (200 mw) are possible.

Both PN and PIN devices can also be used as small-signal amplifiers, mixers, and converters, and may well replace conventional Klystrons, varactor chains, and other microwave sources in many applications. Their full capabilities are still far from being fully exploited, and the future holds much promise for their utility in high-speed circuitry and microwave circuit applications.

GUNN OSCILLATORS

Along with the development of avalanching diode oscillators, a good deal of effort has been expended on improving the performance of bulk effect devices. One such device, first described by Gunn (26), occurs in GaAs. The operation of the device depends on the peculiar band structure of this material which permits, in effect, two different kinds of conduction band electrons to exist. One of these, the normal one, which is generally responsible for charge conduction in the material when used for diodes, tunnel diodes, and transistors, is a so-called "light" electron, with an effective mass perhaps $0.08m$, where m is the normal free electron mass. The other is a "heavy" electron with mass of the order of $1.2m$. Under the normal electric fields applied in standard devices, this material behaves in an ordinary manner, and the light electron is responsible for current flow. When, however, high electric fields, on the order of 10^4 volts/cm, are applied to thin samples of the material, some of these light electrons can be excited into the heavy electron band. This causes a drastic change in the conductivity of the material, since the mobility of the heavy electrons is much smaller than that of the light ones. Thus, the current, which is proportional to the product of conductivity and electric field, decreases even though the electric field increases. We have thus a so-called negative differential mobility, resulting in an *I-V* curve similar in shape to that for the tunnel diode (25).

The device can produce significant microwave power as an oscillator from 1–70 GHz, can act as an amplifier at these frequencies, is partly voltage- and current-tunable (as well as mechanically tunable over a wide band), and is easy to fabricate. It requires low voltage, of the order of 20 volts, to produce output powers in the milliwatt range, and has an efficiency of conversion of dc to ac power of the order of 2 to 3 per cent, which is rather better than, for example, comparable Klystron sources.

The entire field of solid-state microwave device development is currently in a state of turmoil as we discover more and more devices which can generate, modulate, or amplify high-frequency signals, and all with simpler power supplies and more efficiency than comparable vacuum devices.

GUNN BASICS

Gunn diodes are one-port devices that exhibit negative resistance at microwave frequencies due to growing carrier waves. Carrier waves are associated with the drift motion of carriers and may be thought of as being the equivalent of space-charge waves on an electron beam. Although they are heavily attenuated in extrinsic semiconductors with only one type of carrier present, their loss can be reversed by the Gunn effect in the III-V compound of GaAs.

At sufficiently high dc fields, the carriers in GaAs exhibit a negative-differential mobility. Thus, a small increase in field produces a decrease in velocity and a growing carrier wave rather than a decaying one results. Under these conditions, the diode exhibits an rf impedance with a negative real part at frequencies near the point where the transit angle equals 2π. For a sample of GaAs with a length of approximately $10 \mu m$ this frequency is approximately 10 GHz. When placed across the terminals of a microwave source, the reflected power is greater than the incident rf power and the diode functions as a negative-resistance amplifier.

A nonreciprocal device, such as a ferrite circulator, must be used to separate the incident and reflected (amplified) wave, as shown in Fig. 13-36. Matching networks are included between the diode and the circulator to provide a stable gain over the desired frequency range. These networks generally contain an impedance transformer to adjust the source impedance to an acceptable level.

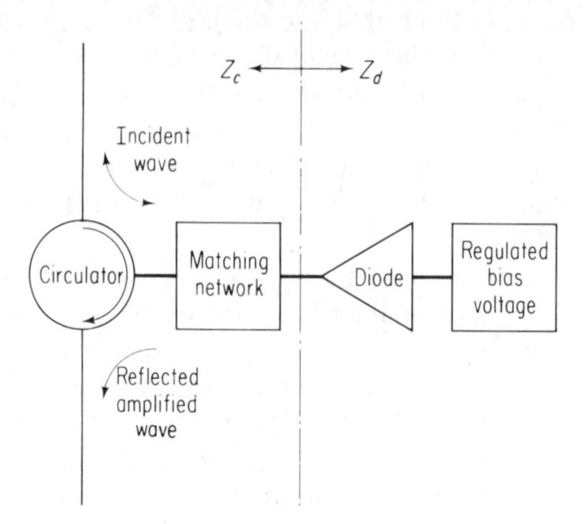

Fig. 13-36. Negative resistance amplifiers require a non-reciprocal circulator to separate the incident and reflected waves.

Supercritically doped Gunn devices can be stabilized under certain bias and circuit conditions. Amplifiers using these diodes will supply approximately one-half their oscillating power capability at 1db compression. Saturated output power will equal oscillation capability. Thus, at X-band, combined diodes have produced 2 W cw and at 35 GHz, 140 mW cw has been obtained. Noise figures ranging from 14 db for X-band to 16 db at Ka-bands are readily achievable.

In general, the Gunn-diode parameters which affect an amplifier's power, bandwidth, noise and stability are very similar to those that affect the performance of a wideband oscillator with the same output power and efficiency. Optimizing the parameters that lead toward wideband-oscillator tunability will also lead toward amplifier stability. As in wideband oscillator design, the product of doping concentration (n) and length (l) must be low enough to avoid the possibility of spontaneous domain triggering. An $n \times l$ product of less than 2×10^{12} and flat-uniform profiles throughout the active layer are required.

The growing space charge waves, along with producing a negative-resistance effect, also cause thermal noise buildup in the Gunn diode. These noise contributions are at a minimum when the internal fields are evenly distributed throughout the device and worst when the field gradients are high and the gain in the active region of the diode is nonuniform and confined to small regions. The noise figure is, therefore, highly dependent on such diode parameters as doping gradients, doping uniformity, type of cathode contact and other elusive but important physical properties.

GUNN AMPLIFIERS

Gunn amplifiers are no longer restricted to prototype systems. With improved diodes and production techniques, their unique properties qualify them for transmitters, drivers or preamps. Gunn-effect amplifiers, sometimes referred to as *Transferred Electron Amplifiers* (TEAs), are gaining rapid acceptance for both wideband and narrowband applications in commercial as well as military systems. Although not able to challenge TWTs power-wise, TEAs have inherently longer lifetimes. By developing careful, consistent control of impurities in wafer growth, contacting and die processing, performance-degrading failures can be nearly eliminated and catastrophic failures reduced to a near vanishing level. Life-test data from randomly selected production diodes gives a MTBF of 200,000 hours, with standard burn-in and processing.

The Gunn-effect amplifier is much less noisy in comparison with Impatt amplifiers (a 16 to 18 db noise figure for the Impatt) and has a much greater gain-bandwidth capability. Instantaneous bandwidths of one octave have been demonstrated compared to 20 to 30% bandwidths for Impatt amplification. The TEA is also more linear, and its third order IM intercept is typically 6 db above the saturated power output contrasted to 1 or 2 db for the Impatt amplifier.

Gunn-effect amps are now being used in prototype satellite communications systems, but their future looks bright in a variety of other applications:

Receiver front-end amplifiers: TEAs find use as either the front-end or as a booster after a tunnel-diode amplifier to improve the receiver's dynamic range. An X-Band noise figure of 10.5 db has been demonstrated in the laboratory.

TWT driver: These are generally narrowband amplifiers (5 to 10% BW) with up to 1 W output and 20 to 30 db gain.

Transmitters: Primary emphasis in this application is on high power with either narrow or wideband performance. Amplifiers may be cascaded to deliver saturated power levels of up to 10 W in X-band.

Using an automatic network analyzer, the TEA designer's first task is to characterize the impedance of the diode. Figure 13-37 shows the measured impedance of an X-band diode having a doping concentration (n_0) of 1.9×10^{15} cm^{-3} and a length (1) of 8 μm. The unit was tested in a micro-pill package (with ceramic dimensions of 0.030×0.012) at 600 mA and 12 Vdc.

The diode exhibits a region of negative resistance from 6 to 16 GHz, when measured at the end of a 70-ohm coaxial line. Note that the maximum negative resistance occurs at 13 GHz and that the reactance is capacitive

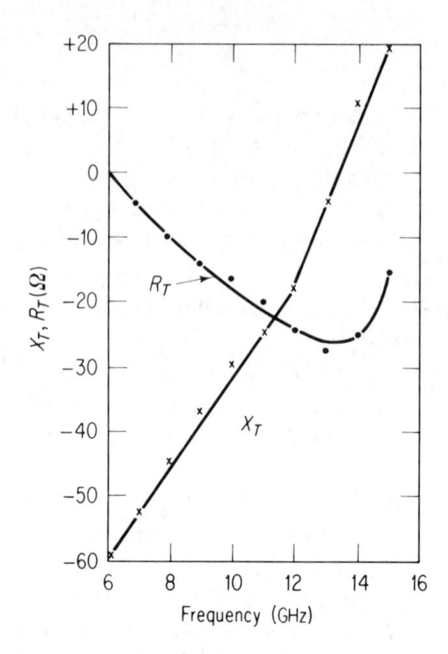

Fig. 13-37. Measured Gunn diode performance shows a negative resistance region from 6 to 15 GHz.

over most of the frequency range, with a resonance point also occurring at 13 GHz. If the package parasitics are removed, the real part of the chip impedance is flat over a relatively wide range and the chip capacity is approximately 0.3 pF. This agrees closely with the value calculated from thickness and area assuming that the active layer is a simple parallel-plate capacitor.

Referring to the impedance data given in Fig. 13-37, some general observations can be made which hold true within certain limits of the package design. The fractional bandwidth over which negative resistance exists is approximately one octave, and the negative resistance is fairly constant over a 45 per cent waveguide bandwidth centered at the frequency where the reactance equals zero. The diode impedance can therefore be modeled as a series RLC circuit and a negative Q can be computed to estimate the gain-bandwidth capability for the device. The negative Q for this example is about 4, and classical negative resistance circuit synthesis indicates that 10 db power gain (voltage gain of 3) can theoretically be realized over an approximate 30 per cent bandwidth, or about 8 db over 40 per cent bandwidth with fairly simple, practical circuitry. The gain-bandwidth capability of this particular device [(voltage gain) × (fractional bandwidth) × (center frequency)] is approximately 10 GHz.

MATCHING FOR STABILITY

The matching networks used in negative-resistance amplifiers are designed to present a reflection coefficient with constant amplitude to the circulator over the frequency range of interest. They are passive, ideally lossless and must be designed from the standpoint of amplifier stability. The voltage gain can be calculated from the impedance

$$I = \frac{Z_c - Z_d^*}{Z_c + Z_d}$$

where Z_d is the diode impedance (Fig. 13-36) and Z_c is the circuit impedance as "seen" by the diode.*

From this gain equation, it can be seen that in order to maintain stability, the function $Z_c + Z_d$ cannot have zeros in the right-hand plane of complex frequencies. A sufficient, but not necessary, condition for stability is that $Z_c \neq Z_d$ for real frequencies, implying that the two loci of Z_c and Z_d can never intersect.

Classical broadband-matching techniques call for a broadbanding shunt resonator and are therefore impractical. If the broadbanding resonator were realized as a shorted-quarter wave-length stub at f_0, the locus of Z_c would intersect $-Z_d$ at two frequencies which would lead to instabilities.

Since classical broadbanding theories must be modified, less bandwidth than predicted is actually realized. For octave bandwidths, computer aided circuit design and optimization is used. A number of amplifiers have been constructed at frequencies ranging from C to Ka bands (31).

Microstrip technology is used (Fig. 13-38) to combine the circulator with the matching and bias networks on a single ferrite or alumina substrate, thereby reducing manufacturing costs considerably.

Failure to provide the correct matching impedance at all frequencies of interest can affect the amplifier's gain flatness. With careful matching, however, the designer can expect to reduce these variations to ± 0.25 db for two-stage, narrowband (5 to 10%) amplifiers and ± 1.0 db for a four-stage unit. For wideband (waveguide bandwidth) amplifiers, larger gain variations can be expected, typically ± 1.5 db depending upon the number of stages, temperature extremes and other factors.

Although a negative-resistance amplifier requires only a three-port circulator for coupling in and out of the diode, either a four-port circulator, which has an isolator on the output, or a five-port circulator, which has isolators on both the input and output can be used for improved stability.

*Z_d^* is the complex conjugate of the diode impedance.

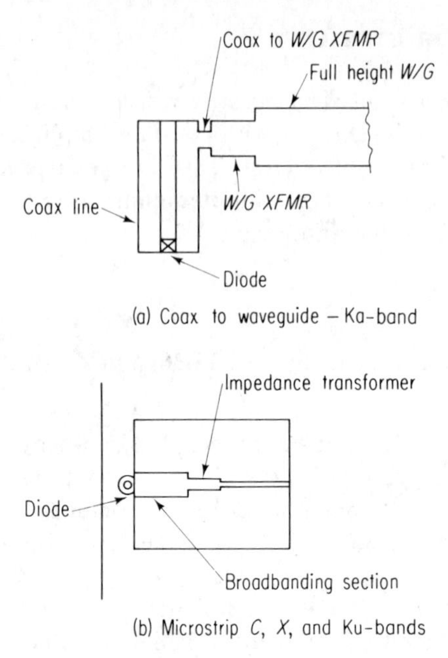

Fig. 13-38. Either coax or waveguide designs are possible. A Ka-band coax to waveguide matching section (a) and a microstrip section suitable for *C, X* or Ku-band (b) are shown.

TEMPERATURE COMPENSATION NEEDED

TEA power-supply circuitry usually consists of a series voltage regulator and a pass transistor. Although Gunn-effect devices are environmentally rugged with respect to shock, vibration and high humidity, they are sensitive to temperature changes. Changes in temperature cause both an overall gain change and a tilt in the gain-frequency response with gain decreasing at the high frequency end as the temperature increases. If necessary, the diode voltage is temperature programmed with thermistor-sensistor bias networks. Even if the bias voltage is compensated at each temperature for optimum gain flatness, variations of approximately 0.04 db/°C over the range of 0° to 50°C can be expected. These gain variations can be minimized by adding heater elements or temperature-programmed PIN attenuators.

Nominal diode bias voltages range from as high as 20 Vdc at C-band to less than 7 Vdc at Ka-band. Bias current is a function of the required gain and saturated power output. Less than 1 A is required for 15 to 20 db gain and 100 mW output.

LARGE SIGNAL PERFORMANCE

Intermodulation products vary considerably from one amplifier to another and depend, not only on the power level, but also on the number of stages, gain, frequency and gain distribution among stages. The third order intermodulation intercept point for multistage amplifiers is usually 2 to 5 db above the saturated output level, which compares favorably to Impatt amplifiers. At the 1 db gain-compression point, third-order intermodulation products can be 15 to 25 db below the carrier; at full saturation, these modulation products are generally 8 to 15 db below the carrier. In comparison with circulator-coupled amplifiers, the hybrid-coupled amplifiers provide greater linear power output. The bandwidth characteristics of the hybrid-coupled amplifiers are essentially identical to the bandwidth capabilities of the individual stages.

Other large signal amplifier effects include harmonic generation and amplitude-to-phase modulation conversion. Harmonic generation is generally not a serious problem. Even at full saturation the second harmonic is usually more than 20 db below the carrier level. Simple filters to further reduce harmonic levels can be included as part of the bias circuit at relatively little cost. Amplitude-to-phase modulation conversion at the 1 db compression point is usually about 1 degree/db or less and at full saturation about 3 degrees/db.

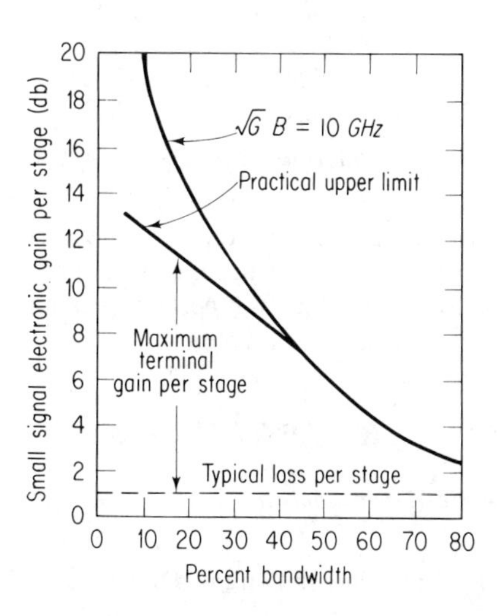

Fig. 13-39. Maximum stage gain is practically limited to about 13 dB. The 10 GHz gain-bandwidth product is indicative of a high quality GaAs device.

In some very narrowband high-gain amplifiers, 5 degree/db can be observed, especially near the band edges.

The far over-drive region of reflection Gunn amplifiers is distinctly different from that of traveling-wave tubes. The slope of power in-power out curve always remains positive, never peaking and never negative.

The gain-bandwidth limitations shown in Fig. 13-39 are based on broadband measurements in X-band. Due to circulator VSWR and temperature effects, the maximum practical electronic gain per stage is limited to 13 db (terminal gain 12 db) for a nominal 5 per cent bandwidth. When ferrite and circuit losses are taken into account, a small-signal terminal gain of approximately 7 db is obtainable over 45 per cent fractional bandwidth.

Of course, several Gunn-amplifier stages can be combined to achieve high gains, but an isolator must be used between each pair of stages to provide stability and minimize interstage coupling and detuning. There is practically no limit to the number of stages that can be cascaded in this fashion. The design of a multistage amplifier involves several trials to obtain the proper balance of gain and gain-compression among a number of stages.

EXERCISES

1. Derive Eqs. 13-4 and 13-5.
2. A certain germanium tunnel diode has the following parameters:

$$\begin{aligned} I_P &= 5 \text{ ma} & C &= 5 \text{ pf} \\ I_V &= 0.5 \text{ ma} & L_S &= 3 \times 10^{-9} \text{ h} \\ V_F &= 475 \text{ mv} & R_S &= 1.5\ \Omega \\ V_P &= 50 \text{ mv} & r_d &= 20\ \Omega \end{aligned}$$

   Calculate the resistance cut-off frequency, the self-resonant frequency, and the switching time.
3. The tunnel diode described in Exercise 2 is to be used as a video amplifier. Calculate the allowable source and load resistance that will allow a voltage gain of 20 db to be achieved. Calculate a general expression for the voltage gain-bandwidth product of such a video amplifier.

REFERENCES

1. E. SPENKE, *Electronic Semiconductors.* New York: McGraw-Book Company, 1958, p. 91.
2. F. H. MITCHELL, JR., "Deriving the Tunnel Diode Curve," *Electr. Ind.*, October, 1961.
3. E. SPENKE, *op. cit.*, p. 4.
4. C. A. BURRUS, "Millimeter Wave Esaki Diode Oscillators," *Proc. IRE*, **48** (December, 1960), 2024.

5. A. BLICHER *et al.*, "Temperature Dependence of the Peak Current of Germanium Tunnel Diodes," *Proc. IRE*, **49** (September, 1961), 1428.

6. J. W. EASLEY and R. R. BLAIR, "Fast Neutron Bombardment of Germanium and Silicon Esaki Diodes," *J. Appl. Phys.*, **31** (October, 1960), 1772.

7. R. STRATTON *et al.*, "Tunnel Diodes, A Special Report," *Electr. Design News*, May, 1960, p. 50.

8. W. J. CUNNINGHAM, *Introduction to Nonlinear Analysis*. New York: McGraw-Hill Book Company, 1958, p. 107.

9. J. ZORZY, "Measurements of the Equivalent Circuit Parameters of Tunnel Diodes," *Genl. Radio Experimenter*, July–August, 1960, p. 3.

10. E. G. NIELSEN, "Noise in Tunnel Diode Circuits," *Proc. Natl. Electr. Conf.*, 1960, p. 285.

11. B. G. KING and G. E. SHARPE, "Low-Gain Wide-Band Esaki-Diode Amplifiers," *Digest Tech. Papers*, 1961 Intl. Solid State Circuits Conf. (Phila., Pa.), p. 98.

12. R. F. TRAMBARULO, "Esaki Diode Amplifiers at 7, 11, and 25 KMC," *Proc. IRE*, **48**, 2022.

13. G. DERMIT, "High Frequency Power in Tunnel Diode Oscillators," *Proc. IRE*, **49**, 1033.

14. K. K. N. CHANG, "Low Noise Tunnel Diode Down Converter Having Conversion Gain," *Proc. IRE*, **48**, 854.

15. F. STERZER *et al.*, "Microwave Tunnel Diode Autodyne Receiver," *Digest Tech. Papers*, 1961 Intl. Solid State Circuits Conf., p. 88.

16. F. D. NEU, "A Tunnel-Diode Wide-Bank Frequency Doubling Circuit," *Proc. IRE*, **49**, 1963.

17. A. G. JORDAN and R. ELCO, "The Tunnel Diode Super-Regenerative Amplifier," *J. Electronics and Control*, **XI** (July, 1961), 65.

18. J. F. BOGUSZ and H. H. SCHAFFER, "Superregenerative Circuits Using Tunnel Diodes," *Digest Tech. Papers*, 1961 Intl. Solid State Circuits Conf., p. 102.

19. P. SPIEGEL, "A Tunnel-Diode Function Generator," *IRE Intl. Conv. Record*, **9**, Part 2 (1961), 164–74.

20. R. A. KAENEL, "High Speed Analog-to-Digital Converters Utilizing Tunnel Diodes," *IRE Trans. Electronic Computers*, **EC-10** (June, 1961), 273–84.

21. B. RABINOVICI and C. A. RENTON, "Full Binary Adder with One Tunnel Diode," *Proc. IRE*, **49** (July, 1961), 1213–14.

22. P. SPIEGEL, "High Speed Scalers Using Tunnel Diodes," *Rev. Sci. Instr.*, **31** (July, 1960), 754–55.

23. W. C. FOLLMER, "Low Frequency Noise Figure in Backward Diodes," *Proc. IRE*, **49** (December, 1961), 1939–40.

24. K. K. N. CHANG, *Parametric and Tunnel Diodes.* Englewood Cliffs, N.J.: Prentice-Hall, Inc., 1964.

25. *IEEE Trans. Electron Devices*, Special Issue, **ED-13**, No. 2, January, 1966. (See articles on avalanche devices and Gunn effect.)

26. J. B. GUNN, *IBM J. Res. Devel,* **8**, No. 2 (1964), 141.

27. M. F. UMAN, *Introduction to the Physics of Electronics.* Englewood Cliffs, N.J.: Prentice-Hall, Inc., 1974.

28. M. P. RISTENBATT, *Semiconductor Circuits, Linear and Digital*, Englewood Cliffs, N.J.: Prentice-Hall, Inc., 1975.

29. C. BELOVE, H. SCHACHTER and D. L. SCHILLING, *Digital and Analog Systems, Circuits and Devices: An Introduction.* New York: McGraw-Hill Book Company, 1973.

30. C. L. UPADHYAYULA and B. S. PERMAN, "Design and Performance of Transferred Electron Amplifiers Using Distributed Equalizer Networks," *IEEE J. of Solid State Circuits, SC*-8, (1973), pp. 29–36.

31. R. E. GOLDWASSER and F. E. ROSZTOCZY, "35 GHz Transferred Electron Amplifiers," *Proc. IEEE* **61**, (1973) pp. 1502–1504.

32. A. A. SWEET and J. C. COLLINET, "Multistage Gunn Amplifiers For FM-CW Systems," *Solid State Circuits Conference Digest Technical Papers* (1972), pp. 42–43.

33. J. F. CALDWELL and F. E. ROSZTOCZY, "Gallium Arsenide Gunn Diodes for Millimeter Wave and Microwave Frequencies," *Proc. Fourth Int. Symp. on GaAs and Related Compounds*, pp. 240–248, The Institute of Physics, London (1973).

14

Integrated Circuits

INTRODUCTION

This chapter presents an introduction to an exciting new field of semiconductor electronics. During the last several years the semiconductor industry has developed the technology for fabricating complete electronic circuits within and on a single silicon "die." The resulting component is referred to here as an *integrated circuit* (IC).

The IC represents one approach that the components industry has taken to make smaller, more reliable, less expensive, higher-performance components. It is, indeed, not the only approach in use today. World War II provided significant motivation for microminiaturizing components. Smaller vacuum tubes were developed and replaced the large bulky designs of the 1930's. Effort to reduce the size of discrete component parts has continued to this day. Much effort has been directed toward fabricating microminiature arrays of passive networks using deposited metallic films or silk screening. Thin magnetic films are being employed for making small, high-performance memories for use in digital computers.

The IC is becoming more and more important as a component to be used in the design of electronic equipments, not only in equipment that must be small and light in weight, but where reliability and performance are demanded. In many areas of application, particularly in digital computers, the IC provides more economical equipment designs.

To apply this new component intelligently, one should understand the rudiments of the technology by which such devices are fabricated, the characteristics of the components of which they are comprised, and some of the circuit design technology being utilized. This chapter presents an introduction to these subjects.

Designing and fabricating integrated circuits require a combination of skills, including those of semiconductor device design, semiconductor process design, and circuit design. Needless to say, the design and development of integrated circuits is a team effort. The "boundary conditions" which apply in the design of IC's differ from those which apply to the design of discrete components. These differences in design philosophy are indicated later in this chapter.

This book has dealt with circuit design principles and the application of those principles to practical circuits. From the standpoint of developing new circuit analysis principles, this text has little need for a chapter on integrated circuits. The impact of the IC technology on the manner in which circuit functions are performed has, however, been so profound from a circuit topology viewpoint, that a chapter such as this is mandatory in a modern text on semiconductor circuits.

We begin with a concise presentation of the basic IC fabrication processes and show how they are used in circuit fabrication. Next, we provide component characteristic information in order to show the differences in characteristics between discrete components and integrated circuit components. From these discussions evolve the "boundary conditions" which dictate the need for different circuit topologies. Several examples of digital and analog circuits are shown to illustrate the impact of these boundary conditions on circuit topology.

TYPES OF MICROCIRCUIT FABRICATION

The initial approach to miniaturizing electronic circuitry was through successive refinements of printed-circuit techniques (1).* Such techniques generally used an insulating substrate (glass or ceramic) on which film-type resistors or capacitors were painted through a mask (in this case, a photographically prepared stencil). Such "thick film" (0.001 in. thickness) circuits have remained the most economical way to prepare many types of circuits and are widely used in both commercial and military equipment (2, 3, 4). Later refinements involved the deposition of components as thin metal films (0.000004 in. thickness) by vacuum evaporation, or sputtering techniques (5). Such a circuit is shown in the photograph of Fig. 14-1. The capacitor and resistor regions are defined by a process of photoengraving whereby the deposited film is selectively removed in unwanted regions. Whereas resistors and capacitors are formed on an insulating substrate, active elements, such as transistors and diodes, must be fabricated separately and fastened onto the substrate. Thick and thin film circuits which contain both active and passive

*Numbers in parentheses refer to material listed at the end of the chapter.

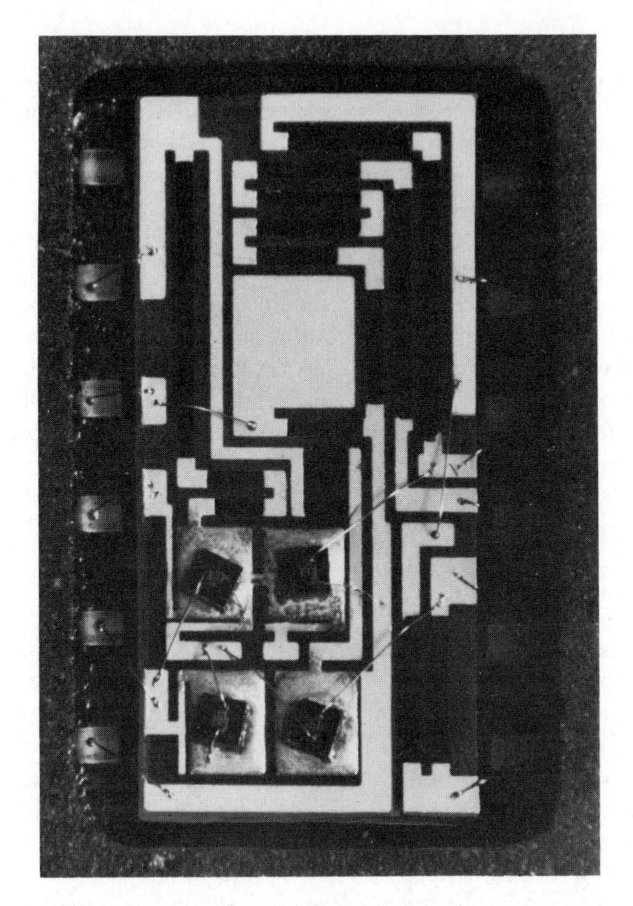

Fig. 14-1. Microelectric circuit made with thin-film passive and attached active components.

components are called *hybrid* circuits. Such circuits are relatively easy to design and build, since the component behavior is similar to that of their discrete counterparts and the equipment investment is modest by today's microelectronics industry's standards. Such circuits are inherently larger than those obtained by more refined procedures and cannot be fabricated at a high enough density to provide the ultimate in cost reduction.

Planar Transistors

The key to a further evolution in microelectronics technology was the development of the silicon planar transistor in 1958. Differing electronic properties are imparted to various regions in the silicon by introducing trace amounts of impurities, typically phosphorus, arsenic, or antimony, to produce n-type or electron-conducting silicon; or boron, aluminum, or gallium, to produce p-type or hole-conducting material. For optimum performance, the collector and emitter areas of the transistors must be quite small, typically measured in tenths of a thousandth of an inch. The thickness of the region between the emitter and collector (the base) is even smaller, typically measured in hundredths of a thousandth of an inch. An important aspect of the planar transistor is the "benign" condition that is established at the point where the active elements (the emitter and collector junctions and the base region) reach the surface. Structures made by the planar process have a silicon–silicon oxide interface formed at a high temperature. Such a surface appears to provide a nearly ideal reflection interface for the holes and electrons that move within the device, an important factor in achieving good transistor performance (6). This oxide covering further serves to protect the sensitive transistor surface from the harmful effects of external contaminants. In addition, the oxide acts as a mask for subsequent diffusion steps; this is of major importance in working with small geometries.

Silicon Integrated Circuits

The planar transistor has another key feature; that is, the insulating oxide on the surface makes it possible readily to interconnect a large number of individual devices to form a "monolithic"* silicon integrated microcircuit. Previous designs had required the attachment of thin (fragile) leads directly to the small active areas of the transistor in order to connect one component to another (as in Fig. 14-1). In a silicon integrated circuit, a number of planar transistors and other silicon planar components (diodes, resistors, and capacitors) are formed in a single piece of silicon, and suitable interconnections are made by thin-film metal regions located on top of the oxide-covered surface. These components and interconnections can be seen greatly magnified in the photograph of a complex silicon microcircuit as shown in Fig. 14-2. Such a circuit contains about 70 individual electronic components in a piece of silicon 0.055 in. by 0.055 in.

An entire silicon wafer is processed at one time to produce a large number of circuits, as shown in Fig. 14-3. These are then separated and individually mounted in metal packages, as shown in Fig. 14-4.

Silicon Integrated Circuit Fabrication Technology

The fabrication of silicon integrated circuits requires processing the silicon wafers through a series of complex manufacturing processes. These include silicon purification, crystal growth, wafer cutting and polishing, epitaxial

*The term *monolithic* is another term commonly used to describe a silicon circuit, emphasizing that it is fabricated within a single block containing continuous regions of different materials that are in intimate contact, with no wired interconnections.

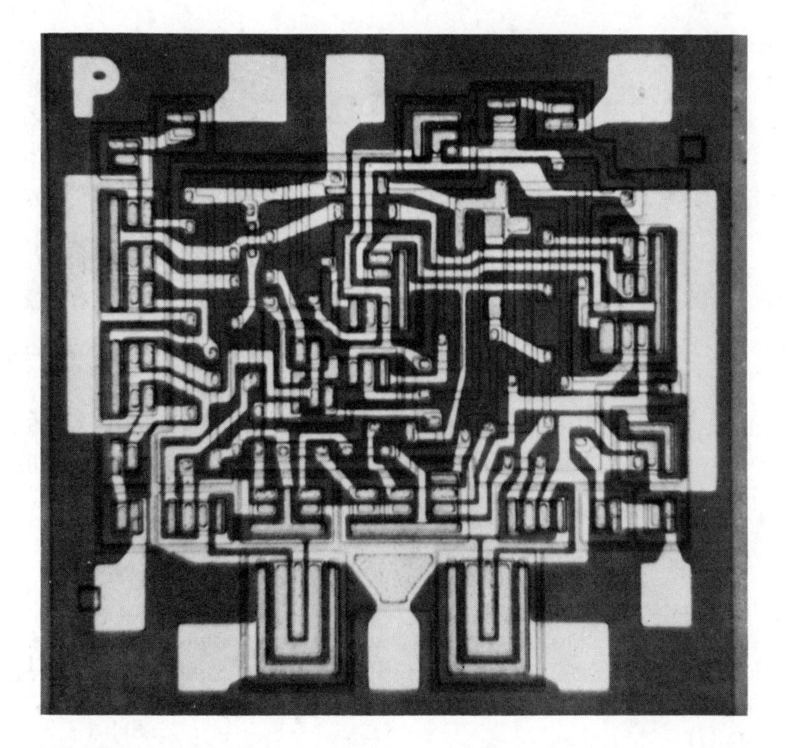

Fig. 14-2. Complex silicon microcircuit. The area covered by this circuit is only 0.055 in. × 0.055 in.

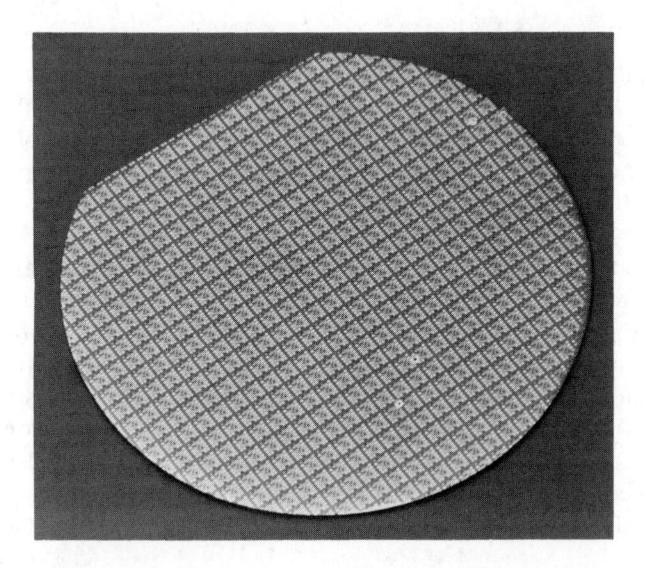

Fig. 14-3. Completed silicon circuits on wafer before cutting.

Fig. 14-4. Silicon circuits mounted in metal and flat packages. (Lid removed to expose mounted die.)

growth of silicon on a silicon wafer, formation of a silicon oxide layer, photoengraving the oxide layer, solid-state diffusion, and photoengraving of metallic layers.

Silicon Preparation and Crystal Growth

Silicon used in solid-state devices must initially be extremely pure (less than one part per billion of impurity atoms). It is commonly obtained by chemical reaction from carefully purified silicon compounds. Common reactions utilized for this purpose are the high-temperature decomposition of silane (7),

$$SiH_4 \longrightarrow Si + 2H_2$$

and the hydrogen reduction of either silicon tetrachloride, or trichlorosilane (8),

$$SiHCl_3 + H_2 \longrightarrow Si + 3HCl$$

The equipment used in purification and deposition is expensive to build and operate, and quality control procedures are rigorous. Hence, high-purity silicon was relatively expensive ($125/kg) in 1966. The amount of material required for each microcircuit, however, is so small that the cost of raw silicon per circuit (including cutting, waste, and defective circuits) is less than one cent.

The vapor-grown silicon is typically converted to single crystal form by growing onto a suitably oriented seed crystal from the molten material (silicon melting point = 1420°C) contained in a crucible. During this operation, a small controlled amount of impurity is added to the silicon to impart the desired conductivity type and resistivity properties. The process is shown diagrammatically in Fig. 14-5(a). Typical single-crystal ingots prepared in this fashion are about an inch and a quarter in diameter and several inches in length. When extreme purity (freedom from dissolved oxygen) is required, other procedures are used so that the molten silicon is not in contact with a crucible.

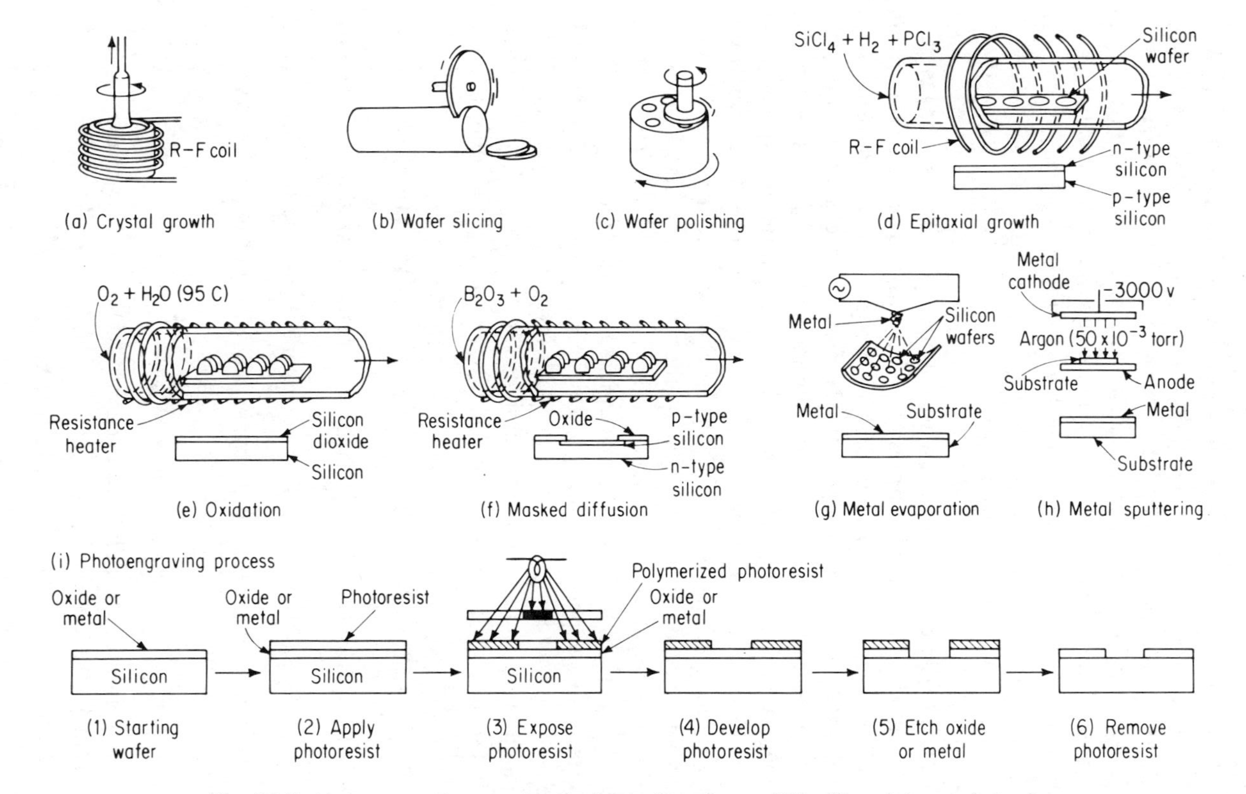

Fig. 14-5. Various processes used in the fabrication of monolithic silicon integrated circuits.

Wafer Preparation

The silicon single-crystal ingot is sawed into wafers (typically 0.010 in. thick) with diamond tipped saws. Subsequently, at least one side of each wafer is lapped and polished until it is smooth and flat. The wafers are then chemically treated to smooth the surface further and to remove that portion of the wafer surface in which the crystal structure has been damaged by the lapping and polishing operations, as in Fig. 14-5(b, c).

Epitaxial Growth and Vapor Deposition

The term *epitaxial growth*, as used in microcircuit fabrication, generally refers to the growth of an additional layer of silicon on a single-crystal silicon substrate by one of the two types of reactions previously described for the preparation of silicon [Fig. 14-5(d)] (9, 10). Suitable care is taken so that the grown layer will be single-crystal material having the same crystal structure and orientation as the substrate and being as free from growth defects as possible. In this case, an appropriate quantity of n- or p-type impurity is normally added during the growth process to produce a region having a conductivity type different from that of the substrate. Epitaxial growth is one of two methods used to produce such a structure, the other being diffusion of impurities into the substrate as indicated subsequently.

Certain other vapor-deposition processes are now being used in the development of advanced microcircuits to deposit layers of metals and insulators in a manner very similar to that used in silicon epitaxial growth. For example, silicon dioxide may be deposited in thin layers by any of the following reactions (11-14):

$$SiCl_4 + 2H_2 + 2CO_2 \rightleftharpoons SiO_2 + 2CO + 4HCl \qquad (14\text{-}1)$$

$$C_2H_5Si(OC_2H_5)_3 \xrightarrow{\Delta} SiO_2 + (\text{organic byproducts}) \qquad (14\text{-}2)$$

$$Si(OC_2H_5 \xrightarrow{\Delta} SiO_2 + (\text{organic byproducts}) \qquad (14\text{-}3)$$

Thin layers of various refractory metals may be deposited by the hydrogen reduction of a suitable halide (14):

$$2MoCl_5 + 5H_2 \longrightarrow 2Mo + 10HCl \qquad (14\text{-}4)$$

At the temperatures generally used to deposit metals by this process, a further reaction may take place:

$$Mo + 2Si \longrightarrow MoSi_2 \qquad (14\text{-}5)$$

Many metal silicides are metallic-type conductors, however, and can be used as such in fabricating a silicon microcircuit.

Oxidation

During the fabrication processing of microcircuits, as well as in the completed device, the silicon wafer is covered with a thin layer of amorphous silicon oxide. This layer has a composition and structure similar to that of a fused quartz. It is most generally formed by flowing either wet oxygen (saturated with water at 95°C) or dry oxygen over the silicon at temperatures in the range of 900° to 1200°C. [Fig. 14-5(e) (15).]

Under these conditions, the initial growth rate is very high. The oxidation rate decreases according to a parabolic law as the oxide becomes thicker, and growth rate is determined by the length of time required by the reactants to diffuse through the already formed oxide. The process of oxidation invariably changes the number of impurities in the silicon at the surface. The impurities that were contained in the silicon may be relatively insoluble in the oxide, so that as silicon is used up during oxidation they remain and become concentrated in the silicon surface. On the other hand, a given impurity may be more soluble in the oxide—and may diffuse out of the silicon into the oxide—leaving a surface layer depleted of the impurity in the silicon. These effects are well understood and are sometimes used to advantage in fabricating a device.

A rather important effect associated with the silicon oxide interface exists because the oxide may become positively charged either by the presence of unwanted impurity ions or by virtue of constitutional defects in the oxide itself. Where this happens, extra "induced" electrons are concentrated in the silicon near the silicon-oxide interface, and the electrical properties of the silicon are strongly modified (generally undesirably) in this region. Troublesome ionic contaminants include the alkali metals (lithium and sodium), and insofar as possible these should be prevented from entering during manufacturing (16).

Diffusion

Diffusion of selected impurities into a semiconductor has been used for many years to form successive n- and p-layers having the necessary precise geometry required for a high-frequency transistor. Diffusion is similarly used to form any or all of the circuit elements in a silicon microcircuit. Typically the impurities are introduced through a preshaped hole in the oxide, to determine the size of the diffused region. The oxide masking effect takes place because, for most impurities (gallium being a notable exception), diffusion through the oxide is much slower than it is in silicon at the same temperature.

To control the exact concentration and the depth of penetration of impurity atoms, the process normally requires two steps: (1) The wafers contained in a quartz boat are placed in a "deposition" furnace where they are exposed to a high concentration of impurity vapor in a suitable carrier gas flowing over the wafers for a predetermined time at a given temperature [Fig. 14-5(f)]. This step deposits a precise quantity of impurities on the silicon surface. (2) The wafers are then placed in a "drive-in" furnace when the impurity diffuses to the desired depth while a layer of oxide is formed simultaneously to close the hole in the oxide mask, thereby preparing the wafer for the next step.

Metal Deposition

Microcircuit interconnections are typically formed from a thin film of metal deposited over the wafer surface. For most microcircuit uses, the metal film is applied after formation of the active devices. It is, therefore, desirable to maintain a relatively low wafer temperature during the deposition process. Lower boiling metals, such as aluminum or gold, are deposited by vacuum evaporation of the molten metal contained in a heated tungsten coil [Fig. 14-5(g)]. Refractory metals, such as tantalum and molybdenum, can also be evaporated, but the higher metal-source temperature required (1500° to 3000°C) is more easily obtained by bombarding the metal with a high-energy stream of electrons (that is, electron beam evaporation). Refractory materials can also be readily deposited by a sputtering process in which the metals are bombarded with ions of an inert gas that remove metal from the source electrode, which is then deposited on the substrate [Fig. 14-5(h)].

Photoengraving Process

A photoengraving process is used to remove selectively layers of oxide or metal in specific areas in order to fabricate the individual microcircuit elements. Thus it is used to cut holes in the surface oxide mask that defines the dimensions of the diffused regions. This process is also used to remove metal from precisely controlled areas to form thin film components and interconnections. The steps involved in the process are shown schematically in Fig. 14-5(i). The photoresist solution is applied to the wafer while the latter is held in a suitable fixture and spun on its axis at several hundred revolutions per minute, to remove most of the resist material except for a thin smooth coating. A mask (similar to a photographic negative) is then placed over the wafer, and the photoresist is selectively exposed to ultraviolet light. After the unmasked resist is polymerized by the action of the light, the nonpolymerized regions can be dissolved, leaving the underlying material unprotected in these regions. This material can then be removed by a selective chemical reaction in which the polymerized photoresist is not attacked. Silicon is removed with hydrofluoric acid. Great care must be taken at this step in order to obtain sharp definition.

From a yield standpoint, photoengraving is one of the most important steps in the entire fabrication process since it is extremely difficult to create and maintain a layer of resist that is free of those defects that degrade its masking qualities. Wherever the acid penetrates the polymerized resist, pinholes are produced in the oxide that impair the diffusion mask so that tiny diffused regions known as *pipes* are produced. These act as unwanted electrical short circuits between elements of the microcircuit. In other cases, a pinhole in the oxide allows the metal interconnection layer to be shorted to the region beneath the oxide insulating layer.

Since component values, for example, the resistance of a resistor, are a function of size, the definition obtainable in the photoengraving process directly affects circuit performance. As the size of the pattern decreases, various difficulties arise, such as those due to light diffraction and poor photoresist removal, so that the limit of the process may be expected to be somewhere in the range of a 0.00001-in. line width.

As mentioned earlier, the technique for producing circuit elements in wafers of silicon is derived from that previously used in the fabrication of planar silicon transistors. The process depends on the diffusion of electrically active impurities into localized regions of the silicon wafer in a controlled fashion. This technique was originally made possible by the discovery by Frosch and Derick (17) at Bell Laboratories that a surface layer of silicon dioxide over the silicon greatly impedes the diffusion of most impurities into the silicon. Thus, if openings are cut in the oxide (by means of the photoengraving process), diffusion will take place only in the exposed areas. Using this technique, regions are formed in the silicon that have electrical properties (conductivity type and resistivity) different from those in the original substrate.

A typical series of processing steps used for silicon microcircuit fabrication is illustrated in Fig. 14-6. The individual regions in the silicon substrate where separate components are to be formed must be isolated. This is commonly accomplished as shown in steps 4, 5, and 6. An epitaxial layer of n-type silicon is grown on a p-type substrate, followed by the diffusion of a p-type impurity through the oxide in a pattern that provides a number of "isolated" n-type regions in which the individual circuit components are to be placed. The boundaries between the p and n regions formed in this operation are PN junction diodes having high blocking resistance to provide electrical isolation between n-type regions.

Further steps in the process include oxidation and then photolithographic processing to expose the areas in which the appropriate impurities are diffused. Between diffusions, the wafer is reoxidized to cover the previously exposed areas, and new openings are made. Finally, contact cuts are made through the oxide as required to form electrical connections to the various regions. In the final circuit, all the elements have been interconnected by

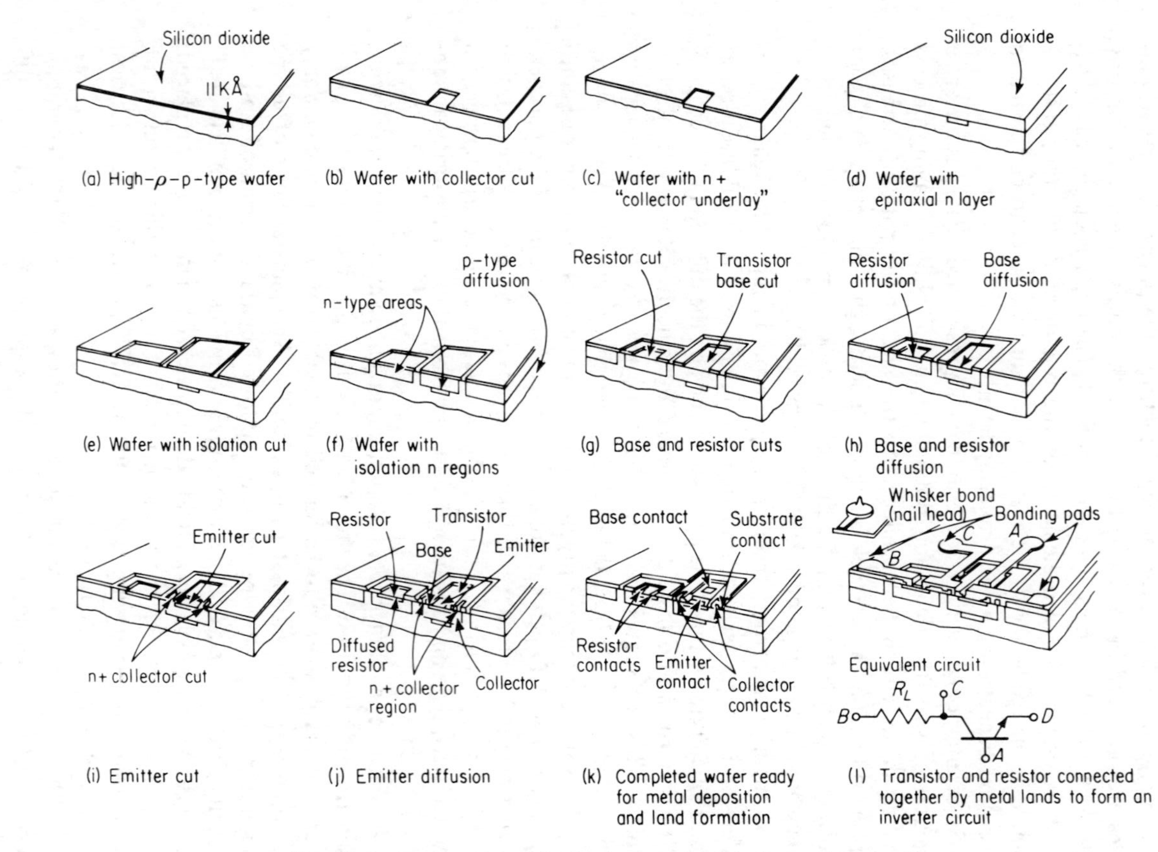

Fig. 14-6. Processes used for fabrication of solid silicon microcircuits.

means of a deposited and photoengraved metal pattern over the oxide, with bonding pads for lead attachment.

Such a sequence of operations can be used to fabricate many different types of circuit elements. The process lends itself better to the formation of active elements (transistors and diodes) than of passive elements (resistors and capacitors).

PASSIVATION

One of the most important fabrication steps in terms of long-term stability and reliability of the monolithic circuit is passivation of the surface. By *passivation* we mean the protection of the surface, particularly the intersections of the various junctions with the surface, from outside contaminants.

The SiO_2 layer used in the diffusion-masking process is a good protection for the surface against most contaminants. Since the final step before opening the contact windows is an oxidation of the surface, all junctions will be protected by at least one layer of oxide. Diffusion through such an oxide window takes place laterally as well as normal to the surface and the junction meets the surface under the oxide instead of at the edge of the oxide. Further SiO_2 protection is provided by one or more additional oxide layers grown in the subsequent fabrication steps. The final window is made smaller than the original diffusion mask for metallization.

Whereas SiO_2 provides good protection from water vapor and other contaminants, an important problem exists due to the possibility of migration of certain metal ions, particularly Na^+ ions, through the SiO_2 layer. This problem can be reduced greatly by depositing a thin layer of phosphorus glass ($P_2O_5 + SiO_2$) on top of the SiO_2.

A particularly effective passivation technique is the use of a thin silicon nitride (Si_3N_4) layer on top of the SiO_2. This type of protective layer, called a *sealed junction method*, eliminates the migration of all important impurities to the surface of the Si. The use of Si_3N_4 passivation requires several extra processing steps, but the resulting structure is sufficiently protected to permit greatly simplified packaging techniques. After the final diffusion step, an additional thin oxide layer is grown to include the diffusion windows. The silicon nitride layer is deposited after this oxide growth. Another layer of SiO_2 is next deposited over the Si_3N_4, since special etching techniques must be used to open windows in the silicon nitride. After an HF etch to open windows and a boiling phosphoric acid etch, a final rinse in HF removes the thin oxide layer in the contact windows and the top oxide layer over the silicon nitride. Therefore, the final passivation medium is a sandwich structure of SiO_2 and Si_3N_4. Metallization can then be performed over the nitride layer and in the windows.

NEWEST TECHNIQUE

Reliable protection from environmental conditions, previously available only with expensive electronic packaging, is now possible for a line of ICs assembled in dual-in-line plastic packages. Called *Gold CHIP* (chip *h*ermetically *i*n *p*lastic), the devices are made with a silicon nitride passivated tri-metal process. Hermeticity is achieved on the IC chip itself rather than with an external package. The protected chip is enclosed in inexpensive plastic (epoxy) only for ease in handling.

Experience has shown that moisture at the surface of an IC causes metallization corrosion and that semiconductor plastic molding compounds are inefficient barriers to that moisture. Ingression of moisture in plastic packages occurs through the molding compound as well as by capillary action at the epoxy/lead frame and mold parting-line interfaces. The result is the creation of open circuits.

Although a coating of phosphorus-doped, chemically-vapor-deposited (CVD) silicon dioxide (SiO_2) glass is widely used for device passivation and mechanical protection during IC fabrication, too heavy phosphorus concentrations in the SiO_2 can drastically increase aluminum corrosion rates and related open-circuit defects. A tri-metal technology, developed at RCA, illustrated in Fig. 14-7, eliminates the corrosion mechanism by substituting deposited layers of titanium, platinum, and gold for the susceptible aluminum film.

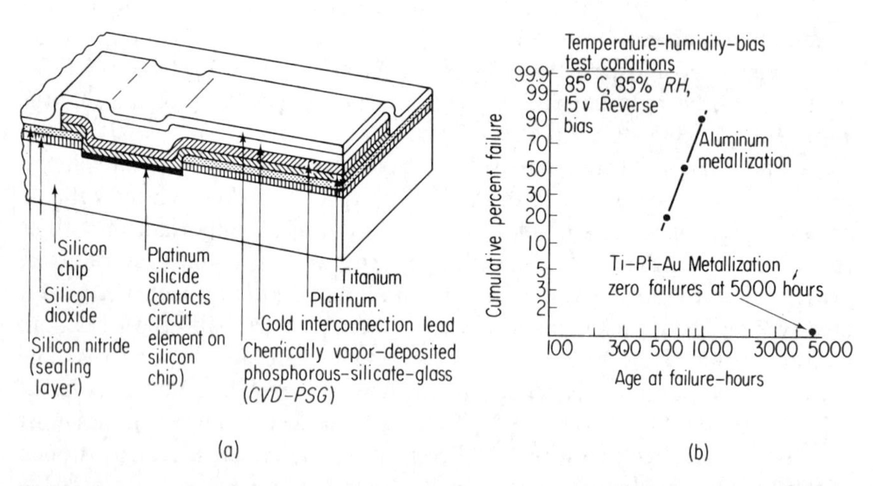

Fig. 14-7. (a) Cross-section of RCA's GoldCHIP structure shows initial silicon nitride passivation layer, platinum silicide contact area, tri-metal layers (titanium, platinum, and gold), and final passivation layer of CVD-PSG. Fabrication process permits ICs to be packaged in plastic, yet remain moistureproof for use under extreme environmental conditions. (b) Test results for hermetic trimetal IC vs aluminum metallization. All 63 tri-metal devices passed 5000 hr of reverse bias/humidity life testing under jungle and seashore conditions without a single failure.

Similar to the beam-lead system developed by Bell Telephone Laboratories, except for electroforming steps for gold beams and anchor pads, the RCA hermetic-chip process follows normal IC fabrication procedures until the final oxide step. At that point a passivation layer of silicon nitride is deposited to hermetically seal the junctions and to protect the silicon surface against moisture and contaminants. Then contact windows are opened by a standard masking operation and platinum is sputtered over the wafer and sintered in the contact areas to form platinum silicide.

A layer of titanium, to provide maximum adherence to contact regions and silicon nitride surfaces, and then a layer of platinum, to act as a diffusion barrier between the titanium and the following gold conductor, are sputtered sequentially on the wafer surface. Then, after using photoresist mask techniques to define an interconnect pattern, a gold layer is electrolytically plated on the platinum interconnect runs. As the final step, a second passivation layer of phosphorus-doped CVD SiO_2 glass is deposited for protection. Packaging in plastic is essentially similar to that for conventional aluminum, nonhermetic chips except that gold wires are used to interconnect gold-surfaced bonding pads on the IC chip and the gold-plated bonding surfaces on the package lead terminals.

ADDITIONAL COMPONENTS

The fabrication sequence previously described shows the techniques for making double-diffused NPN transistors and diffused resistors. Conventional PNP transistors can be made in the same batch processing to provide PNP and NPN circuits in the same chip. Additional photoengraving, etching, and diffusion steps are required to form the additional diffused regions. This additional processing has limited the application of the conventional PNP transistor because of added manufacturing cost.

Two types of unconventional PNP transistors may be readily fabricated using the basic process described. The first of these is called the *substrate* PNP transistor. (Fig. 14-8 shows a cross-sectional view of such a device.) The substrate forms the collector of the transistor. The substrate PNP collector is tied to the most negative potential in the circuit since the substrate must be so connected. Also, all substrate PNP transistors in the same chip will have their collectors in common because the substrate is the collector. The emitter for the PNP can be formed from the base diffusion of the NPN. Alternatively, a separate diffusion step can be used. In this latter case, higher dc gain and higher-frequency response result because a narrower transistor base width may be obtained from a separate emitter diffusion. Despite the common-collector limitation, limited frequency response, and dc gain, the substrate PNP has found significant application in IC's.

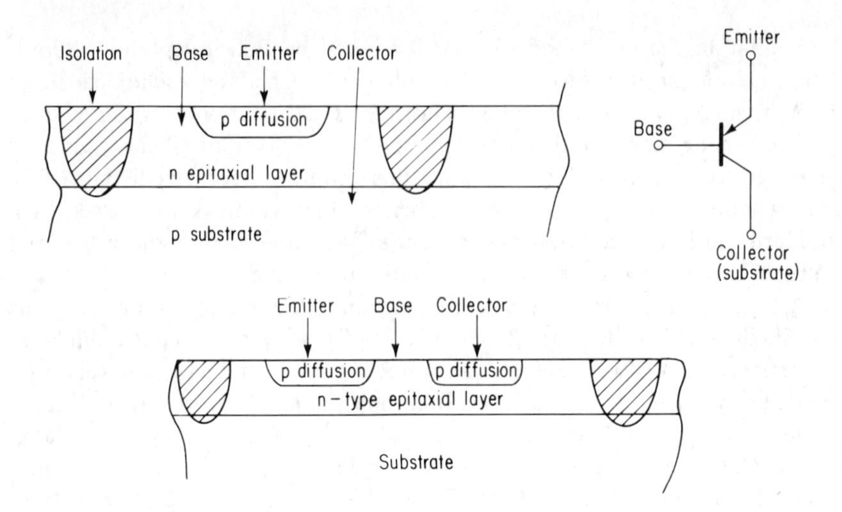

Fig. 14-8. (a) Substrate PNP transistor. (b) Structure of the lateral PNP transistor.

Another easily fabricated form of PNP is termed the *lazy* or *lateral* transistor. (The structure for this device is shown in Fig. 14-8.) The PNP transistor is formed by two closely spaced p-type diffusions. The base width, which is the spacing between these two diffusions, is rather large. This large base width makes the common-emitter dc gain rather low. Typically an h_{FE} of 2–3 may be obtained. The wide base width also limits the frequency response. The f_T of such PNP devices is in the order of 25 MHz. The low dc gain is not a particular drawback in many applications, since an equivalent high-gain PNP may be made by combining a lateral PNP with a conventional NPN. (The composite PNP transistor is shown in Fig. 14-9.) The composite dc gain is approximately the product of the individual transistor gains. The output characteristic of the composite differs from that of a conventional transistor because the base-emitter diode of the NPN transistor appears in series with the output.

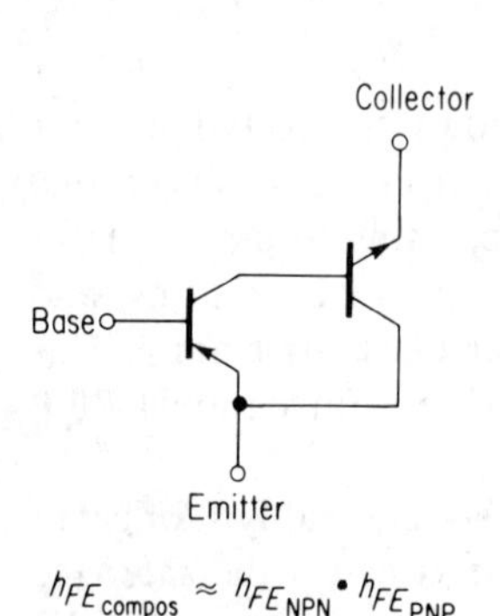

$$h_{FE_{compos}} \approx h_{FE_{NPN}} \cdot h_{FE_{PNP}}$$

Fig. 14-9. Composite PNP transistor.

Two types of capacitors are in general use in IC's today. The capacitance associated with a diffused junction provides a technique for making a polar capacitor. The junction must be reverse-biased to provide a diode with a reasonable Q. The diode capacitance is also voltage sensitive. Alternatively the SiO_2, an integral part of each IC, may be used as the dielectric material.

Diodes are fabricated in IC's by using the base-emitter or collector-base junctions of a transistor diffusion, or by interconnection of the transistor as

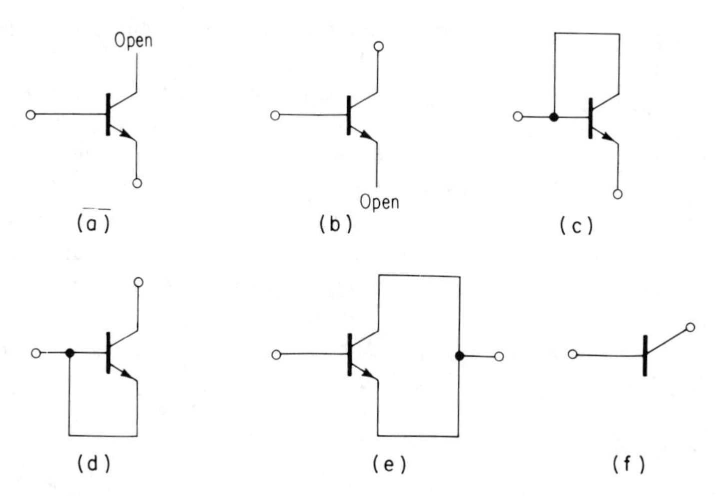

Fig. 14-10. Six possible diode connections that may be fabricated in an IC: (a) E-B, collector open; (b) C-B, emitter open; (c) E-B, collector shorted to base; (d) C-B, emitter shorted to base; (e) C-E, shorted; (f) C-B, no emitter.

a diode. Six possible diode configurations are readily fabricated using the IC process outline previously described. These are summarized in Fig. 14-10. Each configuration provides slight differences in forward voltage drop, switching time, and breakdown characteristics.

SEMICONDUCTOR INTEGRATED CIRCUIT COMPONENT CHARACTERISTICS

In this section, we consider the characteristics of the components formed as an integral part of the IC.

Integrated Circuit Transistors

The NPN transistors used in the IC differ from their discrete-component counterpart in three ways. The IC transistor has a significantly larger parasitic collector resistance because the collector contact is brought to the surface of the silicon chip. There is a significant collector-to-ground parasitic capacitance due to the capacitance of the isolation diode on each collector. The leakage current of this isolation diode is not insignificant in low-power circuits.

These three parasitic elements are shown in Fig. 14-11. The collector resistance and isolation capacitance are distributed elements. In a typical small-area transistor for low-level digital circuits, $R_c \approx 20\ \Omega$ and $C_I \approx 3$ pf. This represents a cut-off frequency in excess of 1000 MHz. This time constant

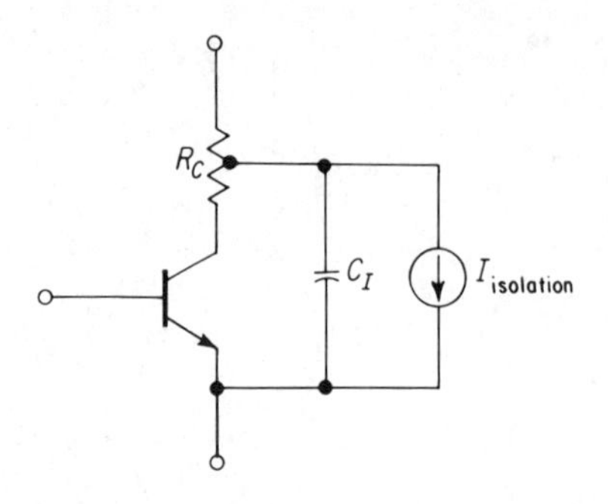

Fig. 14-11. Parasitic elements in an NPN transistor in an IC.

has an insignificant effect on most circuits. In circuits using relatively high collector load resistors (for example, low-power logic circuits), the isolation capacitance represents a significant frequency-limiting element.

The isolation diode leakage is in the range of 10–100 nanoamperes at room temperature increasing to several microamperes at 125°C. This is a negligible current except in very low-power circuits.

Integrated Circuit Resistors

The diffused resistor fabricated as an integral part of the IC differs markedly from its discrete-component counterpart. The diffused resistor has a distributed capacitance associated with it—the distributed capacitance of the PN junction of which it is formed. (A model of the resistor is shown in Fig. 14-12.) The resistor is analogous to a transmission line, except that it has

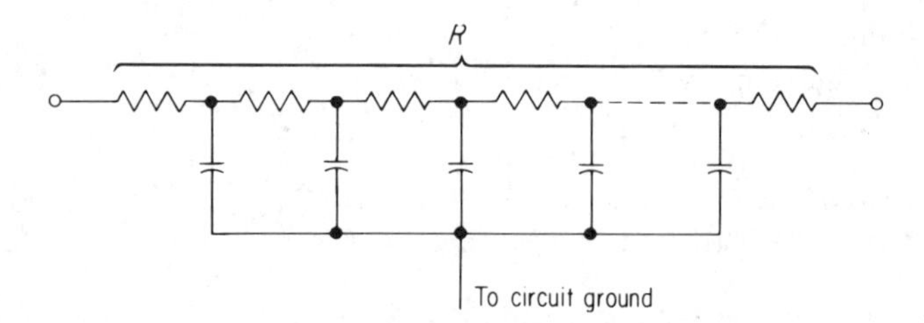

Fig. 14-12. Distributed model of diffused resistor.

negligible inductance. The frequency response of a diffused resistor is shown in Fig. 14-13. The low-frequency or dc resistance for this particular resistor is 1000 Ω. The low-frequency capacitance is approximately 1 pf. The ac resistance has decreased 3 db at 150 MHz. This particular resistor was formed with a line width of 1 mil (0.001 in.). By using narrower line widths, the frequency response will be improved, since the total parasitic capacitance is proportional to the area of the resistor. For a fixed resistor value, the capacitance scales as the square of the resistor width. The cut-off frequency of the resistor therefore varies approximately as the square of the resistor width. This can be shown as follows: The design value for a diffused resistor is

$$R = R_S \frac{L}{W} \tag{14-6}$$

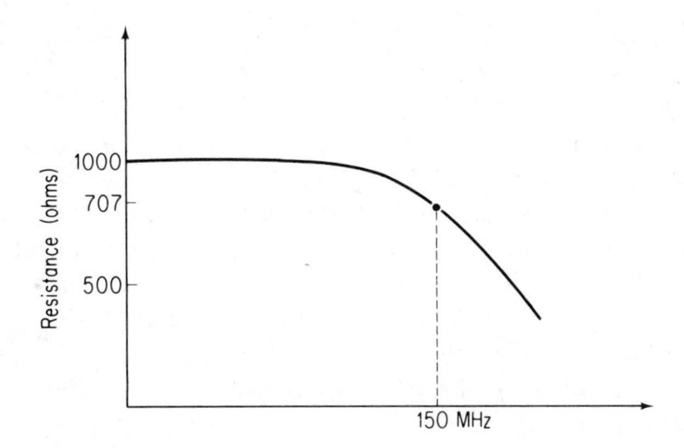

Fig. 14-13. The frequency response of a diffused resistor.

where R_S = shunt resistivity of the base diffusion in Ω/unit sq
L = length of the resistor
W = width of the resistor.

The total parasitic capacitance of the resistor is proportional to the resistor area:

$$C_R = k_R L W \tag{14-7}$$

where k_R is a constant which depends on the diffusion process. Eliminating L between 14-6 and 14-7 we have

$$C_R = k_R \frac{R}{R_S} W^2 \tag{14-8}$$

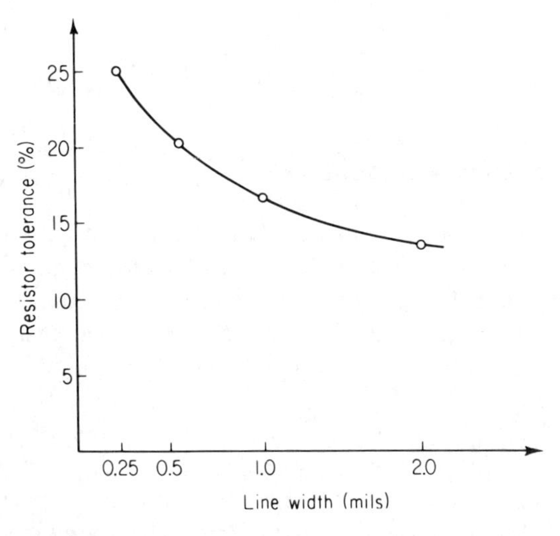

Fig. 14-14. Resistor tolerance dependence on line width.

Therefore, a two-to-one reduction in resistor width will increase the frequency response by a factor of four.

The tolerance of the diffused resistor is a function largely of two variables: the sheet resistivity, which is controlled by the diffusion process, and the resistor width, which is controlled by the photoengraving process. The overall tolerance that can be maintained is primarily a function of the resistor line width. Figure 14-14 shows the present tolerance dependence on line width which is representative of the present manufacturing state of the art.

The temperature coefficient of resistance of a diffused resistor is larger than that of most of its discrete-component competitors. The resistor variations with temperature are due primarily to the variation of the mobility with temperature. (The temperature dependence of a typical diffused resistor is shown in Fig. 14-15.)

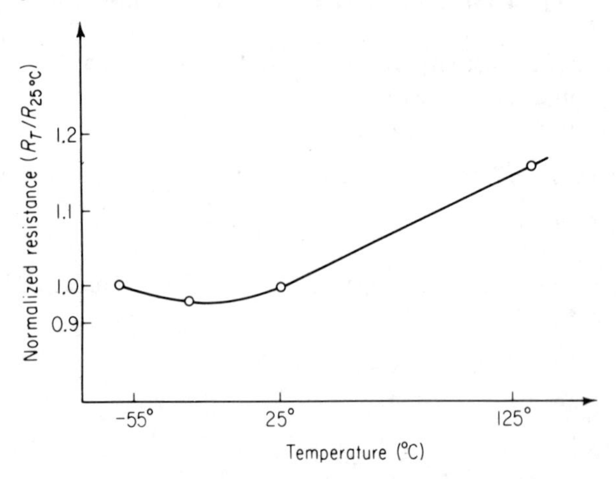

Fig. 14-15. Temperature dependence of diffused resistor.

CIRCUIT DESIGN PHILOSOPHY

The classical circuit design function has always consisted of an iterative process of optimization to provide a circuit form which will provide the desired performance. Here *performance* is used in the general sense to mean electrical performance, cost, and reliability. The circuit design process with discrete components has usually consisted of a closed-loop process as depicted schematically in Fig. 14-16. The process of optimization consists of traversing the steps in this loop until the desired levels of performance are obtained. Generally, the *optimization* is defined as maximum performance at minimum cost with an acceptable level of reliability. Typically, the design process begins with selection of a circuit form which the engineer knows from past experience may perform the necessary electrical function. The circuit is

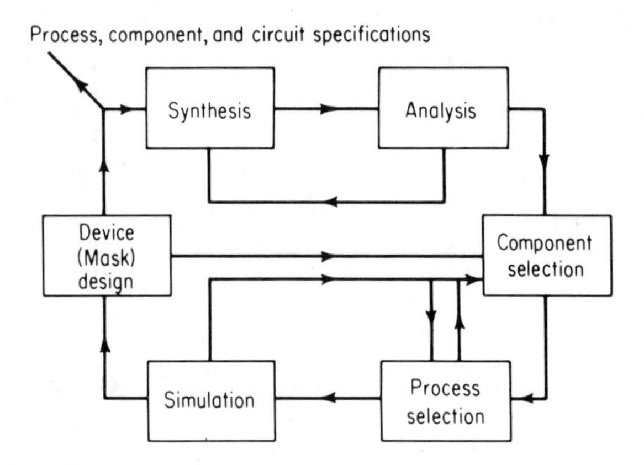

Fig. 14-16. Circuit design process when employing IC technology.

then subjected to a paper analysis or computer simulation. If analysis shows that the circuit form selected does not meet the required performance, further synthesis work is needed. Appropriate modifications must be made to the circuit and their effects analyzed. Once a satisfactory circuit form is selected, a tentative decision is made on component types and required characteristics. The circuit is then built in "breadboard" form, in a fashion to simulate as closely as possible the physical layout of components when the circuit is placed in production. Appropriate tests are then performed on the circuit to make sure the electrical performance requirements are met. If the electrical performance falls short of these requirements, additional synthesis and/or component selection is required. The outputs from the design process once a circuit design is selected are (1) component specifications, (2) circuit performance specifications, (3) specifications covering the physical implementation of the circuit (for example, printed-circuit board layout with component placements specified).

Designing IC's differs substantially from designing discrete-component circuits in two general ways: First, the design process is more complex because the component-selection task present in the discrete-component design process is replaced with three tasks: (1) component specification, (2) process selection, (3) device design. (This is shown in Fig. 14-17.) Secondly, the "ground rules" for selecting the optimum circuit form are different when one is designing an IC. For example, in a discrete circuit, production economy is usually synonymous with (1) minimum use of semiconductor components, (2) maximum use of discrete components, (3) minimum number of component parts. In discrete-component designs, the ratio of active element cost to passive element cost is in the range of 10–1 to 100–1. In other words, a discrete active element is between ten and one hundred times as

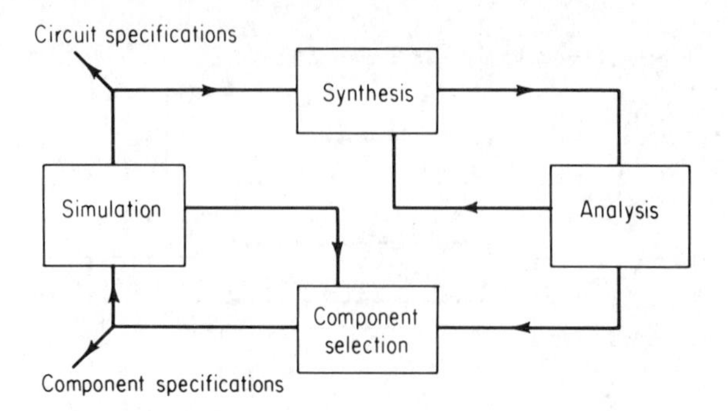

Fig. 14-17. Circuit design process when employing discrete components.

expensive as a passive element. Therefore, circuits implemented with discrete parts tend to use more passive elements than active elements in the interest of economy.

The direct costs associated with producing a discrete-component circuit can be computed from a linear model. Let C_D be the direct costs of circuit production. Then,

$$C_D = C_t + C_d + C_r + C_c + C_l + C_m \tag{14-9}$$

where C_t = cost of transistors
C_d = cost of diodes
C_r = cost of resistors
C_c = cost of capacitors
C_l = cost of inductors
C_m = miscellaneous costs, including assembly labor, rework labor, testing, printed-circuit board costs, etc.

In the IC production, economy is obtained by (1) selecting circuits which make maximum use of active elements; transistors and diodes generally occupy less area on the chip than passive components; (2) limiting use of large-value resistors, since these require large chip area; (3) limiting use of capacitors because these require large area; (4) selecting circuit forms which can achieve critical electrical performance by component matching instead of relying on the absolute tolerance of component values; (5) selecting circuits which do not require high-performance PNP transistors, as these are expensive to fabricate.

Needless to say the direct costs of producing an IC bear no relation to the direct costs of producing a discrete-component design as the manufacturing technologies are completely different. A priority prediction of production costs of an IC is difficult to accomplish because of the complexity of the manufacturing process. A simple linear cost model is presented here to clarify a point. No claim is made for its precise accuracy, and in fact, it would be difficult to obtain cost information in the form required by the model.

In the preceding section, we have seen that hundreds of integrated circuits are fabricated on a wafer of silicon. If at the end of the fabrication sequence each circuit on the wafer met the performance specifications, obviously the cost of a circuit (at that point in the process) would be the cost of processing a wafer to that point divided by the number of circuits on a wafer. Since the cost of processing a wafer is relatively independent of the number of circuits on the wafer, the more circuits per wafer, the less the cost. From this argument we see that the area occupied by the circuit (the "chip area") is a key cost factor.

Unfortunately in a practical manufacturing process not all the circuits on a wafer will meet performance specifications. In other words, some circuits will be defective in some respect. The reasons for defective circuits are myriad but they can be assembled into one of two categories: process defects and design defects. *Process defects* include such things as oxide defects, which may result in shorting between components, and scratches in the interconnection metallization which result in a portion of the circuit not being connected. *Design defects* include those defects which are caused by the circuit design requiring component parameter values that the process cannot provide. For example, suppose for a circuit to meet its performance specification a certain transistor must have $40 \leq h_{FE} \leq 120$. Now suppose that the process used to produce the circuit containing this transistor produces transistors with $30 \leq h_{FE} \leq 150$. Then clearly some of the circuits will not meet the performance.

We propose a linear direct-cost model for the IC as

$$C_D = C_a + C_p + C_d + C_m \tag{14-10}$$

where C_D = direct manufacturing cost
C_a = cost per circuit if all circuits met the performance requirements
C_p = cost increment due to processing defects
C_d = cost increment due to design defects
C_m = miscellaneous costs, such as those associated with testing and packaging. (These are miscellaneous costs, not necessarily insignificant ones.)

Clearly C_a is proportional to the circuit area. The circuit area is approximately proportional to the sum of the number of transistors and diodes (N_{AE}) plus the sum of the resistors (R) in the circuit and the sum of the circuit capacitances (C). Then,

$$C_a = K_A(k_{AE}N_{AE} + k_R \sum R + k_C \sum C) \tag{14-11}$$

in which K_A = factor which converts circuit area to cost
k_{AE} = proportional factor between number of active elements and component area
k_R = proportional factor between total circuit resistance and area occupied by the resistors
k_C = proportional factor between total circuit capacitance and chip area occupied by the capacitor.

C_p should be proportional to the number of active elements and the total resistance and capacitance because an increased number of components results in an increase in area and a higher probability of process defect. Hence

$$C_p = K_p(k_{AE}N_{AE} + k_R \sum R + k_C \sum C)$$

where K_p is a proportionality constant.

Similarly, C_d should be proportional to the number of active elements and the total resistance and capacitance:

$$C_d = K_D(k_{AE}N_{AE} + k_R \sum R + k_C \sum C) \tag{14-12}$$

Therefore, the direct manufacturing cost is

$$\begin{aligned} C_D &= (K_A + K_p + K_D)(k_{AE}N_{AE} + k_R \sum R + k_C \sum C) \qquad (14\text{-}13) \\ &\equiv (K_A + K_p + K_D)\cdot(\text{circuit area}) \end{aligned}$$

The IC should therefore be designed for minimum area if cost is to be minimized.

The cost-weighting factor between active and passive elements in integrated circuits is far different than in discrete-component circuits. The cost of a low-level switching transistor or diode is approximately equivalent to the cost of a 5 KΩ resistor or a 10 pf capacitor.

This influence of circuit area and of the different relative component costs has led to the evolution (and the need) for different circuit topologies in integrated circuits. Inability to make close-tolerance components has also led to the evolution of new circuit form, as we shall see later in this chapter.

EXAMPLES OF INTEGRATED CIRCUITS

In this section we illustrate some of the circuit forms which are advantageous for fabrication as an IC. The particular circuits discussed show, by example, some of the circuit design principles being applied with this new technology. Examples have been drawn from the major circuit categories described earlier in this text, including biasing, low-frequency amplifiers, video amplifiers, RF amplifiers, and various types of digital circuits.

BIASING

Techniques for biasing transistors were presented in Chap. 3. Here we consider an example of biasing a common-emitter amplifier stage to be made as part of an IC. Figure 14-18(a) shows a conventional biasing method. A biasing scheme more appropriate for IC fabrication is shown in Fig. 14-18(b).

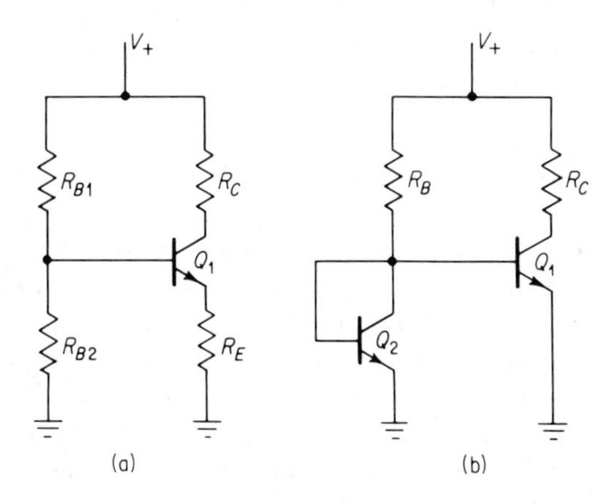

Fig. 14-18. Comparison of techniques for biasing transistors. (a) Resistor bias; (b) diode bias.

The IC biasing scheme relies on the base-emitter voltage drop of transistor Q_2 to provide the base-emitter bias for Q_1. If Q_1 and Q_2 are fabricated with identical geometry and if the input characteristics of both transistors are identical, the collector currents of Q_1 and Q_2 will be identical. The dc transfer characteristic of the common-emitter connection may be approximated as

$$I_C = I_0 e^{qV_{EB}/kT} \tag{14-14}$$

Let the collector current of Q_2 be denoted as I_{bias}. Then,

$$I_{\text{bias}} = I_0 e^{qV_{EB}/kT} \approx \frac{V_+ - V_{EB}}{R_B} \tag{14-15}$$

$$\approx \frac{V_+}{R_B} \tag{14-16}$$

Therefore, the base-emitter bias of Q_1 is,

$$V_{EB} = \frac{kT}{q} \ln \frac{V_+}{I_0 R_B} \tag{14-17}$$

and the collector current of Q_1 is

$$(I_C)_{Q_1} = I_0 \exp \ln \frac{V_+}{I_0 R_B} \equiv \frac{V_+}{R_B} \tag{14-18}$$

This diode-biasing scheme provides a more economical biasing technique than the conventional resistor biasing method because the diode method requires less chip area when fabricated as an IC. In the diode-biasing method, one resistor and one diode perform the same function as three resistors. From an economy standpoint the diode-biasing technique is better. From the standpoint of temperature stability the diode-biasing method is somewhat superior because the characteristics of Q_1 and Q_2 "track" with temperature.

The diode-biasing scheme has a disadvantage for some applications since it provides a lower circuit input impedance. This is a result of the shunting of the diode formed by Q_2 at the input of the circuit.

LOW-FREQUENCY AMPLIFIERS

The subject of low-frequency, small-signal amplifiers was covered in Chap. 4. There it was shown that a low-frequency ac amplifier consists of cascaded stages capacitively coupled and appropriately biased to provide the proper gain, bandwidth, input and output impedances, and output power. It is not feasible at the present state of the art of IC technology to fabricate large values of capacitance for bypassing and coupling. By necessity, therefore, an IC low-frequency amplifier must rely on dc coupling or the use of external capacitors.

An IC designed for low-frequency amplifiers provides a "gain block," a functional element which, with coupling capacitors at input and output, provides the ac amplifier function. (This concept is illustrated in Fig. 14-19.) Here the IC amplifier with transfer function $e_2/e_1 = G(s)$, input impedance Z_{in}, and output impedance Z_{out}, is considered as part of the over-all low-frequency ac amplifier. In general, the coupling capacitors C_i and C_o will be required to isolate the dc output level of the amplifier from the load (C_o) and to allow the amplifier to be biased properly (C_o and C_i).

The specific circuit topology of the IC amplifier will depend on the specifications required by the low-frequency amplifier. Gain, gain stability, frequency response, input and output impedance, and output voltage swing are

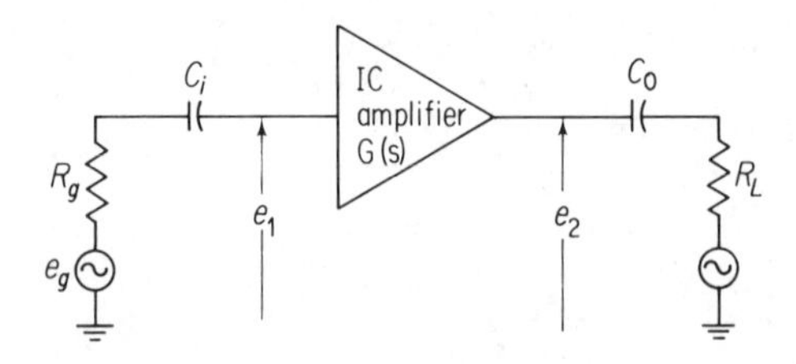

Fig. 14-19. IC "gain block" used to implement an ac amplifier. The frequency response is set by coupling capacitors C_i and C_o and the frequency response of the IC amplifier.

some of the parameters to consider in the design. Generally the amplifier function itself will be implemented with dc circuitry because of the lack of large-value capacitors in IC fabrication technology.

An example of a low-frequency IC amplifier is shown in Fig. 14-20. This is a dc coupled amplifier with internal feedback which is used to control the value of $G(s)$, the voltage transfer function. The over-all amplifier has two common-emitter voltage amplifier stages, Q_1 and Q_3. States Q_2 and Q_4 are emitter followers used for isolation. Diode D_1 is a diode clipper which limits the level of input voltage by clipping. The low-frequency gain may be set to specific values by connecting point A, B, or C to the "feedback return" point. This provides "negative feedback" and degenerates, or decreases, the "open loop" gain, that is, the gain with no feedback. Stage Q_5 and its associated resistors provide temperature stability of the operating points of the rest of the amplifier. The compensation capacitor, C_1, is used to control the frequency response and stability of the amplifier. Provisions are made at the emitter of Q_3 for use of a (large) external bypass capacitor to increase the low-frequency gain. With the circuit constants shown in this design, voltage gain with no feedback and no bypass at the emitter of Q_3 is 50 db.

What features distinguish this IC amplifier from a discrete-circuit design? First, the amplifier is completely dc coupled and avoids the use of large coupling capacitors. Emitter followers are used to provide isolation. A transistor feedback stage is used to provide temperature stability. Versatility of the amplifier is achieved by providing internal feedback components which may be used to control the gain.

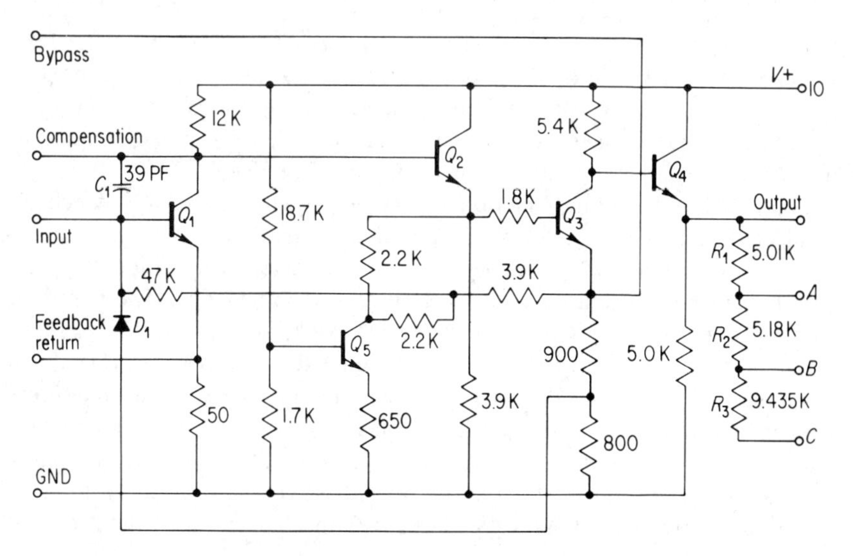

Fig. 14-20. An example of an IC amplifier which may be used as a "gain block" in forming a low-frequency ac amplifier.

TUNED AMPLIFIERS

Design and performance of transistor-tuned amplifiers were covered in detail in Chap. 5. How can an IC be used in a tuned amplifier? At the time this text is being written there is no practical and economical manner for fabricating components as an integral part of an IC that will function as tuned circuits because of the difficulty of fabricating inductance. Therefore the poles and zeroes required in the over-all transfer function of a tuned amplifier must be implemented with discrete components. The IC can, however, provide the required power gain, AGC capability, and isolation between source and load that are among the requirements of circuits to be used in tuned amplifiers. This is illustrated conceptually in Fig. 14-21, where again the IC amplifier is serving the function of a "gain block" as in the previous section on low-frequency amplifiers.

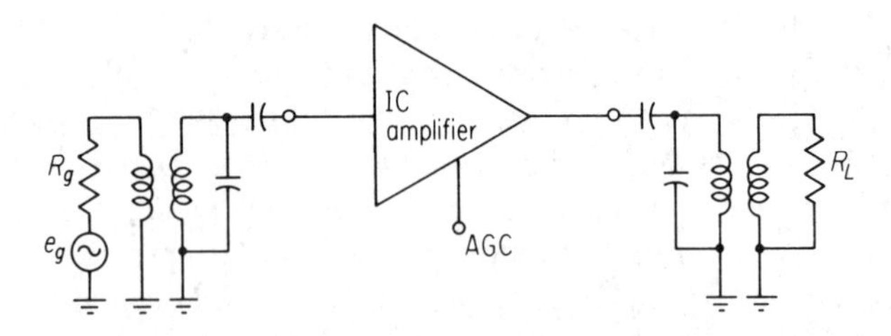

Fig. 14-21. An IC amplifier functioning as a "gain block" in a tuned amplifier. The IC amplifier provides gain, AGC capability, and isolation between source and load.

The specific form of the IC amplifier depends on the over-all requirements of the tuned amplifier, such as gain, noise figure, impedance level, and AGC capability. A specific design useful in tuned amplifiers from low frequencies up through several hundred megahertz is shown in Fig. 14-22.

The basic amplifying function is achieved with Q_1 and Q_3 which are connected as a *cascode* amplifier (common-emitter stage driving a common-base stage). The base of Q_2 is effectively at ac ground because diodes D_1, D_2, and D_3 are heavily biased and represent a low impedance. The common-emitter stage Q_1 is biased via negative feedback from the output dc level through Q_3 to the base of Q_1. This negative feedback also serves to stabilize the operating point of Q_1 and Q_2 against changes in temperature, supply voltage, and transistor parameters. The bias of Q_2 is achieved by the dc level established across diodes D_1, D_2, and D_3 in series. Automatic gain control is achieved by varying the dc voltage at the base of Q_3, thus controlling the dc emitter current of Q_1 and Q_2. The AGC mechanism is of the "reverse" type discussed in Chap. 5.

When this amplifier circuit is incorporated into a tuned amplifier, a tuned circuit is placed at the input, point *A*. A dc blocking capacitor as shown in

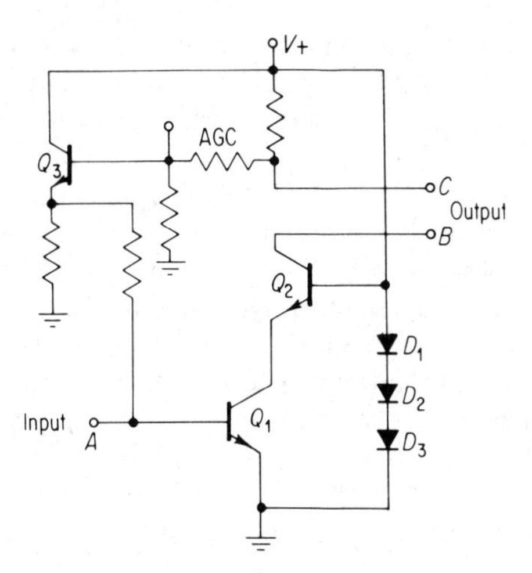

Fig. 14-22. An IC amplifier for application in tuned amplifiers.

Fig. 14-21 must be used to prevent upsetting the bias of Q_1. The output-tuned circuit is placed between points B and C at the output. Here no blocking capacitor is used.

The *cascode* connection is a good choice for this application as it provides (1) high gain because of the two amplifying stages, (2) low noise figure because of the common-emitter input stage, (3) good isolation between input and output because of the two amplifier stages in series. This amplifier is capable of providing 25 db of stable power gain at 60 MHz and requires no neutralization.

Compare this IC amplifier circuit with the discrete-component designs in Chap. 5. The manner in which this circuit maximizes the use of active elements while minimizing the use of passive elements is apparent. Again, dc coupling is used throughout the IC amplifier to avoid large capacitors which are not economic to fabricate. Use of diode biasing in this circuit in lieu of a resistor voltage divider results in no bypass capacitor being needed in the bias network. Using a transistor for stabilizing the operating point (Q_3) is a feature not often found in discrete-component designs because there are other more economical methods for achieving operating point stability in these circuits.

VIDEO AMPLIFIERS

Video amplifiers represent another area where the IC technology can provide useful devices for many applications. As discussed in Chap. 6 on video amplifiers, this type of amplifier is not unlike the low-frequency amplifier (Chap. 4). The primary difference is in the area of frequency response, where the frequency response of video amplifiers extends at least several orders of magnitude beyond that of low-frequency audio amplifiers. In fact, in some applications, the frequency response of video amplifiers is required to go beyond 100 MHz.

The function performed by an IC amplifier when designed and used in a video amplifier is identical in concept to its use in the low-frequency and tuned amplifiers described in previous sections. The IC amplifier provides gain, perhaps some bandpass shaping, isolation, and the required output drive level. The specific IC amplifier circuit topology used is dependent on the requirements of the specific application. Here we present a representative IC amplifier configuration which may be used as a video amplifier in the frequency range from low frequency to 150 MHz.

The circuit schematic is shown in Fig. 14-23. The various points have been numbered to facilitate discussion of the circuit operation. The amplifier has

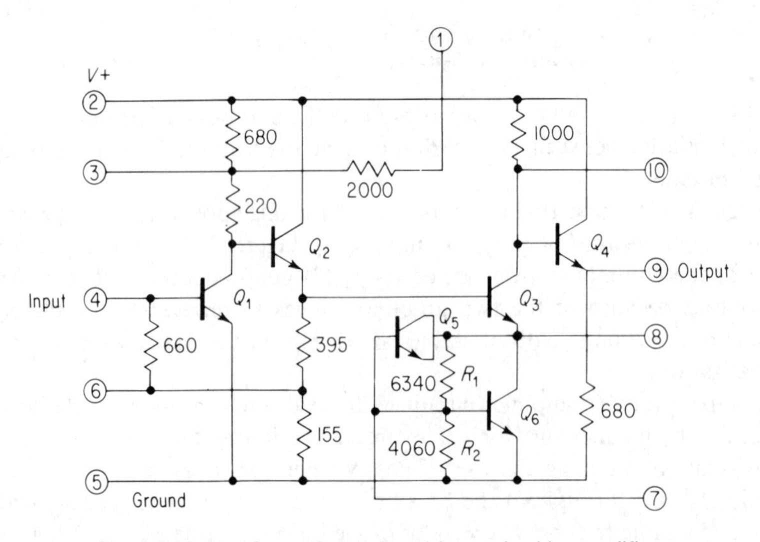

Fig. 14-23. An IC amplifier designed for use in video amplifiers.

two common-emitter gain stages, Q_1 and Q_3. Stages Q_2 and Q_4 are emitter followers which prevent loading of the gain stages. Bais of Q_1 is achieved by negative feedback from the split emitter resistor of Q_2 to the base of Q_1. This biasing technique allows the amplifier to maintain a reasonably high input impedance and provides temperature stabilization of the operating point of Q_1. The gain stage, Q_3, and the associated circuit in its emitter require some explanation. The network in the emitter of Q_3, consisting of Q_6, R_1, R_2, and Q_5, serves the dual role of providing biasing for the emitter of Q_3 and frequency compensation for the gain stage.

Transistor Q_6 and R_1 and R_2 determine the dc potential at the emitter of Q_3. The various dc bias currents and voltages for this network are defined in Fig. 14-24.

$$I_1 = I_B + I_2 \tag{14-19}$$

$$I_2 = \frac{V_{BE}}{R_2} \tag{14-20}$$

The potential at the emitter of Q_3 is

$$V_E = I_1 R_1 + I_2 R_2 \tag{14-21}$$

Substituting Eqs. 14-19 and 14-20 into Eq. 14-21 yields

$$V_E = I_B R_1 + V_{BE}\left(1 + \frac{R_1}{R_2}\right) \tag{14-22}$$

For high h_{FE},

$$V_E \approx V_{BE}\left(1 + \frac{R_1}{R_2}\right) \tag{14-23}$$

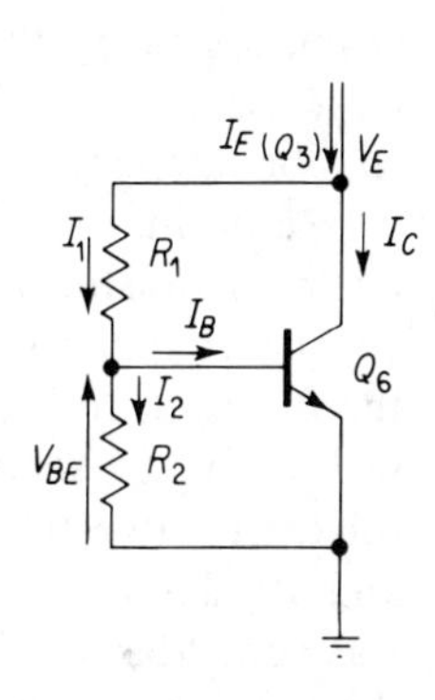

Fig. 14-24. Definition of various dc bias currents and voltages in the transistor bias network.

This bias network therefore fixes the emitter voltage.

The other function served by this bias network is to act as frequency compensation in order to extend the frequency response of the amplifier. Transistor Q_5, in shunt with R_1, with its collector and emitter shorted, acts as a capacitor (C_{TE} and C_{TC} in parallel). The impedance from the collector of Q_6 to ground may be shown to be equivalent to a parallel RC network (Fig. 14-25) where

$$R = \frac{R_1}{\beta_0} + r_e\left(1 + \frac{R_1}{R_2}\right) \tag{14-24}$$

and

$$C = \beta_0(C_{TE_{Q_5}} + C_{TC_{Q_5}} + C_{TC_{Q_6}}) \tag{14-25}$$

Here r_e is the emitter resistance of Q_6, and β_0 is the common-emitter current gain of Q_6. Note that the relatively small capacitance

$$(C_{TE_{Q_5}} + C_{TC_{Q_5}} + C_{TC_{Q_6}})$$

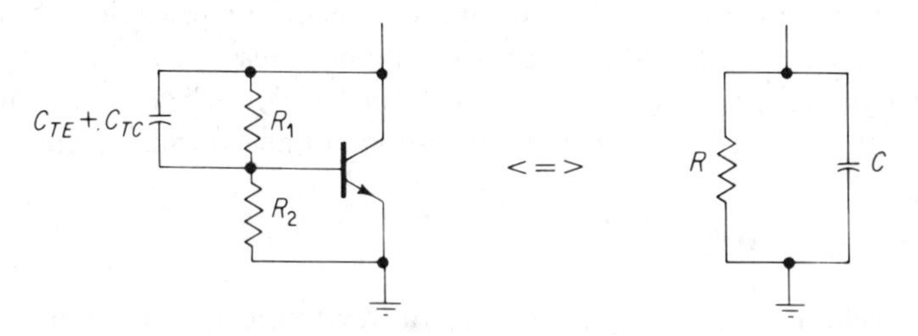

Fig. 14-25. Equivalent output impedance of bias network. This network acts as a frequency compensating circuit.

is multiplied by the transistor current gain and hence appears as a much larger capacitance. This is a distinct advantage for an IC where it is difficult to fabricate large values of capacitance.

Returning now to the description of the amplifier operation, note the feedback resistor between points 1 and 3. When point 1 is connected to point 10, the amplifier gain is reduced and the bandwidth extended. By lowering the feedback resistance between points 3 and 10, different gains and bandwidths can be achieved. Figure 14-26 shows relative gains and bandwidths for three values of feedback resistance, R_F, the resistance between points 3 and 10. To achieve the bandpass flatness noted here, a trimmer capacitor was added between pins 5, 7, and 8 and adjusted for maximum flatness. For the circuit constants shown in Fig. 14-23 the open loop gain is 51 db; with $R_F = 2000\ \Omega$, the gain is 43 db; and with $R_F = 180\ \Omega$, the gain is 28 db. The bandwidth for these gain levels is shown in Fig. 14-26.

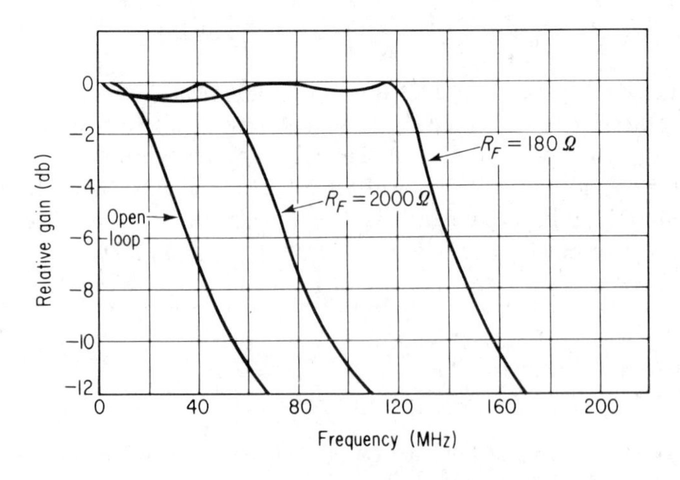

Fig. 14-26. Gain and bandwidth trade-off in the IC video amplifier.

This amplifier maximizes the use of active elements to achieve performance. Use of a transistor for biasing and frequency compensation is a unique feature not generally found in discrete-component designs. As in the previous amplifier designs, dc coupling is used. Using negative feedback provides a convenient way to change the gain by addition of an external component.

LOGIC CIRCUITS

The IC technology has proved very useful for fabricating a variety of logic circuits. (For a review of the design of logic circuits see Chap. 9.) Logic circuits need be designed for only OFF-ON operation; hence they have rather "loose" component tolerance requirements compared with linear circuits.

Logic circuits are therefore ideal for fabrication as IC devices. A variety of IC logic circuits are currently being manufactured and used in data processing equipment. In this section we consider one of these to illustrate some of the differences between IC logic elements and discrete logic circuits.

In Chap. 9 we considered the design of DCTL (Direct Coupled Transistor Logic) as a form of discrete-component logic circuit. This form of logic element is natural for consideration as a candidate to fabricate as an IC because it uses only transistors and resistors. To make this circuit form practical as an IC device, a simple modification must be made to the basic DCTL logic form. A resistor must be put in series with each transistor input to alleviate the base-current "hogging" problem discussed in Chap. 9 and shown in Fig. 9-8. Because of the relatively large spread in input characteristics (V_{BE} versus I_B) experienced in IC transistors, this "hogging" problem is severe. Addition of a resistor at the input of each transistor reduces this spread in circuit input characteristics. Figure 14-27 illustrates the basic circuit modification of DCTL. This is a modified form of RTL. Although DCTL is today a relatively uneconomical way to build discrete-component logic circuits because of the large number of transistors used (one for every input), the modified RTL circuit is a very economical circuit to fabricate as an IC.

More complex circuits may be fabricated from the basic gate element. These include the combinational logic circuits, flip-flops, and shift registers described in Chap. 9. An example of a modified RTL flip-flop is shown in Fig. 14-28. This circuit performs functions similar to that of the *RC*-coupled flip-flop in Fig. 9-50. In the IC flip-flop, use of capacitors is avoided at the expense of a topologically complex circuit. Economy and performance have been obtained by maximizing the use of active elements.

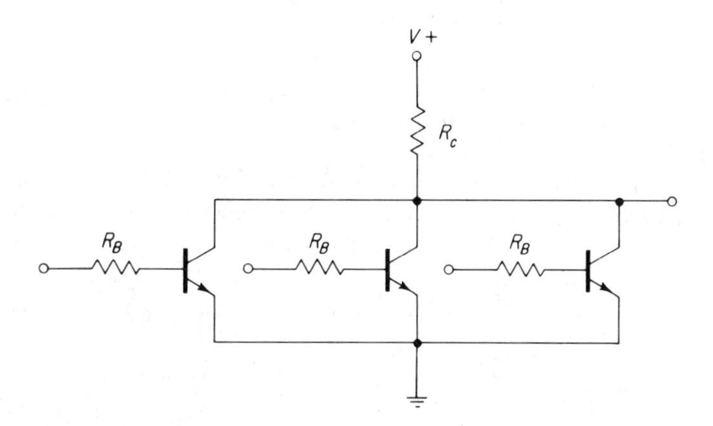

Fig. 14-27. Basic three-input logic gate showing modified form of DCTL. This logic form is modified RTL.

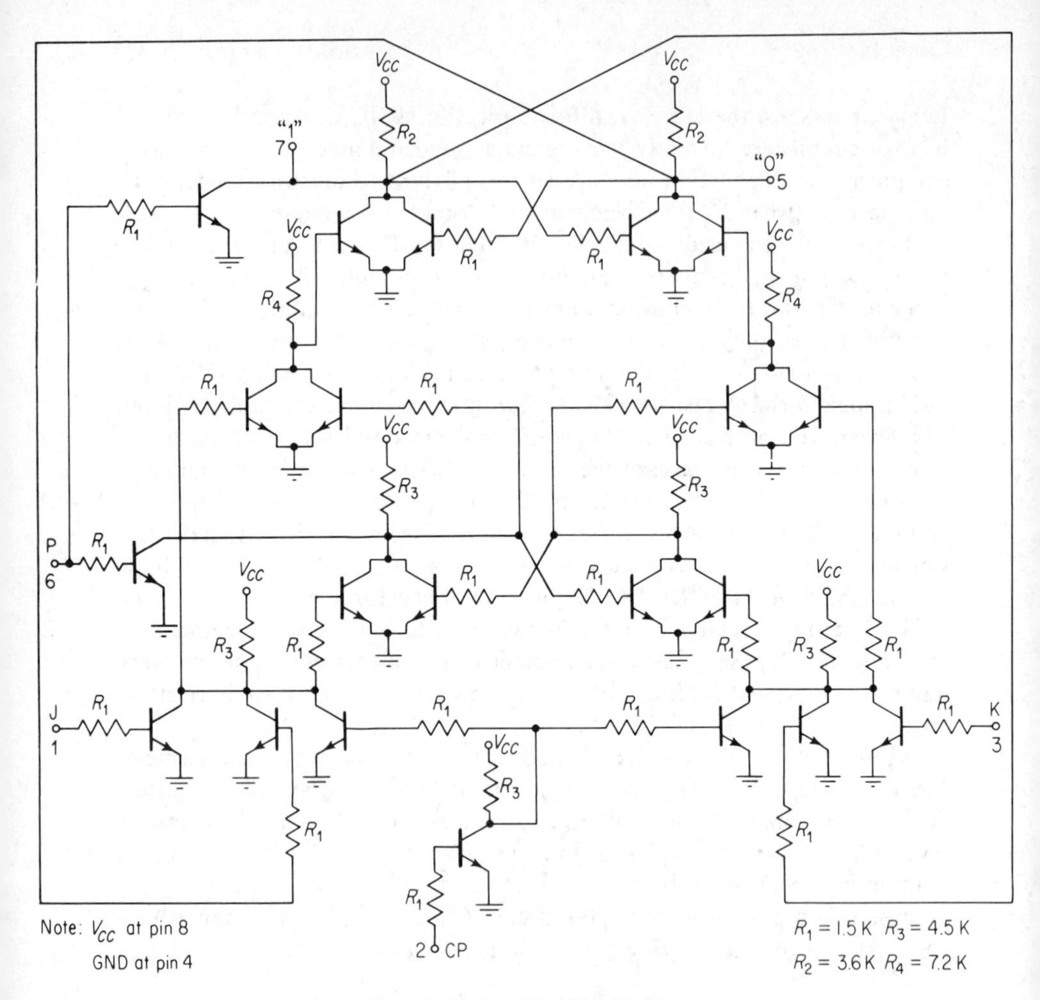

Fig. 14-28. Modified RTL flip-flop.

NEW TRENDS IN MICROELECTRONICS

Thus far, this chapter has dealt with a particular class of silicon integrated circuits—namely, those in which the active elements are bipolar transistors and in which all of the elements are fabricated within a single block of silicon with a total complexity of up to about 50 or 60 elements. Such circuits include over 99% of those in common use today. Nonetheless, a number of important new developments are being evaluated both in the laboratory and in limited product usage. Some of these promise to bring about significant changes in the way microcircuits are designed and used. These will be the subject of this section.

New Types of Transistors

By way of review, recall that the "bipolar" transistors that have been previously discussed are minority carrier devices; that is, the important charge carriers in the active region of the device (the base) are of opposite conductivity type to those majority carriers which would normally characterize the conductivity of this region. In the NPN transistor, the minority carriers in the base are electrons even though the base is p-type.

A second, and in fact older, category of transistor which is becoming increasingly important depends for its amplifying action on changes in the number of majority carriers in the active region of the device. Hence, they are called majority carrier devices. Changes in the number of majority carriers require the application of much higher fields than in the case of minority carrier modulation. Hence, the name *field effect transistor* has been given to such devices.

Field effect transistors can, in turn, be broken down into two basic types. These are junction gate field effect transistors (FET), and insulated gate field effect transistors. The most commonly used insulated gate device is the metal-oxide-silicon (MOS) transistor. The junction FET and two types of MOS transistors are illustrated in Fig. 14-29.

In using field effect devices, a suitable potential is normally applied between the source and drain electrodes. Current flows in the channel region between the electrodes and, as already stated, the current consists of carriers having the same conductivity type as the conductivity type of the source and drain electrodes. Current flowing in the channel is modulated by the application of a voltage to the gate electrode. At this point, however, the action of the various types of FET transistors is seen to vary.

In the junction device, the application of a field to the gate electrode causes the depletion region of the back-biased gate junction to widen into the channel—thus constricting the cross section of the channel and decreasing its conductance. Thus, with zero gate bias, source-drain conduction will be at a maximum. The conductance will decrease with the application of higher gate bias.

In the MOS transistor, on the other hand, carriers in the channel are induced or depleted by the application of a field across the gate oxide, that is, mobile carriers are either attracted to or repelled from the surface.

If the device has been deliberately fabricated with a thin channel of the same conductivity type as the source or drain [Fig. 14-29(b)], then the action of the transistor depends on the ability of the field induced across the gate oxide to repel all of the mobile carriers out of this channel. Such a normally-on device, which is turned off by depleting the carriers in the channel, is also referred to as a depletion mode device. On the other hand, if such a channel

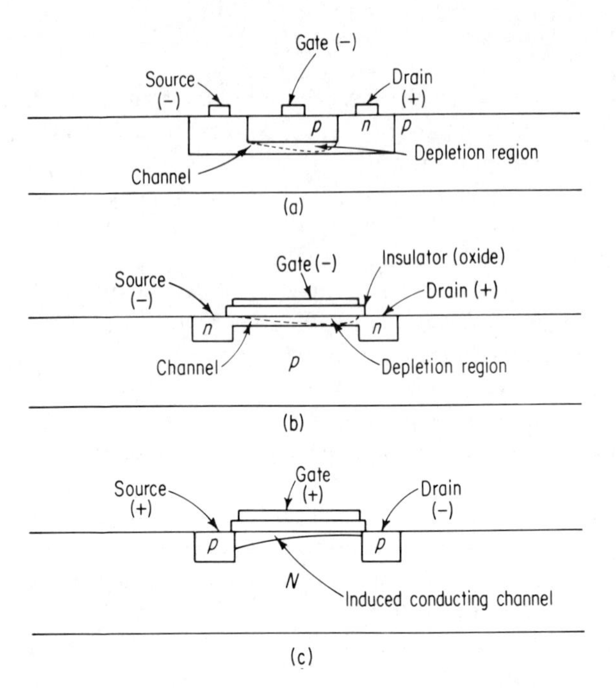

Fig. 14-29. (a) Junction FET; (b) normally-on (depletion mode) insulated gate FET; (c) normally-off (enhancement mode) insulated gate FET.

has not been formed as part of the fabrication procedure [Fig. 14-29(c)], initial conduction cannot take place regardless of the sign of the applied source drain voltage since the current path will include a back-biased junction. The device will be "normally-off", and must be turned on by attracting a large enough quantity of carriers of the correct sign to the surface to form a channel between the source and drain electrodes. Such normally-off devices are also called enhancement mode devices. The electrical characteristics of FET's are shown in Fig. 14-30.

To understand the action and the detailed electrical characteristics of these devices more fully, one must take into account the fact that the field being applied to the channel at any point along its length depends not only on the voltage applied to the gate electrode, but also on the potential drop occurring in the channel region as a result of the current flowing through it between the source and drain. Thus, the depletion or enhancement regions as shown in Fig. 14-29 are not uniformly wide along the length of the channel. At low source-drain voltages, this lateral effect is minimal and the device resembles a variable resistor. As source-to-drain voltage is increased, however, there is sufficient lateral voltage drop in the channel so that next to the source electrode, the channel may be heavily conductive, while next to the

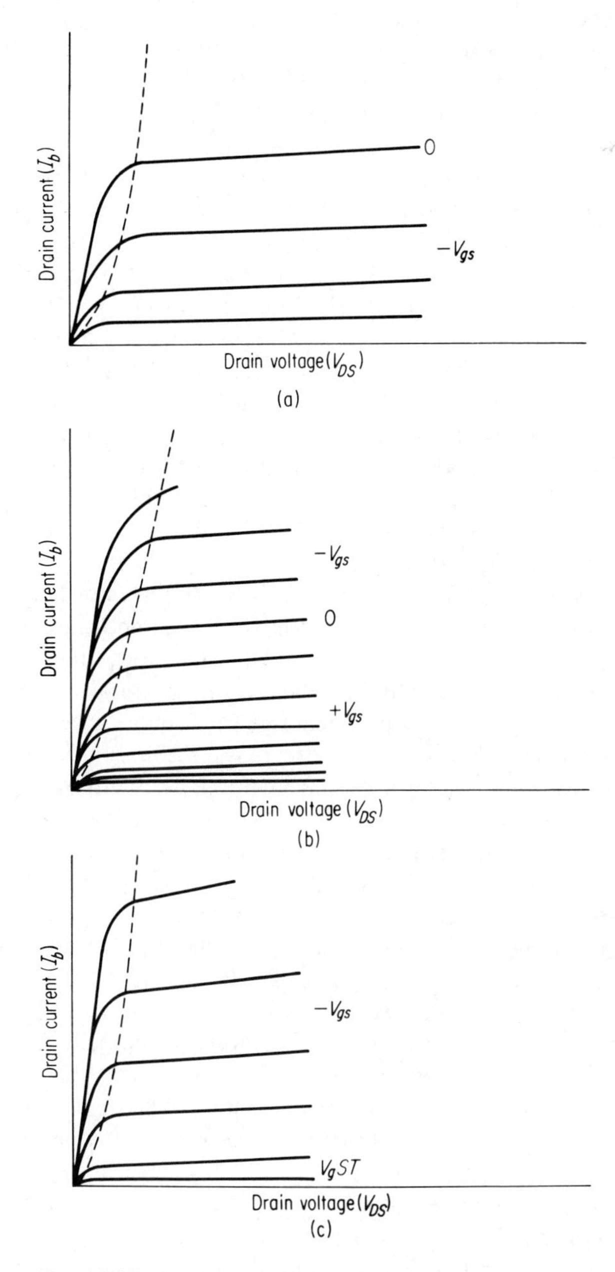

Fig. 14-30. (a) Junction FET; (b) normally-on (depletion) insulated gate FET; (c) normally-off (enhancement) insulated gate FET.

drain electrode, the channel may be almost completely pinched-off. Actually, no matter how much current flows between the source and drain, the channel will never completely pinch off, since if this were to happen the lateral voltage drop would decrease, hence widening the channel in the pinch-off region to restore current flow. An equilibrium condition is thus established in which the current saturates, so that over a wide range of source-drain voltage there is very little change in source-drain current. The point at which this occurs is known as the pinch-off point. From Fig. 14-30, we see that such points form a continuous curve separating the characteristics into two regions known as the pinch-off and prepinch-off regions. In the pinch-off, or normal region of operation, the device has a high output impedance. In fact, the characteristics closely resemble the conventional vacuum tube pentode (and similar circuit design principles apply). Of perhaps greater significance, however, is the fact that field effect devices also have a high input impedance at low frequency. That is, in both classes of devices essentially no gate current (or very little in the case of junction field effect devices) flows. The typical input impedances for an MOS transistor is $10^{15}\ \Omega$.

Junction FET's and normally-on insulated gate field effect transistors find their greatest use and application where a higher input impedance than that available in a bipolar transistor is required. A typical application would be as the input to an amplifier for a ceramic transducer output, since such devices have an inherently high output impedance. FET's are also used as input devices in r-f tuned amplifiers since their high input impedance prevents them from loading tuned circuits. FET's also may exhibit somewhat improved cross modulation and intermodulation characteristics although equivalent results can generally be obtained from properly designed bipolar circuits.

From Fig. 14-30, the effect of gate voltage is also apparent. Both the junction FET and the normally-on insulated gate FET are conducting at zero base bias. The characteristics of the insulated gate device can be varied linearly on both sides of the zero gate bias point since the application of a negative bias will cause further enhancement of the conductivity in the channel in the same manner as the normally-off device shown in Fig. 14-30(b). The junction device can only be operated for a single gate bias polarity, since for positive bias the gate junction will be forward-biased and gate current will flow, thereby reducing the field effect to zero.

An important parameter for the normally-off device is the voltage which must be applied to the gate before current will flow. This voltage is called the threshold voltage and is designated V_{gst}. This voltage is commonly about 4 volts for MOS transistors, although it may be reduced (or, in fact, made positive thereby forming an enhancement device) depending on the specific manufacturing process.

The unique properties of the normally-off MOS device have made it possible to formulate a new class of circuits, which are particularly adaptable to silicon integrated circuits, called "dynamic multiphase logic circuits." The best example of such a circuit is found in the very large serial shift registers which utilize this circuit technique. In these shift registers, information (1 or 0) is stored at each stage in the shift register by charging the input capacitance of an MOS FET. An MOS device is used as a bilateral coupling element to charge and discharge this input capacitance, and an MOS device is also used for the node so that the flow of information can be directed by turning the node device on and off with a multiphase clock.

The basic principles of MOS dynamic logic, as applied to a serial shift register, are shown in Fig. 14-31. A negative voltage (logical 1) clocked to Data In at ϕ_2 time is stored on the node capacitance there. At ϕ_1 time, this information is clocked through the first inverter and stored on the node capacitance at A as a "0". At ϕ_2 time, it is clocked through the second inverter to Data Out as a "1." Thus, the condition at Data In appears at the output delayed by one clock period, or 1 bit.

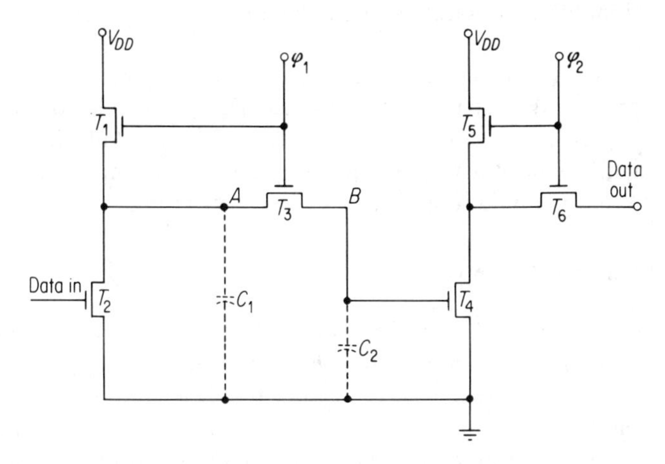

Fig. 14-31. Two-phase implementations of delay stage.

A coupling device, T_2, uses the bilateral properties of the MOS FET to charge and discharge the node capacitance at A. This basic delay stage can be used as a shift register stage or, by adding feedback and appropriate gating circuit, it can be transformed into an RS, JK, or other type of flip-flop. Large shift registers, based on this principle, are further discussed in the next section on complex silicon integrated circuits.

Looking Ahead

A number of significant new developments in silicon monolithic microcircuit fabrication technology are taking place. These include progress in the techniques already described above, as well as the introduction of radically new fabrication concepts. These developments can be placed into two broad categories:

1. Those developments whose impact is largely economic in terms of improved yield and reduced cost per circuit function. Among the more significant of these are:
   a. Greater circuit complexity including, but not limited to, MOS;
   b. New interconnection and packaging techniques;
   c. Computer design of microcircuit chips;
   d. Greater fundamental knowledge of electronic effects in silicon planar oxides.
2. Those techniques which are being studied as means of overcoming performance barriers. These include:
   a. Microcircuit transistors with considerably higher performance;
   b. Thin film/silicon monolithic circuits;
   c. Monolithic insulators and conductors.

It should be noted that in the case of the second group a significantly increased processing cost poses a major barrier to general introduction of these techniques into widespread application. However, sufficiently important new methods have a way of becoming cheaper to implement when no other route can be found to achieve an equivalent result.

New Cost-Directed Technology

Greater Circuit Complexity. Increasing complexity in terms of the number of circuit functions per unit area on a wafer is the most powerful cost reduction tool available to the microcircuit industry. It is an important test that once a given circuit can be produced at reasonable yield (more than 20% "good" circuits on a wafer), product redesign simply to produce more circurity per unit area can result in considerably greater cost reduction than can either yield improvement or process simplification.

Consider the circuits on the wafer shown in Fig. 14-3 where the defective circuits have been marked by an ink dot as part of the in-process testing. The yield, shown here to be about 50%, may be considered typical for production of circuits that are not pushing the limits of the technology and which have been in production for some time. Note that the defective units are so grouped that large areas of the wafer produce good units.

The amount of circuitry per chip can be increased either by increasing the size of the chip or by using smaller geometry techniques to put more components on a given circuit chip. The effect on cost can be seen by reference to Table 14-1. Note that with present technology, it might cost 50¢ to produce a good microcircuit chip and an additional $3.00 to assemble the circuit into a

TABLE 14-1 Relative Effect on Circuit Cost of Increased Complexity versus Increased Manufacturing Efficiency

Cost Factor	Present	Assume 2 Times Yield and 2 Times Productivity	Assume 6 Times Complexity	Assume 100 Times Complexity
Cost/chip	\$0.50	\$0.15	\$1.00	\$25.00
Cost/assembly, testing, and certification	\$3.00	\$3.00	\$6.00	\$10.00
	\$3.50	\$3.15	$\frac{\$7.00}{6} = \$1.16$ cost per equivalent amount of circuitry	$\frac{\$35.00}{100} = \$0.35$ cost per equivalent amount of circuitry

package, test it, and certify its reliability level, for a total cost of \$3.50. An improvement factor of 4 in combined yield and productivity (100% yield for the circuit illustrated in Fig. 14-3) would only reduce the cost by 10%. On the other hand, assume that six times the amount of circuitry is placed on the chip, albeit with twice the original cost. In this case, the cost for an equivalent amount of circuitry is reduced to a third of the original amount. The trend continues even when the cost of the chip (due to higher design costs and smaller production runs) becomes greater than the assembly, test, and certification cost.

In general, as the number of components on a chip is increased, the number of external device package leads goes up rapidly until a complete subsystem function is obtained, at which point the lead count drops back to a practical level. The ability to design large complex arrays is currently limited by the degree to which the systems designer can or will partition his over-all system into suitable size blocks, each with a minimum number of input and output leads.

The complexity concept was initially introduced through the use of the MOS transistor arrays (18). The inherent small size of the components, the elimination of the need for "isolation" regions, and the ability to create simple circuit forms using the built-in capacitor storage mechanisms and easier-to-implement manufacturing process, combine to make it possible to manufacture, for the first time, single-chip microcircuits containing hundreds of components. The high frequency performance of the best MOS transistor is below that of the better bipolar transistors, and the yield of circuits containing the latter type of device is rapidly approaching a level where it is feasible to fabricate much more complex bipolar arrays. Both types of arrays will doubtlessly be used depending on the particular application.

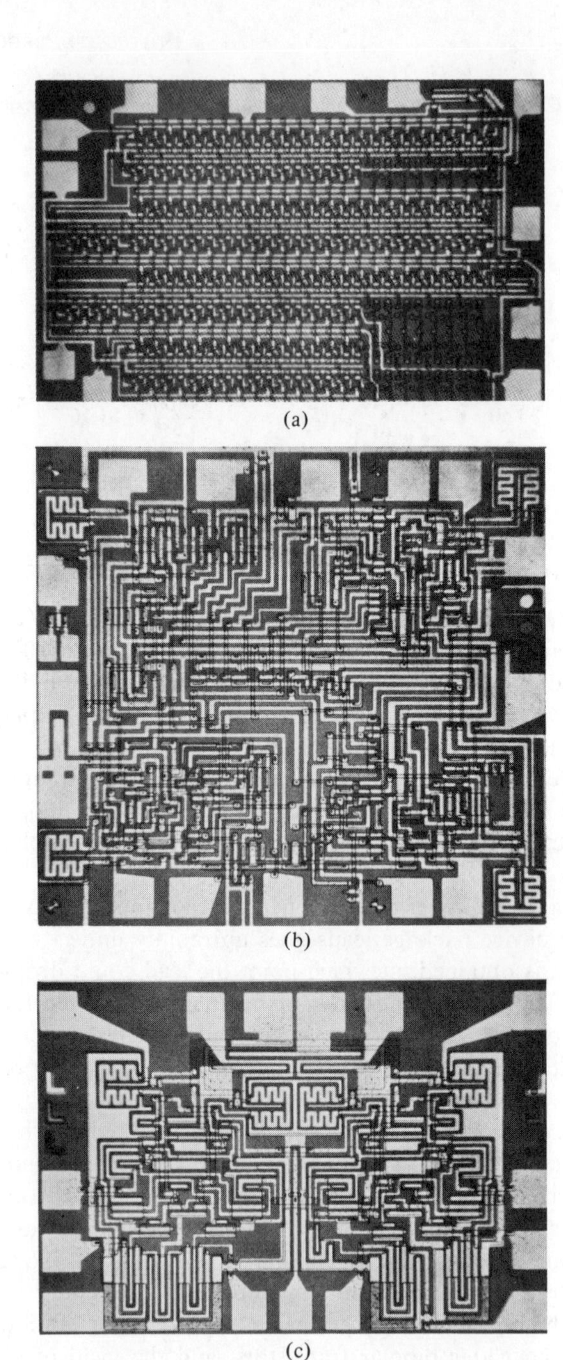

Fig. 14-32. (a) 48-bit shift register with 44 devices; chip size is 66 × 102 mils; (b) BCD counter with 118 devices; chip size is 78 × 90 mils; (c) dual JK flip-flop with 51 devices; chip size is 56 × 86 mils.

To illustrate the considerations involved in array design, it is worth examining a few commercial MOS circuits, as shown in Fig. 14-32, in greater detail.

The advantages of MOS devices are seen most directly in serial input-output blocks having repetitive functions, such as the 48-bit shift register of Fig. 14-32(a), where stages are interconnected in series using a minimum of complex interconnections. This device is based on the stage delay concept of Fig. 14-31 already discussed. One-hundred-bit shift registers of this type containing 100 to 800 MOS devices are available and 200-bit shift registers have been built; however, 200 to 300 devices are considered to be more practical based on acceptable yields (19).

In a circuit such as the BCD counter of Fig. 14-32(b), which has a large number of interconnections on two levels between the various active elements, the interconnections rather than the number and size of the individual components determine the chip area. This situation is typical of the majority of higher speed parallel design computer microcircuits in use today.

A still different situation arises where a circuit requires a number of large-area MOS devices to provide a necessary current drive capability, as in the dual JK flip-flop of Fig. 14-32(c). In this case, bipolar transistors having the required current handling capability would occupy less space than the MOS transistors shown.

In Fig. 14-33, a photomicrograph of a typical complex bipolar circuit (a master-slave JK flip-flop) containing 60 separate components is shown for comparison. This 60-component circuit has about one half the functional

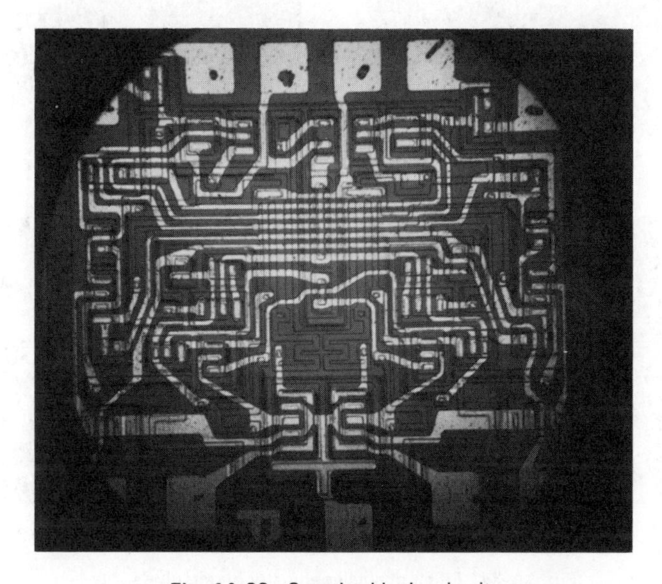

Fig. 14-33. Complex bipolar circuit.

capability of the circuit of Fig. 14-32(c), but about 10 times the speed, with somewhat increased functional capability per flip-flop.

The type of function to be performed and the choice of circuit or system method can strongly affect both the required number of components and the achievable component density.

Small Geometry and Complexity

The level of circuit complexity achievable is being strongly affected by the trend toward smaller geometry circuit fabrication. Considerable progress is being made in reducing the size of microcircuit elements. The result is clearly illustrated in Fig. 14-34, in which microcircuits constructed with different linewidth elements and photographed with the same magnification are compared. The smallest circuit shown (approximately 3 mils from side to side) is a complete digital gate which contains a multiple gating input, a high speed inverter, and a choice of three load resistors, all interconnected as a single functional device.

The ability to fabricate smaller microcircuit elements using photoengraving techniques is being pushed far beyond levels thought possible a few

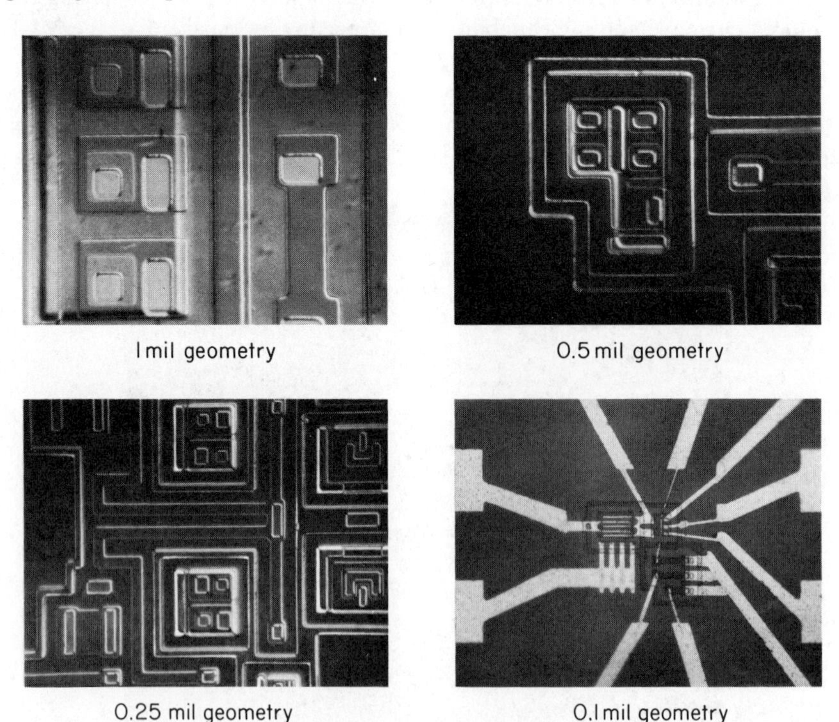

Fig. 14-34. Effect of small geometry on circuit density.

years ago, when it was believed that as circuit dimensions approached a distance equal to several wavelengths of the light used in the photoengraving process, diffraction effects would make it impossible to obtain the desired image. In fact, refinements in the contact printing process have made it possible to fabricate patterns having dimensions equal to 2 or 3 wavelengths of the light used in the photoengraving process. Such a pattern is illustrated in the transistor metalization pattern shown in Fig. 14-35, where micron-wide emitter and base metalization lines with micron spacings are seen.

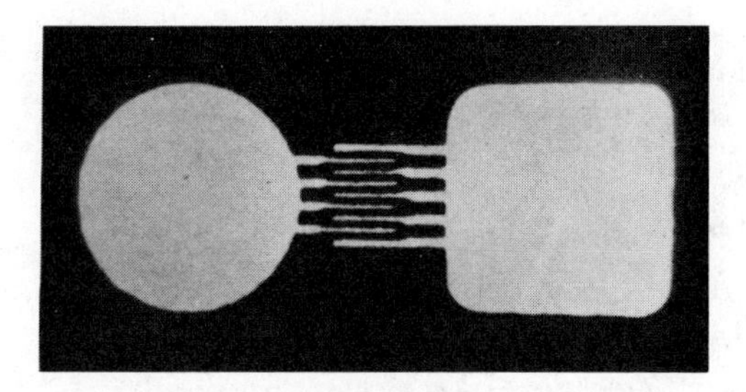

Fig. 14-35. One-micron transistor geometry.

A further insight into the present state-of-the-art and what may be expected is seen in Table 14-2, in which a number of commercially available MOS and bipolar circuits are listed together with normalized component density for chip sizes which permit reasonable production yield. It is noted that except for the special case of the MOS shift register, the component densities are similar for 1-mil bipolar and 0.4-mil MOS, even with the larger size of the components used in the bipolar circuits (the 1 mil and 0.4 mil refer to the smallest oxide cut width). This result is in part due to the space requirements for interconnections in the functionally more complex MOS circuits. The expected impact of small geometry in the bipolar case is seen in the last three developmental circuits where the normalized component count goes up drastically with component density. The larger numbers indicated cannot be taken literally, since a proportionately greater area for interconnections will also be required in more complex bipolar blocks. The impact on computer system design of having available much more complex silicon integrated circuits at no increase in price can readily be envisaged by the reader.

Complexity and Speed

As is noted in a later section on high frequency transistors, we are beginning an era in which the attainment of improved device performance (20) will make it practical to build nanosecond switching circuits. First order considerations, however, indicate that such circuits cannot be operated at these speeds using the components either as discrete elements or as discrete circuits because the RC charging time constant and transmission time delay of the interconnections will severely limit the speed. This problem can only be overcome by assembling major computer subsystems ln a single chip to eliminate long interconnections. Highly complex bipolar arrays with buffer input and output circuits are being designed to operate at nanosecond speeds. An example of such a circuit is the 9-bit parity checker listed in Table 14-2 and shown in Fig. 14-36. This device uses emitter coupled logic circuits with a propagation delay per gate less than 0.1 ns.

Fig. 14-36. Nine-bit parity checker.

In summary, it has been noted that the initial approaches to considerably more complex microcircuits are using the MOS technology. The major reasons are that (1) availability of new circuit forms based on MOS dynamic logic makes it possible to perform a larger number of circuit functions per unit area, and (2) at the present time, the yield for a given chip area is higher for MOS circuitry than for bipolar circuitry. Thus, it is practical to produce circuits of larger area.

The ability to produce complex bipolar circuits at high yield has so improved that bipolar circuits on chip sizes up to about 80 mils square are practical and, with smaller geometry technology, we can expect high-speed

TABLE 14-2 Component Density as Affected by Technology and Circuit Type

Components	Number of Components	Chip Size, Mils	Chip Area Used, Mils^2	Number of Components Normalized to 2200 Mils^2 Useful on 50×80 Chip	Number of Components Normalized to 4100 Mils^2 Useful on 80×80 Chip	
MOS 200 bit shift	1200	90×40	6710		730	
MOS 90 bit shift	540	80×58	3700	320	600	MOS shift registers 0.4 mil
MOS 48 bit shift 0.4 mil	444	66×102	5000	194	360	
MOS 9 bit par. shift 0.4 mil	198	91×118	8700	50	93	
MOS BCD counter	118	78×90	5600	46	86	MOS conventional 0.4 mil
MOS dual JK	51	56×86	3100	36	74	
MOS BCD to dec. conv.	60	70×104	4750	28	53	
DTL JK FF PL941	60	56×56	2240	60	111	
6 input inverter	48	54×62	2500	43	79	Bipolar 1 mil
MW JK 940A	48	62×71	3400	31	58	
High gain diff. amp.	35	50×56	1680	46	86	
Functional receiver circuit	58	53×53	1370	93	174—Bipolar 0.5 mil	
Dual 3 input gate	14	32×35	210	148	276—Bipolar 0.25 mil	
3 bit parity checker (2 level)	58	30×34	(226)	(560)	(1050)—Bipolar 0.10 mil	
9 bit parity checker (3 levels of metalization)	232	45×51	960	540	1000	
27 bit parity checker (two levels of metalization)	754	90×85	3266	510	950	

single chip subsystems comparable in size to those which have already been produced using MOS technology.

New Interconnection and Packaging Techniques

Considerable effort is now being spent on the development of improved electrical interconnections both on the chip and from the chip to the larger system of which it is a part. Reasons for this activity include the following.

Reliability problems that can be caused by metalization defects require expensive visual inspection of the microcircuit chip on the production line to eliminate faulty units. A typical chip metalization defect is illustrated in Fig. 14-37. For this and other reliability reasons (e.g., undesirable metallurgical interactions at the pad-to-wire bond), new metalization techniques are being sought to interconnect elements on a wafer using refractory (high temperature) metals, and various chip coating materials are being evaluated as a means of making chip circuits more scratch resistant.

As already seen in the discussion of Fig. 14-32(b), efficient (parallel) computer subsystem design leads to the need for a large number of interconnections and crossovers which increase chip area. To date, this problem has been partially resolved in MOS circuits through the use of diffused crossunders; however, their inherently high resistance tends to render them unsuitable for low impedance integrated circuits. As a result, various systems for reliable multilayer interconnections are being evaluated in many laboratories.

At present, the expense of putting microcircuit chips into flat packages and attaching individual wire bonds constitutes the major assembly cost in a

Fig. 14-37. Metalization defects.

finished microcircuit. Therefore, techniques are being developed to contact the chip and protect it from its external environment at considerably lower cost.

New Metal Interconnection Systems

In an attempt to solve all of the above problems, there are now in production and under study a number of new metal systems for making interconnections on the microcircuit chip which do not rely only on aluminum. Complete success in this area is hindered because aluminum seems to have provided a nearly optimum combination of properties, including very low contact resistance to silicon, good adherence to the oxide, and high electrical conductivity. Other metal combinations are readily found which have the necessary adherence and conductivity; however, the problem of achieving a good reproducible ohmic contact to silicon is generally a complicating feature. In general, the problem is solvable by using aluminum-doped silicon or a metal silicide to make connection with the silicon where it is contacted with a refractory metal which is used for adherence to the silicon oxide. One or more additional metals may then be superimposed to achieve good electrical conductivity and to provide an external contact to which bonds or solder connections are easily and reliably made. The situation is illustrated in Fig. 14-38 for a two-level interconnection system. For such a system, alloyed aluminum or various conductive silicides are typical materials used to provide the ohmic contact. Molybdenum, titanium, or tantalum are used for the refractory metal; gold, silver, copper, or nickel are typical conductors. In terms of possible metallurgical interactions and resultant reliability problems, these systems are all much more complex than is aluminum (in which the degradation mechanisms are at least understood and allowed for). It is expected that the aluminum–silicon oxide system will be most generally used for some time to come.

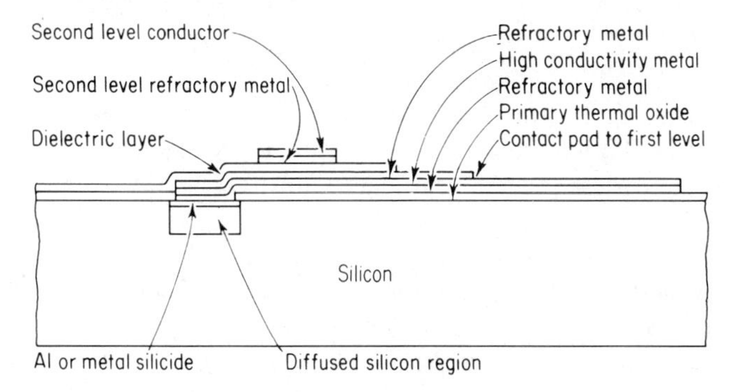

Fig. 14-38. Multilevel interconnection structure.

Packaging by New Techniques

The existing standard process for packaging silicon microcircuits requires the use of fine gold wires to make contact to the microcircuit chip. These leads are clearly seen in Fig. 14-4. The attachment of individual leads is one of the more critical and expensive operations in the present manufacturing process. Moreover, interconnections of this type are a potential cause of poor reliability in finished circuits.

One process which can be used to eliminate such connections is illustrated in Fig. 14-39. Direct connection is made between the lands on the individual silicon microcircuit chips and the thin-film connections on an insulating substrate. Strong connecting leads can be attached directly to the substrate, thereby eliminating the need for any internal package wiring. This basic concept can be extended to the construction of complete subsystem assemblies as shown in Fig. 14-40. Here the thin-film "Mother-board" carries the interactions within the subsystem between individual silicon circuits. The close-up photograph, taken through the glass substrate, shows the chip bonded directly to the thin-film connections. As shown, crossovers under the chip are also obtained by this method. Encapsulation is completed by molding an appropriate plastic case around the subassemblies.

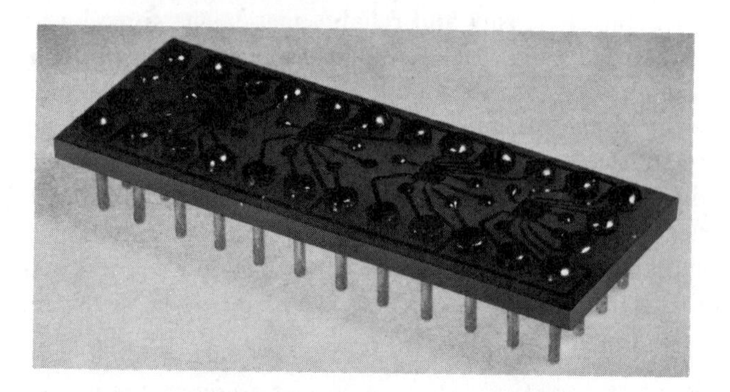

Fig. 14-39. Flip-chip packaging method.

Considering chips containing 200 components each, and bonded to a multilayer thin-film interconnection substrate, a packaging density of 10,000 components on a one-square-centimeter substrate seems achievable. It is interesting to note that at an average of a milliwatt dissipation per component (a practical design value), a power dissipation of 200 mW per chip, or 10 W for the entire package, is required. This is a reasonable level in terms of silicon devices and modern power device packaging technology.

The encapsulation of microcircuits in plastic is now possible because of the considerable improvement in plastic compositions suitable for electronic packaging. One method of constructing a plastic encapsulated IC in what is becoming a standard form factor for some devices is shown in Fig. 14-41.

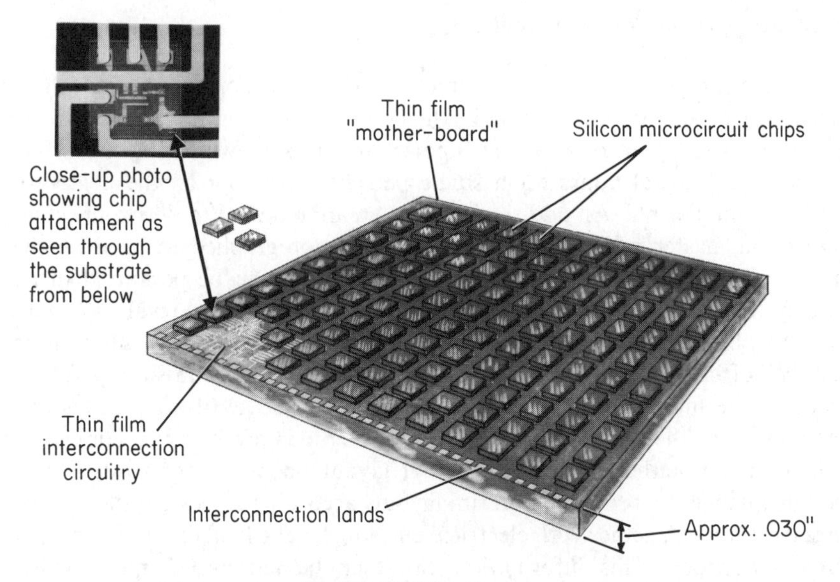

Fig. 14-40. Flip-chip array.

Fig. 14-41. Plastic encapsulated circuit.

Here "flip-chip" techniques have been combined with plastic encapsulation to produce an inherently low cost device for insertion into printed circuit boards. Plastic encapsulated transistors have been used in commercial applications for some time. As additional confidence is gained in the reliability of such devices, increased marketing of plastic encapsulated microcircuits and subsystem assemblies will follow.

Computer Design of Microcircuits

As the complexity of the microcircuit increases, the point is reached where the ability of human beings to make the necessary topological layout is being exceeded. The layout of a typical bipolar circuit as shown in Fig. 14-4 may takes as long as 24 hours for a single person, depending on the degree of optimization desired. An additional 24 hours are required to derive the eight individual master mask drawings which are photographed to produce the masks. Obviously, the number of iterations which can be permitted in an attempt to obtain an optimum configuration is limited. As a result, a circuit is rarely obtained which has the best possible area utilization with a minimum number of crossovers. Moreover, there always exists the possibility for an error in the final design which may not be detected until after the actual fabrication of the circuit. For this reason, methods are being developed to use computers either to perform actual layout operations or to assist the design engineer by providing interim layouts according to design rules stored in the memory, together with electrical checking of the final circuit for proper interconnections. This information can then be fed into a numerically-programmed stencil cutting machine for production of the individual masks.

More desirably, the computer would do the entire layout operation and provide a single optimum configuration with the smallest possible chip area and the minimum numbers of interconnections, thus minimizing parasitics. Unfortunately, the necessary algorithms have not been found for accomplishing this by a straightforward computer procedure. If the circuit has a sufficient number of repetitive features, it is possible to minimize interconnections by programming the computer to test all possible configurations on a trial and error basis. Large amounts of computer memory are required for this purpose although the procedure is presently used on a limited basis in the case of MOS arrays. For bipolar circuits, computer programs have been devised which can take into account all of the mask design rules as stored in the memory and quickly provide the designer with a layout which satisfies the design rules and is electrically correct. The designer can then rearrange components by instructing the computer to make changes which will obviously improve the layout. The chief advantage in this approach is that a number of layouts can be created, tested, and changed through successful iterations with considerably greater ease than in the case of manual layout.

The approach to circuit layout by computer is evolutionary because of the level of conception required to develop new approaches. However, there are several intriguing possibilities for further refinement, some of which are summarized below.

1. Computer-aided circuit analysis and design have been utilized by larger manufacturers for quite some time. Since obvious interactions exist between the design of the circuit itself and the eventual topological

layout chosen, the computer may be used not only to modify the topological layout but also to vary simultaneously the circuit itself to optimize the final design.

2. As more complex circuits are developed, it may be possible to have the computer work with a much more flexible design concept than one which is based on a specific (and limited) set of design rules previously stored in its memory. Rather, the computer could devise its own design limitations from more general principles to fit each individual situation, thereby obtaining a more optimum solution.
3. Developments are under way which will permit the circuit designer to enter information into the computer in the form of a desired network configuration drawn on the face of a cathode ray tube. The trial layout would also be presented graphically using already developed techniques.

Greater Fundamental Knowledge of Electrical Effects in Silicon

Of major importance in the fabrication of integrated circuits at high manufacturing yield is an understanding of the nature of the various electrical effects associated with the planar silicon oxide. It has long been known that under certain conditions, an n-type channel is present in the silicon at the silicon-oxide interface which drastically affects device characteristics. Such induced charge layers or channels have assumed particular importance with the advent of MOS type transistors, since their primary parameters are altered by the presence of such channels. The effects of such an induced layer on both bipolar and MOS transistor circuitry are illustrated in Fig. 14-42. In

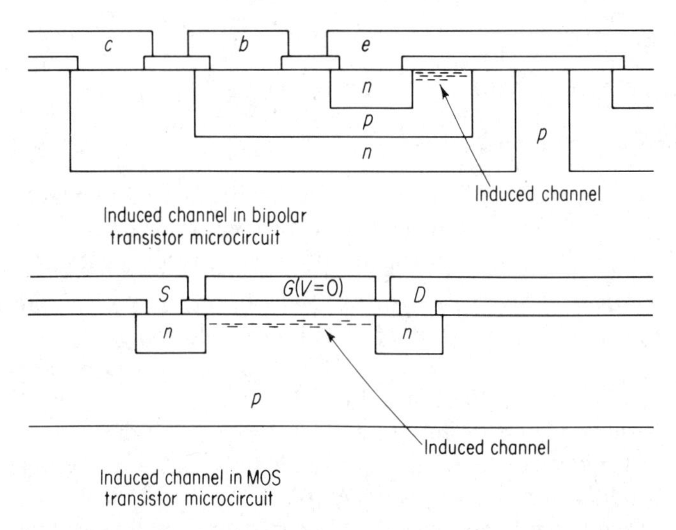

Fig. 14-42. Effect of induced inversion layer on MOS and bipolar microcircuits.

the case of conventional bipolar microcircuitry, a channel appearing across the base region of a transistor essentially produces a collector-to-emitter short or high leakage condition. In MOS transistors, this inversion layer produces source-to-drain leakage in the device, even when no turn-on voltage has been used on the gate.

In addition to the presence of the induced n-type layer, it has been found that the recombination generation rate for mobile carriers at the oxide-semiconductor interface is so variable that the dc leakage of a diode and the low current beta of a transistor can vary by orders of magnitude depending on the details of the planar process. These problems are allowed for in present microcircuit designs, but at a cost in performance. The amount of induced charge is not a stable phenomenon but rather depends on the environment to which the same is subjected. In particular, elevated temperature, presence of an electrical potential across the oxide, exposure to a high energy radiation environment, and interaction with the ambient background can all produce drastic changes in the interface properties.

Fortunately, these effects are now becoming rather well understood as the result of extensive investigations being carried on at various government, industrial, and university laboratories (21, 22). Most of the variability and instability in planar devices can now be explained on the basis of impurity ions which can move in or on the surface of the passivating oxide under the influence of electrical field, particularly at high temperatures. Other more subtle effects also exist, including the presence of certain charged centers resulting from structural or chemical imperfections in the oxide or at the oxide-silicon interface. The subject is much too complicated to cover in detail here. However, considerable material on the subject now exists in the literature.

PERFORMANCE-DIRECTED TECHNOLOGY

Performance Limitations

In spite of the obvious success in the application of standard diffusion technology, fundamental problems exist in conventional silicon integrated circuits when comparisons are made with discrete component circuitry.

1. *Limitations in the active and passive components.* Resistors formed by the diffusion process have a rather wide range of tolerance ($\pm$10–20%) and have large temperature coefficients (1000–2000 ppm/°C). Capacitors formed as PN junctions are voltage sensitive and limited in operating frequency range. These problems are more severe in the case of linear circuits where problems of frequency distribution and tolerance control make it difficult to avoid the limitations by changes in design.

Microcircuit transistors generally show poorer performance than discrete transistors. To a degree this is the result of the transistor art having for a time advanced at a slower pace during the introduction of microcircuit technology. Other limitations are inherent in the incorporation of the transistor into a monolithic circuit. A foremost example is the inability to optimize a single element when a large number of different elements are formed from the same epitaxial or diffused layer. Another limitation results from the uniform diffusion of gold (used to control minority carrier lifetime in high-speed transistors and diodes) throughout the silicon block, making it difficult to fabricate high-speed and low-speed components in the same circuit.

2. *Parasitic limitations.* From the standpoint of over-all circuit design there are further limitations imposed by geometry. These limitations are reflected in the existence of paraistic resistance and capacitance as shown in Fig. 14-43.

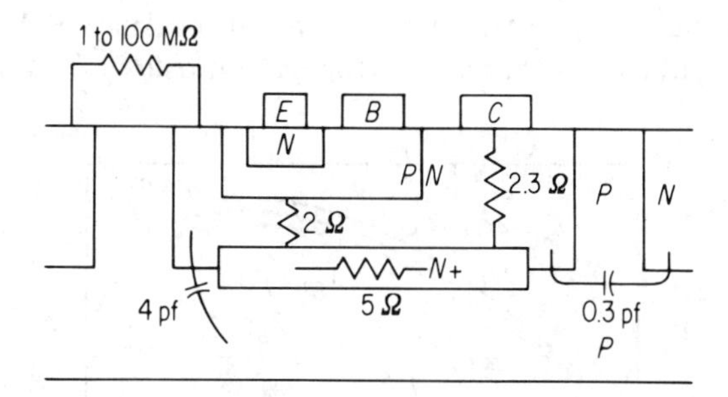

Fig. 14-43. Parasitic resistance and capacitance in conventional silicon microcircuits.

In conventional silicon microcircuits, the series resistance in the collector region is, in most cases, reduced by the introduction of a buried n^+ region as shown by steps 2 and 3 in Fig. 14-6. However, the series resistance is still a factor of 2 to 5 greater than is obtained in discrete transistors designed for low saturation resistance. The problem is intensified by the desire to make a small-top collector contact to the transistor to minimize transistor component size (and capacitance), and to facilitate optimum circuit layout. Typical resistance values are shown in Fig. 14-43, for a particular transistor designed to carry 45 ma with a collector contact along one edge of the unit.

The parasitic capacitance shown in Fig. 14-43 results from the method used to isolate components. The capacitance associated with the isolating PN junctions is about 0.1 pf/mil^2 at 0.5 volt of reverse bias.

High Frequency Transistors in Microcircuits

With the increased demand for both digital and linear circuits to perform at high frequency, the design engineer has again turned his attention to producing higher frequency silicon transistors. The frequency capability and size of the transistor are directly related because frequency response is affected by carrier transit time and capacitance. Although a number of subtle details must be considered in producing high frequency transistors, the basic problem is that of reducing the basewidth and the emitter and base areas to values lower than have heretofore been achieved, while retaining other important characteristics of the device. Recently described transistors have shown a gain-bandwidth product of 9 GHz. The gain-bandwidth product as a function of current and voltage for a 6.4-GHz transistor is shown in Fig. 14-44. Such devices have linewidths and spacings of 0.1 mil or smaller, a total diffused emitter depth of $\frac{1}{3}$ micron, and a base width of 0.1 micron, with a control requirement on basewidth of $\frac{1}{50}$ micron. This is a factor of 4 better than is commonly achieved in conventional microcircuitry. A number of digital circuits have been built to establish the feasibility of using such devices to achieve switching times less than one nanosecond. Results obtained to date include saturating T^2L circuits with a propagation delay of 1.2 ns and unsaturated *ECL* circuits which have a propagation delay of 0.4 ns (20).

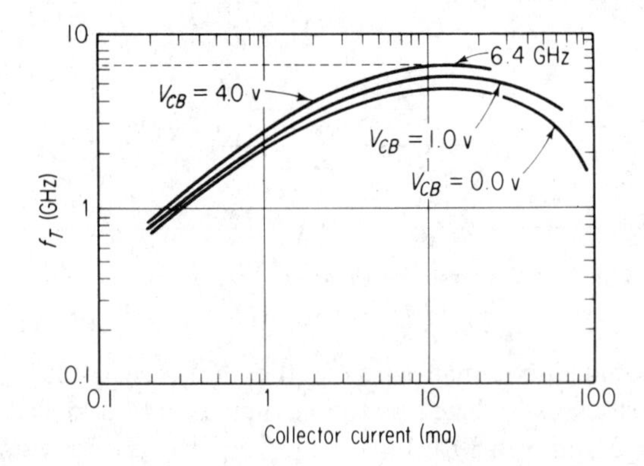

Fig. 14-44. Gain-bandwidth vs voltage and current in experimental VHF transistor.

In the linear field, designers of high frequency amplifiers are beginning to emphasize the construction of general purpose, high gain-bandwidth amplifiers to be tuned by external elements. The complexity of such circuits is directly related to the quality of the transistors used. That is, bandwidth can always be obtained at the expense of gain: a large number of transistor stages with low frequency degenerative feedback can be used to obtain a given gain-bandwidth characteristic in the amplifier. However, the number of stages required is a direct function of the quality of any given transistor. The availability of the high frequency transistors described above has made possible the design of monolithic amplifiers of unusual performance. For example, a single-chip microcircuit with dimensions of 40×40 (mils)^2 is shown in

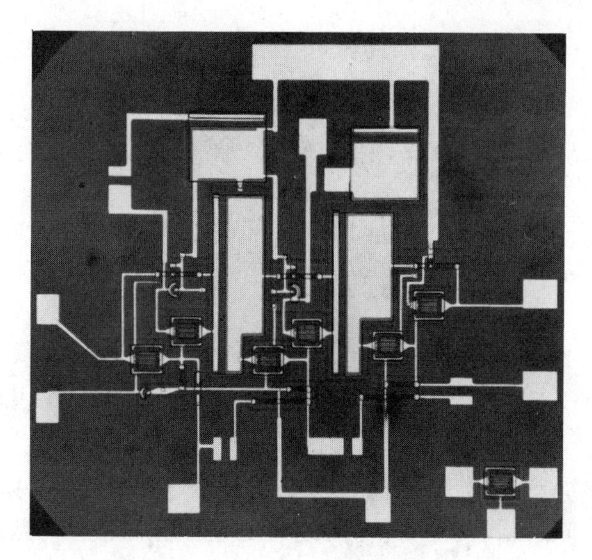

Fig. 14-45. Broadband high-frequency subsystem.

Fig. 14-45. The microcircuit uses 0.1-mil geometry transistors and provides a gain of 60 db over a frequency range of 15 to 150 MHz.

The IC With Thin-Film Passive Components

Because of the limitations in the passive components in all silicon circuits and the desirability of separately optimizing transistors and resistors, other basic types of silicon integrated circuit configurations are being developed. For several years, thin-film type resistors and capacitors with properties very similar to those of discrete components have been satisfactorily produced on insulating substrates. This was done initially by silk screening and printed wiring (thick-film) (23) techniques and later by thin-film methods. Thin-film passive components can be produced to close tolerance ($\pm$ a few percent), show very little temperature variation and, by virtue of the insulating substrate, have good electrical isolation when combined to form a circuit.

In 1948, Brunetti and Curtis (23) suggested that thin-film passive components be formed directly on the insulating envelope of active devices when the latter require a special environment to function properly. They suggested that thin-film components be formed on the glass envelope of a vacuum tube. Such a technique can also be applied to silicon integrated circuits, since thin-film resistors and capacitors of optimum design can be placed on top of the insulating surface of the oxidized silicon. The resultant structure is seen in Fig. 14-46.

In this case, isolation is automatically obtained for these elements by virtue of the silicon oxide layer. The resulting structure is achieved by adding

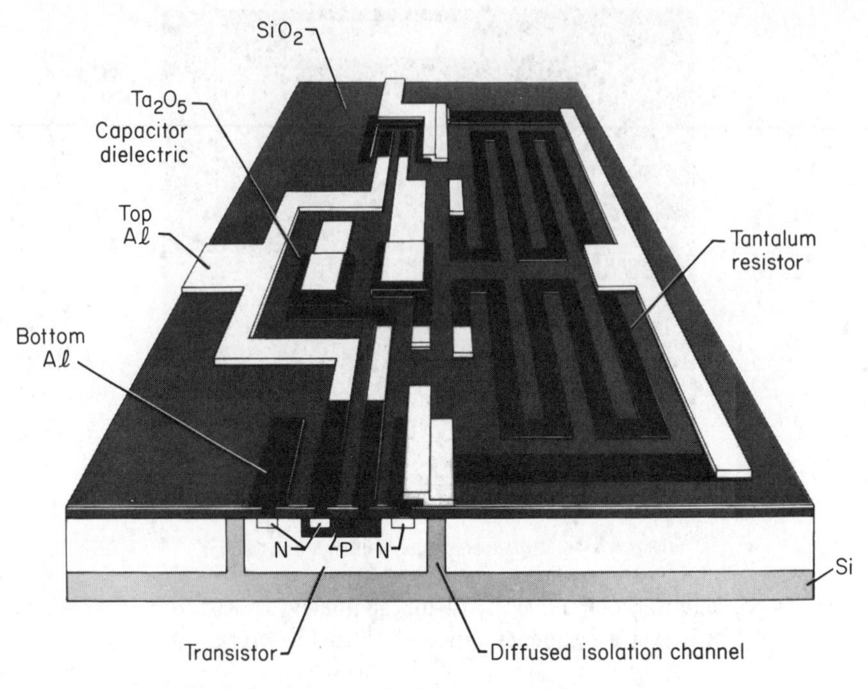

Fig. 14-46. Thin-film/silicon monolithic structure.

thin-film processing steps to the wafer in which the active elements have been previously formed, prior to cutting and assembling the individual circuits. Many different materials have been investigated for use in forming thin-film components which might be compatible with the silicon circuit substrate. Most fully reported, however, are those components obtained using either tantalum (24, 25), or a nickel chromium alloy (26, 27), for resistors, and either tantalum oxide (25) or one of the oxides of silicon (26) as a capacitor dielectric. Tantalum (in some cases partially converted to the nitride) and a nickel-chromium alloy are both strongly adherent to silicon dioxide and relatively stable at elevated temperatures. Tantalum, which can be given a preoxidized surface, is more stable for both resistance and TCR values than is unprotected nickel-chromium alloy. Nickel-chromium resistors are generally stabilized against further oxidation by applying a vapor-deposited oxide coating. As a capacitor dielectric, silicon dioxide is found to be a particularly stable material with high dielectric strength. It may be prepared in a very thin layer by an anodization process with the achievement of a capacitance per unit area as high as 1 pf/mil^2. Tantalum oxide possesses a higher dielectric constant and can be readily produced by anodization of a tantalum thin-film electrode; however, it is subject to failures under high voltage and high temperature stress due to field-induced crystallization of the tantalum oxide.

A new form of high-resistivity sputtered tantalum is currently under investigation as a thin-film resistor material (28). Conventional tantalum or tantalum nitride films are formed with an extremely reproducible sheet resistivity in the range of 40 to 100 ohms at commonly used deposition thickness. After oxide stabilization, a temperature coefficient of −50 ppm is obtained. By modifying the deposition conditions used in the sputtering apparatus (low voltage and high argon pressure), a much higher resistivity form is produced with a choice of 500 to 5000 ohms at a similar thickness (29). Although the temperature coefficient after stabilization is higher (−250 ppm), this material will probably find wide application in silicon microcircuits because resistors of a given value can be made much smaller. Thus, the over-all size of the microcircuits can be reduced to obtain lower cost and/or it will be possible to fabricate more complex circuits in a given area.

Monolithic Insulators and Conductors

The designers of IC's and other solid state devices have long desired to incorporate both insulating and conducting layers within the silicon block. The introduction of electrically insulating material into the IC would improve electrical isolation between components, as compared with PN junction isolation. Moreover, the introduction of metallic conducting layers within the IC could reduce the effects of parasitic resistance which result from the presence of nonmetallic conduction paths from the bottom of the active devices to the upper surface where the metal connection is made.

In efforts to produce isolated silicon active devices on insulated substrates, several investigators have attempted to grow single-crystal silicon on NPN-silicon substrates (30, 31, 32). Moderate success has been achieved in that smooth, approximately single-crystal deposits were obtained (33). Active devices, and in particular MOS transistors, have been produced on silicon obtained in this manner (34). In view of the higher level of dislocations

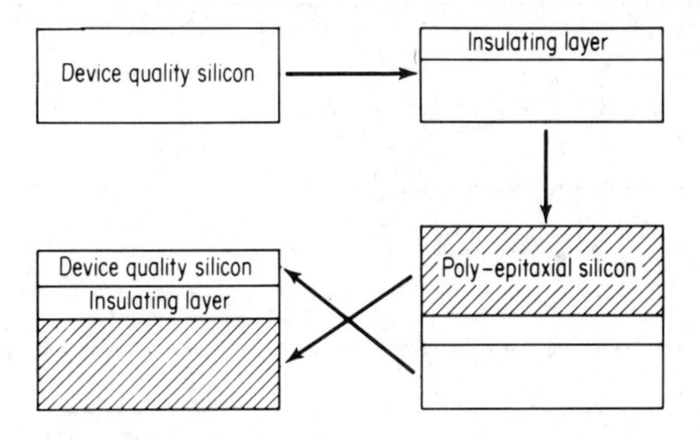

Fig. 14-47. Fabrication concept for single-crystal silicon on insulating substrate.

resulting from the silicon-to-substrate lattice and thermal mismatch, it is unlikely that a high yield bipolar transistor for use in conventional silicon circuits can be formed in this way.

During 1965–66, a conceptual breakthrough developed, making it possible to obtain device quality silicon on an insulating substrate. The concept is illustrated in Fig. 14-47. The technique involves starting with a wafer of device quality silicon, depositing the desired insulation material, and then growing a quantity of polycrystalline silicon to a thickness approximating that of the original wafer. The original silicon side of this composite wafer is then lapped or otherwise reduced in thickness to provide a thin layer of device quality silicon on the insulated substrate. Since the quality of the silicon used

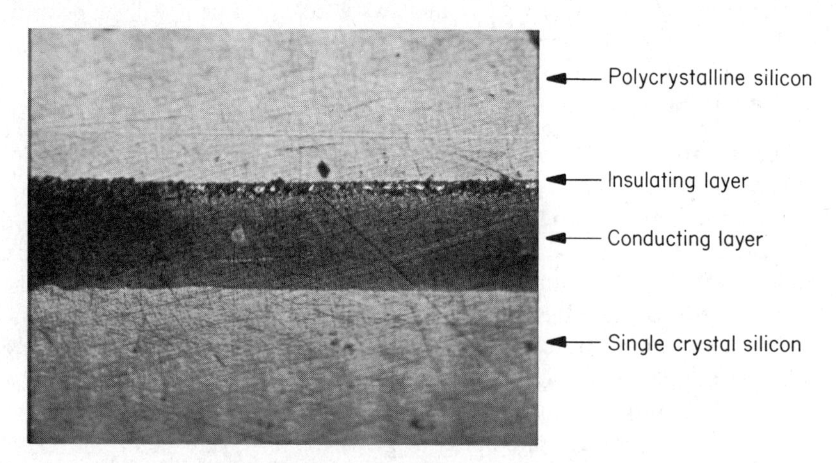

Fig. 14-48. Multi-material monolithic structure.

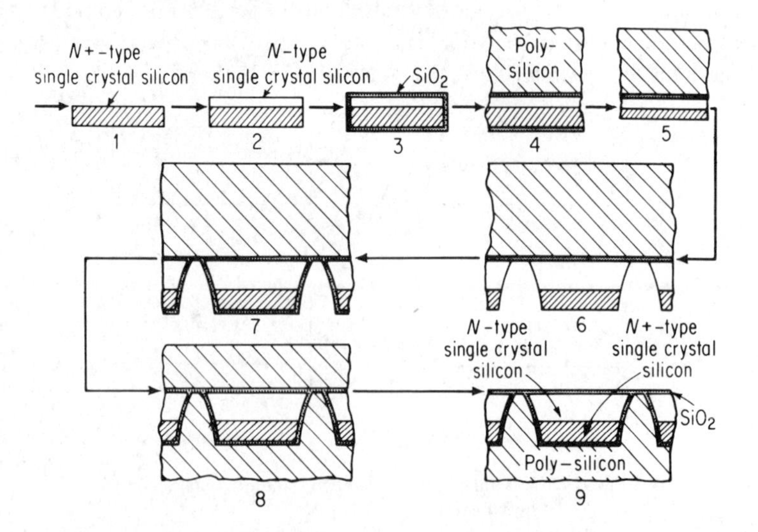

Fig. 14-49. Double-polydielectric isolation process.

to fabricate devices does not depend on the nature of the substrate, any number of intermediate layers can be introduced, including conducting as well as insulating layers. Figure 14-48 shows a cross section of a silicon wafer prepared according to the desired sequence.

Various modifications of this technique, as used to provide improved component isolation have been reported (35). Considerable difficulty arises, however, in implementing such a process on a manufacturing basis. This is because the characteristics of a device are a function of the thickness of the single-crystal silicon layer. After the lapping and polishing to reduce the thickness, the single-crystal silicon must be flat and parallel to a controlled dimension of ± 0.00005 in. Such precision is extremely difficult to obtain on a manufacturing basis. The dimension control is further complicated by the work damage caused by polishing the surface. The damage must be removed by etching to obtain optimum microcircuit yield, and the required etching adds to the thickness variation of the single-crystal silicon layer.

Recently a process was described (36) which largely overcomes these difficulties by means of the sequence of operations shown in Fig. 14-49 and further illustrated in Fig. 14-50. The process involves oxidation of a silicon wafer and deposition of a polycrystalline silicon layer on the oxide. Then, all but a few mils of the single-crystal layer are removed by etching and lapping, or a combination of these. The remaining layer thickness at this operation is not critical. The subsequent steps include oxidation and photoengraving of a pattern of isolation cut-ins into the oxide on the single-crystal side of the wafer, and selective etching with oxide masking down to the layer of oxide inside the silicon structure to form isolated regions. Next, the wafer is oxidized to form a layer of SiO_2 after which polycrystalline silicon is vapor plated on top of the oxide layer on the single-crystal side of the wafer. Finally, the wafer is etched to remove all of the first deposited polycrystalline layer. Etching stops when the layer of SiO_2 underlying the polycrystalline layer is reached. The oxide layer is then used as a mask for fabrication of microcircuits by conventional techniques.

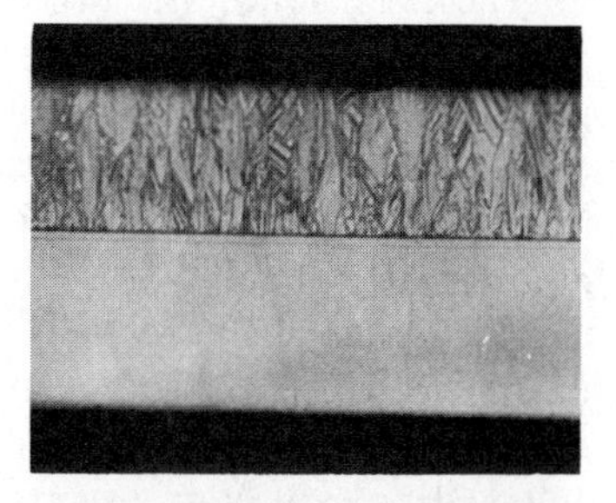

Fig. 14-50. Wafer cross section showing steps in dielectric isolation process.

This process has the following advantages:

1. It eliminates critical lapping steps and therefore can be economically implemented on a production basis.
2. The top surface of the single-crystal silicon is totally planar and is protected by a silicon dioxide layer throughout all processing steps involved in forming the isolated regions.
3. In contrast to other oxide isolation techniques, there is no possibility of work damage existing in the active areas of the microcircuit.

An additional advantage of using the oxide isolation structure is that gold can be introduced into individual regions to produce high speed devices without affecting other components in the same wafer. This is because the oxide acts as a barrier to gold diffusion.

The capacitance associated with the oxide layer is in the range of 0.02 pf/mil^2 for 1 micron of oxide thickness. Thus, a 5-micron thick oxide layer lowers the capacitance coupling by a factor of 25, compared with PN junctions having a slight reverse bias. The capacitance is temperature insensitive. Measured isolation resistance is over 10^{12} ohms at 100 volts. A high frequency amplifier fabricated by this oxide isolation process is shown in Fig. 14-51.

A thick metallic layer can be formed either prior to or after the formation of the oxide layer (37). Thin layers of many metals can be deposited by vapor deposition using apparatus and conditions similar to those used in the silicon epitaxial process. Various investigators have deposited molybdenum by the hydrogen reduction of $MoCl_5$. Molybdenum thus deposited on silicon under

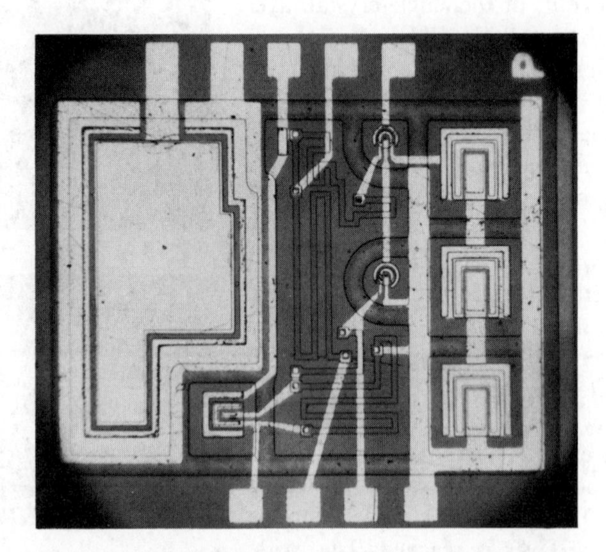

Fig. 14-51. High-frequency amplifier with dielectric insulation.

high temperature conditions readily forms molybdenum disilicide as a metallic layer (electrical resistivity of 21.5 μohm/cm). Other silicides can be similarly formed.

A variety of monolithic structures can be fabricated by a combination of the above techniques, including localized epitaxial growth. Figure 14-52 (top) shows that a metal contact can be made between the bottom of the isolated regions and the upper surface to produce devices having essentially the same saturation resistance as found in two-sided discrete components. In the particular case of the transistors of Fig. 14-43, the extrinsic collector resistance is reduced from 10 ohms to 2 ohms by adding a metallic layer at the isolation barrier interface without making any other changes in the device layout. In this case, the diffused n$^+$ layer is used to assure a good ohmic contact to the metal layer.

When a metal layer is deposited outside the oxide isolation as shown in Fig. 14-52 (center), it can supply a continuous ground plane through the block for shielding purposes or for producing strip-line configurations in very high frequency monolithic circuits. The metallic crossover illustrated in Fig. 14-52 (bottom) will introduce less parasitic resistance than is obtained in the diffused regions used for this purpose in present structures.

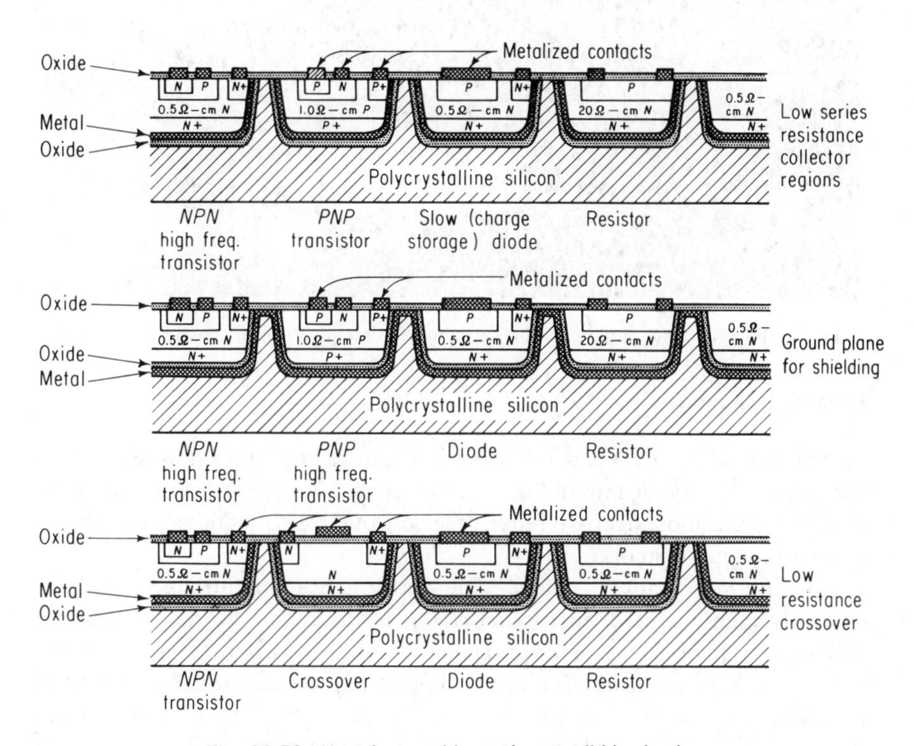

Fig. 14-52. Uses for metal layers in monolithic circuits.

A photomicrograph of an oxide isolation structure containing metal inserts surrounding the isolated component regions is shown in Fig. 14-53.

The feasibility of photoengraving buried metal layers into a desired pattern has recently been demonstrated. This leads to the intriguing possibility of providing a second level of interconnection within the chip beneath the active components. In its simplest form, a high frequency amplifier constructed by this technique contains both a ground plane and bias line within the buried layer so that the upper interconnection level includes only input and output connections.

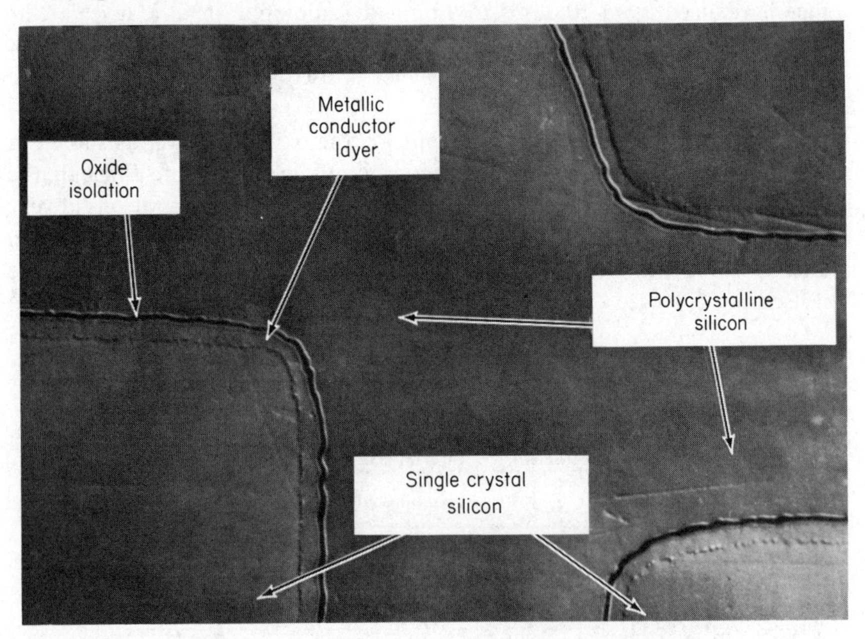

Fig. 14-53. Top view of dielectric isolated region containing metallic conductivity layer.

Expected Changes

Substantial progress is being made in the development of new technology for producing silicon monolithic microcircuits. The greatest effort is being placed in areas most likely to reduce the cost of circuit functions as incorporated in the final system.

In addition, many of the circuit performance limitations existing in currently manufactured microcircuits can be overcome by new processing techniques. Obviously, such improvements will be incorporated into a circuit only when there is a clear circuit performance requirement which can support the additional manufacturing cost that is likely to result from deviations from the standard processes. On the other hand, it appears extremely clear that monolithic silicon technology will eventually handle many of the applications previously thought to require more expensive hybrid approaches.

REFERENCES

1. C. BRUNETTI and R. W. CURTIS, "Printed Circuit Techniques," *Proc. IRE*, **36** (1948), 121–61.
2. E. M. DAVIS *et al.*, "An Approach to Low Cost, High Performance Microelectronics," Paper 13.1, 1963 Western Electronic Show and Convention, San Francisco, Aug. 20–23, 1963.
3. M. KAHN, "Ceramic-Based Microcircuits," Electronic Components Conf., Washington, D.C., May 10, 1960.
4. L. C. HOFFMAN, "Precision Glaze Resistors," *Am. Ceram. Soc. Bull.*, **42**, No. 9 (Sept. 15, 1963).
5. A. E. LESSOR *et al.*, "Thin-Film *R-C* Networks," *IEEE Spectrum*, **1**, No. 4 (April, 1964), 72–80.
6. J. A. HOERNI, "Planar Silicon Diodes and Transistors," Electron Devices Meeting, Washington, D.C., 1960.
7. J. M. WILSON, "High Purity Silicon," *Research*, **10** (1957), 166.
8. C. G. CURRIN and E. EARLEYWINE, "Advances in Elemental Semiconductors," *Semiconductor Products and Solid-state Technology*, **7**, No. 6 (June, 1964), 20–25.
9. S. E. MAYER and D. E. SHEA, "Epitaxial Deposition of Silicon Layers by Pyrolysis of Silane," *J. Electro. Soc.*, **111**, No. 5 (1964), 550–56.
10. H. C. THEUERER, "Epitaxial Silicon Films by the Hydrogen Reduction of $SiCl_4$," *J. Electro. Soc.*, **108**, No. 7 (1961), 649–53.
11. W. STEINMAIER and J. BLOEM, "Successive Growth of Si and SiO_2 in Epitaxial Apparatus," *J. Electro. Soc.*, **111** (1964), 206–209.
12. E. L. JORDAN, "A Diffusion Mask for Germanium," *J. Electro. Soc.*, **108** (1961), 478–84.
13. J. KLERER, "A Method for the Deposition of SiO_2 at Low Temperature," *J. Electro. Soc.*, **108** (1961), 1070.
14. C. F. POWELL *et al.*, "Vapor Plating," John Wiley & Sons, Inc., 1955, pp. **136–43**.
15. B. E. DEAL, "The Oxidation of Silicon in Dry Oxygen, Wet Oxygen, and Steam," *J. Electro. Soc.*, **110**, No. 6 (June, 1963), 527–33.
16. E. H. SNOW *et al.*, "A Study of Ion Migration in Thin Insulating Films Using the MOS Structure," Fall Meeting of the Electrochemical Society, Washington, D.C., 1964, *Elec. Div. Abstr.*, **13**, No. 129 (1964), 17–20.

17. C. J. FROSCH and L. DERICK, "Surface Protection and Selective Masking during Diffusion in Silicon," *J. Electro. Soc.*, **104** (1957), 547–52.

18. D. E. FARINA and J. D. TROTTER, *Electronics*, **138** (1965), 84–98.

19. J. L. SEELY, *Electronic Design*, **114** (1966), 90–93.

20. R. L. LUCE, International Electron Devices Meeting, Washington, D.C., October 20–22, 1965.

21. E. H. SNOW, A. S. GROVE, B. E. DEAL, and C. T. SAH, Fall Meeting of the Electrochemical Society, Washington, D.C., 1964.

22. J. E. THOMAS, JR. and D. R. YOUNG, *IBM J. Res. & Dev.*, **8** (1964), 368.

23. C. BRUNETTI and R. E. CURTIS, *Proc. IRE*, **36** (1948), 121–61.

24. P. G. THOMAS, J. A. EKISS, J. ROSCHEN, and M. J. WALKER, *Proc. NEC*, **18** (1962), 727–35.

25. D. A. MCLEAN, N. SCHWARTZ, and E. D. TIDD, *Proc. IEEE*, **52** (1964), 1450–51.

26. R. S. CLARK and D. W. BROOKS, *Electronic Design*, **12**, No. 26 (1964), 64; **13**, No. 2 (1965), 52.

27. H. J. DEGENHART and I. H. PRATT, Proc. 10th Nat. Vacuum Symp., 480 (1963).

28. M. J. WALKER, Electron Devices Meeting, Washington, D.C., 1964.

29. H. J. SCHUETZE, H. W. EHLBECK, and G. G. DOERBECK, Proc. 10th Nat. Vacuum Symp., 434 (1963).

30. R. G. BRECKENRIDGE, Final Report Contract DA-36-039-SC-90734, Union Carbide Corp., Parma, Ohio, ASTIA AD-418-340.

31. M. MACHA, Final Report Contract NObsr-87634, June 1962 to August 1963, Lear Siegler Res. Labs., Santa Monica, Cal.

32. E. RASMANIS, *Semicon. Prod.*, **6**, No. 7 (1963), 30–33.

33. H. M. MANASEVIT and W. I. SIMPSON, *J. App. Phys.*, **35** (1964), 1349–51.

34. C. W. MUELLER and P. H. ROBINSON, *Proc. IEEE*, **12** (1964), 1487.

35. D. MCWILLIAMS, C. FA, G. LARCHIAN, and O. MAXWELL, JR., *J. Electrochem. Soc.*, III, 153C, July 1964.

36. G. L. SCHNABLE and A. F. MCKELVEY, Electron Devices Meeting, Washington, D.C., 1964.

37. C. G. THORNTON, NEC, Chicago, October 19–21, 1964.

38. R. M. WARNER, JR. and J. N. FORDEMWALT, *Integrated Circuits—Design Principles and Fabrication*. New York: McGraw-Hill Book Company, 1965.

39. M. F. UMAN, *Introduction to the Physics of Electronics*. Englewood Cliffs, N.J.: Prentice-Hall, Inc., 1974.

40. F. C. FITCHEN, *Electronic Integrated Circuits and Systems*. New York: D. Van Nostrand Reinhold Company, 1970.

41. M. P. RISTENBATT, *Semiconductor Circuits, Linear and Digital.* Englewood Cliffs, N.J.: Prentice-Hall, Inc., 1975.

42. *Integrated Electronic Systems.* Westinghouse Defense and Space Center, Englewood Cliffs, N.J.: Prentice-Hall, Inc., 1970.

43. C. BELOVE, H. SCHACHTER and D. L. SCHILLING, *Digital and Analog Systems, Circuits and Devices: An Introduction.* New York: McGraw-Hill Book Company, 1973.

44. J. T. WALLMARK and H. JOHNSON, *Field-Effect Transistors.* Englewood Cliffs, N.J.: Prentice-Hall, Inc., 1966.

45. H. C. LIN, *Integrated Electronics.* Boston, Mass: Holden-Day Publishing Company, 1967.

46. N. HOLONYAK, JR., *Integragted Electronic Systems.* Englewood Cliffs, N.J.: Prentice-Hall, 1970.

15

Modern Applications of Integrated Circuits

In all that has gone before we have analyzed and designed a great variety of simple circuits using discrete components along with some integrated circuits. This final chapter briefly describes some IC circuit/system building blocks such as computer memories and microprocessors, calculators, etc. along with some recent developments in new logic arrangements and devices such as the LED, CCD, etc. Progress continues unabated in device and component technologies. The push to make ICs faster and denser has accelerated the development of:

I^2L (integrated injection logic) devices, with the density of MOS (metal oxide silicon) and the speed of bipolar ICs;

CMOS (complementary MOS) ICs with microwatts of power dissipation needed for consumer electronic products;

Microprocessors that promise to revolutionize the manner in which instruments, systems, and circuits are designed, and to improve information-processing techniques.

At the heart of IC advances are photolithographic and ion-implantation processing schemes. Research into newer schemes, such as electron-beam and ion-beam implantation pattern generation, holds great promise for the future.

Liquid-crystal and LED, *light-emitting-diode*, digital displays continue to improve in power dissipation, efficiency, and cost. LEDs are now available in a wide choice of colors that include red, green, and yellow. Blue LEDs are being investigated. The newer field-effect liquid crystal displays have improved appearance compared with the older dynamic scattering types.

One of the hottest device technologies is the CCD, *charge-coupled device*, which is being readied for use in serial-access memories, image sensors, and signal processing applications. In linear ICs, converters, op amps, multipliers, dividers, multifunction modules, active RC filter networks, and hybrid ICs have also experienced advances.

Upgrading Standard Logic

While the vast majority of electronic circuits make use of TTL, *transistor-transistor logic*, ICs for logic functions, improvements in TTL speed and power-consumption characteristics have resulted in such newer families as low-power Schottky TTL for lower power dissipation and somewhat higher speeds.

SOS, *silicon-on-sapphire*, technology continues to advance. It combines MOS technology with the use of a sapphire substrate. A thin single-crystal silicon film is formed on the substrate by epitaxial growth. Such a technology offers several advantages that include high speeds, low power dissipation, high packing densities, and high reliability, and is potentially cost competitive with bulk-silicon CMOS. Gate delays of a few nanoseconds and microwatts of power dissipation per gate have been possible.

Even though SOS seems to be an almost ideal technology, some problems have yet to be solved. Chief among them are the relatively high-leakage currents at the silicon-sapphire interface, and the high costs of manufacturing, polishing, and cutting the sapphire material.

I^2L—A Bipolar Revival

The newest of IC technologies is that of I^2L, also known as MTL, *merged-transistor logic*, developed in late 1972 by IBM in Germany, and Philips in Holland, and now being pursued worldwide by a number of IC companies. This new approach to bipolar LSI IC design offers high packing densities, low power-delay products, and low fabrication costs. Power-delay products of 0.35 picojoule and gate densities of 100 or more per mm^2 have been reported in I^2L ICs made with conventional bipolar processes.

In an I^2L IC (Fig. 15-1), a pnp transistor acts as a current source and load for an npn inverting transistor operating with multiple-collector outputs. Since the processing required is that of conventional bipolar structures, the low-level I^2L gates may readily be mixed with standard bipolar structures, both digital and linear, on the same chip, to produce interfacing with external circuitry. Circuit functions include logic arrays, control logic, microprocessors, read-only memories, frequency counters and dividers, converters, and linear functions of all types.

One of the promising areas of application is consumer electronics, where low cost and high packing densities become very attractive features. Late in 1974, for example, Texas Instruments teamed up with the Benrus Watch Company to produce a digital wrist-watch with an I^2L IC. The IC has a

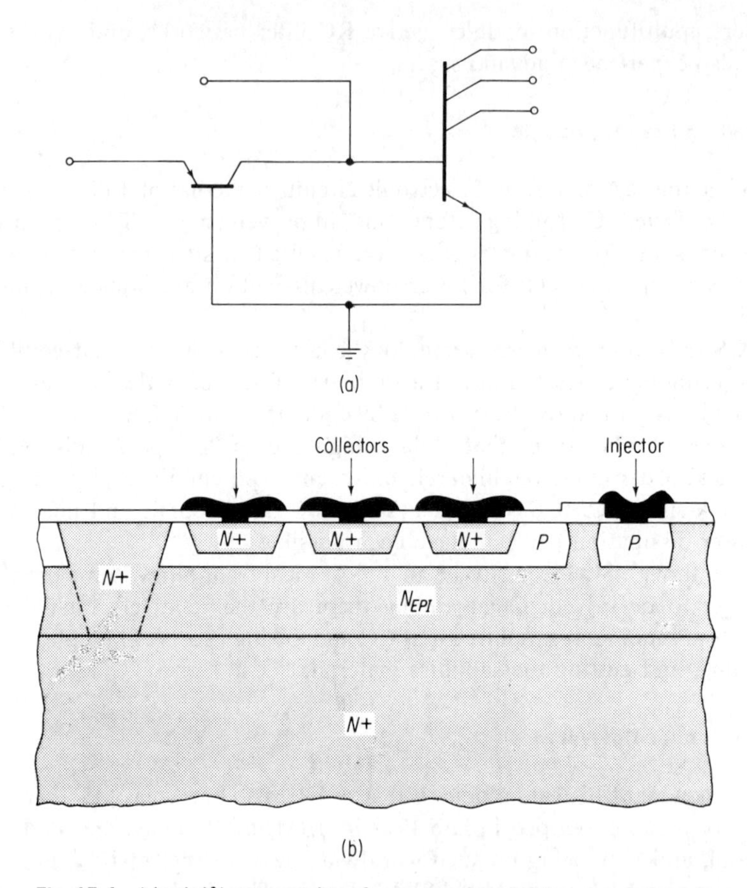

Fig. 15-1. A basic I^2L gate consists of a pnp transistor used as a current source and load for an npn inverting transistor, operating with multiple-collector outputs. This configuration is simple to interface to other types of ICs, as the processing is entirely compatible (a). Furthermore, it allows easy layout of compact circuits (b).

packing density that is 25 per cent less than that of an equivalent IC using CMOS, the IC technology conventionally used for high packing densities.

The real payoff in I^2L is the ability to combine relatively low-speed, low-power devices (for high-density memories or slow logic, for example) with high-speed, high-gain, npn devices (for off-the-chip buffering, driving, and amplification), all on the same chip.

N-channel MOS, or NMOS, is the latest MOS evolution from p-channel MOS and is reported to have better operating characteristics than PMOS, including easier interfacing to TTL levels and better speed-power products. Because NMOS ICs require less silicon (for a given operating speed), they cost less than PMOS devices and are consequently finding their way into high-density memories, like 4-kb RAMs, *random-access memories*, which have now become cost competitive with 1-kb PMOS RAMs.

Processing Advances

Fundamental to all IC device developments are the various processing technologies used to make them, particularly the pattern delineation steps. Today's photolithograph pattern-generation technologies make use of photographic masks that must be produced with almost incredible precision. A typical photomask, for instance, having an active pattern diameter of 7.5 cm, may have a complex pattern with features as small as 2 μm, may be precise within 1 μm, and may have less than 0.5 defect per square cm. Making such a mask is equivalent to laying out a football field to within 0.8 mm, and guaranteeing that within that area there are fewer than a dozen defects as large as a grain of sand. Conventional optical processes are hard pressed to meet such requirements.

Electron-beam masking techniques, which are just beginning to emerge from the laboratory, may well provide the key to improved patterning. Getting better masks is only part of the problem, however, since current contact-printing operations can often fatally damage a mask the very first time it is used on a wafer. Projection printing offers a solution to this problem, but new techniques are necessary if adequate resolution is to be obtained over fields large enough to cover 3-in- (7.62-cm-) diameter or larger diameter wafers. Novel electron-beam projection systems are also being developed to overcome the contact-damage problem while providing high resolution.

In an alternative solution to the mass-degradation problem increasing attention is being given to proximity or *near-contact* printing in which a small, but well-defined gap (typically around 10 μm) is provided between mask and wafer. When used with conventional ultraviolet exposing radiation, diffraction effects associated with the gap inevitably compromise resolution—a problem which can be overcome by exposing the wafer to soft X-rays whose short wave-lengths make diffraction negligible. The special resists and masks necessary to make this process practical have received a great deal of attention—leading contenders being gold on thin-silicon or gold on thin-plastic films.

Advances in ion implantation, which gives tight control of doping impurities, are having a considerable impact on the development of linear and digital-signal bipolar and MOS circuits. Most new processes include at least one ion-implantation step.

A number of device fabrication techniques involving dielectric isolation (for high operating voltages, faster switching speeds, and greater packing densities) include Isoplanar, OXIM, *oxide-insulated-monolithic*, and dielectrically isolated-FET schemes. The leading processes are characterized in

TABLE 15-1 Bipolar and MOS Characteristics

Parameters: Process Type	Parameters: Circuit Type	Nominal Stage Delay (ns)	Power Dissipation per Gate @1 MHz(mW)	Noise Immunity (V)	Area per Gate (sq mils)	Process Steps
TTL	Low Power Static	25	1.2	0.4	—	60–65
TTL	Medium Speed Static	8	7.5	0.4	104–115	60–97
TTL-Schottky	Static	3	18	0.3	126	60–74
High Thrshld PMOS	4-Phase	90–150	0.4	2	10–15	30–40
Low Thrshld PMOS	4-Phase	90–150	0.2	1	10–15	30–40
PMOS Silicon-Gate	2-Phase	40–50	0.16–0.3	0.5–0.8	10–15	29
NMOS-on-Sapphire	Static	15	0.3	1	5–10	30–40
NMOS Silicon-Gate	2-Phase	40–80	0.17–0.28	0.4	12–14	23–34
CMOS	Static	12–17	0.3	3.5–4	15–35	37
CMOS-on-Sapphire	Static	6–10	0.15–0.2	3.5–4	15–25	30–40

Note: Data are based on 2-input NAND gate with fanout of 3.

Table 15-1. Although bipolar has dominated the computer industry during the past decade, a number of MOS computers have recently been introduced, and the pendulum appears to be swinging in that direction. Although bipolar technology clearly has the advantage in speed, MOS is generally superior with respect to power, functional density, and process complexity. Its slower speed is usually adequate for most aerospace applications.

MOS advantages are especially important in achieving large-scale integration (LSI), i.e., more than 100 gates/chip. Typical upper limits of bipolar processes, such as transistor-transistor logic (TTL) medium speed, are about 60 gates/chip, whereas in custom dynamic p-channel MOS (PMOS) logic, densities of 250 gates/chip are common. LSI offers excellent benefits toward achieving high reliability, small size, and light weight—all critical aerospace requirements. Because of its maturity, PMOS represents the lowest risk to the designer. In fact, almost all currently available MOS computers known to the author are implemented with PMOS circuits. Compared to other leading MOS processes, PMOS has the poorest inherent speed-power product;

however, its relatively poor power characteristics are greatly improved through use of dynamic logic. Dynamic logic circuits conditionally transfer stored charge from one capacitive mode to another. They conserve power by drawing current only when the nodes are being charged or discharged (conventional static logic draws current continuously). Dynamic logic relies on multiphase clock signals, either 2- or 4- phase, to define respective charge and discharge time periods.

MOS reliability is aided in two ways. First, LSI reduces the number of inter-chip connections—a major factor contributing to integrated circuit (IC) failure rates. Second, the low power attributes of MOS tend to keep chip temperatures low for a given gate density, minimizing adverse reliability effects of high temperature operation.

Several types of PMOS have been developed, each having unique circuit configurations, substrate crystal lattice orientation, substrate, dielectric, and gate materials. Substrate materials include bulk silicon and sapphire; dielectric types include silicon dioxide and a layered combination of silicon dioxide and silicon nitride; and common gate materials include aluminum and polycrystal silicon, the latter usually identified as silicon-gate. Each process-type possesses its own electrical and geometrical characteristics.

Charge Coupled Devices (CCD)

For the last 3 to 4 years, there has been considerable publicity concerning the charge coupled imaging device. CCDs appear to offer a simple, low cost technique for "transistorizing" the TV camera tube and other such image-sensing devices. Most companies now have the ability to produce laboratory models with excellent characteristics, although no company is claiming mass production capabilities.

Optical character readers may very well be given the impetus needed to achieve their volume growth through the use of Solid-State Imaging Devices—let's call them SSIDs—credit forms, mail sorter readers, credit card verification devices, and many other business machine applications are immediate application areas for SSIDs. The facsimile and copy-machine is another huge potential market. Already a number of companies are working on machines that not only copy material but process it for computer storage or for transmittal to a remote location. Point of sale checkout terminals at supermarkets and mass retail stores are well on their way towards employing SSIDs for speeding the checkout of customer's purchases as well as assist store management in inventory control and bookkeeping. Industrial applications are already being served by SSIDs. These devices are now employed in application where their size, reliability, ruggedness, and, in particular, their dimensional precision is important. Of course, their precision stems from the

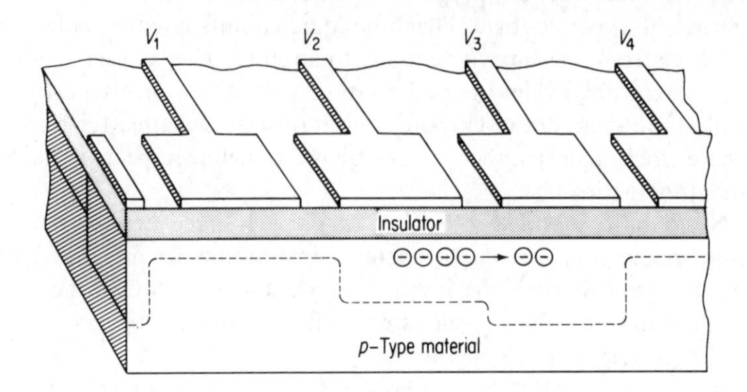

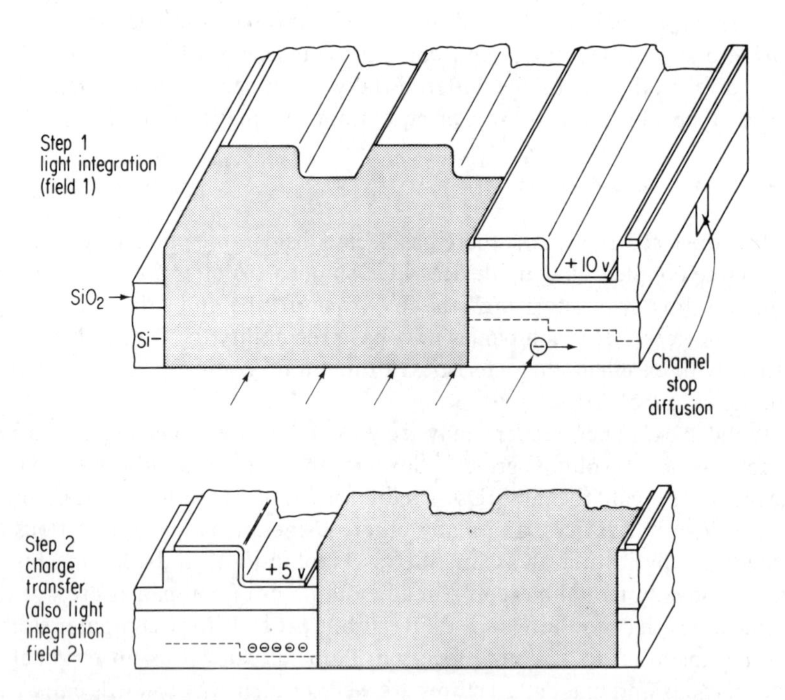

Fig. 15-2. (a) In basic three-phase Charge Coupled Device, V_3 electrode creates "deepest" potential well in which electrons are collected and stored during integration time. During readout time, voltages on electrodes are manipulated so that the deepest potential well is moved under all V_1 electrodes then V_2, and so on until all charges are fed into the readout amplifier; (b–c) dual-level SiO_2 insulation layer allows two-phase charge transfer system to be employed. Charges collect as shown during integration period and during readout alternating voltages causing charge packets to be shifted into readout amplifier.

fact that each sensing element is controlled precisely by the fabrication process, and unlike the vidicon, exact dimensional information can be derived from the video signal.

Just as the pocket calculator has skyrocketed in popularity and the price has plummeted from $395 to $39.50 in just a few years, it is certainly conceivable that the solid state imager may very well follow suit, particularly if a low cost video tape recorder, that is now being worked on by a number of companies, is perfected.

So let us compare the various solid state imaging technologies, and not only with existing devices such as the vidicon. Several of the competitive techniques are diagrammed and described specifically. (See Fig. 15-2)

CAN THEY BE BUILT?

Undoubtedly the most bandied-about word around the solid state imaging business is *manufacturability*. Everyone is confident that they can make the CCD and CID, *Charge Injection Device*, devices a few dozen at a time in the laboratory or in small, closely controlled production facilities. But, if you ask them if they can produce thousands, if not millions, at a low price, then the answer is decidedly different for few, if any, will predict quantity production capabilities any earlier than 1977. Of course, the older self-scanned device technology has proven manufacturability and products and systems are now being delivered.

The one factor that has made the progress in this technology so rapid is that SSIDs are built using the same MOS fabrication technology which has already been perfected for the fabrication of integrated circuits. However, the size of each final SSID chip is much larger than most IC chips. This is why the extensive effort to develop silicon target vidicons has also speeded the development of the CCD. To make these devices, large silicon wafers of extreme purity have to be processed. Not only are near perfect wafers required, but many ions and many impurities that invade the process during the masking, cleaning, oxidizing, and diffusing steps result in a final chip that is either worthless or of poor quality. Fortunately the effort to develop defect-free silicon targets was quite successful. Whereas a few years ago a silicon vidicon target commonly had 20–30 blemishes, these days such targets are usually fabricated defect free, and those with a few blemishes are even difficult to sell at highly discounted prices. Thus, because of this preceding technology, solid state imagers are already well on their way towards becoming a volume produced device. Solid state imaging technology has not only caught up with MOS technology but has surpassed it because the component densities and the chip size are already beyond that considered *state of the art* by some MOS manufacturers.

ZERO SPACE

One of the problems in producing CCDs has been the extremely precise masking processes that are required. Typically, the MOS electrodes involved have been made on a 1.0–1.2 mil centers, which means that the spacing between electrodes is on the order 2 μm or so. Thus any mask, silicon or processing defects could easily have an adverse effect on the quality of the final device. One technique developed at Bell Labs to solve this problem is through the use of a multilayer technique. Instead of forming all capacitor electrodes at the same time, they formed every third electrode in each of three steps and formed an oxide insulating layer between each step. This means that they have 20 μm between electrodes in each layer. (See Fig. 15-3). The resulting device has a zero-space between electrodes and yet the mask and the fabrication tolerances were extremely loose, thus insuring a high degree of manufacturability.

Three such devices have been assembled with a color-separating prism, and appropriate filters, to produce a color TV camera that provides a reason-

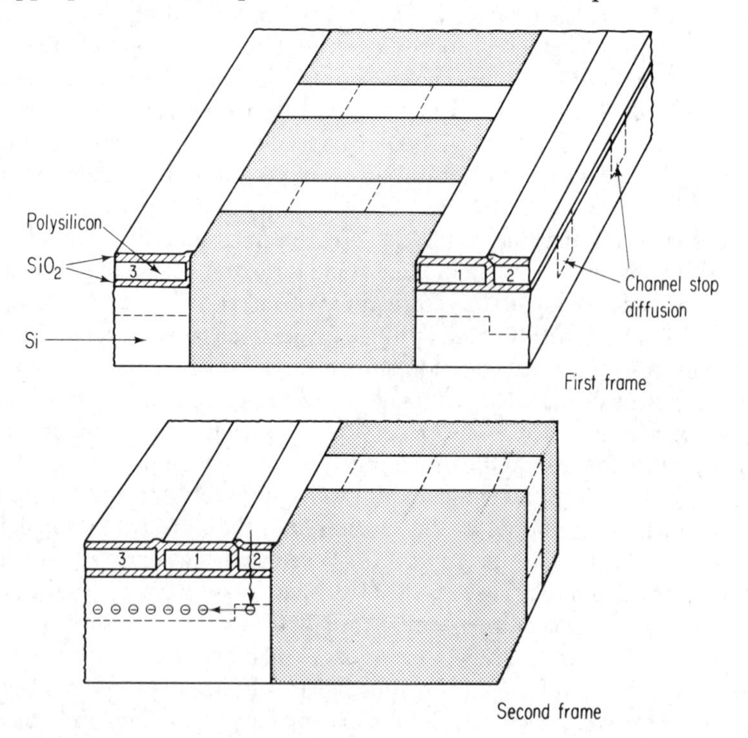

Fig. 15-3. Overlapping electrodes used in the latest Bell Labs system requires three separate electrode forming procedures but improves the yield by reducing tolerances required for each step. Collection areas for first and second frame integration are shown.

able picture quality. In this system, the chip has 220 vertical transfer columns and 128 rows of elements in both the image and the storage areas. By using interlace procedures, a 220 × 256 line picture was achieved that could be displayed on a conventional TV monitor by skipping every other trace. This technology has been used to fabricate other devices as well. For instance, a linear, buried channel sensor was fabricated with 1600 elements for high-resolution slow-scan image sensing as is required in facsimile transmission equipment.

One of the prime reasons General Electric selected the Charge Injection approach (see Fig. 15-4) was because they could employ standard MOS fabrication processes and achieve high yields. One reason for their high yields is that the electrodes can be made much smaller than the total sensing area. This is possible because the sensing elements need not be in extreme close proximity, as in the case of the CCD.

THE STATE OF THE ART

As far as currently *viewable* images are concerned, RCA has apparently overcome manufacturability problems better than anyone. Their chip has 163,840 active elements, (a 512 × 320 array). Its size and associated parameters were selected by the designers at RCA to be compatible with current 525 line TV systems. They started by analyzing the various TV camera tubes available, how they were used in a TV system, and then once the parameters were defined, they used their extensive experience in the development of vidicons to establish their specific design goals for their chip.

One possible offshoot of this effort is the use of the thinned-backside illuminated CCD device in an evacuated cavity with a photo-emissive cathode. By electron-optically focusing the output of the photocathode onto the CCD target, the output signal can be raised considerably above the background signal variation limits. It also should be noted that most silicon devices can be similarly used in an electron-bombarded configuration.

Another feature of CIDs that will be particularly important for some applications is *random access capability*. This may be of particular use in Security and Surveillance applications where, by selective scanning, a degree of electronic zoom-capabilities are achieved for viewing only a portion of the scene where an intruder might be located. Also, this feature can be of use in military applications, such as missile sensors and remote piloted vehicles.

The leading proponent of the buried channel interline transfer system is Fairchild. Much of their early effort was aimed at the development of CCD devices suitable for low light applications. However, LLLTV is not the only application area being aimed at. A variation of the buried channel devices has been recently announced by Philips Research Laboratories.

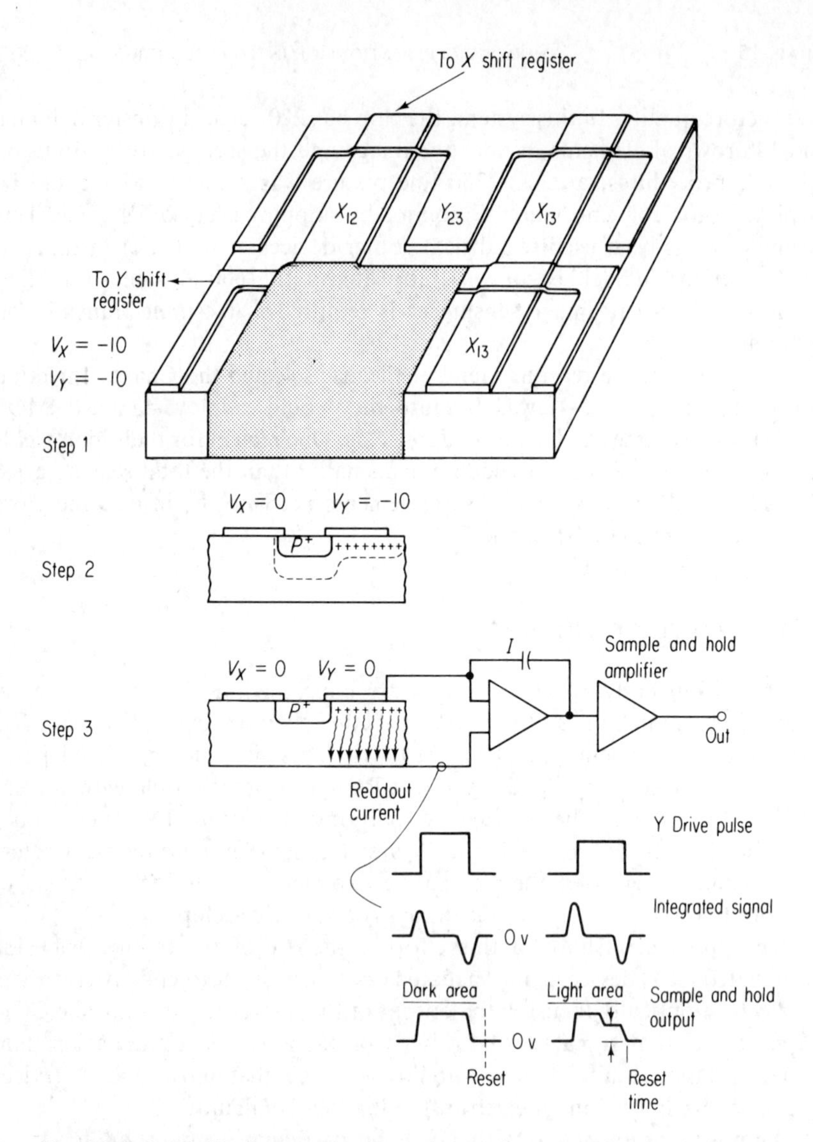

Fig. 15-4. In GE CID System, minority carriers are first moved from under X row to under Y electrodes. Increasing voltage on Y electrodes injects a charge into substrate and causes an output current proportional to charge stored. By integrating leading and trailing edge of readout current, the level of the sample and hold amplifier represents the quantity of stored charges.

They call their device a *peristaltic charge coupled device* (PCCD). The goal of their device is to increase the clock rate capabilities of CCDs by minimizing the time required to collect the tail end of each charge packet. This is achieved by injecting a *charge cloud* in the input and forcing the tail of the charge packet deeper in the silicon where surface effects do not slow down the charge transfer. They report that shift out rates of 100 MHz have been reached with possibilities of reaching even 1 GHz.

At Bell Laboratories, the home of the CCD, the original goal was to develop CCDs so as to be able to produce low cost, reliable Picturephones for person to person communications. However, where the Picturephones have really proven themselves valuable is in such application areas as medical, legal, education, etc. So the reason for working on the CCD imager as a low-cost high reliable picturephone imager has changed, but the need is still there.

From the standpoint of resolution, the self-scanned devices are best suited for linear arrays because the photosensitive elements can be positioned on 1 mil centers and large storage capacitors can be used that are not a part of the photosensitive area. Problems arise when large area arrays are required because the necessary interconnecting lines prevent sensing elements from being closely spaced, and indeed the typical spacing of a Reticon sensor is 4 mils. This means for a given number of photosensitive elements, from 3–4 times the silicon area is required and the fabrication problems skyrocket. (See Fig. 15-5) There is also a debate as to how the high capacitance of the readout lines affect the output amplifier. Also the dynamic range limitation of SSPDs (typically 100: 1) as compared to that of CCDs (typically 1000:1) is causing some debate. It is not an easy comparison because the characteristics demanded by a particular application are important and definitions do not always agree. For instance, a solid state camera that can provide a discernible picture of a scene after an aperture closing of 10 f-stops presents an effective dynamic range of 1000:1 (peak signal to rms noise), yet the S/N ratio (peak signal to peak noise) would be well below that required in a machine reading against discrete thresholds. A concern for the last several years has been whether the CCDs are suitable for low light level imaging.

One question that arises when discussing LLL imaging applications is

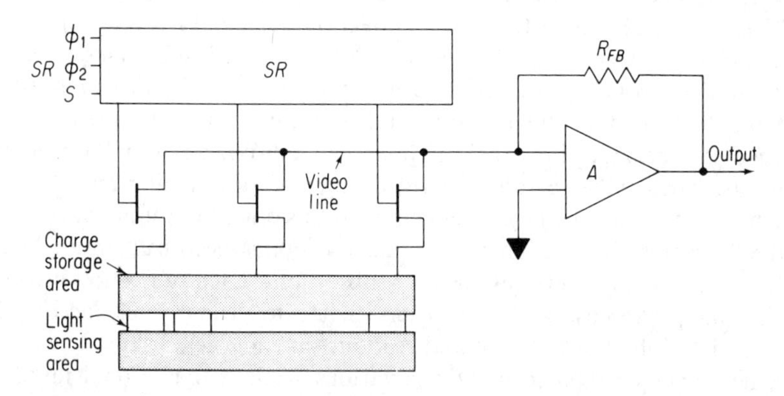

Fig. 15-5. In Reticon's self-scanning photodiode array, photons impinging on light sensing areas discharge stored charge from set level. Self-scanning shift register, SR, consecutively drives FETs ON and recharges individual charge storage areas and produces video output signal proportional to charge replaced.

the effect of aliasing. Aliasing is an effect where the spatial frequency of the image being looked at is greater than the spacing of the detectors. Thus Moire patterns that result may be objectionable to the eye and not tolerable in commercial applications. However, in some military applications the confusion that may occur because of aliasing may be better than having no information at all. This is important because the backside illuminated frame transfer system has a zero MTF at spatial frequencies at the Nyquist frequency (twice the spatial frequency of the available charge sensitive sites) whereas the interline transfer still has a usable MTF even though aliasing may occur at such frequencies.

One problem that arises when trying to use surface channel devices in low light level applications is that at the Silicon-SiO interface charge trapping does occur and degrades the imager results. One technique used to eliminate this problem is to employ a *Fat Zero*. This is achieved by adding a long diffusion, a *great big diode*, across the top of the array. Here, a certain amount of carriers are injected into all channels that are sufficient to fill the trapping areas. Unfortunately, this technique is not yet perfected in that no one has yet been able to inject a uniform *Fat Zero*. The buried channel technique as employed by Fairchild forces the charges away from the surface of the chip and thus eliminates this problem.

Of course we must not forget that the electron-bombarded SSID, previously described, may provide excellent LLLTV results and will definitely be a competitor for such future LLLTV applications.

IS BUCKET-BRIGADE DEAD?

Although CCDs and CIDs have stolen the spotlight, the predecessor to CCDs, the bucket-brigade process, is still of interest. For example, in the IBM laboratories, where the FET production techniques employed in the fabrication of bucket-brigade devices have been refined to a high degree, an area scanner program has been recently implemented. It was intended to explore the potential of wide area optical scanners and the feasibility of defect-free very large chips. This chip, containing over a half-million photoelements, when fully operational will be capable of scanning an $8\frac{1}{2}$ by 11 inch page at the rate of 1 sheet every 4 ms. At this time, 13 contiguous sections of the 16-section chip are functional. (Each section is electrically independent of the other sections and contains four subsections. Each subsection involves 16–512 bit photosensing shift registers that extend completely across the chip, and 2 high speed control and readout shift registers.)

One technique used to make the various sections operational involved *surgical* repair of the circuits. This involved the use of laser milling techniques. Here shorts and bridges were removed using the laser beam much like in trimming resistors. By milling portions of the metallurgy, defects can be corrected, or, if not correctable, prevented from influencing more than a single line in the scanner output. In the latter event, errors in the scanner output caused by a single line can be corrected by circuitry built into the chip that performs an algorithm developed by IBM. This algorithm employs three stages of the shift register and 4 NAND gates and is capable of correcting most errors. It acts by deleting any single white dot in a field of black and any black dot in a field of white. This is done because in the real world of imaging, seldom does a single imaging point exist by itself. An analysis of the types of errors that typically occur indicates that in 10 out of 11 typical images, the error correction will be successful.

A principal objection to bucket-brigade imaging devices that caused researchers to look into CCD still exists and that is the less than desirable transfer efficiency. In a good bucket-brigade device the transfer efficiency is as good as 99.99 per cent efficient, whereas in CCD arrays it approaches 99.999 per cent. This reduced transfer efficiency results in a need for low pass filter compensation of the output signal.

Conclusions

Solid state imaging devices are on the verge of a remarkable growth period. Certainly the technology has been developed, the learning curve is not too long, and many of the problems which seemed significant have been overcome. The main problem now seems to be the effectiveness of the SSID industry in selling their product capability so that they can arrive at the necessary volume. For it is only through volume production that the final knowhow, so necessary to make cost-effective devices in large quantities, can be achieved. Thus there is no question but that in the next few years solid state imagers will be used in hundreds of applications and even start affecting our daily lives. As to whether they will ever invade the home entertainment market, it is still a good question.

MEMORIES

The Choice is Wide with Memories (Commercial)

From mainframe to scratch-pad units, semiconductor memories have experienced rapid growth rates. Anticipated advances in CCD memories, and new developments in magnetic-bubble memories, promise to make the designer's choice of memories the widest it has ever been.

There are several types of binary memory circuit included in a typical computer. We have already discussed the static register and the shift register, both of which are constructed from flip-flop circuits. These memories exhibit the characteristics of high speed and low storage capacity. In recent years,

IC techniques have been improved to the point that large memories are available that, in many cases, compete favorably with core storage memories. As far back as the beginning of the 1970's, some computers were being designed with mainframes that were completely solid state (IBM 370/135 and 370/145) while others combined semiconductor and core memory to advantage (IBM 370/155 and 370/165).

Both bipolar and MOS transistors can be used in flip-flop circuits to provide static storage of binary information. The word *static* in this case means that, once set, each memory cell maintains its state as long as power is supplied continuously to the circuit. The very high gate impedance of the MOS device has made the dynamic register a practical storage unit. In the dynamic register, the binary information is stored as charge on the inherent capacitances of the device gates. Ideally, these stored bits would remain as long as necessary with no further input power required; however, junction leakage slowly drains the charge. Dynamic memories then require the capacitances to be recharged at certain intervals to maintain the stored information. Since power consumption only takes place during recharging of the capacitors, the big advantage of the dynamic memory is low power consumption.

One of the fastest growing markets for semiconductor memories is the read-only memory or ROM. These storage units contain words that are never changed. Each location can be addressed for readout, but this location cannot be written into. There are several small ROM's available in IC form at present. These vary in size from units of perhaps 256-bit capacity up to larger MOS units of 4096 bits. The random-access memory or RAM can also be realized with bipolar transistor IC's or with MOS IC's, although for high speed operation, the MOS unit is not as appropriate as the bipolar memory. If the MOS RAM is dynamic, periodic refreshing of the stored information is necessary to prevent complete discharge of the gate capacitors. Although this requirement for refreshing leads to slightly more complex operation, the packing density of MOS circuitry is quite good, leading to a high storage capacity for the MOS memory.

Memory Technologies (Aerospace)

Choice of memory technology for aerospace applications involves a set of priorities different from those used in selection of memories for commercial applications, i.e., requirements for minimizing power and weight, and for operation in extreme environments. For at least a portion of an aerospace computer memory, nonvolatile random access storage is usually mandatory and, to the extent that nonvolatility and read/write capability are required, semiconductor memories are ruled out except for the *metal-*

nitride-oxide semiconductor, (MNOS) type. Nevertheless, it is likely that evolving architectures will also include wide applications for volatile semiconductor *random-access memories*, (RAMs) as well as *read-only memories*, (ROMs).

With regard to word length and storage capacity, requirements are generally in the range of 16 to 36 bits and from a few thousand words up to 64K words or more. Speed of operation is also important, but often less so than for commercial applications. In recent years, cost has become a major factor. High reliability continues to be a paramount consideration.

Of the many candidate technologies, the ones that warrant close scrutiny for present and near-future aerospace applications are *core*, *plated wire*, and *semiconductor*. Table 15-2 compares typical characteristics of memories in each of these technologies, including the major subdivisions of semiconductor types.

Technological maturity is a major factor in memory selection. Clearly, core memories, which were first suggested in 1951, are more mature than plated wire, which did not begin to see use in actual systems until the mid-1960s. Semiconductor memories (bipolar or MOS) are also based on highly developed technology. Current development concentrates on NMOS and CMOS chips. Single chips with up to 4-kilobit storage are likely to be widely used in the near future, although 1-kilobit chips are now standard. MNOS memories represent an even newer technology, which has yet to progress to field use. It is noteworthy that levels of confidence in the technology base are sufficient to justify development of all candidate memory types for use in the next few years.

The present state of MNOS development necessitates consideration of two varieties, both of which are listed in Table 15-2. One is a RAM, which has read/write speed characteristics comparable to those of core or plated wire. Unfortunately, fully developed units with characteristics equivalent to those in the Table are not yet available, and parameters listed must be viewed as development goals. Basic technical difficulty which remains to be satisfactorily resolved for general applications is the *wear out* phenomenon.

The second type is the electrically alterable ROM (EAROM). Since several milliseconds are required to alter a word in an EAROM, its main use is for cold storage (storage of instructions and constants). The MNOS EAROM is presently available. Considering the status of MNOS development, it would be desirable to mechanize the using system so that a large EAROM could be used for cold storage, and some other type device for parts of the memory requiring random access read/write memory capability. For a short-duration mission, an MNOS RAM could be considered for this purpose.

Effective use of semiconductor memories calls for architectures which

TABLE 15-2 Aerospace Memory Technology

Memory Technology	Readout Type	Storage Type	Read Cycle Time (μs)	Write Cycle Time (μs)	Access Time (μs)	Volume (Cu In./Bit)
Core	DRO	Nonvolatile	1	1	0.4	3.6×10^{-4}
Plated wire	NDRO	Nonvolatile	0.5	0.75	0.4	3.9×10^{-4}
MNOS	NDRO	Nonvolatile	1	2	0.5	1.2×10^{-4}
MNOS EAROM	NDRO	Nonvolatile	1	10,000	0.5	1.2×10^{-4}
Bipolar ROM	NDRO	Nonvolatile	—	—	0.06	1.2×10^{-4}
Bipolar RAM	NDRO	Volatile	—	0.08	0.06	5×10^{-4}
Dynamic NMOS RAM	NDRO[2]	Volatile	0.25	0.3	0.15	3×10^{-4}

Notes:
(1) This table describes complete memory systems using technology presently available or in development. Each design is assumed to provide a maximum of 16K words by 16 bits of storage to facilitate comparison. It is assumed that all are designed to meet severe military environments.

separate the cold storage program portion of memory from the random access read/write segment. Further attention must be given to identifying portions of memory that are to be nonvolatile.

Reasons for increased usage of semiconductor memories can be seen in Table 15-2. ROMs or EAROMs require little power and are characterized by small volume, low cost, and nonvolatility. The bipolar RAM requires considerable power, is relatively large, is volatile, but is extremely fast. MOS RAMs require little power, but are also volatile. The NMOS variety also offers a speed advantage compared to core or plated wire. However, the system designer who specifies a semiconductor memory must recognize the limitations imposed by the volatility of RAMs and the difficulty of changing programs stored in ROMs.

N-channel memories, selling presently at anywhere from 0.25¢ to 6¢ per bit, depending on quantity and performance, are expected to drop down to 0.1¢ per bit by 1977 to become an attractive alternative to core memories, once production levels get high enough. At this moment, many memory experts give CCD the near-term edge over other mass-memory types (Table 15-3). CCDs also have applications as image sensors and in analog-signal processing.

Experimental arrays of 4 kb, 8 kb, and 16 kb have been reported for CCD memories. At the moment, it looks like CCDs will have applications as memory elements somewhere between very-large-capacity-but-slow floppy

Comparison[1] (Typical Characteristics)

Weight (Lb/Bit)	Power 1-μs Cycle (Watt/Bit)	Radiation Hardness	Cost (¢/Bit) (In Production)	Modularity	Technological Maturity
1.4×10^{-5}	3.7×10^{-4}	No[3]	3–5	16K × 16 blocks	Most mature
1.6×10^{-5}	0.8×10^{-4}	Yes	7–12	8K × 16 blocks	Mature
0.4×10^{-5}	0.5×10^{-4}	Yes	—	256 × 4 chips	Not fully developed
0.4×10^{-5}	0.08×10^{-4}	Yes	2–4	256 × 4 chips	Comparatively new
0.3×10^{-5}	0.7×10^{-4}	Yes	0.5–2	1K × 8 chips	Mature
1.4×10^{-5}	6.4×10^{-4}	No[3]	2–4	1K × 1 chips	Mature
1.1×10^{-5}	1.1×10^{-4}	No[3]	1–2	1K × 1 chips	Mature

(2) Dynamic MOS RAMs must be periodically refreshed.
(3) Could be hardened to some extent.

TABLE 15-3 Drum and CCD-equivalent Characteristics (An Example for Low-power Applications)*

Parameter	Drum	CCD
Tracks (loops)	256	1024
Bits per track (net)	32 768	8192
Data rate	2 MHz	2 MHz
Access time		
(maximum)	20 ms	4 ms
(average)	10 ms	2 ms
Usable capacity	8 388 608	8 388 608
Volume	0.084 meter^3	0.0092 meter^3
Weight	56.25 kg	6.75 kg
Power dissipation	300 watts	5 watts (operating) 2 watts (standby)
MTBF	3500 hours	20 000 hours

*Source: J. M. Chambers *et al.* in Session 6, 1974 WESCON Professional Program.

disk and bubble memories, and smaller-capacity but much faster MOS RAMs. Their use as shift registers is also expected to increase.

CCDs for analog-signal processing have found various military and commercial applications, ranging from electronically variable delay lines to transversal filters with up to 500 stages. And CCDs are being readied for use in TV cameras, where their use will make possible very small and light-weight cameras.

Magnetic-bubble memories are now ready for production. While no commerical bubble memories have been made, Bell Telephone Laboratories recently announced the largest-capacity bubble device ever made in the smallest package—460, 544 bits in a package 9.53 by 5.33 by 2.06 cm. The memory has an average access time of 2.7 ms, a data-transfer rate of 700,000 b/s, and a read error rate of less than one error in 630 billion read operations.

Microprocessors—Wave of the Future

There is hardly any circuit or system designer who will not be affected by the microprocessor's role in his future product. Already, over a dozen semiconductor manufacturers have produced commercially available 4-, 8-, and 16-bit microprocessors, with more to come. And microprocessors are expected to take over tasks that were either implemented with much larger and more expensive minicomputers or remained unaccomplished because of cost and size.

CPU microprocessors are self-contained, dedicated circuits on a single chip. Bit slices on the other hand are available in several-chip organizations to synthesize a CPU. The bit-slice organization allows the designer greater flexibility to address system requirements since the basic slice is an ALU (arithmetic logic unit) in 2- or 4-bit form, from which N-by-2 or N-by-4-bit machines can be constructed.

A Boom in Linear Building Blocks

The list of available linear-function building blocks for circuit design continues to grow. The designer's arsenal can include such modular components as A/D and D/A converters, multipliers, dividers, summing circuits, op amps, active filters, and power modules. Many of the above have recently become available in low-cost and miniature monolithic form, while others, such as the op amp, which long ago reached that stage, are being pushed toward the higher-performance levels of higher-cost and larger-size modules.

Almost every major semiconductor manufacturer is active in monolithic A/D converters, where the main thrust has been in the digital voltmeter and panel-meter markets. Despite the fact that many linear building-block components are being made in monolithic form, the modular or hybrid approach, for many functions, dominates when critical performance levels must be met. The future of monolithic products—particularly converters—is in increasingly sophisticated end use of such products, where better performance levels can perhaps be best achieved by combinations of monolithic and other techniques.

DISPLAYS

Solid State in the Lead

Three types of digital displays, for numeric and alphanumeric applications, are in contention: *LED*, *liquid crystal*, and *plasma-* or *gas-discharge*. LEDs, however, are rapidly gaining wider user acceptance and are garnering larger shares of applications.

Improvements continue in LED digital displays, in quantum-efficiency and current-dissipation levels. They are now available, commercially, to operate from 5-volt IC-logic lines, at current-dissipation levels ranging from 3 to 15 mA per segment, for multisegmented displays, in heights of 0.5 inch (1.3 cm).

Vapor-phase nitrogen doping of GaAsP (gallium-arsenide-phosphide) on GaP, instead of the conventional growing of GaAsP on GaAs, with no nitrogen doping, has produced LEDs in different colors with much higher output-intensity levels (Fig. 15-6). Liquid-phase growing of GaP on GaP, produces even brighter and more efficient green LEDs (luminous-flux levels for GaP on GaP are not shown on the graph of Fig. 15-6). Important work is going on in making monolithic LEDs (with decoder/driver on the same chip) with peak current dissipation levels of 5 to 10 mA per segment (300 μA average), resulting in displays for calculators and watches (0.1 in. high).

Despite recent attempts by liquid-crystal-display manufacturers to improve the readability and performance of their product with newer field-effect liquid crystals, compared with the earlier dynamic-scattering type, both still have problems with life expectancies, small viewing angles, a less-than satisfactory operating temperature range, and insufficient multiplexing capability. The liquid-crystal display is dominant in those portable and field applications where very little power is available. Its miniscule power dissipation (down to 0.5 μW per segment) is its main advantage.

Plasma- or gas-discharge displays continue to be used widely, mostly in desk top calculators, but their future is not quite clear. Currently, they offer low-cost and high-readability advantages, along with proven long-life spans. However, the high voltages needed to drive them (about 170 volts) create interface problems for IC circuitry and could be an even worse disadvantage in the future, once expected lower LED prices occur. The real future of plasma displays is probably not in the few-element alphanumeric area, but in monolithic planar displays. These have a fairly large number of resolution elements —about 10,000—and are intermediate between alphanumeric displays and CRTs.

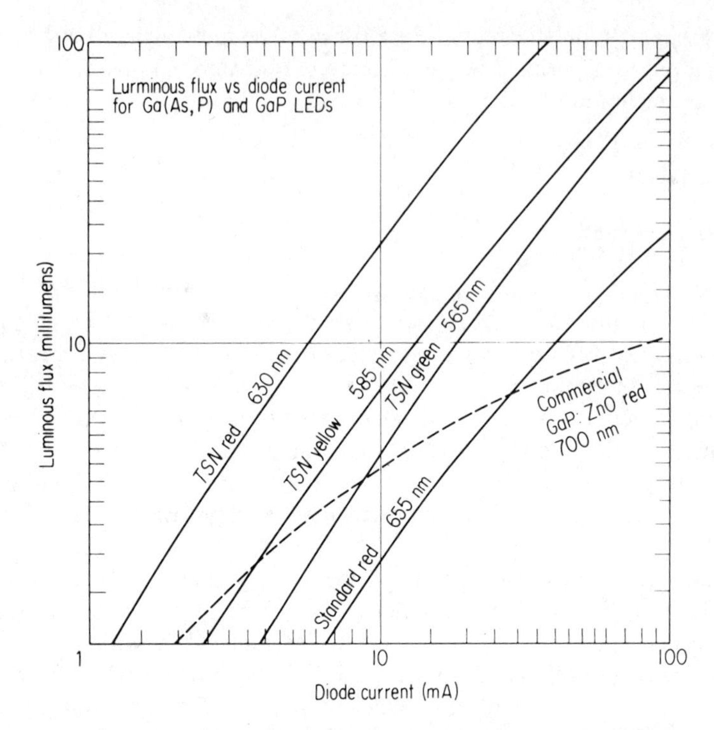

Fig. 15-6. Nitrogen doping (vaporphase) of GaAsp on GaP substrates yields LEDs with luminous flux levels many times higher than conventional GaAsP grown on GaAs substrates. All flux measurements are made with a chip embedded in a medium with an index of refraction of $n = 1.55$. (TSN = Transparent substrate, nitrogen doping.) The dotted curve is an estimate of today's production technology, given as an industry average for ZnO red GaP devices.

REFERENCES

1. M. P. RISTENBATT, *Semiconductor Circuits, Linear and Digital.* Englewood Cliffs N.J.: Prentice-Hall, Inc., 1975.
2. R. D. COMPTON, "Solid State Imaging Revolution," *Electro Optical Systems Design*, April 1974.
3. R. ALLEN, "Circuit System Building Blocks," *IEEE Spectrum*, January 1975.
4. Staff Report, "Modern Memories," *Computer Design*, August 1974.
5. Staff Report, "Gunn Amplifier," *Microwaves*, April 1974.
6. B. G. STREETMAN, *Solid State Electronic Devices.* Englewood Cliffs, N.J.: Prentice-Hall, Inc., 1972.
7. S. D. SENTURIA and B. D. WEDLOCK, *Electronic Circuits and Applications.* New York: John Wiley & Sons, Inc., 1975.

Appendix

BASIC ASSUMPTIONS OF THE CHARGE CONTROL MODEL

The charge control model as developed by Sparkes and Beaufoy (1)* is a first-order theory which describes the dynamic behavior of the bipolar transistor. The origin of this model stems from the basic equations governing current flow in transistors. Transverse current flow in the base of the transistor results from two mechanisms: (1) diffusion of charge carriers because of carrier density gradient; (2) drift of charge carriers because of the existence of an electric field in the base. In analyzing current flow in transistors, a one-dimensional model is assumed in which the quantities of interest vary only in the transverse direction across the base region. This assumption provides a considerable simplification in the resulting model and is a good approximation to the behavior of a number of actual transistor structures.

Figure A-1 illustrates a cross-sectional view of the assumed one-dimensional (PNP) transistor structure. Also shown is the assumed doping density for emitter, collector, and base regions. The emitter and collector depletion region capacitances are represented by C_{TE} and C_{TC}, respectively.

The most important assumptions made in deriving the charge control model are

1. The flow of charge carriers in the transistor is governed by one-dimensional equations. The emitter junction and collector junction are thus considered equipotential planes of the same area.

*Numbers in parentheses refer to material listed at the end of the section.

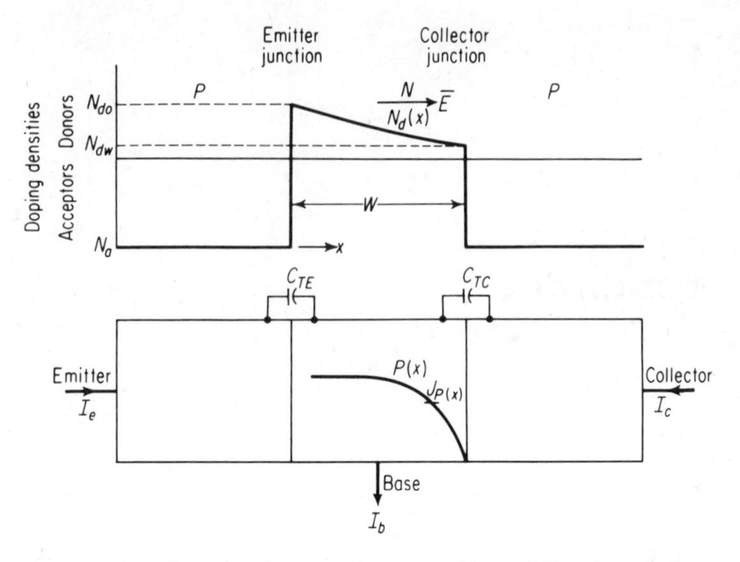

Fig. A-1. One-dimensional transistor structure with nonuniform base doping.

2. The acceptor-doping levels in the emitter and collector regions are much greater than the donor-doping level in the base. Therefore, charge storage takes place only in the collector-depletion region, emitter-depletion region, and base.
3. The voltage dependence of the electrical base width is neglected. This means that the voltage-dependent widths of the emitter- and collector-depletion regions are small with respect to the metallurgical basewidth, W, so that we can use *it* in the relations.
4. The current dependence of the minority carrier diffusion constant is neglected. This condition is violated if the injected hole density is comparable to the donor-doping level in the base.
5. The state of the transistor (cut-off, active, or saturated) is determined by the total stored base charge and not the distribution of the charge.

CURRENT FLOW AND CHARGE STORAGE IN HOMOGENEOUS AND DIFFUSED BASE TRANSISTORS

With these assumptions the equations governing current flow in the base are

$$J_p(x) = q\mu_p p(x)\mathcal{E}(x) - qD_p \nabla p(x) \tag{A-1}$$

$$\frac{\partial p(x, t)}{\partial t} = \frac{p_n - p(x, t)}{\tau_p} - \frac{1}{q} \nabla J_p(x, t) \tag{A-2}$$

$$J_n(x) = q\mu_n n(x)E(x) - qD_n \nabla n(x) \tag{A-3}$$

J_p and J_n are hole and electron current densities, respectively.

With the assumption that $N_a \gg N_d(x)$, $J_n(x)$ is very nearly zero, and the emitter injection efficiency defined by

$$\gamma = \frac{J_p(0)}{J_n(0) + J_p(0)} \approx 1.0 \tag{A-4}$$

From Eq. A-3, with $J_n(x) = 0$:

$$\mathcal{E}(x) = \frac{kT}{q}\frac{1}{n(x)}\frac{\partial n(x)}{\partial x} \tag{A-5}$$

since

$$\frac{kT}{q} = \frac{D_n}{\mu_n} \tag{A-6}$$

We examine a special case in which the electric field, $\mathcal{E}(x) = \mathcal{E}_0$, is a constant throughout the base region. The required doping distribution, $N_d(x) = N(x)$, is determined from

$$\frac{kT}{q}\frac{1}{N_d(x)}\frac{\partial N_d(x)}{\partial x} = \mathcal{E}_0 \tag{A-7}$$

$$\mathcal{E}_0 = -\frac{kT}{qW}\ln\frac{N_{d0}}{N_d(x)} \tag{A-8}$$

The required form of the base doping is exponential:

$$N_d(x) = N_{d0}\exp\left\{\frac{q\mathcal{E}_0}{kT}x\right\} \tag{A-9}$$

Equation A-1 may now be rewritten as a first-order linear differential equation. Noting that $I_p(x) = J_p(x)A$, we have

$$I_p(x) = q\mu_p\mathcal{E}_0 Ap(x) - qD_pA\frac{\partial p(x)}{\partial x} \tag{A-10}$$

We now assume that the lifetime of minority carriers is large (alternatively $h_{FE} \gg 1$) such that $I_p(x)$ is a constant and equal to the collector current

$$I_C = q\mu_p\mathcal{E}_0 Ap(x) - qD_pA\frac{\partial p(x)}{\partial x} \tag{A-11}$$

This equation may be solved for the steady-state spatial distribution of minority carriers in the base corresponding to a certain collector current flow. The solution of Eq. A-11 is

$$p(x) = \frac{I_C}{qA\mathcal{E}_0\mu_p}\left\{1 - \exp\left[\left(1 - \frac{x}{W}\right)\frac{qW\mathcal{E}_0}{kT}\right]\right\} \tag{A-12}$$

The total base control charge required to support the collector current I_C is found by integrating the minority carrier distribution over the base volume:

$$Q_B = qA\int_0^W p(x)\,dx \tag{A-13}$$

The total base charge is

$$Q_B = \frac{WI_C}{\mathcal{E}_0\mu_p}\left[1 - \frac{kT}{qW\mathcal{E}_0}\left\{1 - \exp\frac{-qW\mathcal{E}_0}{kT}\right\}\right] \tag{A-14}$$

For a fixed-collector current, the required base control charge decreases as the electric field increases. The control charge is proportional to the base charge so that the ratio is constant and has the dimensions of time:

$$\frac{Q_B}{I_C} = \frac{W}{\mathcal{E}_0 \mu_p}\left[1 - \frac{kT}{qW\mathcal{E}_0}\left\{1 - \exp - \frac{qW\mathcal{E}_0}{kT}\right\}\right] \tag{A-15}$$

As the "built-in" field decreases to zero ($N_{d0}/N_{dW} \longrightarrow 1.0$)

$$\frac{Q_B}{I_C} = \frac{W^2}{2D_p} \tag{A-16}$$

Equation A-15 gives the required base control charge per unit collector current for a "graded base" transistor in which the base resistivity varies in such a way that the resultant electric field in the base is a constant. Equation A-16 provides the base control charge requirements for a homogeneous base transistor (for example, alloy junction type).

These results illustrate that the total base charge serves as a convenient variable to describe the state of the transistor switch. Turn-on of the transistor requires that a certain base charge be supplied, as dictated by the collector current and voltage to be switched. The time required to complete a change of state is determined (in part) by the time required to supply this base charge.

DERIVATION OF SWITCHING TIME EQUATIONS FROM THE FIRST-ORDER MODEL

Rise and Fall Times

The answer to the switching time problem (2) lies in the analysis of the various charge storage mechanisms in the transistor and of the way in which this charge is routed to and from the transistor. With reference to Fig. A-1, we can write equations for the time-dependent terminal emitter, collector, and base currents in terms of the various components that comprise the terminal currents.

For the emitter current, we have

$$I_e(t) = I_{pe}(t) + \frac{\partial q_b(t)}{\partial t} + C_{TE}\frac{\partial V_{eb}(t)}{\partial t} \tag{A-17}$$

where $I_{pe}(t)$ is the diffusion current injected at the emitter junction. The second term accounts for current which flows to change the base charge; the third term gives the current which flows to charge C_{TE} as a result of a change in V_{eb}.

The collector current is

$$I_c(t) = -I_{pc}(t) + C_{TC}\frac{\partial V_{cb}(t)}{\partial t} \tag{A-18}$$

where $I_{pc}(t)$ is the diffusion current collected at the collector junction and where the second term accounts for charging current through the collector capacitance.

I_{pe} and I_{pc} differ because of recombination in the base. The recombination current is given by

$$I_r(t) = \frac{q_b(t)}{\tau_p} \tag{A-19}$$

Thus, the base current is

$$\begin{aligned} I_b(t) &= I_e(t) + I_c(t) \\ &= \frac{q_b(t)}{\tau_p} + \frac{\partial q_b(t)}{\partial t} + C_{TE}\frac{\partial V_{eb}(t)}{\partial t} + C_{TC}\frac{\partial V_{cb}(t)}{\partial t} \end{aligned} \tag{A-20}$$

In the steady state we have

$$I_B = \frac{Q_B}{\tau_p} \tag{A-21}$$

If we desire to change the state of the transistor (that is, the collector current, or voltage, or both) we must supply additional base current according to Eq. A-20. Thus, the base current represents a mechanism for changing the state of the transistor. If we integrate Eq. A-20 over an increment of time, Δt, we have:

$$\int_{\Delta t} I_b(t)\,dt = \int_{\Delta t} \frac{q_b}{\tau_p}\,dt + \int_{\Delta t} \frac{\partial q_b}{\partial t}\,dt + \int_{\Delta t} C_{TE}\frac{\partial V_{eb}}{\partial t}\,dt + \int_{\Delta t} C_{TC}\frac{\partial V_{cb}}{\partial t} \tag{A-22}$$

This may be written as:

$$Q_{\text{supplied}} = Q_R + \Delta q_b + \Delta q_E + \Delta q_C \tag{A-23}$$

where Q_{supplied} = total charge supplied to the base during the increment
Q_R = total recombination charge lost during the increment
$\Delta q_b, \Delta q_E, \Delta q_C$ = changes in the base charge, emitter capacitance charge, and collector capacitance charge during the increment.

If the transistor is to be maintained in the steady state we must have

$$Q_{\text{supplied}} = Q_R \tag{A-24}$$

If any other value of Q_b is supplied, the state of the transistor must change, hence the term *charge control model.*

All but the recombination term in Eq. A-20 may be readily evaluated as

$$\int_0^{\Delta t} I_b(t)\,dt = \int_0^{\Delta t} \frac{q_b(t)}{\tau_p}\,dt + \Delta q_b + \bar{C}_{TE}\,\Delta V_{EB} + \bar{C}_{TC}\,\Delta V_{CB} \tag{A-25}$$

$\bar{C}_{TE}$ and $\bar{C}_{TC}$ are the average values of emitter and collector capacitance evaluated over the respective change in junction voltages, ΔV_{EB} and ΔV_{CB}.

We now define a time constant τ_T as

$$\tau_T = \frac{\Delta q_b + \bar{C}_{TE}\,\Delta V_{EB}}{\Delta I_C} \tag{A-26}$$

Previously, we have derived the equations for q_b in terms of I_C, namely, Eqs. A-15 and A-16.

For small changes in base-emitter voltage and hence small changes in the collector current, the ratio $\Delta V_{BE}/\Delta I_C$ approaches the dynamic resistance of the base-emitter diode:

$$\lim_{\Delta V_{BE}\to 0} \frac{\Delta V_{BE}}{\Delta I_C} = \frac{kT}{qI_C} \tag{A-27}$$

where I_C is the quiescent dc collector current.

Thus, for this limiting condition of small change in base-emitter voltage, the time constant τ_T (for homogeneous base transistors) is given as

$$\tau_T = \frac{W^2}{2D_p} + C_{TE}\frac{kT}{qI_C} \tag{A-28}$$

This expression for τ_T is recognized as the reciprocal of the common-emitter, current-gain bandwidth product, ω_T. For large signal changes, τ_T may be defined as the reciprocal of an average gain-bandwidth product, $\bar{\omega}_T$. With $\bar{\omega}_T$ defined in this way, a simplification of the switching equation may be effected, since

$$\frac{\Delta I_C}{\bar{\omega}_T} = \Delta q_b + \bar{C}_{TE}\,\Delta V_{EB} \tag{A-29}$$

This average gain-bandwidth product may be considered to be the average of small-signal ω_T over the change of I_C and V_{CE} through which the transistor is switched.

Using these results in Eq. A-25, the time t_r required to switch the transistor through the active region from the edge of conduction ($V_{EB} = 0$, $V_{CB} < 0$) to the edge of saturation ($V_{EB} > 0$, $V_{CB} = 0$) is

$$\int_0^{t_r} I_b(t)\,dt = \int_0^{t_r} \frac{q_b(t)}{\tau_p}\,dt + \frac{I_{CS}}{\bar{\omega}_T} C_{TC}\,\Delta V_{CB} \tag{A-30}$$

where I_{CS} is the collector current flowing when the transistor is at the edge of saturation.

An approximate relationship between the base charge Q_B and collector current I_C during the switching time interval may be derived as follows: The steady-state collector current has earlier been shown to be proportional to the base charge,

$$Q_B = \frac{W^2}{2D_p} I_C \tag{A-31}$$

We shall assume that this equation holds for all time during the switching interval. This means that the base charge builds up during the switch-on process in the manner indicated in Fig. A-2.

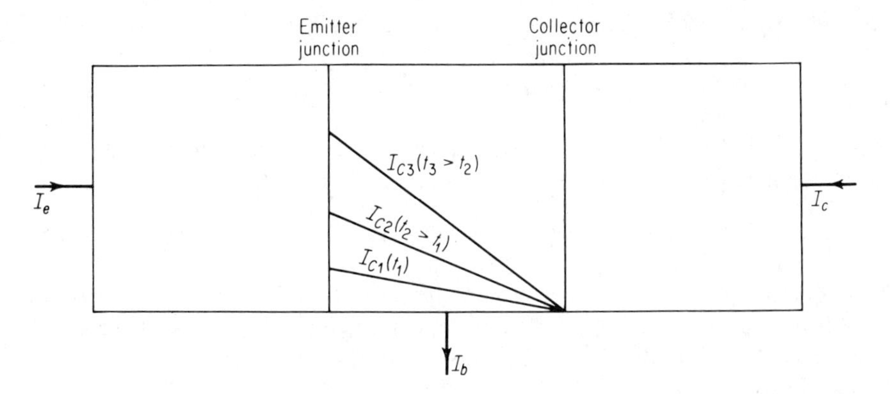

Fig. A-2. Base charge build-up during turn-on.

Substituting Eq. A-31 into Eq. A-30, we obtain

$$\int_0^{t_r} I_b(t)\,dt = \int_0^{t_r} \frac{I_c(t)}{\frac{2D_p}{W^2}\tau_p}\,dt + \frac{I_{CS}}{\bar{\omega}_T} + \bar{C}_{TC}\,\Delta V_{CB} \tag{A-32}$$

Differentiation of this with respect to time gives the general differential equation which describes the switching process in the active region:

$$I_b(t) = \frac{I_c(t)}{\frac{2D_p}{W^2}\tau_p} + \frac{1}{\bar{\omega}_T}\frac{dI_c(t)}{dt} + \bar{C}_{TC}\frac{dV_{CB}(t)}{dt} \tag{A-33}$$

At the end of the switch-on interval, the transistor is driven to the edge of saturation ($V_{EB} > 0$, $V_{CB} = 0$) and

$$I_B = \frac{I_{CS}}{h_{FE}} \tag{A-34}$$

Since Eq. A-32 must also hold in the steady-state condition at the end of the switching interval, we must have

$$\frac{2D_p}{W^2}\tau_p = h_{FE} \tag{A-35}$$

Therefore, the general switching equation, A-32, is

$$I_{b1}(t) = \frac{I_c(t)}{h_{FE}} + \frac{1}{\bar{\omega}_T}\frac{dI_c(t)}{dt} + \bar{C}_{TC}\frac{dV_{CB}(t)}{dt} \tag{A-36}$$

The notation $I_{b1}(t)$ means that the base current is supplying base charge.

If the transistor is caused to switch rapidly, the last two terms of Eq. A-36 will be large. In almost all transistors the first term (recombination term) is small enough so that it can be ignored in such cases. The necessary condition for this approximation can be determined from the integral form of Eq. A-36:

$$Q_{\text{supplied}} = \int_0^{t_r} I_{b1}(t)\,dt = \int_0^{t_r} \frac{I_c(t)}{h_{FE}}\,dt + \frac{I_{CS}}{\bar{\omega}_T} + \bar{C}_{TC}\,\Delta V_{CB} \qquad \text{(A-37)}$$

Assuming that the collector-current increase is reasonably linear during the rise time interval, recombination can be ignored whenever

$$\frac{I_{CS}t_r}{2h_{FE}} \ll \frac{I_{CS}}{\bar{\omega}_T} + \bar{C}_{TC}\,\Delta V_{CB} \qquad \text{(A-38)}$$

Equation A-37 has been derived for the case where the transistor is driven ON from the edge of cutoff to the edge of conduction. The switch-off process from the edge of conduction to the edge of cutoff is the reverse of the turn-on process provided:

1. The transistor is being turned off from a steady-state condition
2. The load impedance is strictly resistive
3. The reverse base current is small enough that the base-emitter diode does not recover (become reverse-biased).

During turnoff, base charge lost by recombination aids in removing charge from the base. The recombination current, therefore, adds to the base current. For the turn-off process, then,

$$I_{b2}(t) = \frac{-I_c(t)}{h_{FE}} + \frac{1}{\bar{\omega}_T}\frac{dI_c(t)}{dt} + \bar{C}_{TC}\frac{dV_{CB}(t)}{dt} \qquad \text{(A-39)}$$

The notation, $I_{b2}(t)$, means that the base current is removing base charge.

Equations A-37 and A-39 may be solved for a specified base current and the time dependence of the collector current thereby obtained. The *rise* and *fall times* may be defined as the time for the collector current to change between two specified levels, if the load is strictly resistive. Alternatively, the rise and fall times may be defined as the time for the collector-to-emitter voltage to change between two specified levels if the load contains capacitance or inductance.

We now compute the rise and fall times under the assumption that the transistor is driven ON and OFF from a constant-current source. It is assumed that the base current, $I_{b1}(t) = I_{B1}$, is large enough to saturate the transistor. Figure A-3 illustrates the circuit being analyzed.

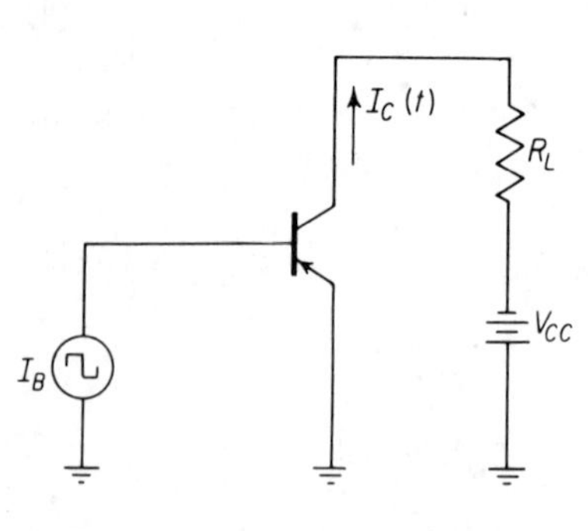

Fig. A-3. Basic switching circuit.

Equation A-35 may be rewritten as

$$I_{B1} = \frac{I_c(t)}{h_{FE}} + \frac{1}{\bar{\omega}_T}\frac{dI_c(t)}{dt} + \bar{C}_{TC}R_L\frac{dI_c(t)}{dt} \qquad \text{(A-40)}$$

The collector current as a function of time is

$$I_c(t) = h_{FE}I_{B1}\left[1 - \exp\frac{-t}{h_{FE}\left(\frac{1}{\bar{\omega}_T} + R_L\bar{C}_{TC}\right)}\right] \tag{A-41}$$

At $t = 0$, $I_c(0) = 0$, whereas at the end of the rise time interval $t = t_r$, $I_c(t_r) = I_{CS}$. The rise time is thus

$$t_r = h_{FE}\left[\frac{1}{\bar{\omega}_T} + R_L\bar{C}_{TC}\right]\ln\left\{\frac{1}{1 - \frac{\beta_C}{h_{FE}}}\right\} \tag{A-42}$$

The quantity, β_C, is the turn-on circuit gain or "forced beta," $\beta_C = I_{CS}/I_{B1}$. The transistor is driven into saturation if $\beta_C < h_{FE}$.

The equation for fall time, A-39, admits a similar solution:

$$t_f = h_{FE}\left[\frac{1}{\bar{\omega}_T} + R_L\bar{C}_{TC}\right]\ln\left\{1 + \frac{\beta_{CO}}{h_{FE}}\right\} \tag{A-43}$$

where $\beta_{CO} = I_{CS}/I_{B2}$, and I_{B2} is the turn-off base current.

Equations A-42 and A-43 are graphed in Fig. A-4. Normalized rise and fall times are shown as a function of normalized overdrive factor β_C/h_{FE} and β_{CO}/h_{FE}. For overdrive factors much less than unity (say, less than 0.2), both the rise and fall time are proportional to their respective overdrive factors. This may be shown by approximating the natural logarithm terms in Eqs. A-42 and A-43 under the assumptions that $\beta_C/h_{FE} \ll 1$ and $\beta_{CO} \ll 1$:

$$\ln\left\{\frac{1}{1 - \frac{\beta_C}{h_{FE}}}\right\} \approx \frac{\beta_C}{h_{FE}} \quad \text{for} \quad \frac{\beta_C}{h_{FE}} \leq 0.2 \tag{A-44}$$

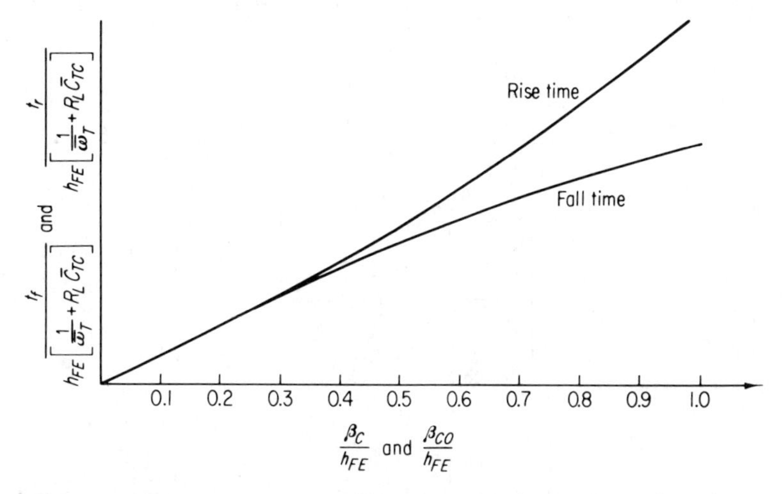

Fig. A-4. Variation of normalized rise and fall times with normalized overdrive factor.

$$\ln\left\{1+\frac{\beta_{CO}}{h_{FE}}\right\}\approx\frac{\beta_{CO}}{h_{FE}}\quad\text{for}\quad\frac{\beta_{CO}}{h_{FE}}\leq 0.2 \tag{A-45}$$

For drive conditions such that these approximations are valid, the rise and fall times are

$$t_r=\frac{I_{CS}}{I_{B1}}\left[\frac{1}{\bar{\omega}_T}+R_L\bar{C}_{TC}\right] \tag{A-46}$$

$$t_f=\frac{I_{CS}}{I_{B2}}\left[\frac{1}{\bar{\omega}_T}+R_L\bar{C}_{TC}\right] \tag{A-47}$$

Thus, for a fixed collector current and fixed load resistance, the rise and fall times can be made arbitrarily small by supplying an arbitrarily large base drive. For drive conditions such that Eqs. A-44 and A-45 are valid approximations, the rise and fall times are predicted to be equal, provided that they are measured with the same overdrive factor. This means that the first-order charge control model predicts switching times through the active region that are independent of the initial state of the transistor (cut-off or saturation).

Storage Time

Next we consider the first-order charge control model for prediction of storage time. The saturated state is achieved whenever emitter-base and collector-base diodes are both forward-biased and serve as both injectors and collectors of minority carriers. The minority carrier distribution in a saturated homogeneous base transistor is shown in Fig. A-5. The saturated collector current, I_{CS}, is supported by the charge $Q_B = I_{CS}/\bar{\omega}_T$. The base current required to support this charge is I_{CS}/h_{FE}. To return the transistor to the unsaturated state, the "excess" charge, Q_{SX}, uniformly distributed throughout the base, must be removed. This charge component builds up

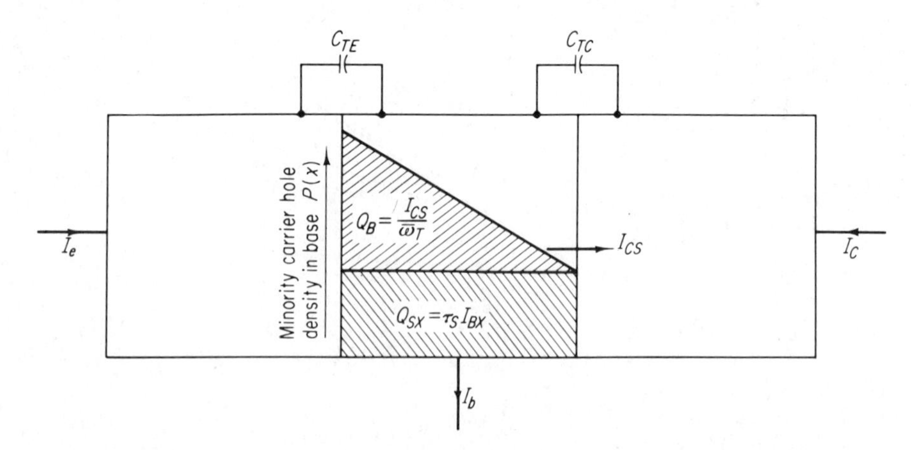

Fig. A-5. Minority carrier distribution in homogeneous base transistor during dc saturation.

as the collector junction becomes forward-biased and begins to inject minority carriers into the base. This excess base charge is controlled by the base current. With reference to Fig. A-5,

$$I_c = -I_{CS} + \bar{C}_{TC} \frac{dV_{cb}(t)}{dt} \tag{A-48}$$

$$I_e = I_{CS} + \frac{Q_B}{\tau_p} + \frac{q_{SX}}{\tau_S} + \frac{dq_{SX}}{dt} + C_{TE} \frac{dV_{eb}}{dt} \tag{A-49}$$

The interpretation of the excess charge lifetime, τ_S, will be covered subsequently. Adding Eqs. A-48 and A-49, the base current is

$$-I_{b2}(t) = \frac{Q_B}{\tau_p} + \frac{q_{SX}}{\tau_S} + \frac{dq_{SX}}{dt} + C_{TE} \frac{dV_{eb}}{dt} + C_{TC} \frac{dV_{cb}}{dt} \tag{A-50}$$

The five components of base current indicated in Eq. A-50 are

1. Current required to support the base charge Q_B (recombination current)
2. Current required to support the excess base charge q_{SX} (recombination current)
3. Current to charge the excess stored base charge
4. Emitter capacitance charging current
5. Collector capacitance charging current.

During the storage time interval, the emitter-base and collector-base voltages change very little so that these charging currents can generally be neglected. The storage time equation thus becomes

$$-I_{b2}(t) = \frac{Q_B}{\tau_p} + \frac{q_{SX}}{\tau_S} + \frac{dq_{SX}}{dt} \tag{A-51}$$

Throughout the storage time interval, the collector current is clamped at its saturated value, I_{CS}. It has been previously shown, Eq. A-35, that under this condition $Q_B/\tau_p = I_{CS}/h_{FE}$. Therefore,

$$I_{b2}(t) = -\frac{I_{CS}}{h_{FE}} + \frac{q_{SX}}{\tau_S} + \frac{dq_{SX}(t)}{dt} \tag{A-52}$$

In the steady-state condition, before turn-off, we have $I_{B2} = -I_{B1}$, where I_{B1} is the turn-on base current. Therefore, just before turnoff (assuming steady-state conditions have previously been reached):

$$I_{B1} - \frac{I_{CS}}{h_{FE}} = \frac{Q_{SX}}{\tau_S} \tag{A-53}$$

Let us define the excess turn-on base current as

$$I_{BX} = I_{B1} - \frac{I_{CS}}{h_{FE}} \tag{A-54}$$

I_{CS}/h_{FE} is the base current just required to hold the transistor on at the edge of saturation. Then,

$$\tau_S = \frac{Q_{SX}}{I_{BX}} \tag{A-55}$$

the ratio of excess stored base charge to excess base current.

For a constant-current turn-off with $I_{b2}(t) = I_{B2}$, the storage time is obtained by solving Eq. A-52 subject to the boundary conditions that at $t = 0$, $Q_{SX} = \tau_S(I_{B1} - I_{CS}/h_{FE})$ and at $t = t_S$, $Q_{SX} = 0$. The storage time for constant-base current turn-off is

$$t_S = \tau_S \ln\left[1 + \frac{I_{B1} - \dfrac{I_{CS}}{h_{FE}}}{I_{B2} + \dfrac{I_{CS}}{h_{FE}}}\right] \tag{A-56}$$

The storage time of the ideal charge-controlled transistor is thus characterized by a single time constant, τ_S.

The derivation of this storage time equation assumed that the turn-off process began from a steady-state situation. We may inquire how long the transistor must be held on for this steady state to be achieved. The increase of excess base charge during turn-on obeys the differential equation:

$$I_{B1} - \frac{I_{CS}}{h_{FE}} = \frac{q_{SX}(t)}{\tau_S} + \frac{dq_{SX}(t)}{dt} \tag{A-57}$$

The excess base charge thus increases exponentially with time according to the relation:

$$q_{SX}(t) = \tau_S\left(I_{B1} - \frac{I_{CS}}{h_{FE}}\right)(1 - e^{-t/\tau_S}) \tag{A-58}$$

Therefore, the steady-state saturation condition is reached in a time $t \approx 3\tau_S$.

The constant-current storage time when the transistor is turned off from a nonequilibrium state may be expected to be shorter than predicted by Eq. A-56 because the excess base charge has not reached its final value. Equation A-56 is easily modified to include this effect:

$$t_S = \tau_S \ln\left[1 + \frac{\left(I_{B1} - \dfrac{I_{CS}}{h_{FE}}\right)(1 - e^{-t_0/\tau_S})}{I_{B2} + \dfrac{I_{CS}}{h_{FE}}}\right] \tag{A-59}$$

The time, t_0, is the length of time the transistor has been held in saturation before turn-off.

Delay Time

We now examine the first-order model of the transistor which characterizes its dynamic behavior in the cut-off region (emitter and collector diodes both reverse-biased). The simple model shown in Fig. A-6 accounts for

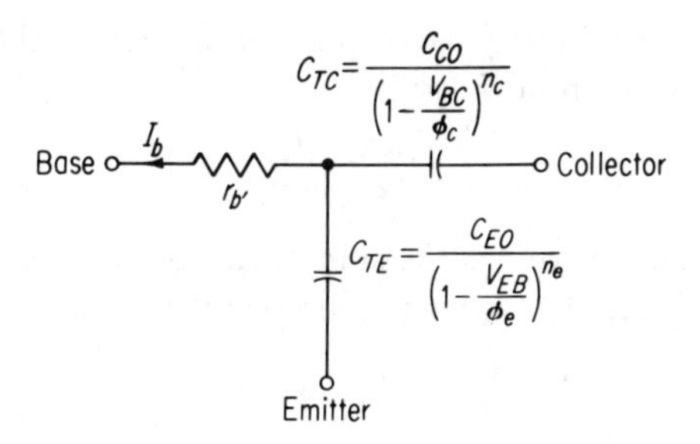

Fig. A-6. First-order transistor model in cut-off state.

the base resistance and the voltage dependence of the diode transition capacitances.

The delay time associated with switching from the cut-off state to the edge of saturation is the time required to change the voltage across the diode capacitances. The base current supplies the charging current to C_{TE} and C_{TC} according to the relation:

$$I_b(t) = C_{TE}\frac{dV_{eb}(t)}{dt} + C_{TC}\frac{dV_{cb}(t)}{dt} \tag{A-60}$$

where the voltage dependence of the capacitances is taken as

$$C_{TE} = \frac{C_{EO}}{\left(1 - \frac{V_{EB}}{\phi_e}\right)^{n_e}} \tag{A-61}$$

$$C_{TC} = \frac{C_{CO}}{\left(1 - \frac{V_{BC}}{\phi_c}\right)^{n_c}} \tag{A-62}$$

The delay time, t'_d, associated with these capacitances charging is the time required to change V_{EB} from its initial value V_{BEI} to the initiation voltage, V_i, and the time required to change V_{BC} from its initial value $(V_{CC} + V_{BEI})$ to the value V_{CC} at the edge of conduction. The initiation voltage is defined to be the intrinsic base-emitter voltage at which appreciable emitter current begins to flow. Thus,

$$\int_0^{t_d'} I_{B1}\,dt = \int_{V_{BEI}}^{V_i} C_{TE}\,dV_{eb} + \int_{V_{BEI}+V_{CC}}^{V_{CC}-V_i} C_{TC}\,dV_{bc} \tag{A-63}$$

Integrating, we have

$$I_{B1}t'_d = \frac{\phi_e C_{EO}}{1 - n_e}\left[\left(1 + \frac{V_{BEI}}{\phi_e}\right)^{1-n_e} - \left(1 - \frac{V_i}{\phi_e}\right)^{1-n_e}\right]$$
$$+ \frac{\phi_c C_{CO}}{1 - n_c}\left[\left(1 + \frac{V_{BEI} + V_{CC}}{\phi_c}\right)^{1-n_c} - \left(1 + \frac{V_{CC} - V_i}{\phi_c}\right)^{1-n_c}\right] \tag{A-64}$$

To this portion of the delay time we must add the inherent *transit time delay*, the time required for minority carriers to diffuse across the base. It has been shown (4) that this inherent delay is

$$t_d'' = \frac{1}{3\omega_T} \qquad \text{(A-65)}$$

The total delay time is taken as the sum of t_d', Eq. A-64, and t_d'', Eq. A-65. In general, the contribution of transit time to over-all delay time is negligibly small.

SUMMARY OF FIRST-ORDER THEORY IN TERMS OF MATHEMATICAL AND EQUIVALENT CIRCUIT MODELS

The result of the analysis to this point is a first-order mathematical model of the dynamic behavior of the common-emitter switch. In summary, the delay, rise, storage, and fall times may be predicted for the common-emitter switch operating with a resistive load with the following set of equations: In all cases $i_b(t)$ refers to a known base drive.

Delay time

$$t_d = t_d' + \frac{1}{3\omega_T} \approx t_d' \qquad \text{(A-66)}$$

where

$$\int_0^{t_d'} i_b(t)\,dt = \frac{\phi_e C_{EO}}{1 - n_e}\left[\left(1 + \frac{V_{BEI}}{\phi_e}\right)^{1-n_e} - \left(1 - \frac{V_i}{\phi_e}\right)^{1-n_e}\right]$$
$$+ \frac{\phi_c C_{CO}}{1 - n_c}\left[\left(1 + \frac{V_{BEI} + V_{CC}}{\phi_c}\right)^{1-n_c}\left(1 + \frac{V_{CC} - V_i}{\phi_c}\right)^{1-n_c}\right] \qquad \text{(A-67)}$$

Rise and fall times

$$i_{b1}(t) = \frac{I_c(t)}{h_{FE}} + \frac{1}{\bar{\omega}_T}\frac{dI_c(t)}{dt} + \bar{C}_{TC}\frac{dV_{cb}(t)}{dt} \qquad \text{(A-68)}$$

$$i_{b2}(t) = -\frac{I_c(t)}{h_{FE}} + \frac{1}{\bar{\omega}_T}\frac{dI_c(t)}{dt} + \bar{C}_{TC}\frac{dV_{cb}(t)}{dt} \qquad \text{(A-69)}$$

Storage time

$$i_{b2}(t) = -\frac{I_{CS}}{h_{FE}} + \frac{q_{SX}(t)}{\tau_S} + \frac{dq_{SX}(t)}{dt} \qquad \text{(A-70)}$$

All standard mathematical techniques available for solving first-order linear differential equations may be used in solving these equations. Operator techniques, such as the Laplace transform, allow these equations to be solved by algebraic manipulations.

For the purposes of analysis, this strictly mathematical model may be replaced by equivalent circuit models. Figure A-7 shows equivalent circuit

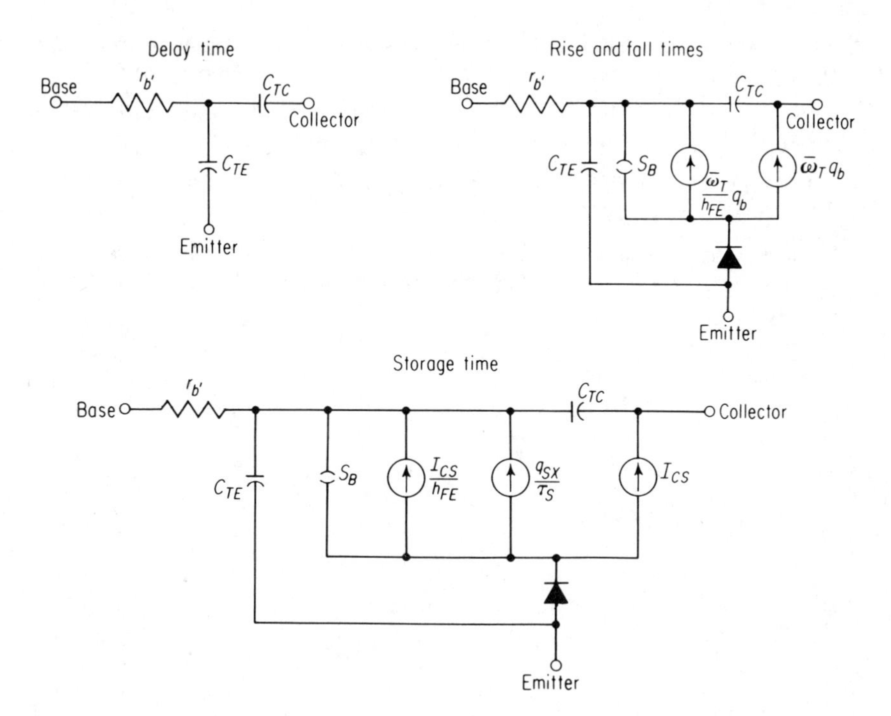

Fig. A-7. Equivalent circuit models derived from the first-order charge control theory.

models (1) for the four transient intervals derived from the first-order charge control theory. The charge storage element, S_B, represents an infinite capacitance since it stores charge without a change in voltage across its terminals. The diode shown in the emitter exhibits the dc characteristics of the base-emitter diode. The diode behavior is ideal with respect to its dynamic behavior (it exhibits no charge storage mechanism).

Either form of representation, mathematical model or equivalent circuit, may be used in circuit analysis. The mathematical description presents a more direct approach, whereas the use of the equivalent circuit provides a better physical picture of the dynamic behavior of the transistor.

REFERENCES

1. R. BEAUFOY and J. J. SPARKES, "The Junction Transistor as a Charge Controlled Device," *A.T.E.J.*, **13**, No. 4 (April, 1957).

2. D. DEWITT and A. L. ROSOFF, *Transistor Electronics*. New York: McGraw-Hill Book Company, 1957, pp. 268–81.

3. G. H. GOLDSTICK, "Comparison of Saturated and Non-Saturated Switching Circuit Techniques," *Trans. Professional Group Electron. Computers*, June, 1960.

4. L. J. Giacoletto, "Study of P-N-P Alloy Junction Transistors from D-C through Medium Frequencies," *RCA Rev.*, **XV**, No. 4 (December, 1954).

5. R. S. C. Cobbold, "The Charge Storage in a Junction Transistor during Turn-off in the Active Region," *Electron. Eng.* (May, 1959), pp. 275–77.

6. J. L. Moll, "Large Signal Transient Response in Junction Transistors," *Proc. IRE*, **42** (December, 1954) 1773–84.

7. J. Ekiss *et al.*, "Characteristic of Switching Transistors," *Final Report U.S. Army Signal Research and Development Laboratory Contract:* DA 36-039-SC88891.

Index

M

T

U

V